T0094300

Practical Computing on the Cell Broadband Engine

Sandeep Koranne

Practical Computing on the Cell Broadband Engine

Sandeep Koranne
2906 Bellevue Court
West Linn, OR 97068
USA
Sandeep.Koranne@gmail.com

ISBN 978-1-4419-0307-5 e-ISBN 978-1-4419-0308-2
DOI 10.1007/978-1-4419-0308-2
Springer Dordrecht Heidelberg London New York

Library of Congress Control Number: 2009928039

Printed on acid-free paper

Springer is part of Springer Science+Business Media (www.springer.com)

I would like to dedicate this book to the early pioneers of high-performance computing, whose untiring effort to extract the most out of computer architectures, continues to improve the quality of our lives to this day.

Preface

It has been just over four years since the Cell Broadband Engine (CBE) was introduced at the International Solid-State Circuits (ISSC) Conference in February of 2005. A number of technical papers at that conference described the inner workings of the architecture, and its first implementation. The chip's computational prowess was (and remains to be) second to none for the sort of applications which matched the architectures intent (and then some[1]). In the years that have passed, the wonderment has been replaced by fascination, and curiosity by admiration. The Cell Broadband Engine Architecture and its first implementation in the Cell processor exhibit exquisite, and mathematical Pareto-optimality (probably by design) across many quantitative (eg., SPE area, power, PPE EMT instruction-queue and TLB design) and other more subjective areas. Here we include the ease of development, software support, open-source availability of system software, and *über coolness* of developing software for the *cell*. It is just fantastic, to see a mailbox message after 2^6 DMA messages have completed processing in a triple-buffer setup. But I get ahead of myself, lets go back to where it started.

During 2001-2003, Sony opened up the Playstation2 console for hobby development using the Sony PS2 Linux kits. Using a cluster of PS2, inexpensive, very high performance clusters could be constructed at a fraction of cost of comparable high-performance computing. The PS2's Emotion Engine chip had a MIPS control processor, attached to two vector co-processors. Even though development had to be done primarily in Vector assembly, the FLOPs were there to be had. During this time I accumulated a number of potential applications, very practical programming problems, each with a compute intensive kernel, which could be elegantly solved on the Emotion Engine, only if some of the restrictions on instruction memory size of the vector co-processors, and development language, could be eased.

With the advent of the Playstation 3 console with the first generation Cell Broadband Engine on November 11 2006, many of these restrictions were removed, and the flawless support of Linux on the CBE (both by Sony, IBM and other third party developers) has made setting up a development environment on the CBE extremely

[1] We refer to Folding@Home `http://folding.stanford.edu`.

accessible, even to those developers who have limited or no prior experience with embedded systems. In this book we shall look at many problem domains. They range from number theory, graph theory, OpenGL visualization, wind-turbine location using digital elevation models, computational biology, and VLSI CAD. Where possible I have included enough information about the subject material as to introduce the problem and its solution, and to make the problem appear in context. Many of these come from areas where I have already seen performance advantages of the CBE, but some of them are exploratory, where we build up a system in layers, building upon methods and code we would have already coded in prior chapters.

I have been careful not to characterize CBE as a processor; this is more a testament to the pioneering nature of CBE's design, which presaged many aspects of now common, high-performance processing units. The novelties included a high-bandwidth Element Interconnect Bus, multiple independent Synergistic Processing Units (SPUs), and instruction set compatibilities. These do not include the exemplary work done by IBM in the enhancement of double precision CBE (Power XCell8i), or the deep sub-micron manufacturing expertise they have brought to bear on CBE's design and manufacture. The leading micro-processor vendors have also eschewed the concept of many core (whether homogeneous or heterogeneous) designs as opposed to a single monolithic block.

The pioneering nature of CBEs advanced functionality also implies a relative dearth of common programming paradigms for effectively using all of CBEs capabilities. Outside of small coding teams at video game developers, system programmers at IBM, Sony and Toshiba, and embedded system developers, it can be safely assumed that most programmers would be more comfortable coding for x86_64 using STL, rather than coding DMA and mailbox sequences. This book may not change it overnight, but it will definitely expand the comfort zone around the CBE. To do this, we shall cover the architecture of the CBE in some detail, especially in programmer friendly constructs and facilities. Moreover, we shall also cover mundane but often essential stepping stones for CBE program development, including installation, programming languages, remote display, and networking support.

This book could have been written in multitude of ways; for example, I had considered writing this book to show how the SPUs of the Cell architecture could model vector supercomputers, or how different parallel programming paradigms mapped on to the Cell architecture. I could also have structured the book focusing on the available software support libraries for the Cell, eg., BLAS, MASS, FFT and so on. Or I could have structured the book around prevalent compiler and program restructuring techniques to write a book on pthreads, OpenMP, MPI or any of the other fine parallelizing libraries. Another orthogonal structuring of this book would have been to present the Cell Broadband Architecture as a state-of-art 90nm, low *k*, Cu (copper) interconnect VLSI chip implementing a high-performance system on a chip (SoC). It contains processing elements, a high-speed interconnect, a high-speed memory controller, a high-speed input/output (I/O) controller, and global control and debug facilities, enough stuff for a complete book on VLSI system design, computer architecture, and compiler optimization. We would still have space to cover a few of the system libraries which are included in the SDK (software development kit), and sample applications. I await such a text on the VLSI system aspect of the chip.

In a way this book touches upon all the above themes, but it is written with the eyes set firmly on the end product; which is a correctly running, maintainable, debuggable program running on the Cell with a higher throughput than achievable on other state-of-art SMP processors. Thus, the major focus of this book is on demonstrating a large number of real-life programs to solve problems in engineering, logic design, VLSI CAD, number-theory, graph theory, computational geometry, image-processing, and others. I do not claim to have presented the best theoretical algorithm for every problem (though I have tried my best to do so), but the wide variety of problems which have been addressed should present the Cell in a new light. The goal is to inculcate in you, the wise reader, a *knack* to looking at a problem, and figuring out where the PPE-SPE cut-line should be, where the double-buffering can occur, where SPE in a pipeline mode can be used. If this book succeeds in convincing you that the Cell is a viable and optimal choice to research, develop and deploy programs in your domain, a goal of this book would have been met. However, if this book prompts you to look at existing programs in a different way, then the goal of the Cell architecture would have been met.

As I mentioned above this book is a programming guide for the CBE, written for another programmer (including the author). Therefore it includes many diagrams, mnemonics, tables, charts, code samples for making program development on the CBE as enjoyable as possible.

But, at the same time, there is almost no description of how to setup Makefiles, build environments, add command line flags to enable debug output, or setting up shell scripts. I expect the serious reader to know much of that already (if you don't know, I have included a comprehensive reading list in this book where such knowledge can be gained in the most productive manner). Secondly, there is no description of how to use the Cell Broadband Engine to develop 3d-graphics intensive applications, games, or frameworks beyond the algorithmic implementations of compute intensive primitives. The reason for this choice is the lack of documentation about the RSX chip (the equivalent of PS2 graphic synthesizer (GS) chip) in the PS3. Moreover, industrial use of the Cell is focusing on compute intensive non-graphical (excluding image processing) applications.

For the non-engineering executives faced with an evaluation of CBE as a candidate for high-performance computing deployment, this lack of detail (referring to my decision to omit details of basic GNU/Linux development) should not be an impediment. They should nevertheless benefit from the material presented in this book to analyze the capabilities of CBE in their application domain. Finally, we present a number of software programming projects, some solved, some partially solved, but also many which can be used by faculty for engaging students into the area of actual code development for practical high-performance computing.

The organization of this book is as follows. In Part-I of this book we shall discuss the following:

In Chapter 1 (Introduction) we introduce the Cell Broadband Engine, the time-line to its development, and its implementation for the PlayStation 3 gaming console. We present computer architecture and VLSI concepts which are required to understand the programming model of the Power Processing Element (PPE) and the Synergistic Processing Elements.

In Chapter 2 (PPE) we discuss the Power Processing Element (PPE). We have spent considerable time and pages to this topic, as not only is the PPE the gatekeeper of the Cell, it is also representative of other POWER processors such as PowerPC 970, 970MP, and Power variants present in a multitude of embedded systems, FPGAs, and game consoles. Gaining familiarity with POWER programming will only enhance your endianness.

In Chapter 3 (SPE) we concentrate on the work horses of the Cell architecture, the synergistic processors. We look at the design trade-offs and understand how they impact current generation software codes. We also look at SPU Channels, DMA support, and Mailboxes are covered as well. We investigate in some depth the SPU ISA (instruction set architecture), as a sound understanding of pipelining and efficient use of the ISA is key to performance. In Chapter 4 (EIB) we briefly discuss the Element Interconnect Bus, the glue which hold the whole system together. Since very little of the bus is part of the problem-state of non-privileged software, we only describe the system aspects of the bus. In Chapter 5 (DMA) we shall examine code to use DMA facilities to transfer data to-and-from SPE and PPE. Since for most applications data transfer is an issue, it is important to understand the DMA concepts very clearly. Thus we have presented many small example which build on sample code. DMA is also the single most common cause of program failure (during development), so its a good idea to develop good defensive programming habits when dealing with DMA.

In Part-II of the book, we go all-out on the code development side, after-all I had promised that this book was written by a programmer, for a programmer. In this part we look at the following aspects:
In Part-II, Chapter 6 presents a homogeneous view of the SIMD development used in this book. I have tried to use SIMD effectively where possible, but I have also been pragmatic about its use, where it complicates code, and the code is not in the the run-time hotspot we have *deferred* (sometimes indefinitely) the choice of rewriting the code in SIMD. We also present alternatives to using C/C++ on SPE to write the book. I wanted to bring this out early on in the book so as to present a choice to the reader upfront, before he/she has spent all the time learning how to program the SPE in detail. We present introductory information on OpenMP, and MPI.

In Part-II Chapter 7 (Software Development Environment) we present information on our development environment. A lot of material on how to install GNU/Linux on PS3, boot-loaders, SDK etc., is readily available on the Internet, and there is a good chance that new versions of Operating Systems (I have used Fedora Core 7 with SDK 3.0), indeed, new versions of PS3 hardware may be available by the time you are reading this book. With that in mind, I only devote that much space to installation and configuration. This is not the book to read on how to build your own custom GNU/Linux kernel for a Cell-blade. However, in this chapter we will cover salient points of the PS3 hardware which you need to be aware of when developing code on it, we cover GCC's behavior on the Power platform. I have also given enough information about the Cell system's development environment so that a developer can form a mental picture even if he/she does not have ready access to the hardware while reading this book.

In Chapter 8 (Hello, World) we write our first compiled code in Chapter 6. GCC pre-processors, big-endian byte shuffles, you name it, its in this chapter (well almost). Chapter 9 presents an overview of the libraries available as part of the Software Development Kit (SDK) ver 3.0. Many of the library functions are used in the sequel of the book and we constantly refer to this chapter for more information on the particular API used.

In Chapter 10, we implement what I call foundational blocks of programming on the Cell. These are kernels of algorithmic routines which I have used most often when solving problems on the Cell. These are not the same as BLAS, FFT or MASS. They are neither framework, nor APIs, but rather a well documented, debugged, and mostly simple way of solving a very specific problem. The fact that they map well to the SPE is incidental, but not required. For example, we present min-cut on graphs as a foundational routine, and we have indeed used that routine extensively in other projects. This chapter presents SPE optimized codes for graph problems, disjoint set-union, number theory and discrete logarithms (using Zech logarithms). The above chapter also contains our implementation of SPE merge-sort which we use extensively in later chapters. Chapter 11 extends on this theme by presenting several graph theoretic algorithms and their SPE implementations.

Chapter 12 presents an overview of the nested data-parallel language NESL and its underlying vector code library model, VCODE. We present NESL as an alternative language in which to code SPU intrinsics. A compiler for VCODE has been developed by me which translates VCODE primitives into 4-way SIMD SPU intrinsics on the SPU. Another alternative parallel programming paradigm in the form of PowerList data-structures is also presented in this chapter. The goal of this chapter is to introduce the rigour of parallel algorithm analysis before we undertake larger projects. Our simplified SPU ISA simulator is also presented therein. The Berkeley *dwarf* paper is also presented in context here; we analyze the requirement of the dwarfs with programming model on the SPE.

We present basic computational mathematics algorithms, eg., series evaluation, polynomial evaluation in Chapter 13. We use polynomial evaluation with steps in our case study on Complex Function solving. We present our heap based polynomial evaluation system which runs in the SPE. No book about the CBE can be complete without some details on dense and sparse matrix algorithms, and mine is no exception. In Chapter 13 we present examples of matrix operations which are computed on the SPU. The ideas presented in this chapter are used later for other matrix and vector algorithms.

Chapter 14 presents a system of using SPEs to perform operations in vector geometry. This is a standard SPE implementation of many computational geometry kernels, and we show that SPEs are good at processing such tasks provided the choice of algorithms be made with SPEs in mind.

Chapter 15 deals with tools and techniques to perform code optimization of SPU codes. We discuss both manual, and automated optimizers. By using the information previously presented on the dual-issue pipeline of the SPU, along-with tools such as *SPU Timing*, *ASMVIS* and *FDPRPRO*, we demonstrate how to improve run-time performance of SPU code (sometimes dramatically).

The last part of this book, Part-III, deals with medium sized projects which we refer to as *Case Studies*. In Chapter 16 we present an implementation of an algorithm to perform line-of-sight estimation on 3d terrain. We use a novel under-sampled SIMD Bresenham's line-drawing algorithm with efficient SPU implementation to compute line-of-sight. We extend this method to perform watershed analysis for facility location (eg., cell-phone towers, wind-turbines).

The chapter *ab-initio* methods for structure recovery using partial distance functions is a compute intensive procedure to discover the structure of a 3d atomic or molecular structure given spectroscopic distance measurements. We have used statistical methods and have run this on the SPE with good performance. Even though the PS3's RSX chip is not part of the current GNU/Linux installation, we can still use remote graphics using OpenGL to display 3d graphics which are formed by some computation performed on the Cell. In Chapter 17 we present a system based on UDP datagrams for communicating data packets from a host machine (running high-performance graphics using OpenGL) with a PS3 which performs the actual computation. This combination is used in our case study on Structure Recovery (Chapter 17).

We present a full polytope exploration system in Chapter 18. This implements a novel polytope enumeration and analysis algorithm which can generate and analyze catalogs of general n,d-polytopes.

Chapter 19 presents an implementation of a simplified (no motion correction, no registration correction) implementation of the core algorithms in Synthetic Aperture Radar. This technique of remote sensing has become vastly popular with increasing compute power, and the SIMD performance of SPEs on this task validates the argument of choosing Cell as the optimal platform for signal-processing applications.

The last couple of chapters are Cell implementations of classical VLSI CAD algorithms such as scheduling, microword-optimization, floorplanning, placement, global routing, coupling length analysis and power estimation. We conclude in Chapter 24.

As with almost all software written on GNU Linux, I acknowledge the large part GNU and Linux have played in not only the writing of this text, but also my programming experience in general. This book is written in Emacs on a Linux box, using LaTeX, xfig, gnuplot and other fine pieces of free software. My heartfelt thanks to all the developers of these projects for burning the midnight oil.

Enjoy programming the Cell Broadband Engine!

West Linn, Oregon *Sandeep Koranne*
May 2009

Accompanying Website for this Book

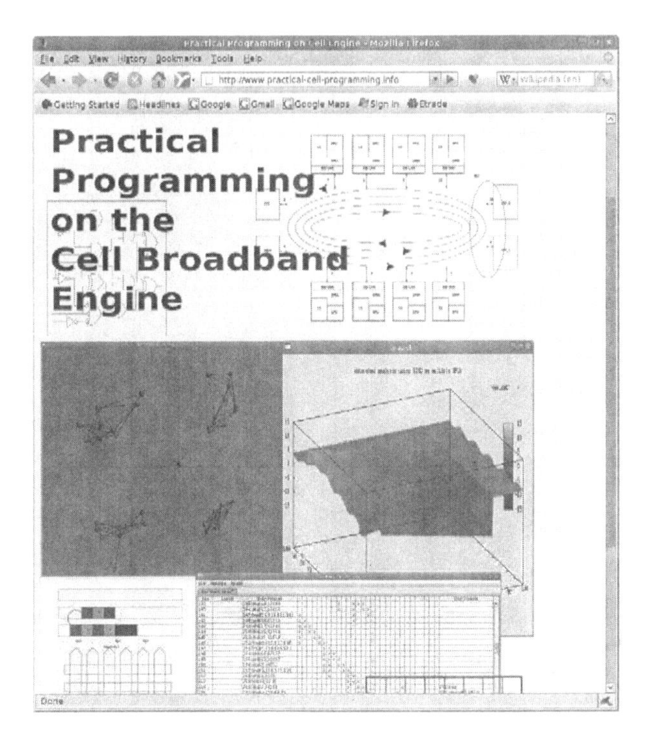

The website `http://www.practical-cell-programming.info`, is a companion to this text. The site contains almost all the source code, and sample-data for the examples presented in this text. The Website will also be continuously updated with further examples, errata and information about latest development around the Cell Broadband Engine.

Contents

List of Tables

List of Figures

Listings

Acronyms

ACCR Address Compare Control Register
ASR Address space register
BAT Block address translation
BHT Branch History Table
BUID Bus unit ID
CR Condition register
CTR Count register
CTRL Control Register
DAR Data address register
DABR Data address breakpoint register
DBAT Data BAT
DEC Decrementer register
DSISR Register used for determining the source of a DSI exception
DTLB Data translation lookaside buffer
EA Effective address
EAR External access register
ERAT Effective Real Address Translation
FPR Floating-point register
FPSCR Floating-point status and control register
FPU Floating-point init
FXU Fixed-point integer unit
GPR General-purpose register
HDEC Hypervisor Decrementer
IBAT Instruction BAT
IEEE Institute of Electrical and Electronics Engineers
IU Integer unit
LPCR Logical Partition Control Register
LR Link register
LSU Load-Store Unit
MMU Memory management unit
msb Most significant bit

MSR Machine state register
NaN Not a number
No-Op No operation
OEA Operating environment architecture
PTE Page table entry
PTEG Page table entry group
PVR Processor version register
RISC Reduced instruction set computing
SDR1 Register that specifies the page table base address for virtual-to-physical address translation
SIMM Signed immediate value
SLB Segment lookaside buffer
SPR Special-purpose register
SPRGn Registers available for general purposes
SR Segment register
SRR0 Machine status save/restore register 0
SRR1 Machine status save/restore register 1
TB Time base register
TLB Translation lookaside buffer
TSRL Thread Status Register Local
TSCR Thread Switch Control Register
TSRR Thread Status Register Remote
TSTR Thread Switch Time-out Register
UIMM Unsigned immediate value
UISA User instruction set architecture
VEA Virtual environment architecture
VXU Vector Media Extension Unit

Part I
Introducing the Cell Broadband Engine

This part of the book focuses on the computer architecture concepts and system design aspects of the Cell Broadband Engine. We analyze the problem-state areas of the Cell in detail to gain a better understanding of the optimizations that can and should be applied for high performance computing.

In Chapter 1 (Introduction) we introduce the Cell Broadband Engine, the timeline to its development, and its implementation for the PlayStation 3 gaming console. We present computer architecture and VLSI concepts which are required to understand the programming model of the Power Processing Element (PPE) and the Synergistic Processing Elements.

In Chapter 2 (PPE) we discuss the Power Processing Element (PPE). We have spent considerable time and pages to this topic, as not only is the PPE the gatekeeper of the Cell, it is also representative of other POWER processors such as PowerPC 970, 970MP, and Power variants present in a multitude of embedded systems, FPGAs, and game consoles. Gaining familiarity with POWER programming will only enhance your endianness.

In Chapter 4 (EIB) we briefly discuss the Element Interconnect Bus, the glue which hold the whole system together. Since very little of the bus is part of the problem-state of non-privileged software, we only describe the system aspects of the bus.

In Chapter 3 (SPE) we concentrate on the work horses of the Cell architecture, the synergistic processors. We look at the design trade-offs and understand how they impact current generation software codes. We investigate in some depth the SPU ISA (instruction set architecture) as a sound understanding of pipelining and efficient use of the ISA is key to performance. We also look at SPU Channels, DMA support, and Mailboxes.

Chapter 1
Introduction

Abstract In this chapter a general outline of the book is presented. An overview of the developments leading upto the design of the Cell Broadband Engine Architecture, and its place in the context of modern processors is given. An overview of the Emotion Engine (EE) architecture of the PlayStation 2 is also given. Key features of the Cell are discussed with respect to their use in high-performance applications.

1.1 About this book

The timing of this book is coinciding with the era of Petaflop computing, with Los Alamos National Laboratory and IBM reporting the sustained Petaflop/s performance of Roadrunner on the LINPACK compute kernel in May 2008 [4], and now the system is already being used for molecular dynamics [120], plasma physics and human visual system modeling [8]. Roadrunner is built using a combination of Cell Broadband Engine processors coupled with AMD Opteron chips connected to form a 12,000 PowerXCell 8i cluster. Each node in this cluster is accelerated to 450 gigaflops/s using the Cell processor. In comparison a general purpose quad-core processor based system would require 32,000 nodes to achieve close to petaflop/s performance. In addition to being the fastest supercomputer at the time of writing this book, Roadrunner is also placed 3rd on the Green500 list.

Cell Broadband Engine Performance Summary

The Cell Broadband Engine (CBE) consists of a PowerPC based core (PPE) supported by 8 Synergistic Processor Elements (SPE) connected together by a high-speed memory coherent Element Interconnect Bus (EIB). The SIMD units in the SPE provide the computational power of the Cell engine; in single-precision mode a fused multiply-add instruction can be executed on each of the 8 SPEs giving a

S. Koranne, *Practical Computing on the Cell Broadband Engine*,

DOI: 10.1007/978-1-4419-0308-2_1, © Springer Science + Business Media, LLC 2009

total of 64 floating point operations per clock cycle. At 3.2 GHz each SPU is capable of performing up to 25.6 GFLOPs/s in single-precision. Additionally it has an integrated memory controller (MIC) providing peak bandwidth of 25.6 Gb/s to an external XDR memory, while the integrated I/O controller provides peak bandwidth of 25.6 Gb/s (for inbound) and 35 Gb/s (for outbound) connections. The EIB provides a peak bandwidth of 204.8 Gb/s [1, 70] for intra-chip transfers among the PPE, the SPEs and the memory and I/O controllers.

Table 1.1 Performance Metrics of the Cell Broadband Engine

Metric	Performance
SPE to SPE Bandwidth	78 – 180 Gb/s
Memory Read/Write	21 Gb/s
I/O Interface	25 (inbound) – 35 Gb/s (outbound)
Optimized Matrix-Matrix Multiple	201 GFLOPs
LINPACK Optimized	155.5 GFLOPs
Single SPU MPEG2 decode	77 frames of HDTV
AES Cryptography on SPE	2.0 Gb/sec

With more than a thousand pages of documentation, APIs and example in the official Cell documentation, one would not expect to see another book on the subject, but this book begins where the HOWTOs end, the APIs finish and the small examples have been run. In a way I started to write this book in 2001, when started to program the Vector Units (VU) of the Emotion Engine chip present in the Playstation 2. I wrote several computation kernels on the VUs [73] (in assembly language). I (along with many other developers) applauded Sony's decision to allow (and make very easy) the installation of Linux on the Playstation 3 console [57], and thats when the concrete idea of a book about practical programming on the Cell Broadband Engine materialized. Recently clusters of PS3s have been used to perform serious supercomputing work and research [82].

Written from the viewpoint of a practising computer scientist and avid programmer, Practical Programming on the Cell (the present text) presents an ideal combination of system level information about the PPE, SPE, EIB, MFC, DMA components of the Cell Broadband Engine, with complete design and implementation details of state-of-art data-structures and algorithms on the SPE. Efficient SIMD and multicore optimized data-structures for hashing, perfect-hashing, union-find, heap, tree and graphs are implemented with full source code available. Key techniques of randomization, probabilistic methods, discrete logarithms and encoding functions are used to map several fundamental algorithms such as searching, sorting, min-cut, minimum-spanning tree, Dijkstra, Floyd-Warshall APSP, graph diameter, graph partitioning and spectral techniques.

Each method is first presented in pseudo-code then implemented on the SPE using compute kernels written in C, SPU intrinsics and SPU assembly [56, 40]. The resulting code is analyzed for instruction scheduling, dual-pipeline issue and load-store optimization. Fundamental methods of SIMD code analysis and optimization are presented using lucid and simple to understand examples as the progression from simple to complete case studies is made.

A common error on the part of the implementation is to not provide any substantial test data along-with the source code for the application. We have tried to not repeat that mistake; in all cases a corpus of data is provided, in many, a program to generate user controlled amount of test data which fits the application parameters has been included.

In addition to mentioning what is covered in this book, we should also candidly state what is *not* covered. We have not documented the API libraries that are installed with the SDK, these include (but are not limited to) SIMD math libraries, BLAS, FFT, MonteCarlo methods, large-matrix libraries, software-managed cache, synchronization, and others. This was done mostly due the ready availability of excellent user documentation for the above libraries. Where we have used the facilities provided by some library we have provided details on its use by the application. I have also not documented ALF[1], and DACS, as these libraries are high-level frameworks, and the objective of this book is to make the programmer intimately familiar with the low level functionality of the Cell. There is also a distinct lean of this book towards optimizing code more from the point-of-view of the SPE, than, the PPE. The rationale, obviously, is the presence of up to 16 SPEs in a Cell blade (as opposed to the rather weak dual-EMT PPEs).

The factual material about the Cell Broadband Engine Architecture and its first implementation has been collected from the series of papers in IBM Journal of Research and Development, [106, 105, 65]. Additionally, source code of the FCC compiler (up to version 4.3.3), especially the PowerPC and SPU targets was consulted for pipelining information. Additional papers which were referenced are cited where needed.

1.2 Background of the CBE Architecture and Processor

During the summer of 2000 it was clear to leading semiconductor design companies that the linear progression of technology at Moore's law alone could not deliver the computational power that content providers of real time entertainment, immersive games and high performance computing for servers, sought for their future needs. Sony in particular wanted to realize a vision to achieve 3 orders of magnitude performance increase compared to the Emotion Engine (the chip in PlayStation 2) [66, 65].

The heart of Sony's PlayStation2 game console was the Emotion Engine (EE) CPU; we briefly describe its architecture for reference, although we shall not attempt to compare implementation of our algorithms on the EE. For more

[1] We briefly discuss ALF in the context of alternative approaches to programming the SPE.

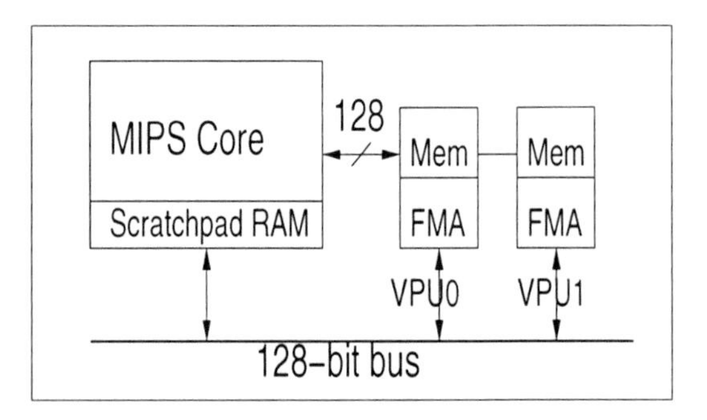

Fig. 1.1 Simplified EE Architecture

details about the EE architecture please refer to [78, 95]. The EE supported SIMD VLIW instruction execution. A simplified block-diagram of the EE CPU is shown in Figure 1.1. The EE contained three processors: the CPU core (which implemented the 64-bit superscalar MIPS IV instruction set), and the VPU0 and VPU1. The CPU core also contained a 16 K-byte scratchpad RAM (SPR). VPU0 and VPU1 were the vector processing units of the EE; each VPU contained a Vector Unit (VU), instruction and data memory, and interface unit. VU0 (the VU of VPU0) included a 4-Kbyte data RAM and a 4-Kbyte instruction RAM; VU1 contained 16-Kbyte of data RAM and 16-Kbyte of instruction RAM. Each VU also had an upper execution unit with four parallel floating point Multiply Accumulators (fMACs$\times$4), a lower execution unit with fDIV, load-store, integer ALU and branch logic, 128-bit$\times$32 floating point registers, 16-bit$\times$16 integer registers. VUs supported 2-slot VLIW, for the upper execution unit and the lower execution unit. The VU had a rich instruction set optimized for high performance graphic calculations, e.g. using a broadcast mechanism a 4$\times$4-matrix geometry transformation could be executed with 4 instructions. Other examples of the VU instructions included ELENG (length function), ERLENG (reciprocal length), ESQRT (square root), ERSQRT (reciprocal square root), ESADD (sum of square numbers) and CLIP (clip judgement). Many of the SPU instructions have one-to-one correspondence with the old instructions.

The Cell/B.E. processor was initially intended for the computer gaming market, and as such, Sony Computer Entertainment, Inc., has shipped millions of PlayStation 3 systems that use this processor. However, when in March 2007 a program to do protein folding (Folding@Home) was made available on the PlayStation 3 system, within a few days it created the world's largest distributed computer for this application. The Cell/B.E. processorbased system contributed more than twice as much application performance as a vastly larger number of PCs could have delivered. IBM and Mercury Computer Systems, Inc., have begun shipping the first generation of Cell/B.E processorbased blade servers. IBM and Los Alamos National Laboratory completed their goal of jointly building a 1-Petaflops supercomputer based on a variant of the Cell/B.E. processor. This supercomputer (the RoadRunner) is currently the world's fastest supercomputer delivering just over a petaflop of computing power.

The Cell Broadband Engine Architecture (CBEA) defines a family of heterogeneous microprocessors that target multimedia and compute-intensive applications [1]. The CBEA resulted from a joint effort among the Sony Group, Toshiba, and IBM to develop the next-generation processor. The following motivations shaped the development of the architecture:

1. Provide outstanding performance on computer entertainment and multimedia applications.
2. Develop an architecture applicable to a wide range of platforms.
3. Enable real-time response to the user and the network.
4. Address the three design challenges facing traditional processors: memory latency, power, and frequency.

The Cell architecture described in this text is the first generation Cell architecture as present in the Playstation 3. An advanced version of the chip the PowerXCell 8i has been recently released which improves the double-precision floating point performance of the SPUs. We shall not use double-precision arithmetic very often in this book. Moreover, the Playstation 3's Cell has only 6 usable SPUs. One SPU is permanently used by the Hypervisor which supervises the GNU/Linux operating system and provides low-level driver functionality. Another SPU is reserved for yield reasons. Thus many of my experiments on the PS3 have run only using 6 SPUs, but the code is generic enough, so that if you have access to systems where more SPUs, even multiple Cell chips are available in a machine, you can change the code easily to make use of all available SPUs.

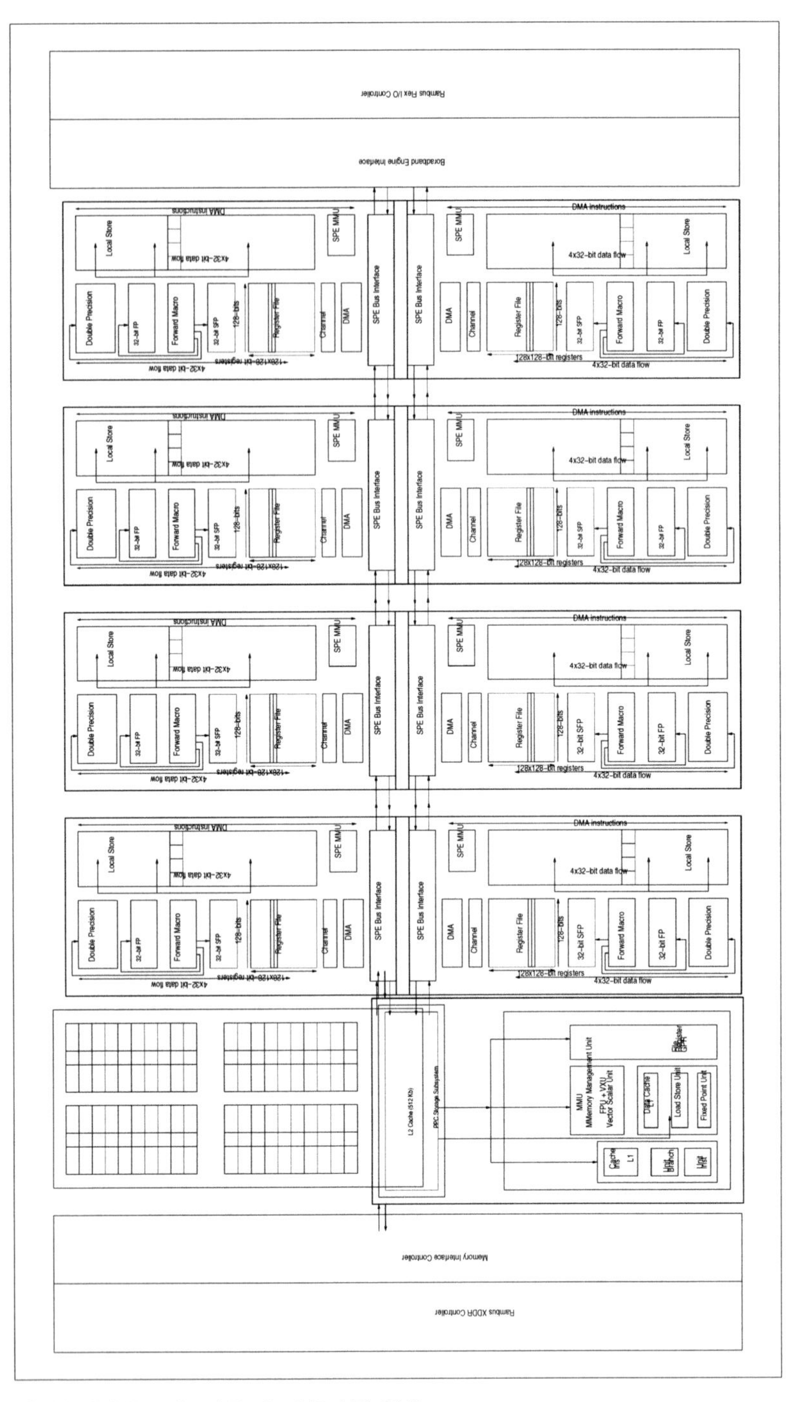

Fig. 1.2 The Cell Broadband Engine [65, 105, 106].

1.3 Design of the Cell Architecture

The Cell architecture is a single-chip multiprocessor with nine (refer to the above discussion for actual PS3 numbers) processor elements operating on a shared, coherent memory. While symmetric multiprocessor systems have been around for a long time, the Cell was the first design to provide two different or heterogeneous processors around a coherent memory. The two processor types are the PowerPC Processing Element (PPE) and the Synergistic Processor Element (SPE). The current Cell architecture has 1 PPE and 8 SPEs per chip. This radical design was conceived of by the design team of Sony, Toshiba and IBM. IBM also contributed leading edge process technology for the manufacturing of the new architecture. The name *broadband engine* refers to the dual faceted nature of the architecture; on one hand it brings together aspects from broadband interconnect entertainment systems, while at the same time, the interconnect used within the system also resembles a supercomputer structure.

Although designed by a committee consisting of designers from Sony, Toshiba and IBM at the STI design center in Austin, Texas, the Cell Broadband Engine provides an unique and refreshing experience in high performance computing. From the design documents subsequently published it is clear that the design team considered a wide variety of conventional multi-core architectures like SMP, CMP and dataflow oriented multiprocessors. The final design of the CBEA combined the 64-bit Power Architecture with coherent memory flow control and added a number of synergistic co-processors to provide the required computational throughput and power efficiency.

In order to meet, and in some cases exceed, the exponential performance requirements of Moore's Law, processor designers had relied exclusively (on the most part) on reducing the clock cycle duration (the reciprocal of which determines the processor's operating frequency). Since the processor needs to complete a finite number of operations during each clock cycle, the clock duration cannot be reduced indefinitely, but the introduction of pipelining in the instruction fetch-decode and execution phases led to designers adopting very deep pipelines. The very large number of pipelines in processors, reaffirmed the logical description of processor frequency and performance as the *maximum performance a processor is guaranteed not to exceed*. The theoretical performance of highly pipelined processors could only be achieved on highly tuned loop codes where the mere presence of a branch miss could lead to 1000s of cycles being wasted. Although programmers were quick to realize the problem and spent significant time tuning kernels (compute intensive part of application which is CPU bound, not to be confused with Operating System kernel), it was clear that increasing pipeline depth to satisfy the clock cycle requirement of the VLSI manufacturing process was resulting in diminishing gains.

Even more serious was the problem of *memory wall*, referring to the problem of increased memory latency as measured in processor cycles, and this latency introduced limits on effective memory bandwidth. While instruction misprediction could be worked around with spending vast VLSI real estate on branch prediction logic, and instruction buffers (which were populated with both conditions of the branch), memory latency could not be solved even by putting large amounts of high speed cache memory close to the processor. In SMP with shared memory, main memory latency was being measured in 1000 processor cycles. The speculative branch prediction made the problem worse, and instruction starvation also became a major problem along-with data starvation. This quickly becomes a vicious cycle, as the processor issues data fetch instructions to satisfy the current data starvation, and then speculatively issues instructions from downstream code. The probability that useful work is being correctly speculated decreases very fast with the speculation depth, it was quite rare in non-tuned work loads to see more than a few speculative memory fetches being used to perform correctly used work. The challenge was therefore to design an architecture which allowed access to main memory effectively by allowing more memory transactions to be in flight concurrently.

The increasing emphasis on global climate change, has prompted IT departments to question the diminishing gains in performance at the cost of large increases in thermal budgets of the compute center. Increasing processor frequency by a factor x, results in a power dissipation increase of x^2 (diluted to some extent by the VLSI process shrink of gate-size which invariably occurs at the same time), which meant that a new *power wall* had also emerged which had to be recognized. Moreover, since software costs are a large fraction of NRE (non-recurring engineering) cost of a product, consumer product companies favor architectures where the same code can be executed on *families* of processors; which implies that architecture designers cannot fore-go the possibility of a hand-held version of their chip. In the case of the CBEA indeed Toshiba has announced a prototype of the Cell processor in a laptop. Indeed the Green500 list of most power efficient supercomputers (available at `http://www.green500.org`) shows that the CBEA powered IBM QS22 cluster based on PowerXCell 8i 3.2 GHz is the most power efficient supercomputer. In the Nov 2008 rankings the CBEA attained the top 7 positions on that list, topping at 536.24 mega-flops per Watt.

The CBEA architecture re-introduced a number of techniques and methods which were known to provide high computational density. These included SIMD (Single Instruction Multiple Data), asynchronous heterogeneous co-processors, RISC (Reduced Instruction Set Computing), super-scalar instruction issue, chip multi-threading, fast interconnection bus and large register files. Later in this chapter we will discuss these feature in detail. It should be noted that there is a distinction between the CBEA Architecture and its implementation in the Cell Broadband Engine chip. The architecture extends the 64-bit Power Architecture with SPEs, MFC (memory flow controller), and DMA engines. The first generation Cell chip uses a 64-bit Power processor element (PPE), with 8 SPEs, an on-chip memory controller along-with an on-chip I/O controller. These units are connected with a coherent on-chip EIB (element interconnect bus). The chip itself was manufactured in a state-of-art 90nm process with SOI (silicon-on-insulator), low k-dielectrics and copper interconnect. Subsequent mask shrinks to 65nm process have also been released, as has been an enhanced CBEA chip with high-performance double precision floating point unit (the PowerXCell 8i Processor).

Another distinction between a traditional processor and the CBEA is the definition of two storage domains: main and local. The main storage domain is the same as that commonly found in most processors. This domain contains the address space for system memory and memory-mapped I/O (MMIO) registers and devices. Associated with each SPU is a local storage address space, or domain, containing instructions and data for that SPU. Each local storage domain is also assigned an address range in the main storage domain called the local storage alias. The SPU can address memory directly only within the associated local storage. To access data in the main storage domain and maintain synchronization with other processors, each SPU and local storage pair is coupled with a memory flow controller (MFC). The combined SPU, associated local storage, and MFC make up the SPE. In addition to the PPE and the SPE, the CBEA includes many features for real-time applications not typically found in conventional processor architectures.

Use of PowerPC technology in the Cell

The use of Power Architecture as the underlying architecture of the PPE maintains full compatibility with 64-bit Power Architecture. Despite its venerable lineage (and to a large extent due to it), the Power architecture is deployed in many embedded and control systems and as such a detailed understanding and experience in writing C/C++ codes for Power is a valuable skill. Note that all 3 game consoles in the market, Nintendo, XBox and PlayStation 3, currently have Power architecture as the main control processor. The Power architecture is also used as a microprocessor with Xilinx Virtex series FPGAs (the Power 440), and in many real-time control systems. The CBEA architecture mandates unmodified execution of Power 970 programs on the Cell processor, and in this regards the PS3 can be used as an embedded Power system running Linux to gain valuable programming skills on this platform. In this chapter we shall analyze the PPE, with regards to its integer, floating point, and multi-threading performance.

1.4 The Cell as a supercomputer

The Cell Broadband Engine has become the processor of choice for high performance computing deployment, especially when scalability of the order of 10,000 computing nodes and beyond is called for. As we mentioned above the RoadRunner project, the fastest supercomputer in the world with Petaflop capabilities, is built using the Cell Broadband Engine Processor. Moreover significant research on using the Cell for a variety of scientific computing tasks is still going on [58, 3]. In this book we shall discuss how to program scientific applications on the CBE. In contrast to other books on this subject we shall not focus on any one narrow aspect of scientific computing, for example, fluid dynamics, or nuclear interaction simulation.

We shall also not focus only on the nuts and bolts of DMA transfer, instruction latency, compiler switches (although all of the above will be needed to some extent). Our goal with this book is to provide the user with enough material that he or she can look at the problem they are trying to solve, and be able to answer the following questions:

1. Is the CBE the right processor and technological solution for this problem ?
2. What are the existing tools, methods and libraries I can leverage to arrive at a solution in the shortest possible time and money ?
3. What are the design patterns I can adopt to arrive at a starting point to think about my problem's solution ?
4. What do I need to get started ?

This book should help you in answering these questions. With the help of many (more than 20) fully worked out examples we hope to show you that the CBE is indeed a versatile compute node, and with the eco-system provided by IBM and other CBE solution providers, it makes a sound technological investment. For the student, who wants to learn how to write high performance computing programs, this book offers a very different perspective from other fine collections on High Performance Fortran, MPI, PVM, for example; we do not expect the student to know which method is best from the very beginning, but by exposing the student to a variety of solution methods, we hope to inculcate in him or her, a knack of choosing a method, and an implementation strategy which is flexible yet efficient to deliver the goods.

Today, there are at least 5 methods of solving high performance demanding applications requirements, they are listed in decreasing order of cost to consumer:

1. Custom Application Specific Integrated Circuit designed exclusively for the problem: eg. IBM Deep Blue Chess move generator, molecular simulation engines (D. E. Shaw's Anton), n-body simulators (grape),
2. Custom FPGA boards with problem specific HDL design code mapped onto hardware,
3. GPU computing using off-the-shelf video card and C-compiler,
4. Cell Broandband Engine as embedded in QS22 racks from IBM/Mercury, and the Playstation3 video game console,
5. Commodity off-the-shelf microprocessors (like Opteron, Xeon, Sun SPARC) connected together using Giga-bit Ethernet clusters to form compute clusters.

Historically, the 1st choice was really the top choice for high performance computing, but with non-recurring costs, and the cost of software dominating the budget, developing custom ASICS and the associated software is no longer affordable (by almost everyone, there are always the exceptions). Custom FPGA boards offer the pedal-to-the-metal advantage of ASICs and definitely lower the ASIC production cost, but writing HDL for complex algorithms (as we discuss later in this book) is non-trivial. Moreover, HDL verification, simulation, synthesis, has to be followed by physical mapping of logic to gates on the hardware which is time consuming and cannot be done in an interactive manner. But nevertheless, FPGAs do offer a good compromise between ASICs and traditional microprocessor based solutions. GPU computing has recently become another player in this space. By utilizing the already installed high performance hardware in the user's machine, GPU computing advocates say, applications can execute 10 to 100 times faster. Writing applications which can take advantage of GPU computing is significantly more difficult than writing straight-line C/C++ code, but this fact alone is not a major problem. The key difference between GPU, CBE, FPGA and traditional microprocessors, are the sweet-spots of application development. Like a choice of baseball bats which offer a trade-off between size of sweetspot and the power of the bat, each of these technical choices of platform comes with its own little sweet-spot of an optimum trade-off between cost, development time, processing speed, and power.

Proponents of each method will be quick to order this list based on the dominant advantage their choice of platform has, and certainly they are right. In this book we shall not focus on the calculation of the benefits of choosing one platform over the other, except to point out that in our opinion (which is shared by a good number of people, both in high-performance computing community, industry and government), the CELL Broadband Engine provides the most flexible platform for long term investment. Why is this important ? as we are going to show in this book writing high-performance code for applications is hard, and the last thing as a developer you want to do is to write the same application over and over again, optimizing for the last 10% on a radically different hardware with its own quirks. Thus we have decided to focus on writing compute bound applications on the CBE. Although the mix of problems we plan to solve in this book covers some problems which are not main stream, they do expose the reader to techniques which we feel the reader will be able to exploit in other, larger problems.

1.5 Major Components of the Cell Broadband Engine

The Cell Broadband Engine Architecture consists of the following:

PowerPC Processing Element (PPE): the PPE serves as the control plane of the architecture. It is a PowerPC Book 2 compliant architecture with some limitations. We discuss the PPE in detail in the next chapter (cf. Chapter 2).

Synergistic Processing Element (SPE): the SPE architecture is the workhorse of the CBEA, and we discuss the design and implementation of the SPE micro-architecture in Chapter 3,

Element Interconnect Bus (EIB): the EIB is a token-ring style communication framework which provides 96 bytes per processor-clock cycle maximum bandwidth. We discuss the EIB in Chapter 4,

Memory Interface Controller (MIC): since the MIC does not provide any user space controllable functionality, we have not analyzed the MIC with the same level of detail as other components. Please refer to the CBEA Architecture Guide for more details on the MIC [17]. In the PS3, the MIC only supports the Rambus XDR memory, but in PowerXCell 8i, DDR2 is also supported. The interleaving strategy of memory banks is of interest to problems where large data sets are stored in EA. Since the PS3 does not have enough system RAM to accommodate such applications, we have not discussed the memory interleaving access strategies, but the interested reader can refer to Flynn [47] for a thorough analysis of memory banks and their impact of memory latency,

Cell Broadband Engine Interface Unit (BEI): the BEI is another component which does not provide user-level programmable features. Its use in single socket Cell hardware (like the PS3) is also limited to the Bus Controller and the Interrupt Controller. The BEI supports the Rambus Flex IO interface which can be configured to use the BIF protocol to connect to another Cell processor directly and coherently,

Other important facets of the CBEA, the DMA facilities, MFC (memory flow controller), and Mailboxes are also covered in Chapters 5.

1.6 Flex I/O, DDR/XDR, Interrupt Controller

Most of the features and facilities provided by these units (which we have not described) are available to privileged mode, or hypervisor software only. Thus, though formally these units do constitute an integral part of the *architecture* of CBEA, for most (if not all) user applications, these units (with the possible exception of EIB) should be a *don't care*. Especially when developing and deploying applications on the PS3 which suffers from the hypervisor's overhead in all system and kernel tasks (eg., SATA-disk access, Giga-bit Ethernet, frame-buffer output etc.), I think it is best to treat these units as a black-box, and concentrate our attention on the PPE, SPE, DMA, MFC and EIB (who said abbreviations don't help).

1.7 Conclusions

This book is not intended to replace the 1000s of pages of excellent reference documentation provided by Sony, Toshiba and IBM on the research, design, development of the Cell Broadband Engine. Nor can this book replace the documentation of the API for the many libraries that comprise the SDK. Even if this could be attempted, the information is bound to get outdated as newer version of the API and SDK are released. A complete overview of every single API call would make this book clumsy and unwieldy to hold comfortably in the hand, which if I may dare say so, should be *the* pleasing attribute of every paper-book in this Laptop, Netbook savvy age. But, where the sample code could not be explained without describing a key API, I have not shied away from it either, and for completeness pointers to standard documentation is given where the API is significantly used.

Thus, the plan of this book is to first explain the workings of the Cell as a computer architecture, without getting bogged down in the programming aspects, and then develop the programming aspects by providing details on the development infrastructure and development libraries.

More details about the Cell Broadband Engine Architecture and its implementation can be found in the following references:

Cell Broadband Engine Architecture [17]:
Cell Broadband Engine Programming Handbook [19]:
PowerPC Architecture [15]:
Synergistic Processor Unit ISA [32]:
Linux Standard Base for PowerPC [12]:
C/C++ Language Extensions for CBEA [33]:
CBE CMOS SOI 65nm Hardware Initialization Guide [18]:
64-bit PowerPC ELF ABI Supplement [14]:
SPU ABI Specification [29]:
SPU Assembly Language Specification [30]:
Cell Broadband Engine Registers [20]:
PowerPC Compiler Writers Guide [13]:
SDK 3.0: ALF for CBE Programmer's Guide and API Reference [26, 27]:
Cell Broadband Engine Processor: Design and Implementation [105]:
Introduction to the Cell Broadband Engine Architecture [106].

Chapter 2
The Power Processing Element (PPE)

Abstract In this chapter we discuss the design, implementation and functionality of the PowerPC Processing Element (PPE), which serves as the control plane of the Cell Broadband Engine; the OS of the Cell runs on the PPE. The PPE contains a 64-bit, dual-thread PowerPC Architecture RISC core and supports a PowerPC virtual-memory subsystem. It has a 32 KB L1 I-cache, as well as a 32 KB D-cache, along with a 512 KB L2 unified cache. We analyze the PPE in detail, as not only does it provide significant facilities for controlling the SPEs, the PPE in Cell can run existing PowerPC 970MP and 970FX compiled applications. Since PowerPC is often used as a control processor in embedded systems, we use the PPE in Cell to understand PowerPC instruction set in detail. The PPE in Cell includes the vector/SIMD multimedia extensions (AltiVec).

2.1 PPE: the Control Plane of the CBEA

Although the Cell Broadband Engine derives much of its touted Gigaflops from the 8 (limited to 6 in the PS3) Synergistic Processor Elements (SPEs), the Power Processing Element (PPE) is the *conductor*, presiding over the computing orchestra. In this chapter we shall focus solely on the PPE, as we examine its design and architecture with the intent of understanding its computational sweet-spot. Thereafter, in this book, we shall mostly concentrate on writing efficient SPE code by optimizing for SIMD, dual-issue and DMA. The role of the PPE shall become like the benevolent elder sibling or parent, watching out, doing its job, and mostly staying out-of-the-way of the blazing SPEs.

As discussed above the PPE performs the control plane functions that typically require a more general-purpose or traditional computing environment which is needed for Operating Systems, and system software. The PPE in CBEA compliant architectures is based on Version 2.02 of the PowerPC Architecture, which offers many of the features required for the application spaces targeted by the CBEA.

The PPE was based on the PowerPC Architecture for four reasons [27]:

S. Koranne, *Practical Computing on the Cell Broadband Engine*,
DOI: 10.1007/978-1-4419-0308-2_2, © Springer Science + Business Media, LLC 2009

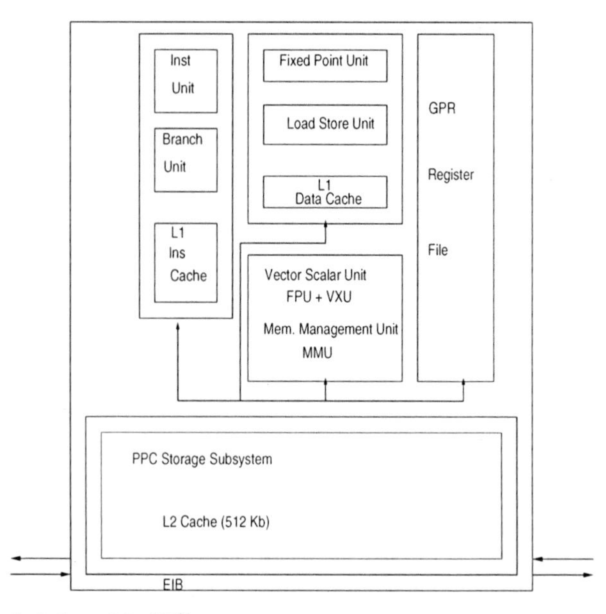

Fig. 2.1 A detailed view of the PPE

1. the PowerPC Architecture is a mature architecture that is applicable to a wide variety of platforms,
2. it supports multiple simultaneous operating environments through logical partitioning,
3. it contains proven microarchitectures that meet the frequency and power challenges of the targeted market segment,
4. use of the PowerPC Architecture leverages IBM's investment in the PowerPC ecosystem.

The PPE complies with the 64-bit implementation of the PowerPC Architecture [13, 14]. It is a dual-threaded processing core that includes an integer unit, a floating-point (FP) unit, a vector multimedia extensions (VMX) unit, and a memory management unit (MMU). See Figure 2.1 for a schematic. The PPE instruction cache (I-cache) is 32 KB and the data cache is 32 KB. In addition, an unified 512-KB Level 2 (L2) cache is included in the PPE, which also has a memory flow controller (MFC) that enables it to perform direct memory access (DMA) transfers to and from main memory, SPEs, or I/O. The MFC also provides memory-mapped I/O (MMIO) transfers to on-chip and off-chip devices. Communication to and from SPEs can be performed via DMA and/or mailboxes. Though based on the PowerPC Architecture, the PPE has made some architecture trade-offs. For example, it is pipelined extensively to allow it to operate at frequencies higher than 3 GHz. To reduce silicon area, program execution is performed in order. It supports allocation management which allows portions of a resource time to be scheduled for a specific resource allocation group. This simplifies real-time application programming. All the caches are covered by an allocation management scheme. The PowerPC Architecture hypervisor extension is also included in the design to allow multiple operating systems to run simultaneously via thread management.

As shown in Figure 2.1, the PPE consists of three main units: the instruction unit (IU), the execution unit (XU), and the vector/scalar execution unit (VSU), which contains the VMX and floating-point unit (FPU). The IU contains the Level 1 (L1) instruction cache (ICache), branch prediction hardware, instruction buffers, and dependency checking logic. The main division between the IU and the rest of the system is at the instruction issue 3 (IS3) stage, which is the main stall point for the PPE. The XU contains the integer execution units (FXUs) and the loadstore unit (LSU). The VSU contains all of the execution resources for FP and VMX instructions, as well as separate VMX and FP instruction queues in order to increase overall processor throughput.

Although the PPE is considered an in-order machine, several mechanisms allow it to achieve some of the benefits of out-of-order execution, without the associated complexity of instruction or memory access reordering hardware. First the processor can make forward progress on a thread even when a load from that thread misses the cache. The processor continues to execute past the load miss, stopping only when there is an instruction that is actually dependent on the load. This allows the processor to send up to eight requests to the L2 cache without stopping. This can be a great benefit to FP and SIMD code, since these typically have a very high data cache miss rate, and it is often easy to identify independent loads. In addition to allowing loads to be performed out of order, the PPE uses "delayed execution pipelines" to achieve some of the benefits of out-of-order execution. Delayed execution pipelines allow instructions that normally would cause a stall at the issue stage to move to a special "delay pipe" to be executed later at a specific point.

2.2 PPE Problem State Registers

The PPE contains the following types of registers (cf. [17]):

1. General-Purpose Registers (GPRs) – the 32 GPRs are 64 bits wide,
2. Floating-Point Registers (FPRs) – the 32 FPRs are also 64 bits wide, with single-precision results maintained internally as double-precision data in the IEEE 754 format,
3. Link Register (LR) – the 64-bit LR can be used to hold the effective address of a branch target,
4. Count Register (CTR) – the 64-bit CTR can be used to hold either a loop counter or the effective address of a branch target,
5. Fixed-Point Exception Register (XER) – the 64-bit XER contains the carry and overflow bits,
6. Condition Register (CR) – conditional comparisons are performed by first setting a condition code in the 32-bit CR with a compare instruction,

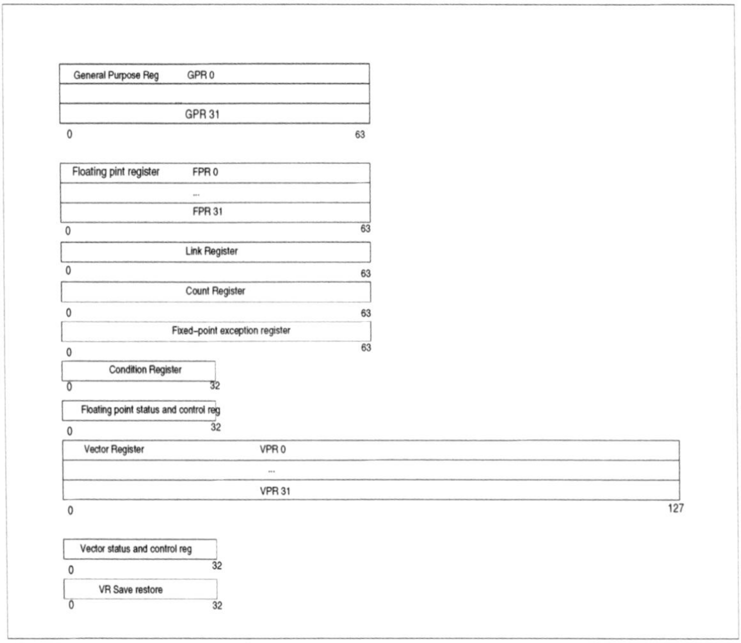

Fig. 2.2 Problem state registers of the PPE.

7. Floating-Point Status and Control Register (FPSCR) – 32-bits, and updated after every floating-point operation by the PPE,
8. Vector Registers (VRs) – 32 128-bit-wide VRs (useful in SIMD operations),
9. Vector Status and Control Register (VSCR) – 32-bit, similar to FPSCR above,
10. Vector Save Register (VRSAVE) – 32-bit, privileged mode only register.

The vector/SIMD multimedia extension defined by the CBEA is very similar to the PowerPC 970 (the major difference being the support for rounding-mode). For compatibility with SPU applications, the vector/SIMD multimedia extension unit in the PPE supports the rounding modes defined by the SPU instruction set architecture. The current PowerPC Architecture supports the base 4-KB page plus one additional large page to be used concurrently. The large-page size is implementation dependent. In the CBEA, many types of data structures are located in main storage, e.g., MMIO registers for the SPEs, local storage aliases, streaming data, and video buffers. If a large-page size of 64 KB is selected, the number of translations needed for MMIO registers and local storage aliases is lower than that for the base 4-KB page size. A large-page size of 1 MB or 16 MB reduces the number of translations required for the streaming and video buffers, but it is too large for mapping the MMIO registers and local storage aliases. To improve the efficiency of the TLBs, the CBEA augments the PowerPC Architecture by providing support for multiple concurrent large-page sizes. The memory management units (MMUs)

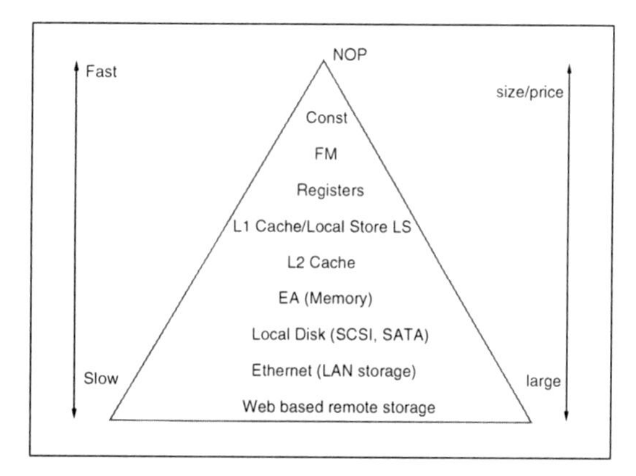

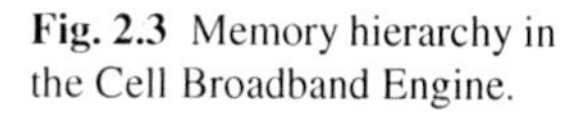
Fig. 2.3 Memory hierarchy in the Cell Broadband Engine.

in the SPE also support the multiple concurrent large-pages extension. We discuss using GNU/Linux commands to enable large page-tables in a later chapter.

2.3 Memory arrays in the CBEA

We first consider the memory hierarchy present in the Cell architecture as shown in Figure 2.3. As we go from bottom to top, the size of the memory decreases, but its price per bit, and speed increase. At the bottom most we have remote storage, perhaps distributed across the Internet. Next up, local area network storage devices and file servers, followed by locally attached disk. The access speeds of these storage devices have almost an order of magnitude difference, and their capacities are also commiserate. Beyond disk storage, we have solid state memory inside the chip itself. These are referred to as SRAM arrays in the next section. They are divided into the Effective Address Memory (512 MB on the PS3), the L2 cache (512 kb), the L1 data cache (32 kb), the L1 instruction cache (32 kb), various queues and cache tables, and the SPU local store of 256 kb.

In the PowerPC and SPU architecture computation is performed between register arguments, and thus registers are also considered as memories, and they are near the top of the pyramid in performance. Only forwarding macro output, which is the result of the previous computation already stored in the arithmetic unit, and immediate operands can be considered faster than register access, but then you are already at the microword level, since all three access take a logical clock cycle to complete. At the top of the pyramid for consistency, is a NOP, meaning, no computation implies no memory access. Many-a-times not doing the computation which requires memory is faster than the fastest cache access.

As mentioned above in the memory hierarchy, the Cell architecture has a number of memories distributed all over the chip. According to one count, the Cell processor contains around 270 arrays. Performance, power, and area constraints drove different designs for the memory arrays. The PPE and SPE contain the largest SRAM devices.

2.3.1 PPE Caches

The PPE has two 32-KB L1 SRAM macros that are organized as two-way associative memory. Interleaving is used with the L1 SRAM macros to improve parity-checking performance. The read latency of the L1 SRAM is three cycles: The first cycle is used by the sum address decoder (SAD) for address generation. The second cycle is used to access the array, and the third cycle is used for parity checking, data formatting, and way select. An L1 SRAM macro is internally constructed as 512 wordlines that access 32 bytes (with parity per byte) per way. Writing of the L1 is performed one cache line at a time and can be 64 bytes or 32 bytes. The writing of one 64-byte-wide cache line addresses two consecutive wordlines. Data that is 16 bytes wide can be read with parity checking on each access. The L1 caches are clocked at full clock rate of the core clock grid.

The L2 cache is a 512-KB eight-way associative cache. The L2 cache is inclusive of the L1 D-cache, but not the L1 I-cache. It supports write-back (copy-back), and allocation on store miss. The cache is constructed of four 1,024 8 140 SRAM macros. Unlike the L1 caches, the L2 cache is clocked at one-half of the clock rate of the core clock grid. Power conservation is achieved by decoding the way selected prior to accessing the array and then activating one-eighth of the L2 macro. The L2 cache has one read-port, and one write-port (but only one read or write per cycle). The L2 array can be accessed as a 140-bit or a 280-bit read. Writes to the array are pipelined and completed in two cycles. Read operations complete in three cycles for the first 140 bits of data. For 280 bits of data, the second 140 bits of data can be completed in the fourth cycle. Writes and reads can be interleaved to provide continuous data transfers to and from the L2 array. It supports ECC on data, parity on directory tags, and global and dynamic power management.

2.3.2 SPU Local Store

The LS array is a 256-KB SRAM array that consumes one-third of the total SPE area. The LS macro consists of a sum address decoder, four 64-KB memory arrays, write accumulation buffers, and read accumulation buffers. The LS completes writes in four cycles and reads in six cycles. The first cycle is used by the SAD to add operands. The second cycle is used to distribute the pre-decoded indices to the four 64-KB subarrays. During the third cycle, decoding of the address is completed and the wordline is selected. The addressed subarray is addressed in the fourth cycle. For reads, the sense amplifier senses the bitline differential signal and holds the value

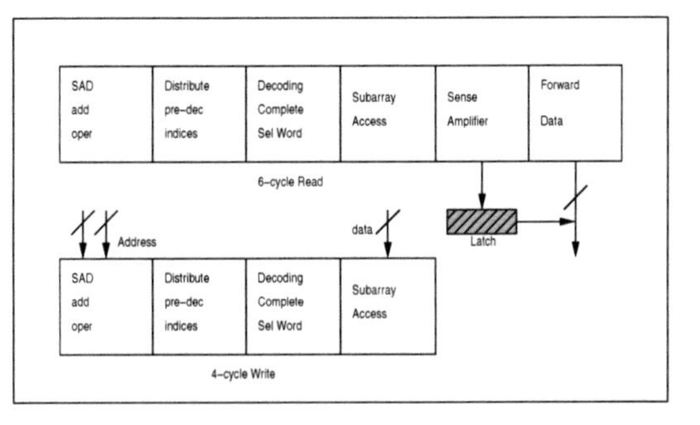

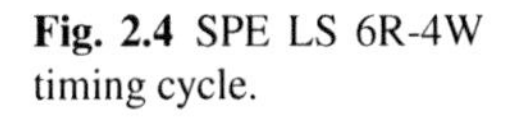
Fig. 2.4 SPE LS 6R-4W timing cycle.

until it is captured in the read latch. In the sixth cycle, the data is forwarded to the executions units. The LS runs at full core clock frequency. A rough sketch of the timing is shown in Figure 2.4.

2.4 Viewing the PPE as a dual-core processor: EMT specifics

The following architected registers are duplicated for multithreading and used by software running in any privilege state (including problem-state):

1. General-Purpose Registers (GPRs) (32 entries per thread)
2. Floating-Point Unit Registers (FPRs) (32 entries per thread)
3. Vector Registers (VRs) (32 entries per thread)
4. Condition Register (CR)
5. Count Register (CTR)
6. Link Register (LR)
7. Fixed-Point Exception Register (XER)
8. Floating-Point Status and Control Register (FPSCR)
9. Vector Status and Control Register (VSCR)
10. Decrementer (DEC)

Thread control registers which are duplicated or thread dependent

Each thread is viewed as an independent processor, complete with separate exceptions and interrupt handling. The threads can generate exceptions simultaneously, and the PPE supports concurrent handling of interrupts on both threads by duplicating some registers defined by the PowerPC Architecture.

The following registers associated with exceptions and interrupt handling are duplicated or are thread-dependent:

1. Machine State Register (MSR)
2. Machine Status Save/Restore Registers (SRR0 and SRR1)
3. Hypervisor Machine Status Save/Restore Registers (HSRR0 and HSRR1)

4. Floating-Point Status and Control Register (FPSCR)
5. Data Storage Interrupt Status Register (DSISR)
6. Decrementer (DEC)
7. Logical Partition Control Register (LPCR)
8. Data Address Register (DAR)
9. Data Address Breakpoint Register (DABR and DABRX)
10. Address Compare Control Register (ACCR)
11. Thread Status Register Local (TSRL)
12. Thread Status Register Remote (TSRR)

In addition, the following thread-independent registers also are associated with exceptions and interrupt handling on both threads:

1. Hypervisor Decrementer (HDEC)
2. Control Register (CTRL)
3. Hardware Implementation Dependent Registers 0 and 1 (HID0 and HID1)
4. Thread Switch Control Register (TSCR)
5. Thread Switch Time-Out Register (TTR)

The following sections describe processor registers that play a central role in controlling and monitoring multithreading activity. The tables that describe register fields have been edited for brevity and clarity (for example, the reserved fields and fields not relevant to multithreading are not described). For complete register descriptions, see the Cell Broadband Engine Registers Specification [20].

Duplicated resources The following arrays, queues, and structures are fully shared between threads running in any privilege state:

1. L1 instruction cache (ICache), L1 data cache (DCache), and L2 cache
2. Instruction and data effective-to-real-address translation tables (I-ERAT and D-ERAT)
3. I-ERAT and D-ERAT miss queues
4. Translation lookaside buffer (TLB)
5. Load miss queue and store queue
6. Microcode engine
7. Instruction fetch control
8. All execution units:
    a. Branch (BRU)
    b. Fixed-point integer unit (FXU)
    c. Load and store unit (LSU)
    d. Floating-point unit (FPU)
    e. Vector media extension unit (VXU)

The following arrays and queues are duplicated for each thread.

1. Segment lookaside buffer (SLB)
2. Branch history table (BHT), with global branch history
3. Instruction buffer (IBuf) queue
4. Link stack queue

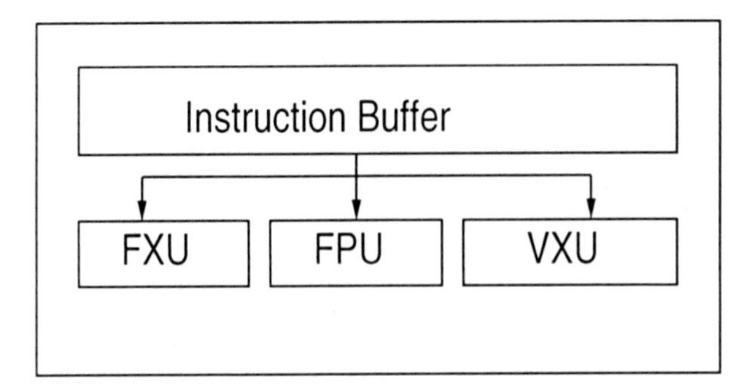

Fig. 2.5 Superscalar instruction execution.

Duplicating the instruction buffer allows each thread to dispatch regardless of any dispatch stall in the other thread. Duplicating the SLB is convenient for the implementation because of the nature of the PowerPC Architecture instructions that access it and because it is a relatively small array. The instruction-fetch control is shared by both threads because the instruction cache has only one read port and so fetching must alternate between threads every cycle. Each thread maintains its own BHT and global branch history (GBH) to allow independent and simultaneous branch prediction.

2.4.1 PPU Intrinsics for managing thread priorities and delays

Table 2.1 Differences between PPE and SPE from a programming point of view.

PPU Intrinsic	Description
`__cctph`	Set current thread priority to high
`__cctpm`	Set current thread priority to medium
`__cctpl`	Set current thread priority to low
`__db[x]cyc`	$x \in 8, 10, 12, 16$ delay current thread for x cycles at dispatch

If the current thread is delayed at dispatch using one of the functions (`__db8cyc`), whether any other thread will get executed depends on the allocation scheme, and relative thread priorities.

2.5 PowerPC Instruction Set

All PowerPC instructions are 4 bytes long and aligned on word (4-byte) boundaries. Most instructions can have up to three operands. Most computational instructions specify two source operands and one destination operand. Signed integers are represented in twos-complement form. The instructions include the following types:

1. Load and Store Instructions: these include fixed-point and floating-point load and store instructions, with byte-reverse, multiple, and string options for the fixed-point loads and stores. The fixed-point loads and stores support byte, halfword, word, and doubleword operand accesses between storage and the 32 general-purpose registers (GPRs). The floatingpoint loads and stores support word and doubleword operand accesses between storage and the 32 floating-point registers (FPRs). The byte-reverse forms have the effect of loading and storing data in little-endian order, although the CBE processor does not otherwise support little-endian order.
2. Fixed-Point Instructions: these include arithmetic, compare, logical, and rotate/shift instructions. They operate on byte, halfword, word, and doubleword operands.
3. Floating-Point Instructions: these include floating-point arithmetic, multiply-add, compare, and move instructions, as well as instructions that affect the Floating-Point Status and Control Register (FPSCR). Floating-point instructions operate on single-precision and doubleprecision floating-point operands.
4. Memory Synchronization Instructions
5. Flow Control Instructions
6. Processor Control Instructions
7. Memory and Cache Control Instructions

2.6 SPE Memory Management and EA Aliasing

All information in main storage is addressable by EAs generated by programs running on the PPE, SPEs, and I/O devices. An SPE program accesses main storage by issuing a DMA command, with the appropriate EA and LS addresses. The EA part of the DMA-transfer addresspair references main storage. Each SPEs MFC has two command queues with associated control and status registers. One queuethe MFC synergistic processor unit (SPU) command queuecan only be used by its associated SPE. The other queuethe MFC proxy command queuecan be mapped to the EA address space so that the PPE and other SPEs and devices can initiate DMA operations involving the LS of the associated SPE.

When virtual addressing is enabled by privileged software on the PPE, the MFC of an SPE uses its synergistic memory management (SMM) unit to translate EAs from an SPE program into RAs for access to main storage. Privileged software on the PPE can alias an SPEs LS address space to the main-storage EA space. This aliasing allows the PPE, other SPEs, and external I/O devices to physically address an LS through a translation method supported by the PPE or SPE. SPE-to-SPE DMA often requires knowledge of SPE LS effective-address. Virtual-address translation is enabled in an SPE by setting the Relocate (R) bit in the MFC State Register. This is done by setting the map problem set area to 1 in using the SPE runtime library API.

2.7 Power Processor Element (PPE) Cache Utilization

Since in this book we don't use self-modifying code, we have no need to discuss the sync operations which are needed when instruction streams are modified by the program. But data-prefetching and *touch* operations on pre-loaded data blocks in the L2 cache can be used to improve performance, especially when streaming data. The following functions are provided as PPE intrinsics for this purpose:

Table 2.2 PPE intrinsics for data prefetching and clear.

Intrinsic	Name	Description
`__sync`	Sync	-
`__lwsync`	Light-weight sync	-
`__dcbt(void*)`	Data cache block touch	data-prefetch hint
`__dcbtst(void*)`	Data cache block touch for store	data-write hint
`__dcbst`	Data cache block store	Write-back cache data to memory
`__dcbf`	Data cache block flush	Flush cache block, write-back if dirty
`__dcbz`	Data cache block zero	set block to 0

In the above functions, the memory pointer is used in a content-addressable way; `__dcbt(x)` provides a hint to the processor that a data block containing x will be loaded in the near future. As the PowerPC does not write-back cache lines to main memory when they are modified, the PPE provides an intrinsic `__dcbst` to perform an immediate write-back to main memory. The `__dcbt` intrinsic acts as a data-prefetch hint only, if the requested data is not found in the L2 cache it is requested from main memory, but not forwarded to the cache. However, for the `__dcbtst`, data is forwarded to the L1 cache. The `__dcbz` instruction is treated as a store instruction from the program order point-of-view, but because the L2 cache processes the write-zero 128-bits at a time it can be significantly faster than writing 8-byte stores. The data cache flush and data cache block store instructions generally finish before the write-back to memory completes. Thus a context-synchronization instruction is inserted after these calls. For more details on the cache instruction, please see Section 6.1.5 of the Cell Broadband Engine Programmer's Handbook [19].

The `__lwsync` instruction can be used by the PPE when it wants to ensure transfer of data from the PPE LS (using SPE initiated DMA) is complete. The `mfc_read_tag_status_all()` function on the SPE only guarantees that the LS is available for use, but not necessarily that the data has arrived at main store. Requiring the SPE to always issue an `mfcsync` command before sending a mailbox notification to the PPE that it has initiated transfer is very inefficient. A better and thus preferred approach is for the PPE to issue `__lwsync` after receiving mailbox notification from SPE, but before accessing any of the computational results.

2.7.1 The `eieio` instruction

The `eieio` instruction (enforce in-order execution of I/O) ensures that all main-storage accesses caused by instructions proceeding the eieio have completed, with respect to main storage, before any main-storage accesses caused by instructions following the `eieio`. An `eieio` instruction issued on one PPE thread has no architected effect on the other PPE thread. However, there is a performance effect because all load or store operations (cacheable or noncacheable) on the other thread are serialized behind the `eieio` instruction. The `eieio` instruction is intended for use in managing shared data structures, in accessing memory-mapped I/O, and in preventing load or store combining. The first implementation of the CBEA (the CBE processor) has a fully ordered and coherent memory interface. This means that the SPE need not issue `mfcsync` or `mfceieio` commands to ensure correct ordering, simple `mfc_read_tag_status_all()` or using a barrier or fence attribute will suffice.

2.7.2 Instruction cache

Both L1 caches are dynamically shared by the two PPE threads; a cache block can be loaded by one thread and used by the other thread. The coherence block, like the cache-line size, is 128 bytes for all caches. The L1 ICache is an integral part of the PPEs instruction unit (IU), and the L1 DCache is an integral part of the PPEs load/store unit (LSU), as shown in Figure 2-2 PPE Functional Units on page 50. Accesses by the processing units to the L1 caches occur at the full frequency of the CBE core clock.

The PPE uses speculative instruction prefetching (including branch targets) for the L1 ICache and instruction prefetching for the L2 cache. Because the PPE can fetch up to four instructions per cycle and can execute up to two instructions per cycle, the IU will often have several instructions queued in the instruction buffers. Letting the IU speculatively fetch ahead of execution means that L1 ICache misses can be detected early and fetched while the processor remains busy with instructions in the instruction buffers. In the event of an L1 ICache miss, a request for the required line is sent to the L2. In addition, the L1 ICache is also checked to see if it contains the next sequential cache line. If it does not, a prefetch request is made to the L2 to bring this next line into the L2 (but not into the L1, to avoid L1 ICache pollution). This prefetch occurs only if the original cache miss is committed (that is, all older instructions must have passed the execution-pipeline pointcalled the flush pointat which point their results can be written back to architectural registers). This is especially beneficial when a program jumps to a new or infrequently used section of code, or following a task switch, because prefetching the next sequential line into the L2 hides a portion of the mainstorage access latency.

When an L1 ICache miss occurs, a request is made to the L2 for the cache line. This is called a demand fetch. To improve performance, the first beat1 of data returned by the L2 contains the fetch group with the address that was requested, and so returns the data critical-sector first. To reduce the latency of an L1 miss, this critical sector of the cache line is sent directly to the instruction pipeline, instead of first writing it into the L1 cache and then rereading it (this is termed bypassing the cache). Each PPE thread can have up to one instruction demand-fetch and one instruction prefetch outstanding to the L2 at a time. This means that in multithread mode, up to four total instruction requests can be pending simultaneously. In multithreading mode, an L1 ICache miss for one thread does not disturb the other thread.

2.7.3 Data cache miss

Loads that miss the L1 DCache enter a load-miss queue for processing by the L2 cache and main storage. Data is returned from the L2 in 32-byte beats on four consecutive cycles. The first cycle contains the critical section of data, which is sent directly to the register file. The DCache is occupied for two consecutive cycles while the reload is written to the DCache, half a line at a time. The DCache tag array is then updated on the next cycle, and all instruction issue is stalled for these three cycles. In addition, no instructions can be recycled during this time. The load-miss queue entry can then be used seven cycles after the last beat of data returns from the L2 to handle another request.

Instructions that are dependent on a load are issued speculatively, assuming a load hit. If it is later determined that the load missed the L1 DCache, any instructions that are dependent on the load are flushed, refetched, and held at dispatch until the load data has been returned. This behavior allows the PPE to send multiple overlapping loads to the L2 without stalling if they are independent. In multithreading mode, this behavior allows load misses from one thread to occur without stalling the other thread. In general, write-after-write (WAW) hazards do not cause penalties in the PPE.

2.8 Understanding the PPE Pipeline

Although the PPE serves as the control-pane for our applications, we have deliberately not used any of the vector functionality of the PPE, instead relying on the

SPUs. But as we promised in the Preface, understanding the PPE architecture in detail also carries over to the other PowerPC based consoles and embedded systems out there. In this section we shall look at the PPE pipeline in some detail.

The PPE is a load-store RISC architecture. We have already presented the system design aspects of the instruction issue, pipelining and caches. In this section we familiarize ourselves with the PPE instruction set and assembly listings. Though the main focus of this book is on SPU optimization, knowledge of the PPE instruction scheduling and pipelining can come in handy when extracting the last clock-cycle worth of optimization on the Cell. Moreover, as we have stated earlier, the PowerPC ecosystem is thriving, especially in the embedded systems space, and the Cell in the PS3 provides an exciting and simple machine to learn PowerPC programming.

Consider the code fragment shown in Listing 2.1.

```
typedef float matrix_t[100][100];
void mult(int size,int row,int col,
matrix_t A, matrix_t B,matrix_t C) {
  int pos;
  C[row][col] = 0;
  for(pos = 0; pos < size; ++pos)
    C[row][col] += A[row][pos] * B[pos][col];}
```

Listing 2.1 Function to perform matrix multiply inner loop

When compiled on the PPE using `gcc -O2` we get:

```
  ;C[row][col]+=A[row][pos]*B[pos][col];
  lfsx 0,9,4
  addi 9,9,4
  lfs 12,0(7)
  addi 7,7,400
  lfsx 13,5,11
  fmadds 0,0,12,13
  stfsx 0,5,11
  bdnz .L4
  blr
```

Listing 2.2 PPE Assembly listing for matrix multiply.

Table 2.3 PowerPC PPE instructions used in matrix multiply inner loop.

Instruction	Description
`lfsx`	load floating-point single (indexed xform)
`addi`	Add immediate
`fmadds`	single-precision floating point multiply add
`stfsx`	store floating point single (indexed xform)
`bdnz`	branch conditional
`blr`	branch to link register

We also compiled the same function on an Opteron system using `gcc -O2`, the output assembly is given below:

```
  ;C[row][col]+=A[row][pos]*B[pos][col];
  movss (%rcx,%r8,4),%xmm0
  incl  %r9d
  mulss (%rax), %xmm0
  incq  %r8
  addq  400, %rax
  cmpl  %r9d, %edi
  addss (%r10,%rdx,4), %xmm0
  movss %xmm0, (%r10,%rdx,4)
  jne .L4
  rep ; ret
```

Listing 2.3 Opteron assembly listing for matrix multiplication.

Consider the floating-point intensive code shown in Listing 2.4.

```
void VMULT( float* X, float* Y, float *Z,
            float t, float r,
            unsigned long int N) {
  unsigned long int i=0;
  for(i=0;i<N;++i)
    X[i] = Y[i] * (X[i]*t + Z[i]*r);
}
```

Listing 2.4 Floating-point computation on PPE.

The above code was compiled using `gcc -O2 -funroll-loops` and the generated assembly is given below in Listing 2.5.

```
        .file   "vmult.c"
 # rs6000/powerpc options: -msdata=data -G 8
        .globl VMULT
        .type   VMULT, @function
VMULT:
.LFB2:
.LVL0:
        mr. 0,6   #, N
        mtctr 0   # N,
        beqlr 0
.LVL1:
        li 9,0    # ivtmp.33,
.L4:
        lfsx 0,9,5        #* Z, tmp136
        lfsx 12,9,3       #* X, tmp134
        fmuls 0,2,0       # tmp135, r, tmp136
        lfsx 13,9,4       #* Y, tmp138
        fmadds 12,1,12,0          # tmp137, t, tmp134, tmp135
        fmuls 13,13,12    # tmp139, tmp138, tmp137
        stfsx 13,9,3      #* X, tmp139
        addi 9,9,4        # ivtmp.33, ivtmp.33,
        bdnz .L4          #
.LVL2:
        blr
.LFE2:
        .size   VMULT,.-VMULT
```

Listing 2.5 PPE floating-point instruction analysis.

A slightly more detailed assembly listing can be generated as well (only relevant portions are shown):

```
   2:vmult.c        ****    unsigned long int i=0;
   3:vmult.c        ****    for(i=0;i<N;++i)
```

```
54                          .loc 1 3 0
55 0000 7CC03379            mr. 0,6   #, N
56 0004 7C0903A6            mtctr 0   # N,
57 0008 4D820020            beqlr 0
58                      .LVL1:
59 000c 39200000            li 9,0    # ivtmp.33,
60                          .p2align 4,,15
61                      .L4:
 4:vmult.c       ****     X[i] = Y[i] * (X[i]*t + Z[i]*r);
63 0010 lfsx 0,9,5        #* Z, tmp136
64 0014 fsx 12,9,3       #* X, tmp134
65 0018 muls 0,2,0       # tmp135, r, tmp136
66 001c fsx 13,9,4       #* Y, tmp138
67 0020 madds 12,1,12,0 # tmp137, t, tmp134, tmp135
68 0024 muls 13,13,12    # tmp139, tmp138, tmp137
69 0028 tfsx 13,9,3      #* X, tmp139
71 002c ddi 9,9,4        # ivtmp.33, ivtmp.33,
72 0030 dnz .L4          #
75 0034 lr
```

Given the PowerPC assembly language and architecture books [15] it is straightforward to follow the assembly language. Another useful reference is the Programming Environments Manual [113] for 32-bit implementations of PowerPC architecture by Freescale Semiconductors. Since the PPE has a single floating-point unit there is no superscalar issue of floating point instructions.

We exercise this code using a simple test wrapper as:

```
extern void VMULT( float*,float*,float*,float,float,unsigned long int);
int main( int argc, char* argv[] ) {
  long time_diff = 0;
  unsigned long int i=0;
  struct timespec t1,t2;
  Setup();
  clock_gettime( CLOCK_PROCESS_CPUTIME_ID, &t1 );
  for(i=0;i<1000;++i)
    VMULT(X,Y,Z,r,t,N);
  clock_gettime( CLOCK_PROCESS_CPUTIME_ID, &t2 );
  time_diff = t2.tv_nsec = t1.tv_nsec;
  printf("\n [%f:%f] it took %ld ns.\n", X[0],X[MAX-1],time_diff);
  return (EXIT_SUCCESS);
}
```

Listing 2.6 Timing PPE floating point performance.

On the PS3, this code performs in 1437120 nano-seconds, and on an Opteron 242 it takes 1876521 nano-seconds. The PS3 PPE is running at 3.2 GHz while the Opteron is running at 1.8 GHz. The Opteron has a dual-issue floating-point pipeline. The assembly code for the Opteron is shown as well:

```
.globl VMULT
        .type   VMULT, @function
VMULT:
.LFB2:
        .file 1 "vmult.c"
        .loc 1 1 0
.LVL0:
        .loc 1 3 0
        testq   %rcx, %rcx      # N
        .loc 1 1 0
        movaps  %xmm0, %xmm3    # t, t
        movaps  %xmm1, %xmm2    # r, r
        .loc 1 3 0
```

```
         je       .L6      #,
15  .LVL1:
         xorl     %eax, %eax        # i
    .LVL2:
         .p2align 4,,7
    .L4:
20       .loc 1 4 0
         movaps   %xmm3, %xmm0      # t, tmp70
         movaps   %xmm2, %xmm1      # r, tmp71
         mulss    (%rdi,%rax,4), %xmm0     #* X, tmp70
         mulss    (%rdx,%rax,4), %xmm1     #* Z, tmp71
25       addss    %xmm1, %xmm0      # tmp71, tmp70
         mulss    (%rsi,%rax,4), %xmm0     #* Y, tmp70
         movss    %xmm0, (%rdi,%rax,4)     # tmp70,* X
         .loc 1 3 0
         incq     %rax     # i
30       cmpq     %rax, %rcx        # i, N
         jne      .L4      #,
    .LVL3:
    .L6:
    .LVL4:
35       .loc 1 5 0
         rep ; ret
    .LFE2:
         .size    VMULT, .-VMULT
```

Listing 2.7 Analysis of floating-point instructions on Opteron.

2.8.1 Discuss assembly language

Flynn [47] provides an excellent quantitative analysis of the many choices which are presented to superscalar pipeline architects. Considering the PPE as a pure load-store architecture with well defined pipeline behavior, we can understand the generated code and the actions of the compiler. This knowledge can guide us when we are writing code for the PPE.

2.9 Measuring and Profiling Application Performance

The Cell Broadband Engine (CBE) processor provides extensive performance-monitoring facilities that assist performance analysis, as well as provide application-optimized and system-optimization features that include:

1. Debugging, analyzing, and optimizing processor-architecture features
2. Profiling the behavior of the memory hierarchy and the interaction of multiple address spaces, as well as tuning system and application algorithms to optimize scheduling, partitioning, and structuring for tasks and data
3. Real-time application-tuning by monitoring bandwidth use and other resource-management behavior
4. Tuning of numerically intensive floating-point applications
5. Tuning of fixed-point applications

The facilities give clear visibility to the details of instruction execution, loads and stores, the behavior of caches throughout the CBE processor, and the entire virtual-memory architecture.

2.10 Conclusion

Since the PPE is PowerPC compliant, much of what is applicable to PowerPC (eg., PowerPC 970MP) is valid and germane to PPE. Thus, we have been able to describe only a fraction of the facilities provided by the PPE. Other excellent references for PowerPC information are the PowerPC Architecture Books [15], and the PowerPC Compiler Writer's Guide [13]. GNU/GCC toolchain for the PowerPC is readily available and investigations of the produced assembly are useful to learn the instruction set of PPE. In the sequel of this book we shall rely on the PPE to provide GNU/Linux facilities, file-system interaction, memory-management and SPE runtime management. We have not presented any VMX code for SIMD processing on the PPE as this topic is relatively mature (due to widespread acceptance of AltiVec), and in our opinion the SPUs provide significantly elegant infrastructure to design SIMD and parallel programs. The SPE, its design, ISA and pipeline is the focus of the next chapter.

Chapter 3
The Synergistic Processing Element

Abstract In this chapter we analyze the Synergistic Processing Elements (SPE) of the Cell Broadband Engine. The 6 identical SPEs (in the PS3, 8 in the general Cell processor) are dual-issue SIMD RISC processors optimized for data-rich operations, running off a 256 KB non-cached local store (LS) memory. The SPE ISA and associated DMA commands are optimized for compute intensive work loads. SPEs communicate to the PPE, MIC and other SPEs using the EIB, which is managed by the Memory Flow Controller in the SPE. VLSI system design of the SPE micro-architecture is also presented as the SPU architecture is radically different from the POWER processor VLSI design (since it needs to make a steeper trade-off between performance and silicon area). Additionally, the SPU is also beginning to be shipped as an independent chip (the Toshiba SPURS engine). The performance of the LS, Register Forwarding Macro, Memory Flow Controller, Multi-source Synchronization Facility, and Channel interface is also discussed, as these are used in the subsequent chapters. The ISA of the SPE is also briefly discussed.

3.1 Introduction to Cell SPE

The Synergistic Processing Element (SPE) is the computational workhorse of the Cell Broadband Engine. The design and implementation of the SPE micro-architecture along with VLSI implementation details is described by Flachs et. al in an IBM Journal of Research and Development article from 2007 [42]. Specific details of SPE presented in this chapter are taken from that paper [42].

The SPE represents the middle ground between graphics processors and general-purpose processors. It is more flexible and programmable than graphics processors but has more focus on streaming workloads than general-purpose processors. The SPE competes favorably with general-purpose processors on a cycle-by-cycle basis with substantially less area and power while running many streaming and HPC algorithms. The efficiency in area and power encourages the construction of a system on a chip using multiple SPEs and offering performance many times that of competitive

S. Koranne, *Practical Computing on the Cell Broadband Engine*,
DOI: 10.1007/978-1-4419-0308-2_3, © Springer Science + Business Media, LLC 2009

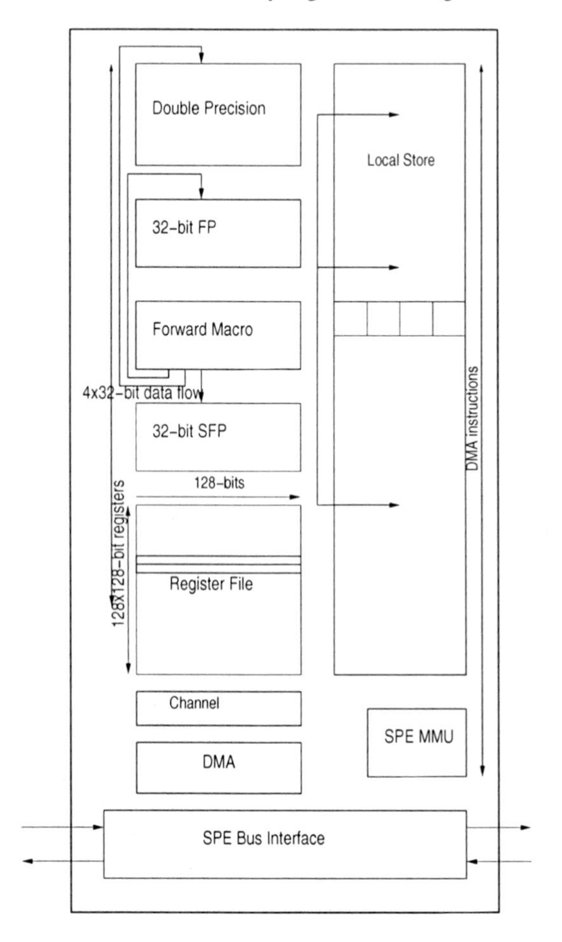

Fig. 3.1 A detailed view of the SPE

general-purpose processors. It is possible to address the memory latency bottleneck and improve application performance through the DMA programming model. This model provides concurrency between data access and computation while making efficient use of the available memory bandwidth. These techniques allow the SPE to have a shorter pipeline, occupy less area, and as a total package, dissipate less power.

The SPU core is a single-instruction multiple-data (SIMD) reduced instruction set computing (RISC) processor [65, 42]. All instructions are encoded in 32-bit fixed-length instruction formats. The SPU features 128 general-purpose registers (GPRs) that are used by both floating-point (FP) and integer instructions. The shared register file allows for a high level of performance for various workloads using only a small number of registers. Loop unrolling, which is necessary to fill functional unit pipelines with independent instructions, is possible with the 128 registers. Most

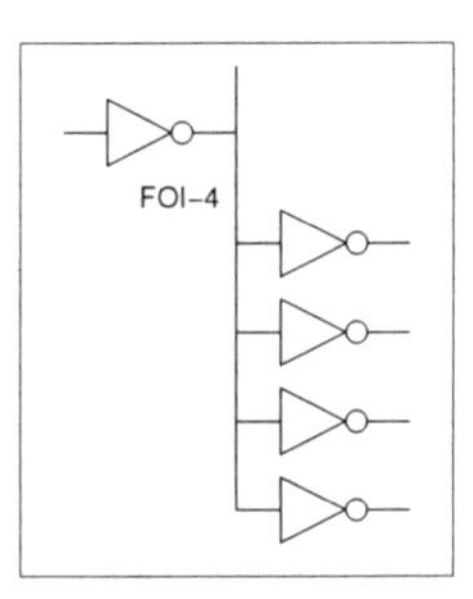

Fig. 3.2 Fan-out-of-4 concept, delay of an inverter driving four copies of itself.

instructions operate on 128-bit-wide data. For example, the FP multiplyadd instruction operates on vectors of four 32-bit single-precision (SP) FP values. Some instructions, such as FP multiplyadd, consume three register operands and produce a register result. The SPU includes instructions that perform SP FP integer arithmetic, logical operations, loads, stores, compares, and branches. The instruction set is designed to simplify compiler register allocation and code schedulers.

The SPE architecture reduces area and power while facilitating improved performance by requiring software to solve difficult scheduling problems, such as data fetch and branch prediction. Software solves these problems by including explicit data movement and branch prediction directives in the instruction stream. SPE load and store instructions are performed within a local address space. The local address space is untranslated, unguarded, and non-coherent with respect to the system address space, and it is serviced by the LS. Loads, stores, and instruction fetch can complete with a fixed delay and without raising exceptions. These properties of LS greatly simplify the SPU core design and provide predictable real-time behavior. This design reduces the area and power of the core while allowing for operation at a higher frequency.

Data is transferred to and from the LS in 128-byte lines by the SPE DMA engine similar to the way a supercomputer uses vector load and store instructions to move data to and from the vector register file. Data moves between system memory and the DMA engine using the on-chip interconnect. The on-chip interconnect sources and sinks 16 bytes of data every other cycle. The SPE DMA engine allows SPE software to schedule data transfers in parallel with core execution.

The micro-architecture of the SPE is designed to support very high frequency operation. A technology independent measure of gate delay is the fan-out-of-4 (FO4) inverter delay (see Figure 3.2), which is the time required for an inverter to drive four copies of itself. In these terms, the SPE is an 11-FO4 design. This allows four to eight stages of logic per cycle, depending on the routing delay between logic. Many signal distribution paths feature about 35% of the cycle time in wire delay. Flachs et al have described several design trade-offs which were considered before the current (first generation Cell) architecture was chosen; especially the choice of a large register file was made consciously to hide instruction latency.

3.2 SPE Local Store

The SPU architecture defines a private memory, also called local storage, which is byte-addressed. Load and store instructions combine operands from one or two registers and an immediate value to form the effective address of the memory operand. Only aligned 16-byte-long quadwords can be loaded and stored. Therefore, the rightmost 4 bits of an effective address are always ignored and are assumed to be zero. This fact is used in many programs for storing meta-data in the pointer, to save space. The size of the SPU local storage address space is 2^{32} bytes. However, an implementation generally has a smaller actual memory size. The effective size of the memory is specified by the Local Storage Limit Register (LSLR). Implementations can provide methods for accessing the LSLR; however, these methods are outside the scope of the SPU Instruction Set Architecture. Implementations can allow modifications to the LSLR value; however, the LSLR must not change while the SPU is running. Every effective address is ANDed with the LSLR before it is used to reference memory. The LSLR can be used to make the memory appear to be smaller than it is, thus providing compatibility for programs compiled for a smaller memory size. The LSLR value is a mask that controls the effective memory size. This value must have the following properties:

Memory Limit Limit the effective memory size to be less than or equal to the actual memory size,

Monotonicity be monotonic, so that the least-significant 4 mask bits are ones and so that there is at most a single transition from 1 to 0 and no transitions from 0 to 1 as the bits are read from the least-significant to the most-significant bit. That is, the value must be 2n-1, where n is log2 (effective memory size). For the PS3 implementation of CBEA, LSLR is $0x0003FFFF$.

The effect of this is that references to memory beyond the last byte of the effective size are wrappedthat is, interpreted modulo the effective size. This definition allows an address to be used for a load before it has been checked for validity, and makes it possible to overlap memory latency with other operations more easily. Stores of less than a quadword are performed by a load-modify-store sequence. A group of assist instructions is provided for this type of sequence.

The LS array is a 256-KB SRAM array that consumes one-third of the total SPE area. The LS macro consists of a sum address decoder, four 64-KB memory arrays, write accumulation buffers, and read accumulation buffers. The LS completes writes in four cycles and reads in six cycles. The first cycle is used by the SAD to add operands. The second cycle is used to distribute the pre-decoded indices to the four 64-KB subarrays. During the third cycle, decoding of the address is completed and the wordline is selected. The addressed subarray is addressed in the fourth cycle. The data is forwarded to the executions units in the 6th cycle. The LS runs at full core clock frequency.

3.3 SPU LS Priority

Although the LS is shared between DMA transfers, load and store operations by the associated SPU, and instruction prefetches by the associated SPU, DMA operations are buffered and can only access the LS at most one of every eight cycles. Instruction prefetches deliver at least 17 instructions sequentially from the branch target. Thus, the impact of DMA operations on loads and stores and program-execution times is, by design, limited. When there is competition for access to the LS by loads, stores, DMA reads, DMA writes, and instruction fetches, the SPU arbitrates access to the LS according the following priorities (highest priority first):

1. DMA reads and writes
2. SPU loads and stores
3. Instruction prefetch

Loads and stores transfer 16 bytes of data between the register file and the LS. To provide good performance while keeping the processor simple, a cycle-by-cycle arbitration scheme is used. DMA requests are scheduled in advance but are first in priority. DMA requests access 128 bytes in the LS in a single cycle, providing more than sufficient bandwidth with relatively little interference with the SPE loads and stores. Loads and stores are second in priority and wait in the issue stage for an available LS cycle. Instruction fetch accesses the LS when it is otherwise idle, again with a 128-byte access to minimize the chances of performance loss due to instruction run-out. Additionally, ECC scrub is also a high priority LS item.

Memory coherence

Memory coherence refers to the ordering of stores to a single location. Atomic stores to a given location are coherent if they are serialized in some order, and no processor or no device can observe any subset of those stores as occurring in a conflicting order. This serialization order is an abstract sequence of values; the real storage location need not assume each of the values written to it. For example, a processor can update a location several times before the value is written to real storage. DMA transfers between the LS and main storage are coherent in the system. A pointer to a data structure created on the PPE can be passed by the PPE, through the main-storage space, to an SPU, and the SPU can use this pointer to issue a DMA command to bring the data structure into its LS. Memory-mapped mailboxes or atomic MFC synchronization commands can be used for synchronization and mutual exclusion. Refer to Section 2.7 on pp. 27 for further details on how PPE-SPE memory coherence works in the CBEA.

3.4 GPRs in the SPE

As mentioned before the SPE has 128 general purpose register, each 128-bit long. Three cycles are required for GPR operation; the first cycle is for address predecoding and decoded signal distribution; the second cycle is for final decoding and array access; and the third cycle is for data-flow distribution. A read operation is performed first, followed by a write operation within an 11-FO4 cycle. Up to eight operations, six reads and two writes (6r2w), are performed independently every cycle. Collisions are avoided by the SPE control logic (SCN). The data from each of three read ports is taken from the true side of a register cell, and the data from the other three read ports is taken from the complement side of a register cell. This leads to three true read ports and three complement read ports at the GPR macro boundary. Dependency checking and data forwarding are performed by several macros: the dependency check macro (DCM), the forwarding macro (FM), and several DPLAs.

Each SPE register contains either a two-way or a three-way mux, depending on the stage. When the FM is operated in shift mode, the register data in a stage is shifted to the next stage. When the data in a stage is forwarded to the destination operand latches, the 16-way mux path is selected. The 16-way mux is implemented with a dynamic 8-way NOR gate followed by a cross-coupled NAND (CCNAND) gate and finally a 2-way static NAND to complete the 16-way multiplexing.

DPLAs are generated with a specially developed program. The program receives an espresso file as an input and generates both schematics and layouts. ANDing is implemented with a dynamic footed NOR followed by a strobe circuit. It is interesting to note that the SPE GPR uses PLAs designed using *espresso* formats, as in Chapter 20 we describe techniques to use the SPEs to perform efficient logic minimization on PLAs and other logic structures.

3.5 SPE Intrinsics and Instruction

The instructions we have left out include spu_shuffle which we describe in our chapter on sorting networks (cf. Chapter 10). The `spus_insert` and `spu_sel` conditional selection instruction are 3 argument instructions which we have already described.

The SPU issues all instructions in program order according to the pipeline assignment. Each instruction is part of a doubleword-aligned instruction-pair called a fetch group. A fetch group can have one or two valid instructions. This means that the first instruction in the fetch group is from an even word address, and the second instruction from an odd word address. The SPU processes fetch groups one at a time, continuing to the next fetch group when the current fetch group becomes empty. An instruction becomes issueable when register dependencies are satisfied and there is no structural hazard (resource conflict) with prior instructions or LS contention due to DMA or error-correcting code (ECC) activity. access priorities.

Dual-issue occurs when a fetch group has two issueable instructions in which the first instruction can be executed on the even pipeline and the second instruction

Table 3.1 Some frequently used SPE intrinsics and their expected behavior.

Intrinsic	Arg1	Arg2	Result
spu_splats	x	-	x,x,x,x
spu_extract	x	element	$x[element]$
spu_insert	x,y	element	$x[element] = y$
spu_promote	x	element	$x[element] = y$
spu_convtf	x	scale (s)	$x/2^s$
spu_convts	x	scale (s)	$x_0 * 2^s$
spu_roundtf	x	-	FRDS rc,x
spu_add	x	y	$rc = x + y$
spu_sub	x	y	$rc = x - y$
spu_and	x	y	$rc = x \& y$
spu_or	x	y	$rc = x \vert y$
spu_andc	x	y	$rc = x \& \neg y$
spu_orc	x	y	$rc = x \vert \neg y$
spu_eqv	x	y	$rc = \neg(x \oplus y)$
spu_xor	x	y	$rc = x \oplus y$
spu_mul	x	y	$rc = x * y$
spu_madd	x	y,z	$rc = xy + z$
spu_msub	x	y,z	$rc = xy - z$
spu_re	x	-	$1/x$
spu_rsqrte	x	-	$1/\sqrt{x}$
spu_absd	x	y	$rc = \vert x - y \vert$
spu_avg	x	y	$rc = (x + y) >> 1$
spu_cmpeq	x	y	$rc = x == y$
spu_cmpgt	x	y	$rc = x > y$
spu_cntb	x	-	$rc = popcnt(x)$
spu_gather	x	-	$rc = LSBs, 0, 0, 0$
spu_sel	x,y	p	$rc = p?x : y$
spu_sl	x	count	$rc = x << count$

can be executed on the odd pipeline. If a fetch group cannot be dual-issued, but the first instruction can be issued, the first instruction is issued to the proper execution pipeline and the second instruction is held until it can be issued. A new fetch group is loaded after both instructions of the current fetch group are issued.

3.6 SPU Instruction Set: Pipelines, Execution Units and Latency

The SPE issues and completes all instructions in program order and does not re-order or rename its instructions. This does not imply that an optimizing compiler is not free to reorder program statements to increase dual-issue rates (as long as the

Table 3.2 SPE Pipeline Information.

Unit	Instructions	Execution pipe	Depth	Latency
Simple fixed	Word arithmetic, logicals	Even	2	2
Simple fixed	Shifts/rotates	Even	3	4
Single precision	MAC	Even	6	6
Single precision	Int MAC	Even	7	7
Byte	Byte avg,sum, pop count	Even	3	4
Permute	QW shift/rotates	Odd	3	4
Local storage	Load and store	Odd	6	6
Channel	Channel read and write	Odd	5	6
Branch	Branches	Odd	3	1–18

compiler can guarantee correct program behavior). The SPE features a dual-issue odd-even pipeline reminiscent of the VU1 dual issue pipeline of the Emotion Engine, but the SPE has nine execution units lined up to service the dual issue pipeline. Table 3.2 contains a description of the pipeline and its division into the functional units.

Table 3.3 Instruction types, latencies and stalls for the SPU Even pipeline

Even Pipeline	Latency	Stalls
Single-precision floating-point	6	0
Integer multiplies, spu_convtf, spu_convts, **fi**	7	0
Immediate loads, logical operations, integer add/sub, select bits	2	0
Double-precision floating-point	7	6
Element rotates and shifts	4	0
Byte operations (pop count, abs. diff., avg, sum)	4	0

Table 3.4 Instruction types, latencies and stalls for the SPU Odd pipeline

Odd Pipeline	Latency	Stalls
Shuffle bytes, quadword rotates/shifts	4	0
Gather, mask, generate insertion control	4	0
Estimate	4	0
Loads	6	0
Branches	4	0
Channel operations, SPRs	6	0

Table 3.3 and Table 3.4 detail the SPE instruction and its associated pipeline. Instruction pairs can be issued if the first instruction (from an even address) is routed

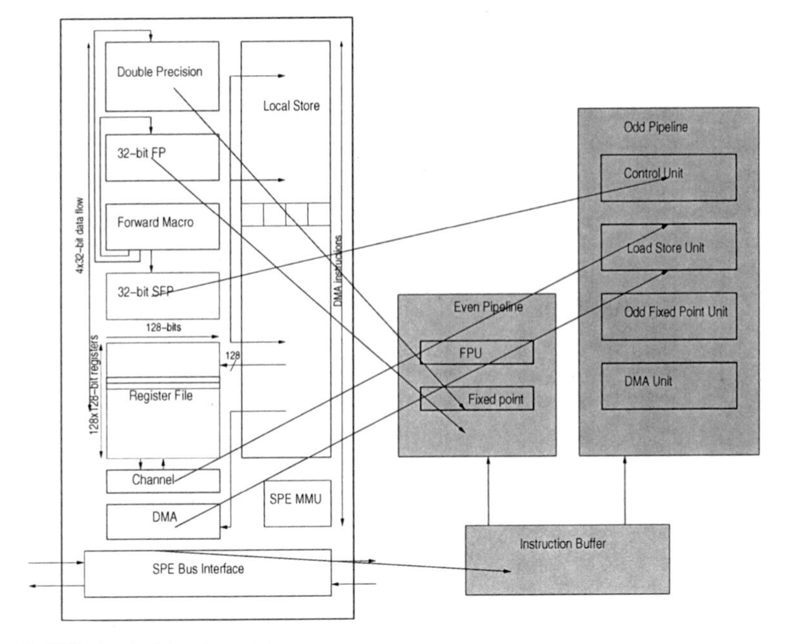

Fig. 3.3 SPE physical implementation

to an even pipe unit and the second instruction to an odd pipe unit. SPE software does not require NOP (nothing-operation) padding when dual issue is not possible, although GNU/GCC provides a NOP in the generated assembly for load-store instructions where clearly no other instruction could be issued. Instruction issue and distribution require three cycles (the details of these 3 cycles are given in the next chapter on instruction buffering). This simplistic instruction issuing implies high performance in minimal area, and simplifies resource and dependency checking.

Operands are fetched from either the register file or the forward network and sent to the execution pipelines. Each of the two pipelines can consume three 16-byte operands and produce a 16-byte result every cycle. The register file has six read ports, two write ports, and 128 entries of 128 bits each, and it is accessed in two cycles. Register file data is sent directly to the functional unit operand latches. Results produced by functional units are held in the forward macro (FM) until they are committed and available from the register file. These results are read from six FM read ports and distributed to the units in one cycle. Short vector SIMD optimizations make use of the features of the SPE and Kurzak et al have been able to achieve close to theoretical peak performance by keeping the above strengths of architecture in mind when developing matrix-matrix multiplication [79].

3.7 Load/Store Optimization on SPU

The goal of this section is to develop an intuitive feel of the load-store issue to minimize extra instructions added by the compiler for scalar operands. The material presented here is based on the research papers by Kahle et al [65], Flachs et al [42], Eichenberger et al [41], and the source code for the GNU assembler for Cell SPU, and the `spu-gcc` compiler, with GNU/GCC version 4.3.3.

Most instructions operate in a SIMD fashion on 128 bits of data representing either two 64-bit doubleprecision floating-point numbers or longer integers, four 32-bit single-precision floating-point numbers or integers, eight 16-bit subwords, or sixteen 8-bit characters. The 128-bit operands are stored in a 128- entry unified register file. Instructions may take up to three operands and produce one result. The register file has a total of six read and two write ports.

The memory instructions also access 128 bits of data, with the additional constraint that the accessed data must reside at addresses that are multiples of 16 bytes. Thus, when addressing memory with vector load or store instructions, the lower four bits of the byte addresses are simply ignored. To facilitate the loading and storing of individual values, such as a character or an integer, there is additional support to extract or merge an individual value from or into a 128-bit register.

An SPE can dispatch up to two instructions per cycle to seven execution units that are organized into even and odd instruction pipes. Instructions are issued in order and routed to their corresponding even or odd pipe by the issue logic, that is, a component which examines the instructions and determines how they are to be executed, based on a number of constraints. Independent instructions are detected by the issue logic and are dual-issued (i.e. dispatched two per cycle) provided they satisfy the following condition: the first instruction must come from an even word address and use the even pipe, and the second instruction must come from an odd word address and use the odd pipe. When this condition is not satisfied, the two instructions are executed sequentially.

The SPE's 256-KB local memory supports fully pipelined 16-byte accesses (for memory instructions) and 128-byte accesses (for instruction fetches and DMA transfers). Because the memory has a single port, instruction fetches, DMA, and memory instructions compete for the same port. Instruction fetches occur during idle memory cycles, and up to 3.5 fetches may be buffered in the instruction fetch buffer to better tolerate bursty peak memory usage. The maximum capacity of the buffer is thus 112 32- bit instructions. An explicit instruction can be used to initiate an inline instruction fetch. Although LS access is relatively cheap (as compared to main memory EA access), `lqx`, does execute on a single pipeline and almost always has associated dependencies which prevent other instructions from executing. Loop-unrolling is discussed in this chapter which is a recommended method to hide LS access behind other computation. The corresponding assembly mnemonic for instruction load is `lqr`. The store instructions are of the form `stqx`, `stqd`, and so on.

Most SPE instructions are SIMD instructions operating on 128 bits of data at a time, including all memory instructions. These instructions expect any scalar operands, such as address offsets, to be present in the *preferred-slot* of the 128-

bit vector register (see Figure 3.6). This is a very important point to keep in mind when writing scalar code as an innocent statement statement like:

```
int LS[1024];
int GeneralAdd(int i, int j, int k) {
  LS[k] = LS[i] + LS[j];
}
```

Listing 3.1 SPU Code for scalar addition.

We compiled this code with `spu-gcc -O2` and examined the assembly:

```
GeneralAdd:
    shli     $15,$3,2            ;# tmp145, i,
    hbr .L3,$lr                  ;#,
    shli     $14,$4,2            ;# tmp147, j,
    ila $6,LS                    ;#, tmp142
    shli     $5,$5,2             ;# tmp143, k,
    lnop
    a    $3,$15,$6               ;# tmp151, tmp145, tmp142
    lqx $13,$15,$6               ;#, tmp152
    a    $4,$14,$6               ;# tmp153, tmp147, tmp142
    lqx $12,$14,$6               ;#, tmp154
    lqx $10,$5,$6                ;#, tmp155
    cwx $9,$5,$6                 ;#, tmp156, tmp143, tmp142
    rotqby   $11,$13,$3          ;# tmp152, tmp152, tmp151
    rotqby   $7,$12,$4           ;# tmp154, tmp154, tmp153
    a    $2,$11,$7               ;# tmp150, tmp148, tmp149
    shufb    $8,$2,$10,$9        ;# tmp155, tmp150, tmp155, tmp156
    stqx     $8,$5,$6            ;#, tmp155
.L3:
    bi   $lr
```

As you can observe, the SPU compiler has to assume the worst case scenario that `LS[i]`, `LS[j]` and `LS[k]` are all not present in the preferred-slot of the vectors. Although this results in some memory overhead, it is insignificant compared to the increase in code size generated by the extra permutation instructions that would otherwise be needed to realign the data. Second, the compiler perform aggressive register allocation of all local computations, such as address and loop index variables, to make good use of the 128-entry register file. These variables thus often reside exclusively in the primary slot of registers and hence require no additional memory storage.

The SPE instruction set natively supports operations on a wide range of data widths, from 8-bit bytes to 64-bit doublewords, unlike most RISC (Reduced Instruction Set Computer) processors, which typically support operations on words or doublewords only. Programming languages such as C/C++ were designed with 8-bit chars as the basic data type granularity, and with short data type of 16-bit and regular words of 32-bits. Consider the case when $i = 3, j = 5$ and $k = 13$. This situation is shown in Figure 3.4.

In this case the relative positioning of the scalars in the vector organized LS is not same, thus a simple vector add operation would not work. The actual location within the 128-bit register is determined by the 16-byte alignment of i, j and k in LS.

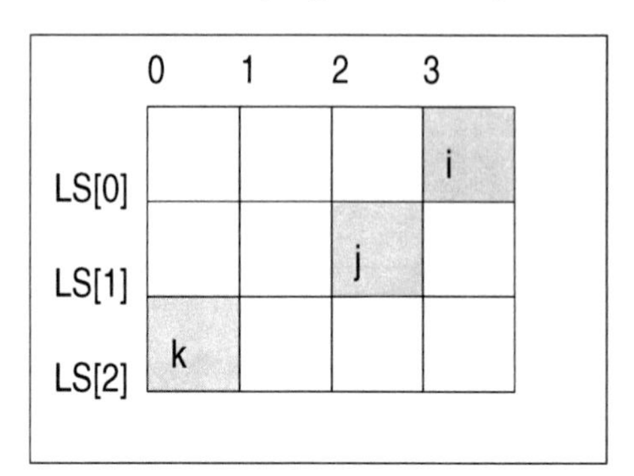

Fig. 3.4 Scalar operands located in Local Store.

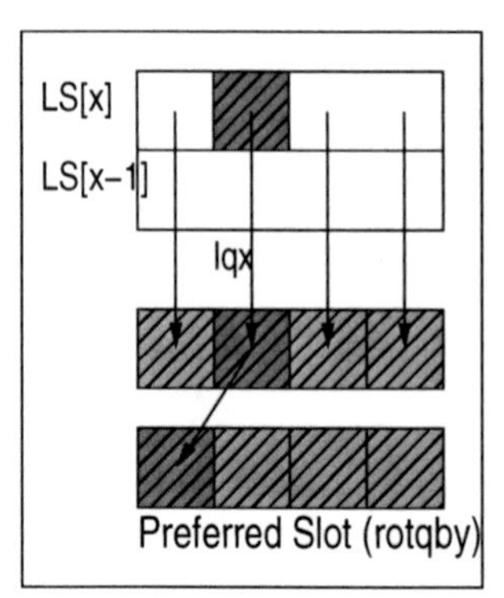

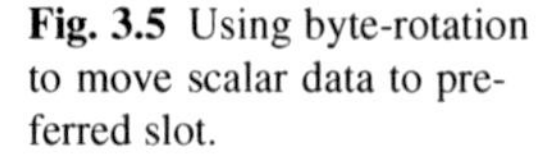

Fig. 3.5 Using byte-rotation to move scalar data to preferred slot.

This is true because the memory subsystem performs only 16-byte-aligned memory requests. Even loading and storing the values from LS must make sure that other elements of the vector are not inadvertently modified by the load/store operation.

The storing of a scalar value is also not as straightforward as on a scalar processor. The compiler must first load the original 128-bit data in which the result resides, variable *LS*[*k* in our example, then splice the new value in the original data, and finally store the resulting 128-bit data to the local memory. An example is shown in Figure 3.5.

To illustrate this, the compiler examine an example of ab+c, where all variables are declared as 16-bit shorts. The integral promotion rule in the C programming language requires any subword types be automatically promoted to the integer type before performing any computations thereon. Effectively, C allows programmers to specify subword data types in terms of alignment, size, and layout in memory, but does not allow specifying subword arithmetic. In essence, shorts and chars in C are simply a compressed representation of integers in memory, which are decompressed when loaded from memory.

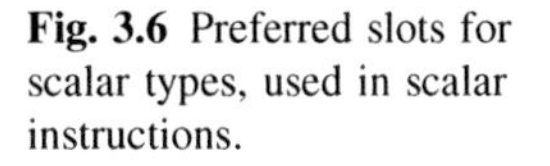

Fig. 3.6 Preferred slots for scalar types, used in scalar instructions.

Another important thing to keep in mind when developing code for SPE is that the SPE, like many other SIMD units, does not support each data type equally. For example, it provides hardware support only for 16-bit integer multiply-add instructions. Hence, a 32-bit multiply instruction is supported in software by using a series of 16-bit multiply-add instructions. This is why MPY 16-bit should be cast wherever possible.

3.8 Branch Optimization on SPU

To save area and power, the SPE omits hardware branch prediction and branch history tables. However, mispredicted branches flush the pipelines and consume 18 cycles, so it is important that software employ techniques to avoid mispredict. The goal of this section is to develop an intuitive feel of the dual-issue pipeline to maximize dual-issue rates. In Chapter 15 we shall look at some advanced methods to measure and then increase dual-issue rates of SPU code. The material presented here is based on the research papers by Kahle et al [65], Flachs et al [42], Eichenberger et al [41], and the source code for the GNU assembler for Cell SPU, and the `spu-gcc` compiler, with GNU/GCC version 4.3.3. An experimental study of sorting and branch prediction has been conducted by Biggar et al [5] which highlights the importance of branch prediction and optimization.

For example consider the following code taken from our program on floorplanning:

```
for(i=0;i<trials;++i) {
  // random permutation of vector by choosing p and q
  if((p==0)||(q==0)||(p==(n-1))||(q==(n-1))) continue;
  int temp_val = A[ p ];
  A[p] = A[q];
  A[q] = temp_val;
}
```

The above code performs very poorly on the SPE, as the branch due to the `continue` cannot be predicted. By using the fact that the code in question is swapping some vector elements to permute the vector, we can write the code as follows:

```
for(i=0;i<trials;++i) {
  // random permutation of vector by choosing p and q
  q=((p==0)||(q==0)||(p>=(n-1))||(q>=(n-1))) ? p:q;
  int temp_val = A[ p ];
5 A[p] = A[q];
  A[q] = temp_val;
}
```

We replace the `continue` by a computation which swaps `A[p]` with `A[p]`, when the condition for swap is not valid. By unrolling the loop further, we can hide the LS latency of the swap itself behind computation performed by the random number generation. When a commonly taken branch is necessary, especially for the backward branch of a loop, software can utilize the SMBTB, which is loaded using the branch hint instructions (`hbr`, `hbra`, and `hbrr`). A correctly hinted branch can execute in a single cycle. These instructions identify both the address of a branch and the probable address of the branch target. When a branch hint instruction is executed, instructions from the branch target are read from the LS and written to the SMBTB. Later, when the indexed branch instruction is sent to the issue logic, the SPE instruction sequencer can send the first instruction of the branch target in the next consecutive cycle.

The SPE hardware assumes that branches are not taken, but the architecture allows for a *branch-hint* (`hbr`) instruction to override the default branch prediction policy. This instruction specifies the location of a branch and its likely target address. When the hbr instruction is scheduled sufficiently early (at least 11 cycles before the branch), instructions from the hinted branch target are prefetched from memory and inserted in the instruction stream immediately after the hinted branch. In addition, the branch hint instruction causes a prefetch of up to 32 instructions, starting from the branch target, so that a branch taken according to the correct branch hint incurs no penalty.

The SPEs are heavily pipelined, making the penalty for incorrect branch prediction high, namely 18 cycles. In addition, the hardware's branch prediction policy is simply to assume that all branches (including unconditional branches) are not taken. This architecture choice reduces hardware complexity, and enables higher clock frequencies.

3.8.1 Branch elimination

The compiler attempts to eliminate taken branches. One method is *if-conversion*, which use compare-and-select (`selb`) instruction to avoid branching. Profile guided techniques to predict and avoid branches are described in Section 15.3, in Chapter 15 on pp. 302. When the hint is correct, the branch latency is essentially one cycle; otherwise, the normal branch penalty applies.

3.9 Instruction Scheduling

The scheduling process consists of two closely interacting subtasks: scheduling and bundling [41, 42]. The scheduling subtask reorders instructions to reduce the length of critical paths. The bundling subtask ensures that the issue constraints are satisfied to enable dual issuing and prevent instruction fetch starvation, i.e., the situation when the instruction buffer is empty and awaiting refill from an explicit instruction fetch (said to be ifetch but, technically speaking, hbr.p) instruction.

The scheduler's main objective is to schedule operations that are on a critical path with the highest priority and schedule the other less critical operations in the remaining slack. However there are some constraints involving numbers of instructions; for example, the constraint that an hbr branch hint instruction cannot be more than 256 instructions nor less than eight instructions from its target branch. In addition the hardware dual-issue constraint states that the first instruction must use the even pipe and reside at an even word address, whereas the second instruction must use the odd pipe and reside at an odd word address. This is satisfied by inserting nop and lnop instructions into the instruction stream.

Given the above discussion on the complexity of instruction issue and scheduling it is no surprise that the SDK authors propose using SPU intrinsics wherever possible. The major advantage of programming with intrinsics is that the programmer has full control over the handling of the data alignment and the choice of SIMD instructions, yet continues to have the benefit of high-level transformations such as loop unrolling and function inlining (i.e, copying subroutine code into the calling routine) as well as low-level optimizations such as scheduling, register allocation, and other optimizations.

In addition, the programmer can rely on the compiler to generate all the scalar code, such as address, branch, and loop code generation; code that is often error-prone when hand-coded in assembly language. In addition, it is much easier to modify code programmed with intrinsics over the course of an application's development than to modify assembly language.

Recent research into BLAS kernel optimization on the CELL SPU has been done by the ICL research group at University of Tennessee and has resulted in a number of technical reports and working software. The publications are available as LAPACK Working Note 184, 189 and 201. The researchers report achieving over 170 gigaflops with a single Cell processor when solving a symmetric positive definite system of linear equations in single-precision.

3.10 Memory flow

Data movement is performed asynchronously with respect to the execution of the SPU program. Therefore, the CBEA provides a tag-group completion facility so that an application can determine when a specific command or set of commands completes. Independent facilities are available for both the MFC proxy and MFC SPU command queues. A tag ID is provided with each DMA command, and commands issued with the same tag ID to the same queue are called a tag group. To determine when tag groups are complete, a mask that selects the tag groups of interest is written to the tag-group mask channel or register. The status of the selected groups is reported in a status register for the MFC proxy command queue. For the MFC SPU command queue, a query type is written to the MFC write tag-group status update request channel. The request can be that either "all selected tag groups are complete" (which is `mfc_read_tag_status_all()`) or "any one of the selected groups is complete", (which is `mfc_read_tag_status_any()`). Once the query type is set, reading from the blocking tag-group status channel causes the SPU to stall until the query condition is met. Like the command issue sequence, the stall can be avoided by first determining the channel count value of the channel.

The CBEA also provides a way to generate an interrupt when a tag group completes. For the MFC proxy command queue, a similar query type is provided to determine when the interrupt is generated. The interrupt is typically routed to the PPE for processing. For the MFC SPU command queue, the event facility provides the interrupt when the condition is met.

As MFC commands are issued, they are placed in the appropriate command queue. MFC commands can be executed and completed in any order, regardless of the order in which they were issued. The out-of-order execution of commands allows the MFC to use system resources efficiently to achieve the best performance. For example, if the EIB supports simultaneous reads and writes, the MFC can simultaneously execute one get and one put command. Command ordering is provided by using `barrier` and `fence` attributes. These are covered in Section 5.1.1, pp. 68.

3.11 Mailbox Facility

A key advantage of the CBEA is the independent nature of the SPEs. Since the SPEs are decoupled from the PPE and other SPEs, the CBEA provides a mailbox facility to assist in process-to-process communication and synchronization. The mailbox facility provides a simple, unidirectional communication mechanism typically used by the PPE to send short commands to the SPE and to receive status in return. This facility consists of an inbound, an outbound, and an outbound interrupt mailbox. Each of these mailboxes can have a depth of one or more entries. The direction of the mailbox is relative to the SPU. For example, the inbound mailbox is written by the PPE and read by the SPU. The SPU accesses the mailboxes using the SPU channel instructions. The mailbox channels are blocking, causing the SPU to stall if the outbound mailbox is full or the inbound mailbox is empty. For the PPE and other devices, MMIO registers provide access to the mailboxes and the mailbox status.

Mailboxes support the sending and buffering of 32-bit messages between an SPE and other devices, such as the PPE and other SPEs. Each SPE can access three mailbox channels, each of which is connected to a mailbox register in the SPUs MFC. Two one-entry mailbox channels the SPU Write Outbound Mailbox and the

SPU Write Outbound Interrupt Mailboxare provided for sending messages from the SPE to the PPE or other device. One four-entry mailbox channelthe SPU Read Inbound Mailboxis provided for sending messages from the PPE, or other SPEs or devices, to the SPE. Each of the two outbound mailbox channels has a corresponding MMIO register that can be accessed by the PPE or other devices.

Mailboxes are useful, for example, when the SPE places computational results in main storage via DMA. After requesting the DMA transfer, the SPE waits for the DMA transfer to complete and then writes to an outbound mailbox to notify the PPE that its computation is complete. If the SPE sends a mailbox message after waiting for a DMA transfer to complete, this ensures only that the SPEs LS buffers are available for reuse. It does not guarantee that data has been coherently written to main storage. The SPE might solve this problem by issuing an mfcsync command before notifying the PPE. But doing so is inefficient. Instead, the preferred method is to have the PPE receive the notification and then issue an lwsync instruction before accessing any of the resulting data. Alternatively, an SPE can notify the PPE that it has completed computation by writing, via DMA, such a notification to main storage, from which the PPE can read the notification. (This is sometimes referred to as a writeback DMA command, although no such DMA command is defined.) In this case, the data and the writeback must be ordered. To ensure ordering, an mfceieio command must be issued between the data DMA commands and the notification to main storage. Although the mailboxes are primarily intended for communication between the PPE and the SPEs, they can also be used for communication between an SPE and other SPEs, processors, or devices. For this to happen, privileged software needs to allow one SPE to access the mailbox register in another SPE by mapping the target SPEs problem-state area into the EA space of the source SPE. If software does not allow this, then only atomic operations and signal notifications are available for SPE-to-SPE communication.

Reading and Writing Mailboxes

Data written by an SPE program to one of the outgoing mailboxes using an SPE write-channel (wrch) instruction is available to any processor element or device that reads the corresponding MMIO register in the main-storage space. Data written by a device to the SPU Read Inbound Mailbox using an MMIO write is available to an SPE program by reading that mailbox with a readchannel (rdch) instruction. An MMIO read from either of the outbound mailboxes, or a write to the inbound mailbox, can be programmed to raise an SPE event, which in turn, can cause an SPU interrupt. A wrch instruction to the SPU Write Outbound Interrupt Mailbox can also be programmed to cause an interrupt to the PPE or other device.

Each time a PPE program writes to the four-entry inbound mailbox, the channel count for the SPU Read Inbound Mailbox Channel increments. Each time an SPE program reads the mailbox, the channel count decrements. The inbound mailbox acts as a first-in-first-out (FIFO) queue; SPE software reads the oldest data first. If the PPE program writes more than four times before the SPE program reads the data, then the channel count stays at 4, and the fourth location contains the last data written by the PPE. For example, if the PPE program writes five times before the SPE program reads the data, then the data read is the first, second, third, and fifth messages that were written. The fourth message that was written is lost.

Mailbox operations are blocking operations for an SPE: an SPE write to an outbound mailbox that is already full stalls the SPE until a entry is cleared in the mailbox by a PPE read. Similarly, a rdch instructionbut not a read-channel-count (rchcnt) instructionfrom an empty inbound mailbox is stalled until the PPE writes to the mailbox. That is, if the channel count is 0 for a blocking channel, then an rdch or wrch instruction for that channel causes the SPE to stall until the channel count changes from 0 to nonzero. To prevent stalling, SPE software should read the channel count associated with the mailbox before deciding whether to read or write the mailbox channel. This stalling behavior for the SPE does not apply to the PPE; if the PPE sends a message to the inbound mailbox and the mailbox is full, the PPE will not stall. Because a wrch instruction will stall when it tries to send a value to a full outbound mailbox, SPE software cannot over-run an outbound mailbox. All outbound messages must be read by some entity outside the SPU before space is made available for more outbound messages. In contrast, the SPU Read Inbound Mailbox can be over-run by an outside entity because the MMIO write used for this purpose does not stall. Although, the stall state is a low-power state.

How Mailboxes are commonly used

For example, the mailbox facility can be used for a command-driven SPU application. In this example, the SPU is typically stalled waiting for a command to be placed in the inbound mailbox. When a command is received, the SPU performs the requested operation given in the mailbox data or sequence of mailbox data. Once the operation is complete, the SPU places a return code in the outbound interrupt mailbox. The write of the outbound interrupt mailbox generates an interrupt for the PPE that indicates the completion of the requested operation. The SPU code then reads the next command from the inbound mailbox and stalls if a new command is not available. Mailbox message values are intended to communicate messages up to 32 bits in length, such as buffer completion flags or program status. In fact, however, they can be used for any short-datatransfer purpose, such as sending of storage addresses, function parameters, command parameters, and state-machine parameters.

Mailbox facility implementation using MFC queues

The MFC provides a set of mailbox queues between the SPU and other processors and devices. Each mailbox queue has an SPU channel assigned as well as a corresponding MMIO register. SPU software accesses the mailbox queues by using SPU channel instructions. Other processors and devices access the mailbox queues by using one of the MMIO registers. In addition to the queues, the MFC provides queue status, mailbox interrupts, and SPU event notification for the mailboxes. Collectively, the MMIO registers, channels, status, interrupts, mailbox queues, and events are called the mailbox facility.

The MFC provides two mailbox queues for sending information from the SPU to another processor or to other devices: the SPU outbound mailbox queue and the SPU outbound interrupt mailbox queue. These mailbox queues are intended for sending short messages to the PPEs (for example, return codes or status). Data written by the SPU to one of these queues using a write channel (wrch) instruction is available to any processor or device by reading the corresponding MMIO register. A write channel (wrch) instruction that can target the SPU Write Outbound Interrupt Mailbox Channel can also cause an interrupt to be sent to a processor, or to another device in the system.

An MMIO read from either of these queues (SPU outbound mailbox or SPU outbound interrupt mailbox) can set an SPU event, which in turn causes an SPU interrupt. One mailbox queue is provided for either an external processor or other devices to send information to the SPU: the SPU inbound mailbox queue. This mailbox queue is intended to be written by a PPE. However, other processors, SPUs, or other devices can also use this mailbox queue. Data written by a processor or another device to this queue using an MMIO write is available to the SPU by reading the SPU Read Inbound Mailbox Channel.

SPU Outbound Mailbox Register SPU_Out_MBox

The SPU Outbound Mailbox Register is used to read 32 bits of data from the corresponding SPU outbound mailbox queue. The SPU Outbound Mailbox Register has a corresponding SPU Write Outbound Mailbox Channel for writing data into the SPU outbound mailbox queue.

SPU Inbound Mailbox Register SPU_In_MBox)

The SPU Inbound Mailbox Register is used to write 32 bits of data into the corresponding SPU inbound mailbox queue. The SPU inbound mailbox queue has a corresponding SPU Read Inbound Mailbox Channel for reading data from the queue.

A read channel (rdch) instruction that targets the SPU Read Inbound Mailbox Channel loads the 32 bits of data from the SPU inbound mailbox queue into the SPU register specified by the read channel (rdch) instruction. The SPU cannot read from an empty mailbox. If the SPU inbound mailbox queue is empty, the SPU stalls on a read channel (rdch) instruction to this channel until data is written to the mailbox.

A read channel (rdch) instruction to this channel always returns the information in the order it was written by PPEs or by other processors and devices.

SPU Mailbox Status Register SPU_Mbox_Stat

The SPU Mailbox Status Register contains the current state of the mailbox queues in the corresponding SPE. Reading this register has no effect on the state of the mailbox queues.

3.12 Use SPU Intrinsics with Mailboxes

This section describes functions that can be used to manage Mailboxes on the SPU side:

spu_read_in_mbox: read Next Data Entry in SPU Inbound Mailbox

```
(uint32_t) spu_read_in_mbox(void)
spu_readch(SPU_RdInMbox)
```

The next data entry in the SPU Inbound Mailbox queue is read. The command stalls when the queue is empty. The application-specific mailbox data is returned. Each application can uniquely define the mailbox data.
spu_stat_in_mbox: get the Number of Data Entries in SPU Inbound Mailbox

```
(uint32_t) spu_stat_in_mbox(void)
spu_readchcnt(SPU_RdInMbox)
```

The number of data entries in the SPU Inbound Mailbox is returned. If the returned value is nonzero, the mailbox contains data entries that have not been read by the SPU.
spu_write_out_mbox: send Data to SPU Outbound Mailbox

```
(void) spu_write_out_mbox (uint32_t data)
spu_writech(SPU_WrOutMbox, data)
```

Data is sent to the SPU Outbound Mailbox, where data is application-specific mailbox data, or the command stalls when the SPU Outbound Mailbox is full.
spu_stat_out_mbox: get Available Capacity of SPU Outbound Mailbox

```
(uint32_t) spu_stat_out_mbox(void)
spu_readchcnt(SPU_WrOutMbox)
```

The available capacity of the SPU Outbound Mailbox is returned. A value of zero indicates that the mailbox is full.
spu_write_out_intr_mbox: send Data to SPU Outbound Interrupt Mailbox

```
(void) spu_write_out_intr_mbox (uint32_t data)
spu_writech(SPU_WrOutIntrMbox, data)
```

Data is sent to the SPU Outbound Interrupt Mailbox, where data is application-specific mailbox data. The command stalls when the SPU Outbound Interrupt Mailbox is full.

spu_stat_out_intr_mbox: get Available Capacity of SPU Outbound Interrupt Mailbox

```
(uint32_t) spu_stat_out_intr_mbox(void)
spu_readchcnt(SPU_WrOutIntrMbox)
```

The available capacity of the SPU Outbound Interrupt Mailbox is returned. A value of zero indicates that the mailbox is full.

Mailbox APIs on the PPE side

The following functions are defined in `libspe2.h` and are available for use on the PPE. In the following API functions `spe` specifies the SPE context pointer of the SPU whose mailbox is to be operated upon.

spe_out_mbox_read: This function reads up to count available messages from the SPE outbound mailbox for the SPE context spe. Its a non-blocking function call and moreover if less than count mailbox entries are available, only those will be read. The return value is the number of 32-bit mailbox messages read.

```
int spe_out_mbox_read (spe_context_ptr_t spe,
                       unsigned int *mbox_data,
                       int count)
```

spe_out_mbox_status: fetches the status of the SPU outbound mailbox of the SPE context,

```
int spe_out_mbox_status (spe_context_ptr_t spe)
```

spe_in_mbox_write: Write up to count messages to the SPE inbound mailbox for the SPE. Depending on the `behavior`, this call may be blocking or non-blocking. The blocking version of this call is useful to send a sequence of mailbox messages to an SPE program, the non-blocking version is better when events are used. The `behavior` values can be any of:

1. SPE_MBOX_ALL_BLOCKING
2. SPE_MBOX_ANY_BLOCKING
3. SPE_MBOX_ANY_NONBLOCKING

```
int  spe_in_mbox_write  (spe_context_ptr_t  spe,
                         unsigned int *mbox_data,
                         int count,
                         unsigned int behavior)
```

spe_in_mbox_status: fetches the status of the SPU inbound mailbox for the corresponding SPE,

```
int spe_in_mbox_status (spe_context_ptr_t spe)
```

spe_out_intr_mbox_read:

```
int spe_out_intr_mbox_read  (spe_context_ptr_t  spe,
                             unsigned  int  *mbox_data,
                             int count,
                             unsigned int behavior)
```

spe_out_intr_mbox_status:

```
int spe_out_intr_mbox_status (spe_context_ptr_t spe)
```

3.13 Synergistic Processor Unit Channels and Memory Flow Controller

In the Cell Broadband Engine Architecture (CBEA), channels are used as the primary interface between the synergistic processor unit (SPU) and the memory flow controller (MFC). The SPU channel access facility is used to configure, save, and restore the SPU channels. The SPU Instruction Set Architecture (ISA) provides a set of channel instructions for communication with external devices through a channel interface (or SPU channels). Although the SPU ISA defines channels as 128-bit (or a quadword GPR sized), the CBEA defines every channel in the system to be only 32-bit; there is no inconsistency as for the SPU GPR the 32-bits are from the preferred-slot of the register.

Channels are unidirectional, function-specific registers or queues. They are the primary means of communication between an SPEs SPU and its MFC, which in turn mediates communication with the PPEs, other SPEs, and other devices. These other devices use MMIO registers in the destination SPE to transfer information on the channel interface of that destination SPE. Specific channels have read or write properties, and blocking or nonblocking properties. Software on the SPU uses channel commands to en-queue DMA commands, query DMA and processor status, perform MFC synchronization, access auxiliary resources such as the decrementer (timer), and perform interprocessor- communication via mailboxes and signal-notification.

A Synergistic Processor Element (SPE) communicates with the PowerPC Processor Element (PPE) and other SPEs and devices through its channels. Channels are unidirectional interfaces for sending and receiving variable-size (up to 32-bit) messages with the PPE and other SPEs, and for sending commands (such as direct memory access [DMA] transfer commands) to the SPEs associated memory flow controller (MFC). Each SPE has its own set of channels. SPE software accesses channels with three instructions: read channel (rdch), write channel (wrch), and read channel count (rchcnt). These instructions enqueue MFC commands into the SPEs MFC SPU command queue for the purpose of initiating DMA transfers, querying DMA and synergistic processor unit (SPU) status, sending mailbox and signal-notification messages for synchronization of tasks, and accessing auxiliary resources such as the SPEs decrementer.

The channels are architecturally defined as blocking or nonblocking. When SPE software reads or writes a nonblocking channel, the operation executes without delay. However, when SPE software reads or writes a blocking channel, the SPE might stall for an arbitrary length if the associated channel count (which is its remaining capacity) is 0. In this case, the SPE will remain stalled until the channel count becomes 1 or more, as shown in Figure 17-1, or the SPE is interrupted. The stalling mechanism allows an SPE to minimize the power consumed by message-based synchronization of tasks, because fewer logic gates are switching during the stall.

Table 3.5 SPU Channel mnemonics and description.

Channel Instruction	Mnemonic	Description
Read channel	rdch	$GPR \Leftarrow channel$
Write channel	wrch	$channel \Leftarrow GPR$
Read channel count	rchcnt	$GPR \Leftarrow channelcount$

Architecturally, SPU channels are defined as read only or write only; channels cannot be defined as both read and write. In addition to the access type, each channel can be defined as nonblocking or blocking. All blocking channels have an associated channel count. Nonblocking channels do not have an associated channel count; a read channel count instruction (rchcnt) that targets a nonblocking channel always returns a count of 1. Channels that are defined as blocking cause the SPU to stall when reading a channel with a channel count of 0, or when writing to a full channel (that is, a channel with a channel count of 0).

Each blocking channel has a corresponding count (that is, depth), which indicates the number of outstanding operations that can be issued for that channel. The channel depth (that is, the maximum number of outstanding transfers) is implementation dependent. Software must initialize the channel counts when establishing a new context in the SPU, or when it resumes an existing context.

Table 3.6 SPU Channel for communicating with Mailbox

Channel	Description	Type
SPU_WrOutMBox	SPU Write Outbound MBox	Write blocking
SPU_RdInMBox	SPU Read Inbound MBox	Read blocking
SPU_WrOutIntrMBox	SPU Write Outbound Interrupt MBox	Write blocking

The key features of the SPE channel operations include:

1. All operations on a given channel are unidirectional (they can be only read or write operations for a given channel, not bidirectional).

2. Accesses to channel-interface resources through MMIO addresses do not stall.
3. Channel operations are done in program order.
4. Channel read operations to reserved channels return zeros.
5. Channel write operations to reserved channels have no effect.
6. Reading of channel counts on reserved channels returns 0.
7. Channel instructions use the 32-bit preferred slot in a 128-bit transfer.

3.14 SPU Timer and Events

The SPU supports an event facility that enables waiting for or polling for specific events. The event facility can also generate an SPU asynchronous interrupt for specific events. If SPU interrupts are enabled, an occurrence of an unmasked event results in an SPU interrupt handler being invoked with the first instruction of the interrupt handler located at local storage address 0. An example of using SPU Timer is given in Section 8.4, pp. 109.

3.15 SPE Context on the PPE

Saving and restoring an SPE context can be very expensive in terms of both time and system resources. The complete SPE context consists of:

1. 256 KB of local storage (LS)
2. 128 128-bit general purpose registers (GPRs)
3. Special purpose registers (SPRs)
4. Interrupt masks
5. Memory flow controller (MFC) command queues
6. SPE channel counts and data
7. MFC synergistic memory management (SMM) state
8. Other privileged state

In all, the SPE context occupies more than 258 KB of storage space.

3.16 Conclusion

The SPE is a both a radical departure from current co-processors, or graphics processors, as well as a reminder of erstwhile floating-point co-processors (eg., FPS-140) which could be loaded with data using and then asynchronously execute their stored program. Once the computation was complete they would send a small message to the host (an IBM mainframe at that time), and result could be read back from co-processor memory to mainframe memory. The SPEs have replaced the memory transfer by the EIB DMA, and the simplistic computations performed by early co-processors by relatively high-level SIMD computations resulting in 100s of gigaflops for single-precision computation. In this chapter we have analyzed the design, implementation and programming facilities of the SPE. In the next chapter we analyze the dual-issue pipeline of the SPE in greater detail and also present some more information on the SPU intrinsics and instruction we use in the remainder of this text.

3.17 What we have not discussed

MFC multisource synchronization facility, (see Section 8.6),
SPU signal notification channels and commands, (see Section 8.4),
SPU event facility, (see Section 8.4),
CBEA privileged mode environment (not covered in this text),
Software management of TLB (except for enabling huge-page tables,
ERAT (effective-to-real-address translation),
Segment lookaside buffer (SLB),
Page Table Entries, reversed page table entries, OS relationship, TLB replacement policies,
Data stream form of the dcbt and associated optimizations (see Section 2.7),
CBEA I/O Architecture (as it is mostly controlled by privileged state only),
PPE Interrupts (except software controlled interrupts),
Logical partitions and hypervisor,
CBEA power and thermal management,
Atomic synchronization primitives (see Section 8.6),
SPE/PPE mutual exclusion using condition variables, or atomic synchronization (see Section 8.6),
SPU event facility and decrementer facility (see Section 8.4),
Software management of TLBs facility,
Cache replacement management facility,
Internal interrupt controller.

Chapter 4
Element Interconnect Bus

Abstract In this chapter we discuss the Element Interconnect Bus (EIB) of the Cell Broadband Engine. We present the design of the EIB as a token-ring, managed communication path for transferring command and data amongst the CBEA components and IO. The EIB supports full memory-coherent and SMP operation. The EIB consists of four 16-byte-wide data rings. Each ring transfers 1 PPE cache line at a time. Each element connected to the EIB has an on-ramp, and an off-ramp, both of which can be active. We provide a traffic-circle analogy with token collection to explain the workings of the EIB which is an important variable in deciding how to partition the problem on the SPEs. Experimental data on EIB performance is also presented in this chapter.

4.1 Introduction to the Element Interconnect Bus

The EIB in the Cell/B.E. processor allows for communication among the PPE, the SPEs, the off-chip memory, and the external I/O. It uses the direct memory access (DMA) paradigm to communicate between any source-destination pairs from the above list. As we have already from Chapter 1, the 3.2 GHz Cell in the PS3 has a theoretical peak performance of 204.8 Gflop/s (single precision) and 14.6 Gflop/s (double precision). To utilize such a high compute rate, the memory latency and inter-processor communication latency has to be very low. The EIB is designed with the sole purpose of providing efficient and low latency data communication in the Cell processor. The EIB has a peak bandwidth of 204.8 Gbytes/s for intrachip data transfers among the PPE, the SPEs, and the memory and I/O interface controllers.

By design, the SPUs can directly access only its local store, every other request to get or put data on the main memory is performed by the MFC which has DMA controllers to perform DMA data transfer operation to/from local store to main memory. The architecture of the SPU and DMA controller has to be understood in conjunction as the design choices made by the architects influence good program design, and data transfer methodologies. When the SPU program is operating in the *sweet-*

S. Koranne, *Practical Computing on the Cell Broadband Engine*,
DOI: 10.1007/978-1-4419-0308-2_4, © Springer Science + Business Media, LLC 2009

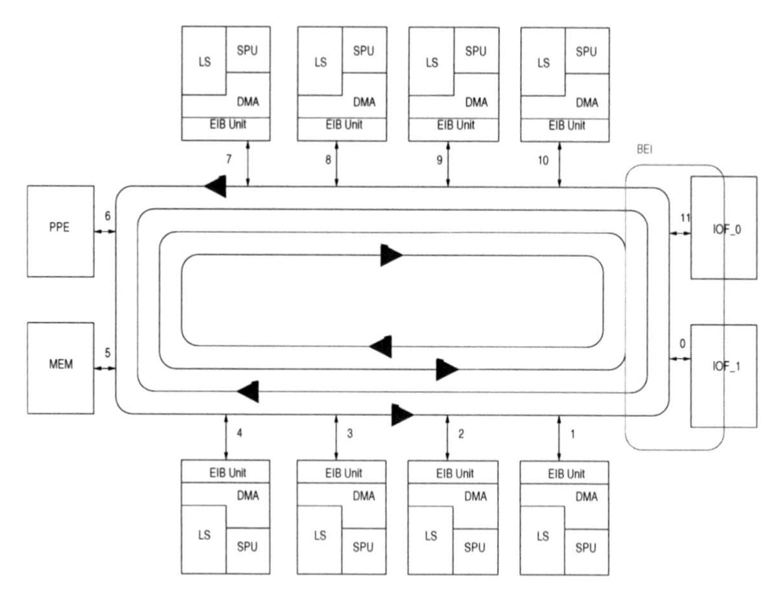

Fig. 4.1 The Element Interconnect Bus (EIB) of the Cell Broadband Engine.

spot of the architecture design, the computation and data-transfers don't interfere with each other, and we can expect to see high computational rates. In the sequel of this book we shall uncover a number of DMA factoids and best practices which have worked for me. This chapter provides some of the rationale for these practices.

These DMA engines also allow direct transfers between the local stores of two SPUs for pipeline mode of operation. We have used the pipeline mode to perform compress-encrypt on the SPUs in a pipeline where data is compressed, and then sent off to the next SPU for encryption.

The EIB can be thought of as the *network* in the Cell network-on-chip (NOC) processor. It is similar to a high-speed token-ring managed bus with address snooping. The EIB consists of one address bus and four 16-byte-wide data rings, two of which run clockwise and the other two counterclockwise. Each ring can allow up to three concurrent data transfers as long as their paths do not overlap. The EIB operates at half the speed of the processor (1.6 GHz for the PS3).

Each requester on the EIB starts with a small number of initial command credits to send out requests on the bus. The number of credits is the size of the command buffer inside the EIB for that particular requester. One command credit is used for each request on the bus. When a slot becomes open in the command buffer as a previous request progresses in the EIB request pipeline, the EIB returns the credit to the requester. When a unit requires access to a data ring in order to send data to another unit, it makes a single request to the data ring arbiter on the EIB. The arbiter processes requests from all requesters and decides, as optimally as it can, which data ring is granted to which requester and the time at which the data ring is granted. The memory controller is given the highest priority to prevent stalling of

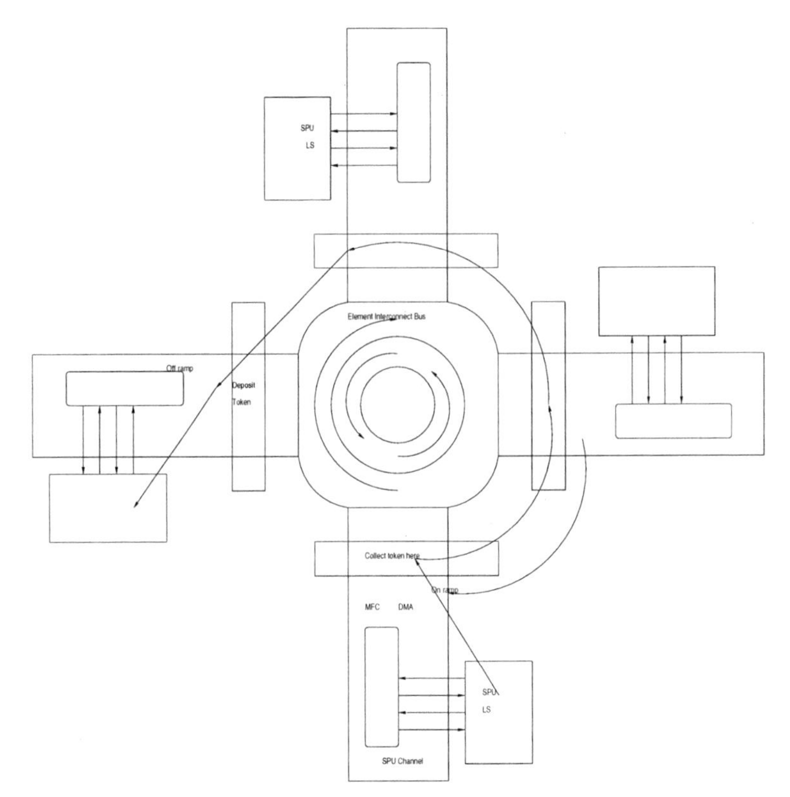

Fig. 4.2 A traffic roundabout analogy for the EIB.

the requester of the read data, while all others are treated equally with a round-robin priority. Any ring requester can use any of the four rings to send or receive data. The data arbiter does not grant a data ring to a requester if the transfer would cross more than halfway around the ring on its way to its destination or would interfere with another data transfer already in progress.

An analogy of the EIB is shown in Figure 4.2. The on and off-ramps are not shown separately but assuming normal road driving consider the on-ramp on the right side of the road. Every vehicle represents a packet of data (16 bytes); before the vehicle can enter the round-about it must possess a valid *token*. It can get a token from the token arbitration engine which ensures key properties of bandwidth, priority, contention and starvation prevention. Token need to be surrendered at the off-ramp and are presumably recycled by the system. With this simple scheme may properties of the round-about and thus the EIB bus can be shown. These include maximum number of simultaneous messages in flight, minimum and maximum bounds on message latency among others.

Each unit on the EIB can simultaneously send and receive 16 bytes of data every bus cycle. The maximum data bandwidth of the entire EIB is limited by the maximum rate, one 16-byte data transfer per bus cycle, at which addresses are snooped across all units connected to the EIB. Since each snooped address request can potentially transfer up to 128 bytes, the theoretical peak data bandwidth on the EIB at 3.2 GHz is 128 bytes 1.6 GHz = 204.8 GB/s. Gonzalez et al have performed extensive analysis of the Cell Broadband Engine for high-memory bandwidth applications [64, 45]. They report that by following some strict programming rules individual SPE to SPE communication almost achieves the peak bandwidth when using the DMA controller when transferring data of atleast 1024 bytes. Moreover, implementing two data streams using 4 SPEs can be more efficient than having a single data stream using 8 SPEs. The main results of their experiments [64] is given below:

1. Load/Store access from PPU to L1 cache: PPU can effectively attain half peak performance when accessing atleast 8 bytes. Since there is 16 bytes/CPU clock cycle link between PPU and L1, the theoretical peak performance is 33.6 Gb/s. Peak bandwidth obtained was 8 Gb/s for 4 bytes, and dropped even further to 4 Gb/s for 2 byte and 2 Gb/s for single byte access,
2. Load/Store access from PPU to L2 cache: from Chapter 2 we know resources are shared between each thread for PPU and thus peak bandwidth increases when EMT multi-threading is used. Peak bandwidth was 8 Gb/s,
3. L1/L2 cache to main memory: bandwidth is close to 44 Gb/s for L2 load, 22 Gb/s for L2 store and 25 Gb/s for L2 copy,
4. PPE to Main Memory: read access to memory achieved same performance as L2 cache reads which 8 Gb/s. Performance for memory store is lower at 2 Gb/s (which quickly saturates the L2 cache for store),
5. SPE to Memory: SPU get average at 16 Gb/s, spu_put at 16 Gb/s and memory get and put was 22.4 Gb/s
6. SPE to Local Store: since Local Store is uncached, carefully written code can achieve the maximum bandwidth of 33.6 Gb/s,
7. SPE to SPE: peak bandwidth for 2 SPEs is 33.6 Gb/s, 67.2 for 4 SPEs and 134.4 Gb/s when all 8 SPEs are used.

The sustained data bandwidth on the EIB will often be lower than the peak bandwidth because of several factors: the locations of the destination and the source relative to each other, the potential for interference between the new transfer and those already in progress, the number of Cell/B.E. chips in the system, whether the data transfers are to or from memory or between local stores in the SPEs, and the efficiency of the data arbiter.

Reduced bus bandwidths can result when all requesters access the same destination (memory or local store) at the same time, when all transfers are in the same direction and cause idling on two of the four data rings, when there are a large number of partial cache line transfers lowering the bus efficiency, or when each data transfer is six hops, inhibiting the units on the way from using the same ring.

The resource allocation groups (RAGs) are groups of zero or more requesters. These requesters are physical or virtual units that can initiate load or store requests or DMA read or write accesses. There are 17 requesters,

1. One PPE (both threads)
2. Eight SPEs.
3. Four virtual I/O channels for physical interface IOIF0
4. Four virtual I/O channels for physical interface IOIF1

Delays occur when outbound and inbound requests are competing for the same resource. For example, accesses from two virtual channels might contend for the same memory bank token; because there is only one physical link, accesses on the link are serialized even if they are to different virtual channels. Accesses on different virtual channels might delay each other due to:

1. EIB-address or data contention
2. SPE LS contention
3. A miss in the I/O-address translation cache for one virtual channel
4. IOIF contention

The last type of contention (IOIF) should be minimal because the IOIF resource is managed by the RAM facility.

4.1.1 EIB Possible Livelock Detection

The EIB possible livelock detection interrupt detects a condition that might cause a livelock. It is a useful tool for programmers to use to detect whether they have created an unexpected bottleneck in their code during development.

4.2 Important things to remember about the EIB

With all the raw numbers presented above, it is no surprise that many people will not retain the finest of the points about peak bandwidth, number of simultaneous DMAs in flight, and so on. To make life easy for all of us, this section contains the bare-minimum information you should keep in mind when reading the subsequent chapters and when writing code:

1. PS3 is running at 3.2 GHz, each clock cycle of the SPE/PPE is 320 pico-second, or 10 cycles are 3.2 nano-second,
2. LS access is 10ns, main-memory is 100ns, DMA start is 10 clock cycles, EIB clock cycles are at 1.6 GHz
3. EIB is fair: true, memory access gets priority, but there is no starvation,
4. EIB is coherent: if you transfer data, and do a MFC `mfc_read_tag_status_all`, you can assume data has arrived where you sent it without error or need for acknowledgement,
5. SPE physical placement is seldom important: unless you know this could be a problem, it wont be for you (i.e., unless you are doing heavy SPE-SPE communication, most data transfers involve either the PPE or memory, which implies a

half-way round-trip for atleast half the SPEs in the computation). If the load balancing is such that you expect a particular SPE to communicate more frequently with the memory physical placement can be a concern,

6. Every transmission is atleast 128 bytes: so try to pack as much as you can even in short messages,
7. Pipe-line mode can be very efficient: SPE-SPE peak is 200 GB/s, while memory bandwidth of 25.6 is shared by all the processing units,
8. Background DMA processing allows overlap of data transfer and communication,
9. Alignment: this wont be the last place where alignment will be discussed in the book. 128 bit alignment is the most efficient and from a programming point-of-view least troublesome (debugging DMA errors is not easy),
10. Maximum single transmission is 16 Kb: DMA list can contain 2048 DMA transfers using a single command. Using DMA list shifts much of the address computation load from the SPEs compute unit to the MFC, making it work like an additional processing unit,
11. Best case scenario:peak performance is achievable for transfers when both the effective address and the local storage address are 128-byte aligned and the transfer size is an even multiple of 128 bytes.

4.3 Conclusions

In this chapter we introduced the Element Interconnect Bus (EIB) which holds the Cell together. Since the EIB does not offer any significant programmable elements which are of use and interest to application software we concluded this chapter with important things to remember about the EIB, which are the restrictions on addresses and packet size optimizations.

Chapter 5
Direct Memory Access (DMA)

Abstract In this chapter we discuss the Direct Memory Access (DMA) functionality of the Cell architecture. Sender and receiver initiated DMA plays a significant role in program optimization. We present the DMA engine, *fence* and *barrier* concept for ordering data transfers. The DMA engine can be used as an additional data movement processor using *dma-list*. Double-buffering and SPE-to-SPE DMA is discussed in the context of examples in the sequel of this text.

5.1 Introduction to DMA in the Cell Broadband Engine

The SPEs DMA engine handles most communications between the SPU and other Cell elements and executes DMA commands issued by either the SPU or the PPE. A DMA commands data transfer direction is always referenced from the SPEs perspective. Therefore, commands that transfer data into an SPE (from main storage to local store) are considered get commands (gets), and transfers of data out of an SPE (from local store to main storage) are considered put commands (puts). DMA transfers are coherent with respect to main storage. Programmers should be aware that the MFC might process the commands in the queue in a different order from that in which they entered the queue. When order is important, programmers must use special forms of the get and put commands to enforce either barrier or fence semantics against other commands in the queue.

The MFCs MMU handles address translation and protection checking of DMA accesses to main storage, using information from page and segment tables defined in the PowerPC architecture. The MMU has a built-in translation look-aside buffer (TLB) for caching the results of recently performed translations.

The MFCs DMA controller (DMAC) processes DMA commands queued in the MFC. For the most part we analyze only the MFC SPU command queue.

In the absence of congestion, a thread running on the SPU can issue a DMA request in as little as 10 clock cycles (we had seen in Chapter 3 that instruction latency for SPU channel read-write is 2 clock cycles). For each of the five parameters

S. Koranne, *Practical Computing on the Cell Broadband Engine*,
DOI: 10.1007/978-1-4419-0308-2_5,

in `mfc_get(&LS, EA, sizeof(cb), tag_id, 0, 0);` data is written to the SPU channels. The local store address, Effective Address, size of data, tag and the actual command itself (get).

The overall latency of generating the DMA command, initially selecting the command, and unrolling the first bus request to the BIUor, in simpler terms, the flow-through latency from SPU issue to injection of the bus request into the EIBis roughly 30 SPU cycles when all resources are available. If list element fetch is required, it can add roughly 20 SPU cycles. MMU translation exceptions by the SPE are very expensive and should be avoided if possible. If the queue in the BIU becomes full, the DMAC is blocked from issuing further requests until resources become available again. Examples of DMA transfers are present in almost every major SPE program presented in the later chapters. SPE-to-SPE DMA without PPE involvement (except for initial LS address gathering) is shown in Section 19.6, pp. 375.

A transfers command phase involves snooping operations for all bus elements to ensure coherence and typically requires some 50 bus cycles (100 SPU cycles) to complete. For gets, the remaining latency is attributable to the data transfer from off-chip memory to the memory controller and then across the bus to the SPE, which writes it to local store. For puts, DMA latency doesnt include transferring data all the way to off-chip memory because the SPE considers the put complete once all data have been transferred to the memory controller.

5.1.1 Difference between fence and barrier synchronization

In particular, there are two flags that can be embedded into a command, *barrier* and *fence*. Both flags affect only commands in the same tag group. Generally, the embedded fence flag will not allow the command to execute until all commands within the same tag group and issued prior to the command with the embedded fence flag are compete. That is, the fence flag requires that all commands within the same tag group issued prior to the command with the embedded fence be completed prior to the execution of the command with the fence flag. The fence flag does not affect subsequent commands in the queue. Thus, a command issued after a command with an embedded fence flag can execute before the command with the fence.

The barrier flag affects all previous and subsequent commands within the same tag group. Generally, the barrier flag will not allow the command with the barrier flag or any subsequent commands within the same tag group to be executed before the execution of every command in the same tag group issued prior to the command with barrier flag. For example, commands within the same tag group and issued after a command with a barrier flag cannot execute before the command with the barrier flag. Typically, when all commands within the same tag group issued prior to the command with the barrier flag are complete, the command with the barrier flag and subsequent commands within the same tag group can be executed.

The *barrier command*, as opposed to the *barrier flag* orders all subsequent MFC commands with respect to all MFC commands preceding the barrier command in

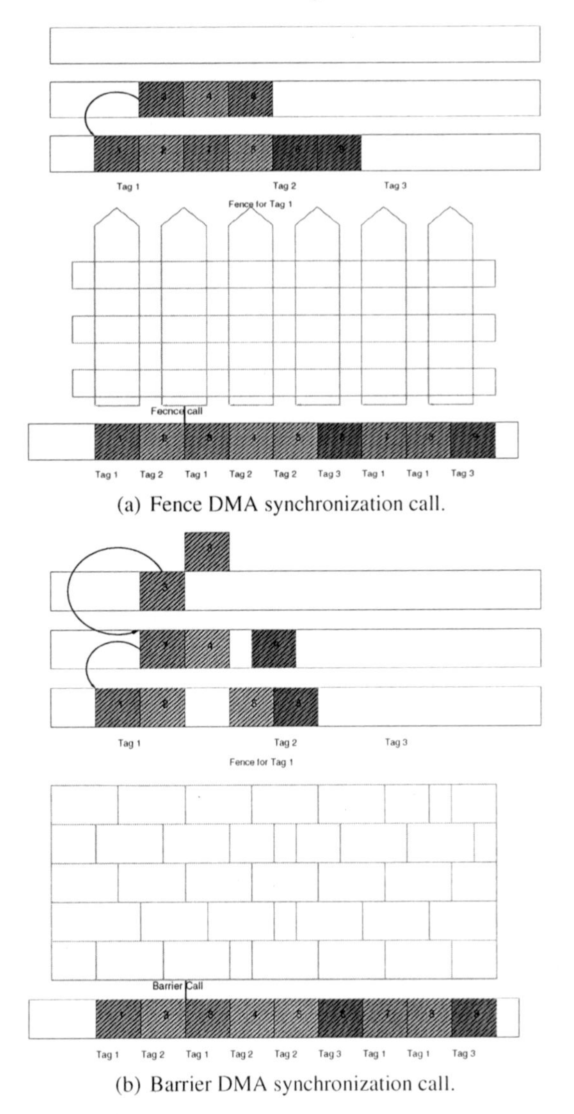

(a) Fence DMA synchronization call.

(b) Barrier DMA synchronization call.

Fig. 5.1 Example of Fence/Barrier DMA Synchronization calls.

the DMA command queue, independent of tag groups. The barrier command will not complete until all preceding commands in the queue have completed. After the command completes, subsequent commands in the queue may be started. The C syntax for issuing a barrier command is:

```
spu_writech(MFC_TagID, tag_id);
spu_writech(MFC_Cmd, 0xC0);
```

As we discussed above, barrier command is valid for all tag IDs, thus the setting of the tag ID is optional and need only be provided to assign an ID to the barrier command for subsequent tag-completion testing.

A barrier option might be useful when a buffer read takes multiple commands and must be performed before writing the buffer, which also takes multiple commands. In this example, the commands to read the buffer can be queued and performed in any order. Using the barrier-form for the first command to write the buffer allows the commands used to write the buffer to be queued without waiting for an MFC tag-group event on the read commands. (The barrier-form is only required for the first buffer-write command.) If the buffer read and buffer write commands have the same tag ID, the barrier ensures that the buffer is not written before being read.

The fence and barrier modifiers also provide stronger consistency of storage accesses in the weakly consistent storage model of the CBEA for several combinations of storage accesses involving commands in the same tag group. Thus, programmers do not need to add additional synchronization commands for these specific combinations if the commands are in the same tag group and either the fence or barrier modifier is used.

DMA Error	Cause of error
Transfer size alignment	Transfer size is not 0,1,2,4,8 bytes or (mod 16) bytes, Transfer size is greater than 16 KB
List transfer size alignment	List transfer size is greater than 16 KB
LS Address alignment	LS(0:14) must equal LSA(14:28) LS address must be on 8-byte boundary LS address LSBs should be multiple of transfer size
EA address alignment	$EA(60:63)(EAH\|\|EAL) \neq LSA(28:31)$
List address alignment	LA(29:31) are not 000
Atomic command pending	Previous atomic command (eg. putllc) is pending
Invalid MFC Command	MFC command type mismatch
Invalid MFC Opcode	Opcode bits are wrongly set
Invalid MFC Tag	Invalid tag-id, may be due to reserved bits

Table 5.1 DMA command errors and possible causes.

There are four MFC atomic update commands: `getllar`, `putllc`, `putlluc`, and `putqlluc`. The getllar, putllc, and putlluc MFC atomic update commands are performed without waiting for other commands in the MFC SPU queue and thus have no associated tag. There is a barrier command which is not the same as the barrier option on put and get. The `barrier` command works independent of tag groups. The fence form ensures that all previous MFC commands of the same tag group are issued prior issuing the fenced command. The barrier form ensures that all previous MFC commands in the same tag group are issued before issuing the barrier command, and that none of the subsequent MFC commands in the same tag group are issued before the barrier command.

5.2 Conclusions

In this chapter we looked at the DMA engine and the associated data ordering primitives of *barrier* and *fence*. We also discussed MFC atomic update commands which are used for SPE-PPE synchronization. Performance Guidelines for MFC Commands:

1. Minimize small transfers. Transfers of less than one cache line consume bus bandwidth equivalent to a full cache-line transfer.
2. Align source and destination addresses of large transfers (128 bytes of larger) to a cache-line boundary.
3. Have SPEs Initiate DMA transfers. Let the SPEs pull their data instead of PPE pushing the data to the SPE. This is beneficial for four reasons: a. There are eight times more SPEs than PPEs. b. An MFC SPU command queue is twice as deep (16 entries) as the PPEs MFC proxy command queue (eight entries). c. Consumer-managed transfers are easier to synchronize. d. The number of cycles required to initiate a DMA transfer from the SPE is smaller than the number to initiate from the PPE.
4. Avoid PPE pre-accesses to large data sets, so that most SPE-initiated DMA transfers come from main storage rather than the PPEs L2 cache. DMA transfers from main storage have high bandwidth with moderate latency, whereas transfers from the L2 have moderate bandwidth with low latency.
5. Minimize the use of synchronizing and data-ordering commands.

Part II
Programming the Cell Broadband Engine

In the previous part we have seen the system level overview of the Cell Broadband Engine. Previous chapters focused on the hardware and computer architecture aspects of the engine, in this part we discuss the corresponding software infrastructure and models we need to develop high-performance computer engines on the Cell.

In Chapter 6 we basically lay the foundation of this book by developing the parallel programming model around *tasks*. We discuss theoretical models and empirical laws which ideally govern parallel application performance. We describe the complexity class *NC* and its realization as a Boolean-circuit. We give a thorough overview of `pthread` library as we make extensive use of it from now on. We also provide some material on OpenMP, MPI and Cell Messaging Library which is not covered elsewhere in this book. We discuss the Berkeley *dwarf* paper [43]. We conclude by listing an unordered collection of programming pitfalls on the Cell which the reader should know about upfront.

In Chapter 7 we describe a simple development environment we have setup on the PS3. Despite going into some details, this chapter is not comprehensive enough as this is one part of the story which will be revised often as new SDK versions appear and newer Linux distributions find their way on the Cell. We discuss parallelism identification, and how to map common algorithmic constructs to high-level parallelism using the SPEs.

In the next chapter (Chapter 8, pp. 103) we put together our first Cell program. We write our first application on the CELL Processor using native development tools, `gcc` and `spu-gcc`. We will put the PPE core through its paces, learning what it can, cannot do. We discuss EMT performance, and where EMT is useful (in the control plane, and memory interfacing). We shall also discuss GNU/Linux system facilities as they can be put to good use to avoid rewriting code which is available in POSIX compliant libraries or other fairly standard GNU distributions. We shall cover EMT, integer performance, compiler capabilities, alignment, using disk, huge-memory page tables, Ethernet, 32-bit and 64-bit development, and memory management. We present simple programs that communicate data between the PPE and the SPE, and how to integrate `pthread` based tasks with PPE/SPE tasks.

Chapter 6
Foundations for Program Development on CBE

Abstract In this chapter we present the foundations of parallel computing. We discuss Amdahl's Law, Gustafson's Law, the *NC* complexity class and Boolean circuits. We present pthreads, the POSIX thread API for expressing parallel programs. The multi-faceted parallelism present in the Cell architecture is introduced. We also briefly discuss OpenMP. This chapter also includes a list of things to watch out for when writing programs for the Cell.

6.1 Theory of Parallel Programming

We can classify parallelism as one of two categories: (a) control-flow parallelism and (b) data-parallel parallelism. Control flow parallelism is the most general model, but suffers from an inherent problem known as *Amdahl's law* which limits the total speedup of the system as a function of its serial (or non-parallel) components. On the other hand, data-parallel is more restricted, but can yield very high levels of parallelism. There are now several models which implement control-flow parallelism on modern computers. The models are implemented using several techniques such as *message passing*, *multiple instruction multiple data* (MIMD), and *locking*. Shared memory multiprocessors use locking, while distributed computing based parallel systems use message passing.

Data-parallel (which is the favored model in this book as well) can be efficiently implemented using single-procedure multiple data (SPMD) or single-instruction multiple data. As we show later in this chapter, and in the sequel of this book, Synergistic Processor Units (SPUs) of the Cell implement data-parallel operations using both SIMD (at the instruction level) and SPMD (using DMA with multiple data packets being operated on by the SPE program)[1]

[1] In this book I use SPU and SPE interchangeably, but mostly when dealing with SIMD or instruction pipeline, floating-point, etc, I use SPU. When dealing with DMA, MailBoxes, synchronization, I refer to the unit as SPE.

S. Koranne, *Practical Computing on the Cell Broadband Engine*,
DOI: 10.1007/978-1-4419-0308-2_6, © Springer Science + Business Media, LLC 2009

In his book *Foundations of Parallel Programming* [81], Lewis describes the difference between control-flow analysis using Amdahl's law and *Gustafson's Law* for analyzing data-parallel parallelism. We describe the analysis below, but we should note, that both laws are derived in ideal settings where communication overheads (for example) are not analyzed in this simple model.

We first define *speedup* as the ratio of T_1 (which is the time taken by the program to complete on a single processor) to T_n (which is the time taken by the program to complete on n processors):

$$speedup = \frac{T_1}{T_n} \tag{6.1}$$

We define *efficiency* E as a measure of the gains made by running it on $n > 1$ processors:

$$E = \frac{speedup}{n}$$

It used to be the case that at best one could hope for *linear* speedup, but with the advent of the *memory-wall*, and power-optimized CPU voltage scaling, we can actually achieve *super-linear* speedup (if the data is divided so that it becomes small enough to fit in the L2 cache, then program run-time can decrease by a factor of 100).

6.1.1 Amdahl's Law

Amdahl's law states that the time to run a parallel program on n processors depends on the fraction of program β which is serially executed and the $(1-\beta)$ that is parallel.

$$T_n = \beta T_1 + \frac{T_1(1-\beta)}{n}$$

Substituting in Eqn. 6.1, we get

$$speedup = \frac{n}{\beta n + (1-\beta)} \tag{6.2}$$

Consider a program which has $\beta = 0.6$ and a speedup factor of 10, then by Amdahl's law the maximum speedup this program can achieve is:

$$s_{max} = \frac{1}{0.6 + \frac{1-0.6}{10}} = 1.5625$$

When we compare the component-wise speedup of 10, with the maximum achievable by the program as a whole, we can appreciate the struggle computer architects have had with Amdahl's law, and the resulting architectures which have favored to speedup all parts, albeit, at the cost of ignoring performance headroom in other areas. SIMD optimization and multiprocessor have only recently been introduced in mainstream microprocessors. As it stands, Amdahl's law paints a pretty bleak picture of the advantages of parallelism, but as we will see in the next section, SIMD (or data-parallel) programs can change the fundamental assumption, and thus arrive at a different conclusion.

6.1.2 Gustafson's Law

A basic assumption made in Amdahl's law is that speedup is constant (at best), but it ignores the fact that in SIMD programs parallelism and hence speedup increases with the size of the vectors. When the program is run on a vector of n, let the time be T_n; set $T_n = 1$ for normalization w.r.t n processors. When the same program is run on 1 processor the serial part takes time β and the parallel part $(1-\beta)n$, the total time is $T_1 = \beta + (1-\beta)n$, substituting in the equation for speedup we get

$$speedup = \frac{\beta + (1-\beta)n}{1}$$

Using the same example as Amdahl's law, we set $\beta = 0.6, n = 10$, then speedup is

$$speedup = 0.6 + 0.4 * 10 = 4.6$$

As we can see with SIMD the speedup is higher, and when we combine the reduced instruction count of SIMD, memory packing and potentially lower cache misses, SIMD can be very beneficial.

6.2 The Class *NC* of complexity analysis

Before we jump into programming, it might be useful to know how much parallelism the problem offers (or alternatively what is the complexity of the best-known parallel algorithm). For this analysis we use a complexity class *NC* which plays the same role as class *P* in algorithm analysis. The underlying machine model is simple, a PRAM with concurrent read-exclusive write (CREW) memory, with a polynomial number of processors. This has been shown to be equivalent to an arithmetic or Boolean-circuit model with the following properties: the number of nodes n is called the *size* of the circuit, each node can perform simple bit-wise operations such as *and*, *or*, and *not*. An example of a Boolean circuit is shown in Figure 6.1, with the longest path marked in red. The longest path from the input to the output of the circuit is called the *depth* and it corresponds to the shortest time output can be produced. The reader should convince himself of the fact, that there exist problems for which throwing more processors will not reduce the depth, or equivalently, the run-time of the algorithm.

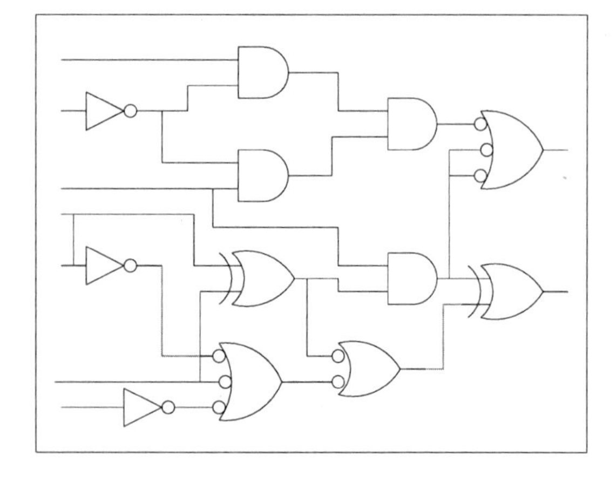

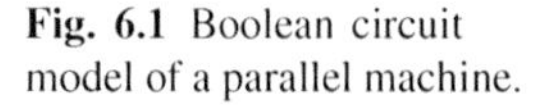

Fig. 6.1 Boolean circuit model of a parallel machine.

A problem is considered to be *efficiently parallelizable* if it known to be in *NC*. The class *NC* is defined as the class of problems which can be solved in polylogarithmic ($(\log n)^{O(1)}$) time on a CREW PRAM using a polynomial ($n^{O(1)}$) number of processors. Using the Boolean circuit model, *NC* corresponds to a Boolean circuit which solves the problem using $n^{O(1)}$ nodes and $\log n$ (upto a constant factor) time.

As an example consider the problem of computing matrix multiplication in parallel. A *NC* algorithm for parallel matrix multiplication can be easily constructed using n^3 processors, and using a $\log n$ addition-tree. Thus the depth or time of parallel matrix multiplication is $1 + \log n$ using n^3 processors.

6.3 Parallel programming concepts

Before we undertake program development on the Cell architecture we should understand the parallel programming model that the architecture provides to us. In this chapter we shall address the terminology and basic principles of communicating parallel programs (mostly for tightly-couple shared memory processors). Since the Cell is a heterogeneous architecture (the ISA of PPE is different from the SPU), and the high EIB bandwidth coupled with autonomous local store allows a unique programming paradigm to flourish. We first discuss parallel programming terminology using `pthreads` a popular POSIX threading library which is available on the PS3 GNU/Linux. Since many SPE run-time library calls are blocking, using multi-threading on the PPE even when there is only a single computation task is the only way to involve multiple SPEs.

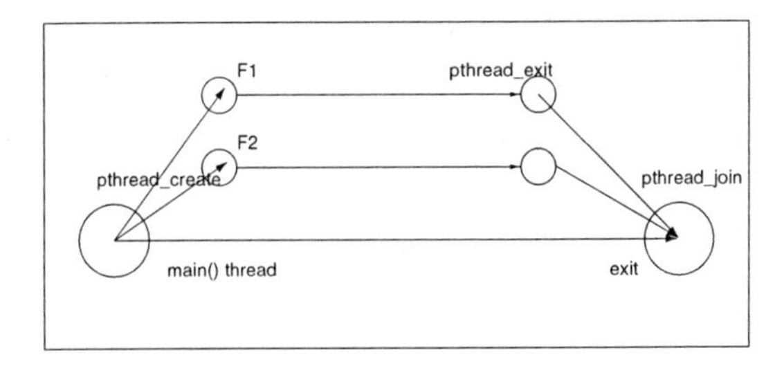

Fig. 6.2 Example of pthread spawning 2 threads for functions F1 and F2.

6.3.1 Parallel programming terminology

We define some commonly used parallel programming terms in this section. For computer architecture terms we have used the terminology as given by Flynn [47].

Critical section: a critical section is a piece of code that can only be executed by one task at a time. It typically terminates in fixed time, and another task only has to wait a fixed time to enter it. Some synchronization mechanism is required at the entry and exit of the critical section to ensure exclusive use. A semaphore (see below) is often used.
Latency: latency is a time delay between the moment something is initiated, and the moment one of its effects begins. In parallel programming, this often refers to the time between the moment a message is sent and the moment it is received. Programs that generate large numbers of small messages are latency-bound.
Load balance: distributing work among processor elements so that each processor element has roughly the same amount of work.
Race condition: a race condition occurs when the output exhibits a dependence (typically unexpected) on the relative timing of events. The term race condition also refers to an error condition where the output of a program changes as the scheduling of (multiple) processor elements varies.
Task: a unit of execution.
Thread: a fundamental unit of execution in many parallel programming systems. One process can contain several (or dozens or hundreds) of threads.

6.4 Introduction to pthreads

Even before we delve deep into the programming model of pthreads, it is useful to see what are the knobs and tuning parameters available to us to manage pthreads. These management methods are independent of the PPE thread scheduling policies or data-cache touching functions, but for Cell, the pthread library scheduling functions use the PPE intrinsics for changing thread priority.

POSIX pthread library allows management of threads by specifying the following:

1. Keys: threads use keys to maintain private copies of shared data items. Since SPE threads use local store this is an excellent model for SPE tasks. A single globally defined key points to different memory location for each thread. We can use keys when making function calls and we want thread specific data to be passed along,
2. Attributes: thread attributes allow us to control scheduling behavior, stack size and initial state,
3. pthread_once mechanism: is like the singleton constructor concept, it ensures that a function or action is performed *exactly once* regardless of how many times (or different threads) attempt to call it. Useful for file or network handling, use this mechanism instead of writing static variables for this purpose,
4. Thread cancellation: allows a thread to self-destruct, for example when searching for a data in a binary-tree, once the data is found, the rest of the threads can be cancelled,
5. Thread scheduling: specifies to the Operating System which thread you (as a programmer, not as a user, unless you make the thread scheduling visible to the end-user) would like the CPU to be scheduled. PPE has extensive control on the scheduling behavior of threads executing on the dual-thread PPU (see Chapter 2),

6.4.1 Understanding pthread programming model

Since pthread is an API library, all of its functionality is exposed using standardized function calls and these calls make up the programming model for pthreads. We discuss some of the important and CBE relevant pthreads call in this chapter for ready reference from other parts of this book, for a more detailed explanation we refer you to the book Pthreads Programming by Nichols et. al [94].

The pthread_create() function creates a new thread, with attributes specified by attr, within a process. If attr is NULL, the default attributes shall be used. If the attributes specified by attr are modified later, the thread's attributes shall not be affected. Upon successful completion, pthread_create() store the ID of the cre- ated thread in the location referenced by thread.

```
int pthread_create(pthread_t *restrict thread,
      const pthread_attr_t *restrict attr,
      void *(*start_routine)(void*), void *restrict arg);
```

Attributes objects are provided for threads, mutexes, and condition variables as a mechanism to support probable future standardization and customization. Attributes objects provide clean isolation of the configurable aspects of threads. For example, stack size is an important attribute of a thread, but it cannot be expressed portably.

The pthread_attr_init() function shall initialize a thread attributes object attr with the default value for all of the individual attributes used by a given implementation. The pthread_attr_destroy() function shall destroy a thread attributes object.

```
int pthread_attr_destroy(pthread_attr_t *attr);
int pthread_attr_init(pthread_attr_t *attr);
```

```
#include <stdlib.h>
#include <stdio.h>
#include <errno.h>
#include <pthread.h>
int r1 = 0, r2 = 0;
static void F1(int *pr1) {
  // some compute intensive function
}
static void F2(int *pr2) {
  // some other compute intensive function
}
int main(int argc, char **argv) {
  pthread_t t1, t2;
  if(pthread_create(&t1,NULL,(void *) F1,(void *) &r1) != 0)
    perror("pthread_create"), exit(1);
  if(pthread_create(&t2,NULL,(void *) F2,(void *) &r2) != 0)
    perror("pthread_create"), exit(1);
  if(pthread_join(t1, NULL) != 0)
    perror("pthread_join"),exit(1);
  if (pthread_join(t2, NULL) != 0)
    perror("pthread_join"),exit(1);
  return (EXIT_SUCCESS);
}
```

Listing 6.1 Introduction to pthreads

```
int pthread_once(pthread_once_t *once_control,
       void (*init_routine)(void));
pthread_once_t once_control = PTHREAD_ONCE_INIT;
```

Usually the controlling thread in the PPE will perform setup tasks, but this can be sub-optimal, especially when the setup is compute intensive or involves dealing with blocking I/O. In this case, the setup can be performed as part of the SPE task. Since SPE run-time directly does not have any function similar to pthread_once, we can simulate this behavior by static variables in the local store. The advantage of using pthread_once is still there, as we can move the function to PPE and easily see if the performance issue has been addressed.

6.4.2 Pthreads Keys: using thread specific data

As threads are created, run and destroyed, their working space consists of their stack space, global variables and memory allocated from the heap. During the course of program execution if you want to associate specific thread specific data to each thread, so that the same variable in different threads points to different memory, you can use *keys* to achieve this. Consider the example of searching for a value in a tree, every thread function will store its own fragment of the binary tree on which its operating. This can be done using keys:

```
static pthread_key_t tree_key;
int search_init() {
```

```
  pthread_key_create( &tree_key, (void*)free_key );
}
```

This function needs the variable for the key as well as a destructor function. Since thread cancellations and exits can leak thread specific data pointed to by keys, implementing this destructor correctly helps prevent memory leaks. Now we can use this key as follows:

```
pthread_setspecific( tree_key, (int*)(binary_tree+offset));
```

We set this threads `tree_key` to `binary_tree`offset+, later on we can use this key to retrieve the binary tree offset as follows:

```
int *tree_offset;
pthread_getspecific( tree_key, (void**)&tree_offset);
```

```
void *pthread_getspecific(pthread_key_t key);
int pthread_setspecific(pthread_key_t key, const void *value);
```

6.4.3 Pthreads Summary

Detailed function description of pthread functions can be found using the `man pthread_create` manpage and then following the related sections. In this section we present the most often used pthread functions, and their description for ready reference.

```
#include <stdlib.h>
#include <stdio.h>
#include <errno.h>
#include <pthread.h>
#define N 100
typedef int matrix_t[N][N];
typedef struct {
  int       id,size,Arow,Bcol;
  matrix_t  *MA, *MB,*MC,dummy;
} arg_t;
matrix_t MA,MB,MC;
void mult(int size,int row,int col,
          matrix_t A, matrix_t B,matrix_t C) {
  int pos;
  C[row][col] = 0;
  for(pos = 0; pos < size; ++pos)
    C[row][col] += A[row][pos] * B[pos][col];
}
void *mult_worker(void *arg) {
  arg_t *p=(arg_t *)arg;
  mult(p->size, p->Arow, p->Bcol, *(p->MA), *(p->MB), *(p->MC));
  free(p);
  return(NULL);
}
int main(int argc, char **argv) {
  int       size, row, col, num_threads, i;
  pthread_t *threads;
  arg_t *p;
  unsigned long thread_stack_size;
  pthread_attr_t *pthread_attr_p, pthread_custom_attr;
```

Table 6.1 Pthread Functions

Name	Description
pthread_create	creates thread (pthread_t, attr, function, arg)
pthread_exit	exits thread (status)
pthread_attr_init	initializes attribute (pthread_attr_t*)
pthread_attr_destroy	destroys attribute (pthread_attr_t*)
pthread_join	blocks parent till *T* terminates (pthread_t *T*, **void****)
pthread_detach	storage can be claimed when thread terminates (pthread_t),
pthread_self	return the thread id of calling thread (pthread_t)
pthread_equal	compare thread id (pthread_t t1, pthread_t t2)
pthread_control	allows 1 call to function (pthread_once_t *, **void*** ())
pthread_mutex_init	initializes mutex with attributes (pthread_mutex_t*, pthread_mutexattr_t*)
pthread_mutex_destroy	destroys mutex (pthread_mutex_t*)
pthread_mutexattr_init	initializes mutex attribute (pthread_mutexattr_t*)
pthread_mutexattr_destroy	destroys mutex attribute (pthread_mutexattr_t*)
pthread_mutex_lock	acquire a lock on a mutex (pthread_mutex_t*)
pthread_mutex_trylock	attemp to acquire a lock (pthread_mutex_t*)
pthread_mutex_unlock	unlock the mutex (pthread_mutex_t*)
pthread_cond_init	initializes cond with attributes (pthread_cond_t*, pthread_condattr_t*)
pthread_cond_destroy	destroys cond (pthread_cond_t*)
pthread_condattr_init	initializes cond attribute (pthread_condattr_t*)
pthread_condattr_destroy	destroys cond attribute (pthread_condattr_t*)
pthread_cond_wait	blocks calling thread until condition (pthread_cond_t*, pthread_mutex_t*)
pthread_cond_signal	wakeun up thread waiting for cond (pthread_cond_t*)
pthread_cond_broadcast	waken up all threads waiting on cond (pthread_cond_t*)

```
    size = N;
    threads = (pthread_t *)malloc(size*size*sizeof(pthread_t));
    for (row = 0; row < size; ++row)
      for (col = 0; col < size; ++col)
35      MA[row][col] = 1, MB[row][col] = row + col + 1;
    num_threads = 0;
    for(row = 0; row < size; row++)
      for (col=0;col<size;col++,num_threads++) {
        p = (arg_t *)malloc(sizeof(arg_t));
40      p->id = num_threads;p->size = size;
        p->Arow = row;p->Bcol = col;
        (p->MA) = &MA;(p->MB) = &MB;(p->MC) = &MC;
        pthread_create(&threads[num_threads],&pthread_custom_attr,
                       mult_worker, (void *) p);
45  }
    for (i = 0; i < (size*size); i++)
      pthread_join(threads[i], NULL);
    return (EXIT_SUCCESS);
}
```

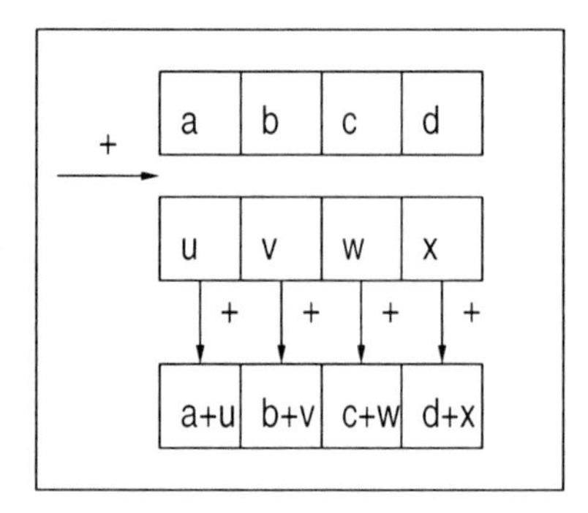

Fig. 6.3 4-way integer SIMD with '+' as the instruction.

Listing 6.2 Parallel matrix multiple

6.5 Multi-faceted parallelism in the CBE Architecture

The CBE processor provides a foundation for many levels of parallelization. Starting from the lowest, fine-grained parallelizationSIMD processingup to the highest, course-grained parallelization networked multiprocessingthe CBE processor provides many opportunities for concurrent computation. The levels of parallelization include:

1. SIMD processing, see Figure 6.3,
2. Dual-issue superscalar microarchitecture,
3. Multithreading (EMT on PPE),
4. Multiple execution units with heterogeneous architectures and differing capabilities,
5. Shared-memory multiprocessing,
6. Networked distributed-memory multiprocessing,

Thus the Cell architecture includes almost all prevalent forms of parallel programming models within it.

6.6 Alignment using compiler attributes

`float particle_mass __attr__ ((aligned (16)));` specifies to the compiler that the variable `particle_mass` should be aligned to a quadword boundary. The `sizeof()` operator for a vector data type always returns a size of 16 bytes. To create a vector literal the C/C++ array notation is used:
`vector unsigned int V = { 0,1,2,3 };`. The `const` and `restrict` qualifiers should

Table 6.2 Vector data types of the CBEA, and their organization as a 128-bit quantity.

Vector Data Type	128-Split	Content	SPU/PPU
`vector (un)signed char`	16 - 8-bit	(un)signed char	Both
`vector (un)signed short`	8 - 16-bit	(un)signed short	Both
`vector (un)signed int`	4 - 32-bit	(un)signed int	Both
`vector (un)signed long long`	2 - 64-bit	(un)signed long long	SPU
`vector float`	4 - 32-bit	single-precision float	Both
`vector double`	2 - 64-bit	double-precision float	Both
`qword`	1 - 128-bit	Quadword	SPU
`vector bool char`	16 - 8-bit	Boolean	PPU
`vector bool short`	8 - 16-bit	Boolean	PPU
`vector bool int`	4 - 32-bit	Boolean	PPU
`vector pixel`	8 - 16-bit	1/5/5/5 pixel	PPU

Vector Keyword Type	Typedef equivalent	SPU/PPU
`vector unsigned char`	`vec_uchar16`	Both
`vector signed char`	`vec_char16`	Both
`vector unsigned short`	`vec_ushort8`	Both
`vector signed short`	`vec_short8`	Both
`vector unsigned int`	`vec_uint4`	Both
`vector signed int`	`vec_int4`	Both
`vector unsigned long long`	`vec_ullong2`	SPU
`vector signed long long`	`vec_llong2`	SPU
`vector float`	`vec_float4`	Both
`vector double`	`vec_double2`	SPU
`vector bool char`	`vec_bchar16`	PPU
`vector bool short`	`vec_bshort8`	PPU
`vector bool int`	`vec_bint4`	PPU
`vector bool pixel`	`vec_pixel8`	PPU

Table 6.3 Single token typedefs for C/C++ macro support.

be used when the code permits. Most readers will be familiar with the `const` qualifier which prohibits the compiler from using *that* pointer to modify the pointed-to variable or memory location. Note this does not mean that the variable or memory location is constant or read-only, the qualifier applies only to the variable handle to which it is applied.

The `restrict` qualifier is intended to assist the compiler in generating better code by making a programmatic assurance to the compiler that all access to the pointed object (in the current scope) is obtained *only* through that particular pointer. For example:

```
void *Children( const N * restrict left, const N * restrict right );
```

Data type T	sizeof(T)	Alignment (x)	lg(x)
char	1	byte	3
short	2	half-word	4
int	4	word	5
long	4	word	5/6
long long	8	doubleword	6
float	4	word	5
double	8	doubleword	6
pointer	4	word	5
pointer(LP64)	8	doubleword	6
vector	16	quadword	7

Table 6.4 Default data type alignment

To minimize the instruction fetch and decode logic, the SPE has simplistic branch prediction and speculation mechanisms. Performance of compute intensive code can be improved significantly by eliminating branches from inner-loops, or where total branch elimination is not possible using `selb` type selection instructions to conditionally select the appropriate result of the branch. Where the instruction to produce the result can produce a side-effect (see why purely functional languages were the pioneers in parallel computing), applying feedback-directed branch optimization or programmer-directed branch prediction can improve performance. To specify the likeliness of a branch use: `__builtin_expect(int expr,int value)`. Note that `__builtin_expect` evaluates the expression; this should not be confused with speculative evaluation of functions. As an example, consider the following:

```
int memory_overfull = get_memory_status();
if (__builtin_expect( memory_overfull, 0 ) {
  // print error to log, and call garbage collection
}
```

As we have described earlier in this chapter, `gcc` defines the `__LP64__` preprocessor define when compiling in 64-bit mode. For SPU targets `spu-gcc` also defines `__SPU__` which can be used to add code exclusively for the SPU target.

6.7 OpenMP

Many compilers support some set of OpenMP directives. OpenMP is a parallel programming model for shared-memory systems. Pioneered by Silicon Graphics, Inc. (SGI), it is now embraced by most of the computing industry and is exhaustively documented at http://www.openmp.org. The book Parallel Programming in OpenMP, by Chandra et al., is another good source, as is Parallel Programming with OpenMP at

```
http://www.osc.edu/hpc/training/openmp.
```

OpenMP directives are instructions to the compiler to perform tasks in parallel. In Fortran, they are expressed as source-code comments; in C and C++, they are expressed as a **#pragma**. All OpenMP directives in C and C++ are prefixed by, **#pragma** omp. A typical OpenMP directive looks like this one, which says that individual iterations of the loop are independent and can be executed in parallel:

```
#include <omp.h>
#include <stdio.h>
int main (int argc, char *argv[]) {
  int th_id, nthreads;
  #pragma omp parallel private(th_id)
  {
    th_id = omp_get_thread_num();
    printf("Hello World from thread %d\n", th_id);
    #pragma omp barrier
    if ( th_id == 0 ) {
      nthreads = omp_get_num_threads();
      printf("There are %d threads\n",nthreads);
    }
  }
  return 0;
}
```

Listing 6.3 Simple Hello,World in OpenMP.

Since OpenMP is a shared memory programming model, most variables in OpenMP code are visible to all threads by default. But sometimes private variables are necessary to avoid race condition and there is a need to pass values between the sequential part and the parallel region (the code block executed in parallel), so data environment management is introduced as data clauses by appending them to the OpenMP directive. Variables can be declared shared, which means they have a single storage location during the execution of a parallel section. In contrast, a variable can be declared to be private. It will have multiple copies: one within the execution context of each thread. There are different flavors of private, including *threadprivate*, *firstprivate*, and *lastprivate*.

OpenMP provides mutual exclusion and event synchronization as ways for multiple OpenMP threads to communicate with each other. The directives critical and barrier can be used. OpenMP allows programmers to specify loops, loop nests, or code sections that may be executed in parallel. More details on OpenMP can be found at the following reference:

```
htttp://www.llnl.gov/computing/tutorials/openMP
http://www.openmp.org
Quinn Michael J, Parallel Programming in C with MPI
and OpenMP McGraw-Hill Inc. 2004. ISBN 0-07-058201-7
```

6.7.1 SPE Programming

1. Use SPU intrinsics where possible
2. Overlap DMA with computation using multiple buffering

3. Optimize for superscalar execution by knowing fetch/issue rules of the even/odd pipeline
4. Design for limited Local Store
5. Eliminate branch instructions as much as possible, use conditional selection, or branch prediction
6. Unroll loops to mitigate loop branches and dependency stalls
7. Be cognizant of the 16x16 bit integer multiplier, avoid 32-bit multiplies (see $GF(2^m)$ Zech)
8. Use shuffle-byte instructions for efficient table lookup and permutations (see Batcher network, sorting example)
9. Prefer to use quadword loads and stores (See under sampled Bresenham for example)

6.8 Double buffering example

```
#define BF_SIZE 4096
volatile unsigned char B[2][BF_SIZE] __attribute__ ((aligned(128)));
void double_buffer_example(unsigned int ea, int buffers)
{
  int next_idx, idx = 0;
  // Initiate first DMA transfer
  spu_mfcdma32(B[idx], ea, BF_SIZE, idx, MFC_GET_CMD);
  ea += BF_SIZE;
  while (--buffers) {
    next_idx = idx ^ 1;
    spu_mfcdma32(B[next_idx], ea, BF_SIZE, idx, MFC_GET_CMD);
    ea += BF_SIZE;
    spu_writech(MFC_WrTagMask, 1 << idx);
    (void) spu_mfcstat(MFC_TAG_UPDATE_ALL);
    use_data(B[idx]); // Use the previous data
    idx = next_idx;
  }
  spu_writech(MFC_WrTagMask, 1 << idx);
  (void) spu_mfcstat(MFC_TAG_UPDATE_ALL); // Wait for last transfer done
  use_data(B[idx]); // Use the last data
}
```

6.9 Software Managed Cache

The SPU compiler provides transparent (though not efficient) access to variables present in the EA memory address space. This is done by using the `__ea` qualifier to a variable. The code also needs to be compiled using the `spu-gcc -mea64` for 64-bit EA addressing, or `spu-gcc -mea32` for 32-bit addressing. Additionally, the size of the cache to be used also needs to be specified as `spu-gcc -mcache-size=8kb`. Modifications to variables with the `__ea` qualifier are not synchronized with the local store, this is the programmers responsibility. Although, DMA is much faster at transferring large amounts of data, and DMA can also be overlapped with computation using double-buffering, having some critical variables managed by the software can be useful.

6.10 Physical Organization

Physical organization refers to the principle of managing files and directories for code, design-documentation and test-cases. These are important subjects, but beyond the scope of this book. As a good introduction to this subject I would recommend *Large-Scale C++ Software Design*, by Lakos [80].

Table 6.5 Differences between PPE and SPE from a programming point of view.

Feature	PPE	SPE
Processing Units	1-4	8-16
No. of Registers	32	128
Register type	Separate FP, fixed, SIMD	Unified
Latency	Cached	Fixed
Size of store	$2^{64}-1$	256 KB LS
Instruction set	PowerPC	New SPU ISA
Compiler	PowerPC GCC	spu-gcc

Moreover the vector/SIMD multimedia extensions provide these operations that are not directly supported by the SPU instruction set:

1. Saturating math
2. Sum-across
3. Log2 and 2x
4. Ceiling and floor.
5. Complete byte instructions

Vector/SIMD multimedia extensions support several vector-pack and unpack operations, although comparable operation can be performed with SPU instructions.l for creating

6.11 Things to watch out for

During the writing of this book and the associated code I collected the following set of problems which were most common. I present them here as things to watch out for when writing code for the Cell.

1. Mis-aligned data in datap,argp when using pthread_data indirection: always use `_malloc_inline` to align data yourself,
2. spu_insert does not modify its argument, in fact none of them do:
3. divide by zero exception: looks like a hang
4. declare shared data **volatile**: compiler may perform optimization otherwise,
5. Exceeding local store capacity:
6. sizeof control block should be a multiple of 64: keeping it a multiple of 64 prevents problems when moving code from 64-bit to 32-bit, and vice-versa,
7. Align address variable of the control block:
8. Keep DMA maximum size in mind when scheduling DMA:
9. Deadlock prevention: in the following code, PPE is waiting for MailBox message, while SPE has already returned context to PPE thread, which has exited. Always make sure that MailBox returns are non-blocking, or are sent even in error checking failures,
10. Avoid magic constants: during testing I reduced the dimension of PDF to 2, and put in a constant 2 in a readahead buffer function to parse floating point data. The system would not work when I moved to 3d (2 Hrs),
11. abs vs fabs: on SPU there is a big and silent difference between abs and fabs, C++ programmers beware
12. correct absolute path to spe_image_open: give absolute path to spe_image_open, as it does not do PATH lookup,
13. SPE return codes at exit are 8 bits only: for larger data use mailbox,
14. When using mmap, make sure file open flags match PROT flags, eg., O_RDONLY only lets you do PROT_READ: this is obvious but when passing file descriptors it is easy to overlook, as the open and mmap calls can be in separate files,
15. unsigned char SIMD is 16-way and not 8-way,: this tripped me up once, as char is often associated with 8 as it has 8-bits (maybe its just me),
16. Using O_DIRECT without `mmap` will cause 0 bytes to be read
17. Don't reserve mfc_tag_reserve for every transfer, you will see a problem after 31
18. Integer Multiplies: The SPU contains only a 16 x16 bit multiplier. Therefore, to perform a 32-bit integer multiply, it takes five instructionsthree 16-bit multiplies and two adds to accumulate the partial products. To avoid extraneous multiply cycles, observe the following rules:

    a. make array-element size a power of 2, thus making indexing equivalent to shifts,
    b. if the operands are less than 16 bits, cast them to unsigned short (see below), For example (this is given in CBE Programmers Guide):

    ```
    #define MULTIPLY(a, b)\
    (spu_extract(spu_mulo((vector unsigned short)spu_promote(a,0),\
    (vector unsigned short)spu_promote(b, 0)),0))
    ```

    c. always cast constants if they are shorter than the default (int),

6.12 Strategies and Paradigms for Identifying Parallelism

In this book we have analyzed several computational tasks, and have implemented their parallel implementation on the SPU. A general guiding principal that I have learnt to follow on the Cell is to recognize high-level constructs in the algorithm pseudo-code and immediately realize an efficient SPU mapping. Some of the models that I use in this book are as follows:

1. **SPU Competition**: many algorithms have the following structure: In many cases

```
1 while bestCost >= threshold do
2     PerformComputation();
3 end
```

   (see Section 11.12) we implement the `while` loop in parallel on multiple SPEs, and choose the result based on the lowest cost. This can be thought of as a competition between the multiple SPEs. Similar strategies work when doing a search-space exploration with a cost metric (consider the example of *perfect hash functions* given in Section 10.6 on pp. 164). There also, a competition between 6 SPEs is held to determine the size of the smallest perfect hash table,
2. **Randomized algorithm**: when studying randomized algorithms for the first time it is disquieting to grasp the notion of randomization; but once the algorithm has been rigorously analyzed, the same attributes can be used to put in place a simple, elegant parallel algorithm. Since by definition (and by design) randomized algorithms cannot depend on initial state (except to guarantee enough randomness in the system), they are stateless in their starting point or number. We have used this to good effect in solving `min-cut` (cf. Algorithm 14, pp. 179) and in writing our PDF (partial distance function) solver in Chapter 17. If all the state you are communicating is random, why communicate at-all, and especially the phrase *random trials* should immediately bring a picture to your mind with all 6 SPEs humming along in parallel running *trials*.
3. **Divide and Conquer**: this is the simplest and most popular parallel paradigm, and for good reason. It usually works well, and with the EIB providing excellent cover for your data communication needs, with a little-bit of care, even a simple text-book style implementation of a compute intensive task using divide-and-conquer will yield positive results,
4. **Data parallelism**: although easy to describe (just do the same computation on different data), in practice, this model is valid only for data-streaming models. Nevertheless, we have used it in solving the line-of-sight (LOS) problem (cf. Chapter 16) as we compromised in keeping enough of the *state* about the problem which is disconnected from the ephemeral data-stream, on each of the SPEs. For image processing, and audio streaming applications, the computation state is very light weight, and can easily be replicated on the compute nodes, but for gen-

eral purpose applications (eg., doing a timing update on a VLSI circuit), careful attention should be devoted to the storage of persistent problem-state (computation side), as opposed to the ephemeral data-state.

5. **Restart Mode**: the SPUs program-counter is disconnected from the main thread, and they cannot inflict damage to the main memory (unless the application writer directed it using some variant of `mfc_put`), as a consequence SPU threads can be killed and restarted with impunity. This is not as easy to do using `pthread_cancel` as you would imagine it to be. Moreover, since SPU kernels cannot (where they can, they should not) consume and hold system resources (eg., `malloc` from main memory, file descriptors, sockets) resource leaking is significantly reduced. These properties can be used to write algorithms which destructively modify data-structures with the goal of reducing run-time. When the current task is done, rather than performing the expensive book-keeping associated with undoing the destructive writes to the data-structure, the SPE context or atleast the data-structure local store is reloaded from known-good initial contents.

6.13 Conclusion

In this chapter we touched upon the theory of parallel algorithms. This is an active area of research, both from an algorithimic point of view, as well as from an implementation point of view. The Cell offers an unique platform to experiment with ideas in loosely coupled, asynchronous, but coherent processors, connected by a high-speed interconnect. In the sequel of this book we shall see how to use the novel features of the architecture to leverage the theoretical benefits such architectures have promised for decades, but only now have they been realized in cost effective compute devices.

I also collected a list of frequent problems, or unusual occurrences, which caused me either a great deal of time or code, or sometime both to fix, in this chapter on pp. 89, Section 6.11. A revised and updated list will be maintained on the website for this book as well.

Chapter 7
The development environment

Abstract In this chapter we discuss the following (a) how to setup a development environment for programming the CBE, (b) for PS3 owners details instructions for installing Fedora Core 7 on PS3 and setting up IBM SDK 3.0, tutorials to follow, (c) for Linux users how to setup the MySim IBM Full System Simulator and cross compiler to write, develop and debug applications on the CBE, (d) for Windows users, how to download VMWare Image containing IBM Full System Simulator with IBM SDK 3.0 installed, (e) how to get a remote shell account on POWER.ORG servers at OSU (Oregon, USA), Germany and China, (f) detailed presentation of the CELL Broadband Engine, PPC core, SPU, Element Interconnect Bus, DMA, Mailbox, SIMD, dual-instruction pipeline.

7.1 How to setup an PS3 for Cell development

Detailed instructions and HOWTOs to install GNU/Linux on the PS3 hardware are available on the Internet. Since operating system versions, and firmware versions keep on changing almost on a monthly basis, you should make sure that the OS version and your PS3 firmware version are compatible with these instructions. In the worst-case you could end-up with a PS3 with your disks formatted (although nothing worse than that should happen), you can always go back to the Game OS and reformat your disk and try again, maybe with a different OS version. In my development setup I have used Fedora Core 7 with Cell SDK 3.0 running on one of the first 20 Gb Japanese model PS3.

Fedora 7 installs on the PS3 out of the box, and if you are used to installing any distribution of GNU/Linux, you should find most of the installation fairly routing. The PS3's base specs are 256MB of main memory, a Cell/B.E. processor, wired Ethernet, 802.1 WiFi (on some PS3 models), PS3 GNU/Linux can access either the gigabit wired Ethernet or the 802.11g wireless, but not both at once. In this book we have not used the wireless Ethernet (did I mention I have a 20 Gb first Japanese model, it does not have a WiFi or card readers). I have not installed any of the

S. Koranne, *Practical Computing on the Cell Broadband Engine*,
DOI: 10.1007/978-1-4419-0308-2_7, © Springer Science + Business Media, LLC 2009

X-window packages or GNOME or KDE as I don't intend to run any graphics application on the Cell. I have installed SSH, screen and all the development packages, and any other packages demanded by the Cell SDK 3.0 installation script.

7.2 Alternatives to installing GNU/Linux on the PS3

Although mentioned here, none of these alternatives provide the same ease of use, or bang-for-the buck as the PS3, *caveat emptor*. There do exist the following alternatives for those who do not have ready access to a PS3 (which they can install GNU/Linux on):

1. **Install IBM Full System Simulator and the SDK Cross compilers**: the IBM Full System Simulator is able to simulate a single Cell processor (with 8 SPEs), a dual-Cell configuration with upto 16-SPEs, and with customization, you can control many aspects of the system under design. The simulator gives cycle accurate data for PPE and SPE execution, and can in many cases pin-point exact cause for DMA errors, whereas the PS3 would simply say `Bus error`. This added visibility comes at a cost of runtime, for all but the simplest of applications, using the full system simulator as a development machine is frustrating. Even using cross compilers only gets half-way, as applications run atleast 100x slower than the native Cell (eg., on the PS3). Once you have got used to native execution speeds on the PS3, the simulator looks and feels even more slower; but its use as a debug device and system configuration explorer tool lives on (despite the fact that it was originally written as a software test platform before the actual chip arrived). I have not used the Full System Simulator as I installed GNU/Linux on the PS3, and none of the examples in this book have use it either,
2. **Get a VMWare image containing IBM Full System Simulator and SDK**: the IBM full system simulator mentioned above only runs on GNU/Linux, so if you are a Windows developer wanting to try out Cell development, and you don't have or want to install GNU/Linux on the PS3, then you can download VMWare images for Fedora 7/8 with the IBM Simulator installed. Registering on IBM's Alphaworks website and downloading the actual simulator is also recommended (even if you don't install it, so that you are abreast of latest news about the simulator and also that you have perused their licensing). So with the VMware images, you are running a simulator (the IBM Full System Simulator) inside another (the VMWare *simulator*, which run GNU/Linux on Windows). Whats the performance like ? Well, I have not tried this method, but it seems to work fine for many people. The following links may be useful if you decide to go this route:
VMWare Images

```
http://sti.cc.gatech.edu/programming.html
http://www.pad.lsi.usp.br/cell/
http://www.ibm.com/developerworks/power/cell/
```

3. **Get an account on the Open Source PPC servers**: from `www.power.org`, it is possible to get an account on a PowerPC machine if you are using to develop

Open Source software. PowerPC machines at OSU OSL (Oregon State University, Open Source Labs) have access to Cell blades.

To provide software developers exposure to the CBEA with and without access to hardware, a Software Development Kit (SDK) is available on the IBM alphaWorks Web site
`www.alphaworks.ibm.com/topics/cell` with open-source content distributed on the Barcelona Supercomputing Center Web site
`www.bsc.es/projects/deepcomputing/linuxoncell`.
The SDK is continuously being enhanced with additional components and features. The key foundational components of the SDK are as follows:

1. A Cell/B.E. processorenabled Linux kernel supporting the unique CBEA features.
2. The IBM full system simulator for the Cell processor that supports simulation of either uni-processor or dual-processor systems. As mentioned above both functional and cycle accurate simulation modes are provided,
3. A system root image (sysroot) that provides a standard set of Linux operating system utilities and services for use in the simulated Linux operating system environment.
4. Standard CBEA-specific libraries including the newlib C languages standard library, SIMD math libraries, and an SPE runtime management library that exposes the SPEs as heterogeneous POSIX threads.
5. Eclipse-based integration development environment (Cell IDE): an example of this IDE is shown in Figure 7.2 however none of the examples in this book need any feature provided by Eclipse. We write our own Makefiles and scripts, in that way we gain better understanding of what the system is doing (or expected to do), and we can make required changes more easily. But, nevertheless, if you are familiar with Eclipse, please feel free to use that as the editor, nothing in the book uses GNU/Emacs either,
6. GNU and IBM XL C/C++ compilers: in this book only the GNU GCC compilers are used. The `ppu-gcc`, `ppu-g++` and the `spu-gcc` compilers in particular. These compilers are available as open source projects, and especially for the SPE looking at the assembler source code can provide insights into register naming, stack space and local store memory. I have not analyzed the code generation for SPE in detail, but as we see later in this book, a thorough understanding of the SPE dual-pipeline and instruction latency is very important to write high-performance code; and the open source availability of these compiler makes the process that much more understandable to someone devoting time and energy towards it,
7. debug tools including the GNU debugger (gdb): the debugger has been enhanced to enable combined PPU and SPU debugging. The only caveat here is that cross-compiled debug binaries do not play nicely with system gdb on the PS3. Thus I found it necessary on many occasions to abandon my cross compiled debug binary and perform a debug build natively on the PS3 to continue debugging. An example of debugging with SPU code is given later in this chapter.

8. Performance analysis tools including the OProfile system profiling tool, the Cell/B.E. processor performance counter, which provides access to Cell/B.E. processor performance monitoring facilities, code analyzer, Feedback-Directed Program Restructuring technology of FDPR-pro, and a static performance analysis tool called spu_timing. The output from many of the tools can be visualized using the Eclipse-based visual performance analyzer (VPA).
9. An SPE executive for invoking SPU executables directly from a Linux shell. The executive provides standard system services including file I/O, shared memory, memory map, and time of day.
10. Many code samples that demonstrate programming techniques, optimized libraries with source code to jump-start application development, and workloads to demonstrate the computational capabilities of the Cell/B.E. processor.
11. Programming model frameworks such as the Accelerated Library Framework (ALF), which provides services to assist in the development of parallel applications and libraries on multicore architectures with hierarchical memory such as the CBEA.

The SDK is supported on the x86, PowerPC 64, and Cell/B.E. processor blade systems running Fedora[1] core for the Linux operating system. These configurations minimize system costs, ensuring wide access to anyone who wants to evaluate Cell/B.E. technology, learn Cell/B.E. programming, or develop Cell/B.E. applications and tools.

The tool chains for both the PPE and SPE processor elements produce object files in the ELF format. ELF is a flexible, portable container for relocatable, executable, and shared object (dynamically linkable) output files of assemblers and linkers. The terms PPE-ELF and SPE-ELF are used to differentiate between ELF for the two architectures. CESOF is an application of PPE-ELF that allows PPE executable objects to contain SPE executables. GNU/Linux thread application programming interfaces (APIs) allow programmers to focus on application algorithms instead of managing basic tasks such as SPE process creation and global variable sharing between SPE and PPE threads. The SPE ABI is also very well documented in the SDK documents.

7.3 The CESOF Format

The CBEA Embedded SPE Object Format (CESOF) standard defines a portable, extensible object-file format for executable object files (both static and dynamically linked). As we have seen in the previous chapters (and should know readily by now), PPE and SPE architectures and instruction sets are completely different, and currently separate toolchain components (`ppu-gcc` for PPE and `spu-gcc` for SPE) are used to compile and link object code. Since ELF provides no direct way of com-

---

[1] I have also installed the SDK on an Gentoo system running x86_64 without any problem.

bining object code of different ISAs, CESOF format was designed to provides three key capabilities for developers:

1. SPE programs have their own section in PPE elf-binaries
2. SPE and PPE programs can share global variables
3. `ld` or other PPE linkers can consume object code containing either PPE or SPE code

The software-development tool chains for CBEA processor programs conform to three standards:

1. Tool Interface Standard (TIS) and Execution specification
2. Executable and Linking Format (ELF) specifications
3. PPE and SPE Application Binary Interface (ABI) specifications

CESOF uses only the facilities defined in the TIS-ELF standard. CESOF files can be archived, linked, and shared just like other PPE-ELF object files, and can be combined with other PPE ELF objects. Standard ELF tools for introspecting the file work just file (infact running `objdump -fh` on the file produces globs of interesting information).

The SPU sync Instruction

The SPU sync instructionwhich differs from the PPE sync instruction discussed in Chapter 2. SPU sync causes the SPU to wait until all pending store instructions to its LS have completed before fetching the next sequential instruction. It orders instruction fetches, loads, stores, and channel accesses. The sync instruction affects only that SPUs instruction sequence and the coherency of that SPUs fetches, loads, and stores, with respect to actual LS state. The SPU does not broadcast sync notification to external devices that access its LS, and therefore the sync does not affect the state of external devices.

1. volatile: add volatile to SPU data, else compiler will do constant propagation
2. non quad loads: edge_table[i] load of an integer takes 4 loads, making it a vector unsigned int, get 4x * (4-1) reduction in memory loads
3. check assembly: always check assembly to make sure instruction count and clock cycle in loops are reasonable, not too high and not too low
4. pipeline: understand and memorize the instruction class of various common instructions, such as shift, arithmetic, loads
5. double-precision: if you don't need the extra precision, it is much faster to stick with single-precision.

Fig. 7.1 Dual pane setup with PPE and SPE code.

7.4 Basic development environment

Although every developer has his own preference for a suitable development environment I can describe my current setup which has worked for me in writing the example applications in this book. My current development environment consists of a GNU/Linux desktop running Gentoo on an AMD Opteron processor box. I have installed the cross compiler for the Cell SDK, and thus am able to syntax-check, and build binaries. I have a headless PS3 which I connect to using SSH and SCP. I use a simple build script which builds the binaries and copies them to `/tmp` on the PS3 and runs them at the same time. This is shown in Figure 7.1. A significant number of algorithms presented in this text were first implemented using Common Lisp. I have included the Lisp code, and Lisp generated test-cases for most applications. A short summary of the functions of Common Lisp which are heavily used is given in the Appendix.

An Eclipse based Cell IDE is also available with the SDK and some people (especially those having prior familiarity with Eclipse) find that more convenient. An example setup with Eclipse is shown in Figure 7.2. Even if you use Eclipse to setup the project and the remote host, it it useful to know the basic steps which are required to compile, link and embed SPU objects in PPE ELF files to create a Cell binary. The following is a step-by-step guide to achieve this result:

1. Compile SPU code: use `spu-gcc` with required options to compile a C code into object code using the SPU ISA, eg

```
spu-gcc -O2 file.c -o file.exe
```

2. Embed SPU object code in a shared library:

```
ppu-embedspu spe_func spe_func.exe spe_func.so
```

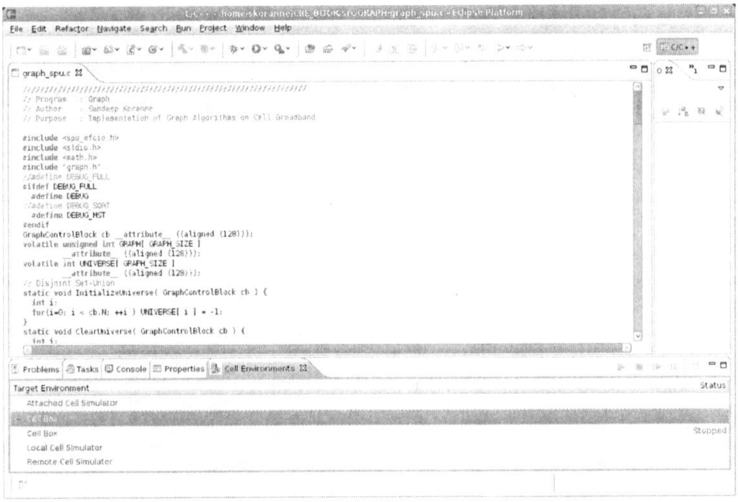

File Edit Refactor Navigate Search Run Project Window Help

graph_spu.c

```
////////////////////////////////////////////////////////////////////
// Program    : Graph
// Author     : Sandeep Koranne
// Purpose    : Implementation of Graph Algorithms on Cell Broadband

#include <spu_mfcio.h>
#include <stdio.h>
#include <math.h>
#include "graph.h"
//#define DEBUG_FULL
#ifdef DEBUG_FULL
  #define DEBUG
//#define DEBUG_SORT
  #define DEBUG_MST
#endif
GraphControlBlock cb __attribute__ ((aligned (128)));
volatile unsigned int GRAPH[ GRAPH_SIZE ]
         __attribute__ ((aligned (128)));
volatile int UNIVERSE[ GRAPH_SIZE ]
         __attribute__ ((aligned (128)));
// Disjoint Set-Union
static void InitializeUniverse( GraphControlBlock cb ) {
  int i;
  for(i=0; i < cb.N; ++i ) UNIVERSE[ i ] = -1;
}
static void ClearUniverse( GraphControlBlock cb ) {
  int i;
```

Problems | Tasks | Console | Properties | Cell Environments

Target Environment	Status
Attached Cell Simulator	
Cell Box	
Cell Box	Stopped
Local Cell Simulator	
Remote Cell Simulator	

Fig. 7.2 Eclipse based Cell IDE, part of Cell SDK.

3. Compile PPE Code:

```
ppu-gcc -O2 ppe_code.c -lspe2 -lpthread -o ppe_code -lspe_func
```

4. Copy binary to remote cell host:

```
scp ppe_code skoranne@cell:/tmp
```

5. Launch binary on remote host:

```
ssh skoranne@cell /tmp/ppe_code
```

7.5 Debugging code on the SPU

For more details refer to the latest GNU/gdb documentation, it contains a section called "Cell Broadband Engine SPU architecture", which describes specific command for the SPU. Some of the useful commands are described below:

info spu event
info spu signal
info spu mailbox
info spu dma
info spu proxydma

The following commands are not specific to the SPU, but are useful nevertheless,

info reg $< r >$:

backtrace/bt:
disassemble: default is function closes to program counter, if you give an argument, then the function closest to that location is disassembled and shown,
info target: prints information about the ELF object,

```
(gdb) b Collatz
Breakpoint 1 at 0x188: file collatz_spulet.c, line 14.
(gdb) run
Starting program: /tmp/collatz
Sweeping Collatz from 1 to 1<<28
Breakpoint 1, Collatz (a=1, b=268435456) at collatz_spulet.c:14
14        ULL sum=0;
(gdb) info reg r1
r1  {uint128 = 0x000..e0, v2_int64 = {0x3..eee0,
    v4_int32 = {0x3fee0, 0x3eee0, 0x3fee0, 0x3fee0},
(gdb) info reg r2
r2     {0x0, 0x0, 0x0, 0x0}}
(gdb) list
11      typedef unsigned long long int ULL;
12      ULL Collatz(ULL a, ULL b) {
13        ULL i;
14        ULL sum=0;
(gdb) info spu mailbox
SPU Outbound Mailbox
0xc0006c00
SPU Outbound Interrupt Mailbox
0xc0006c00
(gdb) info spu signal
Signal 1 not pending (Type Or)
Signal 2 not pending (Type Or)
(gdb) info spu dma
Tag-Group Status  0x00000000
Tag-Group Mask    0x00000000 (no query pending)
Stall-and-Notify  0x00000000
Atomic Cmd Status 0x00000000

Opcode  Tag TId RId EA LSA  Size LstAddr LstSize E
0       0   0   0       0x00000 0x00000
Breakpoint 1, Collatz (a=1, b=268435456) at collatz_spulet.c:14
14        ULL sum=0;
(gdb) disassemble
Dump of assembler code for function Collatz:
0x00000160 <Collatz+0>: stqd    $1,-128($1)
0x00000164 <Collatz+4>: ai      $1,$1,-128
0x00000168 <Collatz+8>: lqd     $2,80($1)       # 50
0x0000016c <Collatz+12>:        cdd     $5,0($1)
0x00000170 <Collatz+16>:        shufb   $2,$3,$2,$5
0x00000174 <Collatz+20>:        stqd    $2,80($1)       # 50
0x00000178 <Collatz+24>:        lqd     $2,96($1)       # 60
(gdb) info target
Symbols from "/tmp/collatz".
Unix child process:
        Using the running image of child process 10321.
        While running this, GDB does not access memory from...
```

```
Local exec file:
        '/tmp/collatz', file type elf32-spu.
        Entry point: 0x28
        0x00000000 - 0x00000024 is .init
        0x00000028 - 0x00000b80 is .text
        0x00000b80 - 0x00000b9c is .fini
        0x00000ba0 - 0x00000c20 is .rodata
        0x00000c80 - 0x00000c88 is .ctors
        0x00000c88 - 0x00000c90 is .dtors
        0x00000c90 - 0x00000c94 is .jcr
        0x00000ca0 - 0x00000fd0 is .data
        0x00000fd0 - 0x00001000 is .bss
```

7.6 Conclusions

In this chapter I have described by development environment which I used to code the examples and case-studies presented in this book. I have also listed installation information in this chapter, along with alternative methods of developing on the Cell, in case you don't have physical access to a Cell machine. The GNU debugger with enhancements for SPU debugging was also introduced.

Chapter 8
Hello World

Abstract In this chapter we discuss the following (a) writing first application on the CELL Processor using native development tools, and gcc, (b) putting the PPC core through its paces, learning what it can, cannot do, (c) understanding EMT, integer performance, compiler capabilities, alignment, using disk, huge-memory page tables, Ethernet, (d) 32-bit and 64-bit development, memory management, (e) remote program launching on the CELL, (f) using the SPUs for simple tasks, communications, pthreads.

8.1 Writing simple programs on the Cell Broadband Engine

We first describe some techniques to gather information about the operating system and the compiler. A nice trick to get `gcc` to dump all of the predefined preprocessor defines it knows about, is to run:

```
[skoranne@cell ~]%echo | gcc -E -dM -
```

To generate the list of 32-bit defines add `-m32` to the argument, conversely, to get the list of all 64-bit defines, add `-m64` to the argument. The salient preprocessor defines which are present on `gcc 4.1.2` on the PS3 are: `__linux__`, `__BIG_ENDIAN__` and `__powerpc__`. In the 64-bit mode, the preprocessor definition of `__LPC64__` is also present (all of the above are defined with a value of 1, in case you were wondering).

Another useful command line combination is:

```
gcc -fverbose-asm -O2 -c -g -Wa,-a,-ad foo.c > foo.lst
gcc -Wl,-q #tells linker not to strip executable
```

this writes out assembly code along with original C/C++ source code to better understand the assembly.

```
#include <stdio.h>
#include <unistd.h>
#include <limits.h>
```

S. Koranne, *Practical Computing on the Cell Broadband Engine*,
DOI: 10.1007/978-1-4419-0308-2_8, © Springer Science + Business Media, LLC 2009

```
    union {
5     long l;
      unsigned char c[ sizeof(long) ];
    } u;
    int main( ) {
      int i=0;
10    u.l = LONG_MAX;
      #ifndef __LP64__
        u.l = 0x12345678;
      #else
        u.l = 0x12345678ABCDEF21;
15    #endif
      for( i=0; i < sizeof( long ); ++i ) {
        printf( " c[%d] = 0x%x  ", i, u.c[i] );
      }
      printf( "\n" );
20    return (0);
```

Listing 8.1 Big-endian, little-endian code

We compile the code shown in Listing 8.1 using $-m32$ for 32-bit binary, and as expected on the PS3, we get

```
c[0] = 0x12   c[1] = 0x34   c[2] = 0x56   c[3] = 0x78
```

On an Opteron, we get:

```
c[0] = 0x78   c[1] = 0x56   c[2] = 0x34   c[3] = 0x12
```

We have already see above that gcc includes a preprocessor `__BIG_ENDIAN__` to detect the endian-ness of the machine. When we write code which either needs to communicate with other machines over the network, or may use disk resident binary data, we need to be cognizant of the endian-ness of the PowerPC vis-a-vis x86_64 machines.

Since the PS3 is running GNU/Linux, we can use standard techniques to probe for information. On my system running

```
[skoranne@cell ~]% uname -a
Linux cell 2.6.21-1.3 [date] ppc64 ppc64 ppc64 GNU/Linux
```

produces the following information.
More details can be found by running `cat /proc/version`.

Table 8.1 General system information for the PS3 running GNU/Linux

Kernel name	Linux
Node name	cell.home
Kernel release	2.6.21-1.3194.fc7
Kernel version	#1 SMP Wed May 23 22:13:52 EDT 2007
Machine	ppc64
Processor	ppc64
Hardware platform	ppc64
Operating System	GNU/Linux

Table 8.2 Contents of /proc/cpuinfo on the PS3

processor	0
cpu	Cell Broadband Engine, altivec supported
clock	3192.000000MHz
revision	5.1 (pvr 0070 0501)
processor	1
cpu	Cell Broadband Engine, altivec supported
clock	3192.000000MHz
revision	5.1 (pvr 0070 0501)
timebase	79800000
platform	PS3

Table 8.3 Contents of /proc/meminfo on the PS3

MemTotal	217068 kB	Total amount of RAM, in kB
MemFree	3300 kB	RAM left unused by system
Buffers	1232 kB	RAM used by file buffers
Cached	52768 kB	RAM used as cache memory
SwapCached	16260 kB	Swap space used as cache memory
Active	152072 kB	RAM recently used, not reclaimed
Inactive	35604 kB	Amount of buffer or page cache available
SwapTotal	524280 kB	Total amount of swap available
SwapFree	502716 kB	Total amount of swap free
Dirty	0 kB	Memory size waiting to be written to disk
Writeback	0 kB	Memory size actively being written to disk
AnonPages	133572 kB	See discussion below
Mapped	9836 kB	Memory size used to map devices, files or libs
Slab	16300 kB	Kernel data structures
SReclaimable	5956 kB	
SUnreclaim	10344 kB	
PageTables	1844 kB	Memory size used by lowest level page table
NFS_Unstable	0 kB	Memory used by NFS
Bounce	0 kB	
CommitLimit	632812 kB	
Committed_AS	225520 kB	
VmallocTotal	8589934592 kB	Total allocated virtual address space
VmallocUsed	4648 kB	Used virtual address space
VmallocChunk	8589929120 kB	Largest contiguous block in virtual add. space
HugePages_Total	0	-NA-
HugePages_Free	0	-NA-
HugePages_Rsvd	0	-NA-
Hugepagesize	16384 kB	Size of each hugepage unit. See below.

Since the PS3 system is fairly low on physical RAM you may run into situations where the system starts to *thrash*. Judicial use of on-disk memory management using *mmap* can alleviate this problem. We shall discuss examples of using disk based memory store later in this book. A related concept, which is shown in the above table, is of *SwapCached* memory: memory that was once swapped out to swap space, was swapped back in, but a copy still resides in the swapfile or swap partition. Imagine the case where this memory table was swapped in, since the kernel employs copy-on-write, and lets assume that this particular page was not written to; then, when the kernel wants to swap this page out to swap, since the page already exists in SwapCached, there is no need to physically write the contents. The page is simply freed from physical memory. Also of interest are *anonymous pages*.

The memory used by the kernel to manage its DMA buffers can be seen by running *cat/proc/slabinfo*. The kernel uses *slabs* as a higher level mechanism to manage system memory. It is interesting to run `cat /proc/slabinfo |grep "size-32(DMA)"` every so often during program execution to see the effect of PPE-SPE DMAs.

```
#include <unistd.h>
#include <stdio.h>
int main() {
  long sz = sysconf(_SC_PAGESIZE);
  long num_phys_pages = sysconf( _SC_PHYS_PAGES );
  long num_avphys_pages = sysconf( _SC_AVPHYS_PAGES );
  printf("\n PAGE_SIZE=%ld NUM_PHYS_PAGES=%ld "
         "NUM_AV_PHYS_PAGES=%ld\n",
         sz, num_phys_pages, num_avphys_pages );
  return 0;
}
PAGE_SIZE=4096 NUM_PHYS_PAGES=54267 NUM_AV_PHYS_PAGES=1210
```

Listing 8.2 PPE page size

The file `/proc/buddyinfo` is used to diagnose memory fragmentation issues. Running this command on the PS3 yields:

```
Node 0,zone DMA 210 19 1 4 1 1 0 1 1  0 0 0 0
```

The entries of this file denote that (at the time this command was run), the system had 210 of $2^0 * PAGE\_SIZE$ pages used by the DMA zone; the same zone also had 19 of $2^1 * PAGE\_SIZE$ pages, and so on.

8.2 SPE Program Development

The `spu-gcc` compiler is used to compile programs for the SPEs. In this chapter we will show example of simple SPE programs in the C language. They will introduce concepts of SIMD, vector operands, mailbox features and MFC I/O facilities. Control of the SPE is done using the libspe2 SPE run-time management library. This library constitutes the low-level API for application access to SPEs. We shall show simple examples which do not need PPE support using the *elfspe* scaffolding program provided along with the libspe2 library. This program (elfspe) is registered with the *binfmt_misc* Linux utility, and allows the Linux kernel to execute simple SPE elf binaries without a corresponding PPE program. The kernel creates a simplistic PPE binary internally and launches the SPE binary in the same thread. Using the elfspe program we can write simple scripts to demonstrate and test SPE constructs. Obviously, the SPE runtime management facilities are not used in this case, as they are designed to be used when the data dependent PPE control is required.

```
////////////////////////////////////////////////////////////////////
// Program: hello_world.c
// Author : Sandeep Koranne
//
////////////////////////////////////////////////////////////////////

#include <stdio.h>
#include <stdlib.h>
extern int _etext;
extern int _edata;
extern int _end;

register volatile unsigned int *RETADDR asm("0");
register volatile unsigned int *SP asm("1");
register volatile unsigned int *SSIZE asm("2");
static void MemoryInfo(void) {
    printf("     &_etext= %p\n", &_etext);
    printf("     &_edata= %p\n", &_edata);
    printf("       &_end= %p\n", &_end);
    printf("RETADDR      = %p\n", SP);
    printf("SPTR         = %p\n", SP);
    printf("SSIZE        = %p\n", SSIZE);
}

int main( unsigned long long spuid /* __attribute__ ((unused)) */,
    unsigned long long argp )
{

  printf(" Hello, World! from %d with %d \n", spuid, argp );
  MemoryInfo();
  return (0);
}
```

Listing 8.3 SPE hello world program

We compile this program as `spu-gcc hello_world.c`, and running this program on the PS3 produces the well known output, along with some other information, as shown below.

```
Hello, World! from 1 with 262128
     &_etext= 0x93c
     &_edata= 0xdd0
       &_end= 0xdf0
RETADDR      = 0x3ff70
SPTR         = 0x3ff70
SSIZE        = 0x4c0
```

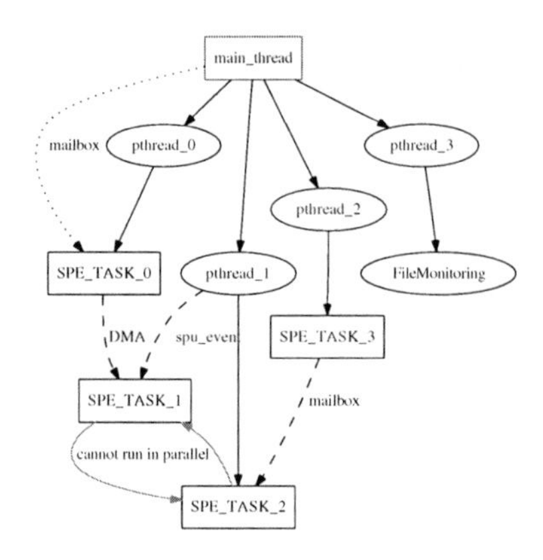

Fig. 8.1 PPE/SPE tasks and threads with communication.

8.3 Using pthread with SPU tasks

Since the SPE context run call from PPE is blocking, to execute multiple SPE tasks we have to use multiple threads in the PPE. This can be done using pthread library as shown in the next example. An example of what an application might setup for parallel computation is shown in Figure 8.1. In the figure, the main thread launches four threads, labelled `pthread_1` to `pthread_4`. Each thread presumably would have its own designated role, but we can already see that `pthread_3` does not participate in SPU management, instead it runs a function called `FileMonitoring`. The other three threads perform SPE management. It is interesting to observe that `pthread_1` manages both `SPE_TASK_1` as well as `SPE_TASK_2`, since SPE contexts are blocking when they are running, we can see from the figure that these two contexts don't run in parallel.

The communication facilities of DMA, mailbox and spu_event shall be discussed in the sequel of this chapter, and in other chapters of this book, in detail.

```
    #include <stdlib.h>
    #include <stdio.h>
    #include <libspe2.h>
    #include <pthread.h>
 5  extern spe_program_handle_t hello_from_spu;
    #define NUM_MAX_SPU    6
    void* spe_task( void *argp ) {
      spe_context_ptr_t ctx;
      unsigned int E = SPE_DEFAULT_ENTRY;
10    ctx = *(( spe_context_ptr_t * )argp );
      spe_context_run( ctx, &E, 0, NULL, NULL, NULL );
      pthread_exit( NULL );
      return NULL; //
    }
15  int main() {
      int i, spu_threads;
      spe_context_ptr_t ctxs[NUM_MAX_SPU];
      pthread_t threads[NUM_MAX_SPU];
      for(i=0; i<spu_threads; ++i) {
20      ctxs[i] = spe_context_create (0, NULL);
        spe_program_load (ctxs[i], &hello_from_spu);
```

```
      pthread_create (&threads[i], NULL, &spe_task, &ctxs[i]);
    }
    for (i=0; i<spu_threads; i++) {
25    pthread_join (threads[i], NULL);
      spe_context_destroy (ctxs[i]);
    }
    return ( EXIT_SUCCESS );
  }
```

Listing 8.4 SPE multiple thread program

8.4 Developing applications using libspe2

In this section we will look at `libspe2` functions for loading and running SPE contexts. We shall present simple example to query for system resources and calculate SPE properties. Towards the end we shall discuss a concrete example of SPE usage; we shall develop a file converter program (this one converts the characters in the file to lower-case). We shall see examples of pre-loading files with PREFAULTING, and how SPU events can be used in an asynchronous manner.

8.4.1 Major functions of the libspe2 library

We give the most common libspe2 functions below for easy back-reference when following the code in this chapter, and for the remainder of this book

The `spe_mfcio_put` and other PPE driven SPE MFC commands are not described in the above table as these commands are not recommended in compute-intensive applications. We have architected our solution so as to always use SPE initiated DMA.

```
   ////////////////////////////////////////
   // Program: spe_demo
   // Author : Sandeep Koranne
   ////////////////////////////////////////
5  #include <iostream>
   #include <libspe2.h>
   #include <stdlib.h>
   #include <string.h>
   #include <errno.h>
10 #define MAX_SPES 16
   static pthread_mutex_t ping_pong_mutex = PTHREAD_MUTEX_INITIALIZER;
   struct SPE_THREAD {
     spe_context_ptr_t ctx;
     pthread_t tid;
15 };
   void* run_spe_thread( void* argp ) {
     struct SPE_THREAD *spt = (struct SPE_THREAD*)argp;
     int flags = 0;
     unsigned int entry = SPE_DEFAULT_ENTRY;
20   spe_context_run( spt->ctx, &entry, flags, spt->ctx, NULL, NULL );
     return NULL;
   }
```

Table 8.4 Common API functions of the libspe2 library.

API Function	Arguments	Error
spe_context_create	flags, Gang context	EINVAL, ENOMEM
	SPE_EVENTS_ENABLE,	EPERM, ESRCH, ENODEV
	SPE_MAP_PS	
spe_context_destroy	SPE Context	ESRCH, EAGAIN, EFAULT
spe_gang_context_create	flags	ENOMEM, EINVAL, EFAULT
spe_gang_context_destroy	gang ctx	ESRCH, EFAULT
spe_cpu_info_get	info mask, cpu_node (-1)	EINVAL
	SPE_COUNT_PHYSICAL_CPU_NODES	
	SPE_COUNT_PHYSICAL_SPES	
	SPE_COUNT_USEFUL_SPES	
spe_image_open	(r)spe_program_handle_t	EACCES,EFAULT
	const char* filename	
spe_image_close	spe_program_handle_t*	EINVAL
spe_program_load	spe_context_ptr_t	ESRCH,EINVAL,ENOEXEC
	spe_program_handle_t*	
spe_context_run	spe_context_ptr_t	ESRCH,EINVAL,EIO
	unsigned int* entry	EFAULT,EPERM
	unsigned int flags	
	void* argp, **void*** envp	
	spe_stop_info_t*	
spe_top_info_read	spe_context_ptr_t	ESRCH,EAGAIN,ENOTSUP,EINVAL
	spe_stop_info_t*	
spe_event_handler_create	(r)spe_event_handler_ptr_t	ENOMEM,EFAULT
spe_event_handler_destroy	spe_event_handler_ptr_t	EAGAIN,EFAULT
spe_event_wait	spe_event_handler_ptr_t	ESRCH,EINVAL,EFAULT
	spe_event_unit_t*	
	int max_events,**int** timeout	
spe_out_mbox_status	spe_onctext_ptr_t	ESRCH,EIO
spe_out_mbox_read	spe_onctext_ptr_t	ESRCH,EIO,EINVAL
	unsigned int* data, count	
spe_in_mbox_status	spe_onctext_ptr_t	ESRCH,EIO
spe_in_mbox_write	spe_onctext_ptr_t	ESRCH,EIO,EINVAL
	unsigned int* data, count	
	SPE_MBOX_ALL_BLOCKING	
	SPE_MBOX_ANY_BLOCKING	
	SPE_MBOX_ANY_NONBLOCKING	
spe_ls_area_get	spe_onctext_ptr_t	ESRCH,ENOSYS
spe_ls_size_get	spe_onctext_ptr_t	ESRCH
spe_ps_area_get	spe_onctext_ptr_t	ESRCH,EACCES,EINVAL,ENOSYS

Table 8.5 SPE Stopinfo `spe_stop_info_t` structure.

Stop Reason	Description
SPE_EXIT	Normal exit with code
SPE_STOP_AND_SIGNAL	SPE ran a stop-and-signal instruction
SPE_RUNTIME_ERROR	SPE stopped due to run-time error condition
SPE_RUNTIME_EXCEPTION	SPE stopped asynch. due to exception
SPE_RUNTIME_FATAL	OS caused due to insufficient resources
SPE_CALLBACK_ERROR	-
SPE_ISOLATION_ERROR	-

```
   int main( int argc, char* argv [] ) {
25   spe_context_ptr_t ctx;
     spe_program_handle_t *spe_prog = NULL;
     spe_program_handle_t *spe_ping = NULL;
     spe_program_handle_t *spe_pong = NULL;
     spe_stop_info_t stop_info;
30   int rc;
     int info=-1;
     int flags = 0;
     int num_cpu = spe_cpu_info_get( SPE_COUNT_PHYSICAL_CPU_NODES, -1 );
     std::cout << "There are " << num_cpu << " CPUs.\n";
35   for(int i=0; i<num_cpu; ++i) {
       info = spe_cpu_info_get( SPE_COUNT_PHYSICAL_SPES, i );
       std::cout << "There are " << info
                 << " physical SPUs on CPU " << i << ".\n";
       info = spe_cpu_info_get( SPE_COUNT_USABLE_SPES, i );
40     std::cout << "There are "
                 << info << " usable SPUs on CPU " << i << ".\n";
     }
     info = spe_cpu_info_get( SPE_COUNT_PHYSICAL_SPES, -1 );
     std::cout << "There are " << info << " total SPUs.\n";
45   spe_prog = spe_image_open("/tmp/spe_func.exe");
     spe_ping = spe_image_open("/tmp/spe_ping.exe");
     spe_pong = spe_image_open("/tmp/spe_pong.exe");
     if( !spe_prog || !spe_ping || !spe_pong) {
       perror("spe_image_open");
50     exit(-1);
     }
     struct SPE_THREAD CTX[ MAX_SPES ];
     CTX[0].ctx = spe_context_create( SPE_EVENTS_ENABLE|flags, NULL );
     rc = spe_program_load( CTX[0].ctx, spe_prog );
55
     CTX[1].ctx = spe_context_create( SPE_EVENTS_ENABLE|flags, NULL );
     rc = spe_program_load( CTX[1].ctx, spe_ping );

     CTX[2].ctx = spe_context_create( SPE_EVENTS_ENABLE|flags, NULL );
60   rc = spe_program_load( CTX[2].ctx, spe_pong );

     rc = pthread_create( &CTX[0].tid, NULL, run_spe_thread, &(CTX[0]) );
     rc = pthread_create( &CTX[1].tid, NULL, run_spe_thread, &(CTX[1]) );
     rc = pthread_create( &CTX[2].tid, NULL, run_spe_thread, &(CTX[2]) );
65   // throw a ball to func, who will return it
     unsigned int ball=1;
     while( spe_in_mbox_status( CTX[0].ctx ) <= 0 );
     spe_in_mbox_write(CTX[0].ctx,&ball,1,SPE_MBOX_ALL_BLOCKING);

70   while( spe_out_mbox_status( CTX[0].ctx ) <= 0 );
     spe_out_mbox_read(CTX[0].ctx,&ball,1);

     // PING
```

```
  while( spe_in_mbox_status( CTX[1].ctx ) <= 0 );
  spe_in_mbox_write(CTX[1].ctx,&ball,1,SPE_MBOX_ALL_BLOCKING);

  while( spe_out_mbox_status( CTX[1].ctx ) <= 0 );
  spe_out_mbox_read(CTX[1].ctx,&ball,1);

  // PONG
  while( spe_in_mbox_status( CTX[2].ctx ) <= 0 );
  spe_in_mbox_write(CTX[2].ctx,&ball,1,SPE_MBOX_ALL_BLOCKING);

  while( spe_out_mbox_status( CTX[2].ctx ) <= 0 );
  spe_out_mbox_read(CTX[2].ctx,&ball,1);

  pthread_join( CTX[0].tid, NULL );
  pthread_join( CTX[1].tid, NULL );
  pthread_join( CTX[2].tid, NULL );

  return 0;
}
```

Listing 8.5 libspe2 example spe_demo.C

```
/////////////////////////////////////////
// Program: LIBSPE demo
// Author : Sandeep Koranne
/////////////////////////////////////////
#include <spu_mfcio.h>
#include <stdio.h>
int main(unsigned long long speid,
   unsigned long long argp,
   unsigned long long envp __attribute__ ((__unused__))) {
  volatile unsigned int ball=0;
  ball = spu_read_in_mbox( );
  printf("\nSPE (0x%llx) says Hello to %d.\n",speid, ball);
  ball++;
  spu_write_out_mbox( ball ); // back to PPE
  return 217;
}
```

Listing 8.6 libspe2 example spe_func.c

```
/////////////////////////////////////////
// Program: LIBSPE demo
// Author : Sandeep Koranne
/////////////////////////////////////////
#include <spu_mfcio.h>
#include <stdio.h>
int main(unsigned long long speid,
   unsigned long long argp,
   unsigned long long envp __attribute__ ((__unused__))) {
  volatile unsigned int ball=0;
  ball = spu_read_in_mbox( );
  printf("\nSPE (0x%llx) says Ping to %d.\n",speid, ball);
  ball++;
  spu_write_out_mbox( ball ); // back to PPE
  return 0;
}
```

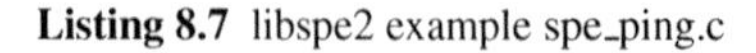

Listing 8.7 libspe2 example spe_ping.c

```
/////////////////////////////////////////
// Program: LIBSPE demo
// Author : Sandeep Koranne
/////////////////////////////////////////
```

```
5  #include <spu_mfcio.h>
   #include <stdio.h>
   int main(unsigned long long speid,
      unsigned long long argp,
      unsigned long long envp __attribute__ ((__unused__))) {
10   unsigned int ball=0;
     ball = spu_read_in_mbox( );
     printf("\nSPE (0x%llx) says Pong to %d.\n",speid, ball);
     ball++;
     spu_write_out_mbox( ball ); // back to PPE
15   return 0;
   }
```

Listing 8.8 libspe2 spe_pong.c

The next code listing in Listing 8.9 shows event monitoring in PPE code.

```
   ////////////////////////////////////////
   // Program: spe_demo, event monitoring
   // Author : Sandeep Koranne
   ////////////////////////////////////////
5  #include <iostream>
   #include <libspe2.h>
   #include <stdlib.h>
   #include <string.h>
   #include <errno.h>
10 #define MAX_SPES 16
   static pthread_mutex_t ping_pong_mutex = PTHREAD_MUTEX_INITIALIZER;
   struct SPE_THREAD {
     spe_context_ptr_t ctx;
     pthread_t tid;
15 };
   void* run_spe_thread( void* argp ) {
     struct SPE_THREAD *spt = (struct SPE_THREAD*)argp;
     int flags = 0;
     unsigned int entry = SPE_DEFAULT_ENTRY;
20   spe_context_run( spt->ctx, &entry, flags, spt->ctx, NULL, NULL );
     return NULL;
   }

   int main( int argc, char* argv [] ) {
25   spe_context_ptr_t ctx;
     spe_program_handle_t *spe_prog = NULL;
     spe_program_handle_t *spe_ping = NULL;
     spe_program_handle_t *spe_pong = NULL;
     spe_stop_info_t stop_info;
30   int rc;
     int info=-1;
     int flags = 0;
     int num_cpu = spe_cpu_info_get( SPE_COUNT_PHYSICAL_CPU_NODES, -1 );
     std::cout << "There are " << num_cpu << " CPUs.\n";
35   for(int i=0; i<num_cpu; ++i) {
       info = spe_cpu_info_get( SPE_COUNT_PHYSICAL_SPES, i );
       std::cout << "There are " << info
                 << " physical SPUs on CPU " << i << ".\n";
       info = spe_cpu_info_get( SPE_COUNT_USABLE_SPES, i );
40     std::cout << "There are " << info
                 << " usable SPUs on CPU " << i << ".\n";
     }
     info = spe_cpu_info_get( SPE_COUNT_PHYSICAL_SPES, -1 );
     std::cout << "There are " << info << " total SPUs.\n";
45   spe_prog = spe_image_open("/tmp/spe_func.exe");
     spe_ping = spe_image_open("/tmp/spe_ping.exe");
     spe_pong = spe_image_open("/tmp/spe_pong.exe");
     if( !spe_prog || !spe_ping || !spe_pong) {
       perror("spe_image_open");
50     exit(-1);
```

```
    }
    struct SPE_THREAD CTX[ MAX_SPES ];

    CTX[0].ctx = spe_context_create( SPE_EVENTS_ENABLE|flags, NULL );
55  rc = spe_program_load( CTX[0].ctx, spe_prog );

    CTX[1].ctx = spe_context_create( SPE_EVENTS_ENABLE|flags, NULL );
    rc = spe_program_load( CTX[1].ctx, spe_ping );

60  CTX[2].ctx = spe_context_create( SPE_EVENTS_ENABLE|flags, NULL );
    rc = spe_program_load( CTX[2].ctx, spe_pong );

    spe_event_handler_ptr_t evnt_hdl;
    spe_event_unit_t event;
65  evnt_hdl = spe_event_handler_create();
    event.spe = CTX[0].ctx;
    event.events = SPE_EVENT_SPE_STOPPED | SPE_EVENT_IN_MBOX;
    spe_event_handler_register( evnt_hdl, &event );

70  rc = pthread_create( &CTX[0].tid, NULL, run_spe_thread, &(CTX[0]) );
    rc = pthread_create( &CTX[1].tid, NULL, run_spe_thread, &(CTX[1]) );
    rc = pthread_create( &CTX[2].tid, NULL, run_spe_thread, &(CTX[2]) );

    // throw a ball to func, who will return it
75  unsigned int ball=1;
    unsigned int count_events = spe_event_wait( evnt_hdl, &event, 1, -1 );
    std::cout << std::endl << "We have " << count_events
              << " events waiting for us.\n";
    if( event.events & SPE_EVENT_IN_MBOX ) {
80    spe_in_mbox_write(CTX[0].ctx,&ball,1,SPE_MBOX_ALL_BLOCKING);
    }
    while( spe_out_mbox_status( CTX[0].ctx ) <= 0 );
    spe_out_mbox_read(CTX[0].ctx,&ball,1);
    count_events =
85    spe_event_wait( evnt_hdl, &event, 1, -1 ); // in-box empty
    count_events =
      spe_event_wait( evnt_hdl, &event, 1, -1 ); // spe stop
    std::cout << std::endl << "We have " << count_events
              << " events waiting for us.\n";
90  if( event.events & SPE_EVENT_SPE_STOPPED ) {
      spe_stop_info_read( CTX[0].ctx, &stop_info );
      if( stop_info.stop_reason == 1 )
        std::cout << std::endl << "SPE stopped with code = "
                  << stop_info.result.spe_exit_code << ".\n";
95  }
    // PING
    while( spe_in_mbox_status( CTX[1].ctx ) <= 0 );
    spe_in_mbox_write(CTX[1].ctx,&ball,1,SPE_MBOX_ALL_BLOCKING);

100 while( spe_out_mbox_status( CTX[1].ctx ) <= 0 );
    spe_out_mbox_read(CTX[1].ctx,&ball,1);

    // PONG
    while( spe_in_mbox_status( CTX[2].ctx ) <= 0 );
105 spe_in_mbox_write(CTX[2].ctx,&ball,1,SPE_MBOX_ALL_BLOCKING);

    while( spe_out_mbox_status( CTX[2].ctx ) <= 0 );
    spe_out_mbox_read(CTX[2].ctx,&ball,1);

110 pthread_join( CTX[0].tid, NULL );
    pthread_join( CTX[1].tid, NULL );
    pthread_join( CTX[2].tid, NULL );
    spe_event_handler_destroy( evnt_hdl );
    return 0;
115 }
```

Listing 8.9 Event monitoring demo spe_event_demo.C

```
////////////////////////////////////////
// Program: spe_demo
// Author : Sandeep Koranne
////////////////////////////////////////
#include <iostream>
#include <libspe2.h>
#include <stdlib.h>
#include <string.h>
#include <errno.h>
#include <sys/types.h>
#include <sys/stat.h>
#include <fcntl.h>
#include <unistd.h>
#include <sys/mman.h>
#ifndef __SPU__
#include "../align.h"
#endif
#define MAX_SPES 16
typedef struct _control_block {
  unsigned int N;
  unsigned int M;
  unsigned long long addr;
} ControlBlock;
static pthread_mutex_t ping_pong_mutex = PTHREAD_MUTEX_INITIALIZER;

struct SPE_THREAD {
  void* argp;
  spe_context_ptr_t ctx;
  pthread_t tid;
};

void* run_spe_thread( void* argp ) {
  struct SPE_THREAD *spt = (struct SPE_THREAD*)argp;
  int flags = 0;
  unsigned int entry = SPE_DEFAULT_ENTRY;

  spe_context_run( spt->ctx, &entry, flags, spt->argp, NULL, NULL );
  return NULL;
}

struct FILE_THREAD {
  ControlBlock* cb;
  spe_context_ptr_t ctx;
  pthread_t tid;
  int filedes;
  char* mem;
};

void* page_fault_file( void* arg ) {
  struct FILE_THREAD* fth = (struct FILE_THREAD*) arg;
  spe_event_handler_ptr_t evnt_hdl;
  spe_event_unit_t event;
  unsigned int ball=1;
  spe_stop_info_t stop_info;
  struct stat file_information;
  off_t file_size = 0;
  int rc;
  size_t page_size = sysconf( _SC_PAGE_SIZE );
  std::cout << std::endl << "PAGE_SIZE = " << page_size << std::endl;
  // file handling if requested
  if( fth->filedes != -1 ) {
    rc = fstat( fth->filedes, &file_information );
    file_size = file_information.st_size;
    file_size = (1+(file_size/page_size))*page_size; // round up
    std::cout << std::endl << "File size = " << file_information.st_size
```

```
            << " bytes, rounded = " << file_size << ".\n";
    _free_align( fth->mem );
    void* ret = mmap( NULL, file_size,
                      PROT_READ|PROT_WRITE,
70                    MAP_SHARED, fth->filedes, 0 );
    if( ret == MAP_FAILED ) {
      perror("mmap");
      exit(-1);
    }
75  std::cout << std::endl << "Address = " << std::hex << ret;
    fth->mem = (char*)ret;
    fth->cb->addr = (unsigned long long int)ret;
    fth->cb->N=file_information.st_size;
    fth->cb->M=file_size;
80 }
  evnt_hdl = spe_event_handler_create();
  event.spe = fth->ctx;
  event.events = SPE_EVENT_SPE_STOPPED | SPE_EVENT_IN_MBOX;
  spe_event_handler_register( evnt_hdl, &event );
85 unsigned int count_events =
    spe_event_wait( evnt_hdl, &event, 1, -1 );
  std::cout << std::endl << "We have "
            << count_events << " events waiting for us.\n";
  if( event.events & SPE_EVENT_IN_MBOX ) {
90  spe_in_mbox_write(fth->ctx,&ball,1,SPE_MBOX_ALL_BLOCKING);
  } else {
    std::cerr << std::endl << "protocol mismatch. exiting.";
    exit(-1);
  }
95 while( spe_out_mbox_status( fth->ctx ) <= 0 );
  spe_out_mbox_read(fth->ctx,&ball,1);
  std::cout << std::endl << "Got the ball back: "
            << ball << std::endl;
  count_events =
100 spe_event_wait( evnt_hdl, &event, 1, -1 ); // mbox Q empty
  count_events =
    spe_event_wait( evnt_hdl, &event, 1, -1 ); // spe stop
  std::cout << std::endl << "We have "
            << count_events << " events waiting for us.\n";
105 if( event.events & SPE_EVENT_SPE_STOPPED ) {
    spe_stop_info_read( fth->ctx, &stop_info );
    if( stop_info.stop_reason == 1 )
      std::cout << std::endl << "SPE stopped with code = "
                << stop_info.result.spe_exit_code << ".\n";
110 } else {
    std::cerr << std::endl << "protocol mismatch. exiting.";
    exit(-1);
  }

115 spe_event_handler_destroy( evnt_hdl );
  std::cout << std::endl << "Starting write_back......\n" << std::endl;

  if( fth->filedes ) {
    munmap( fth->mem, file_size );
120 close( fth->filedes ); fth->mem = NULL;
    fth->cb->addr=0;
  }
  std::cout << std::endl << "Done\n" << std::endl;
  return NULL;
125 }

int main( int argc, char* argv [] ) {

  spe_context_ptr_t ctx;
130 spe_program_handle_t *spe_prog = NULL;
  spe_program_handle_t *spe_ping = NULL;
  spe_program_handle_t *spe_pong = NULL;
```

```
    int rc;
    int i;
135 int info=-1;
    int flags = 0;
    ControlBlock *cb[MAX_SPES];
    int filedes = -1;

140 if( argc > 1 ) {
      filedes = open( argv[1], O_RDWR | O_DIRECT | O_LARGEFILE );
      if (filedes == -1 ) {
        perror("open");
        exit(-1);
145   }
    }
    int num_cpu = spe_cpu_info_get( SPE_COUNT_PHYSICAL_CPU_NODES, -1 );
    std::cout << "There are " << num_cpu << " CPUs.\n";
    for(i=0; i<num_cpu; ++i) {
150   info = spe_cpu_info_get( SPE_COUNT_PHYSICAL_SPES, i );
      std::cout << "There are " << info
                << " physical SPUs on CPU " << i << ".\n";
      info = spe_cpu_info_get( SPE_COUNT_USABLE_SPES, i );
      std::cout << "There are " << info
155             << " usable SPUs on CPU " << i << ".\n";
    }
    info = spe_cpu_info_get( SPE_COUNT_PHYSICAL_SPES, -1 );
    std::cout << "There are " << info << " total SPUs.\n";
    for(i=0; i < info; ++i ) {
160   cb[i] = new ControlBlock();
      #define TEXT_SIZE 1 << 14
      cb[i]->addr = (unsigned long long int)
        (_malloc_align( TEXT_SIZE,10 ));
    }
165 char* location = (char*)(cb[0]->addr);
    sprintf(location, "Hello from Sandeep Koranne");
    cb[0]->M = strlen(location)+1;
    spe_prog = spe_image_open("/tmp/spe_lowercase.exe");
    spe_ping = spe_image_open("/tmp/spe_ping.exe");
170 spe_pong = spe_image_open("/tmp/spe_pong.exe");
    if( !spe_prog || !spe_ping || !spe_pong) {
      perror("spe_image_open");
      exit(-1);
    }
175 struct SPE_THREAD CTX[ MAX_SPES ];
    CTX[0].ctx = spe_context_create( SPE_EVENTS_ENABLE|flags, NULL );
    rc = spe_program_load( CTX[0].ctx, spe_prog );
    CTX[0].argp = cb[0];
    CTX[1].ctx = spe_context_create( SPE_EVENTS_ENABLE|flags, NULL );
180 rc = spe_program_load( CTX[1].ctx, spe_ping );
    CTX[2].ctx = spe_context_create( SPE_EVENTS_ENABLE|flags, NULL );
    rc = spe_program_load( CTX[2].ctx, spe_pong );
    FILE_THREAD fth;
    fth.mem = location;
185 fth.ctx = CTX[0].ctx;
    fth.cb  = cb[0];
    fth.filedes = filedes;
    rc = pthread_create( &CTX[0].tid, NULL, run_spe_thread, &(CTX[0]) );
    rc = pthread_create( &CTX[1].tid, NULL, run_spe_thread, &(CTX[1]) );
190 rc = pthread_create( &CTX[2].tid, NULL, run_spe_thread, &(CTX[2]) );
    rc = pthread_create( &fth.tid, NULL, page_fault_file, &fth );
    // throw a ball to func, who will return it
    unsigned int ball=1;
    // PING
195 while( spe_in_mbox_status( CTX[1].ctx ) <= 0 );
    spe_in_mbox_write(CTX[1].ctx,&ball,1,SPE_MBOX_ALL_BLOCKING);
    while( spe_out_mbox_status( CTX[1].ctx ) <= 0 );
    spe_out_mbox_read(CTX[1].ctx,&ball,1);
    // PONG
```

```
200    while( spe_in_mbox_status( CTX[2].ctx ) <= 0 );
       spe_in_mbox_write(CTX[2].ctx,&ball,1,SPE_MBOX_ALL_BLOCKING);
       while( spe_out_mbox_status( CTX[2].ctx ) <= 0 );
       spe_out_mbox_read(CTX[2].ctx,&ball,1);
       pthread_join( CTX[0].tid, NULL );
205    pthread_join( CTX[1].tid, NULL );
       pthread_join( CTX[2].tid, NULL );
       pthread_join( fth.tid, NULL );

       std::cout << std::endl << "Completed processing\n Text \n";
210    for(int i=0; i < info; ++i ) {
         _free_align( (void*)(cb[i]->addr) );
         delete cb[i];
       }
       return 0;
215  }
```

Listing 8.10 File transformation example asynch.C

The dependency tree of the `ProcessText` function are shown in Figure 8.2.

The above program can be used to illustrate a number of common SPU programming practices. Refer to the listing shown in Listing 8.11. In this example we use 16-way SIMD operating on **`char`** data-type to convert a given text file *in-situ*. The SPE initiate DMA, on an Effective Address which is the result of `mmap` of the file; the DMA access by the SPE causes page-faults on the PPE's OS, which cause data to be read in from disk and transported to SPE LS using DMA. Once the data is in the local store, the SPU converts it to lowercase using SIMD, and then puts the data back in EA using DMA. An example of this program running on a portion of the file is shown in Listing 8.12. The variable `orig` refers to the contents of the local store indexed using `vector` **`unsigned char`**, the variables in the vector registers `GTA` and `LTZ` compute select bits depending on whether the element of the vector is greater-than-or-equal to *A*, and less-than-or-equal *Z*, respectively. Once the select bits are computed an *offset* is added to the `UPC` variable; this offset is 0 if the character is already lowercase, else, it is the distance between 'A'-'a', or 32 bytes. Every uppercase letter in the vector is offset by this selected amount, rendering every character in the vector lowercase. This vector is then written back to the local store. At the end of the current batch of characters, we put the data from LS to EA, and simultaneously download the next batch for processing.

```
    ///////////////////////////////////////////
    // Program: LIBSPE demo
    // Author : Sandeep Koranne
    ///////////////////////////////////////////
 5  #include <spu_mfcio.h>
    #include <stdio.h>
    #include <simdmath.h>
    #define TEXT_SIZE 1024
    typedef struct _control_block {
10    unsigned int N;
      unsigned int M;
      unsigned long long addr;
    } ControlBlock;
    ControlBlock cb __attribute__ ((aligned (128)));
15  #define USE_SIMD
    #ifdef USE_SIMD
    volatile vec_uchar16 TEXT_BLOCK[ TEXT_SIZE ]
    __attribute__ ((aligned (128)));
    #else
```

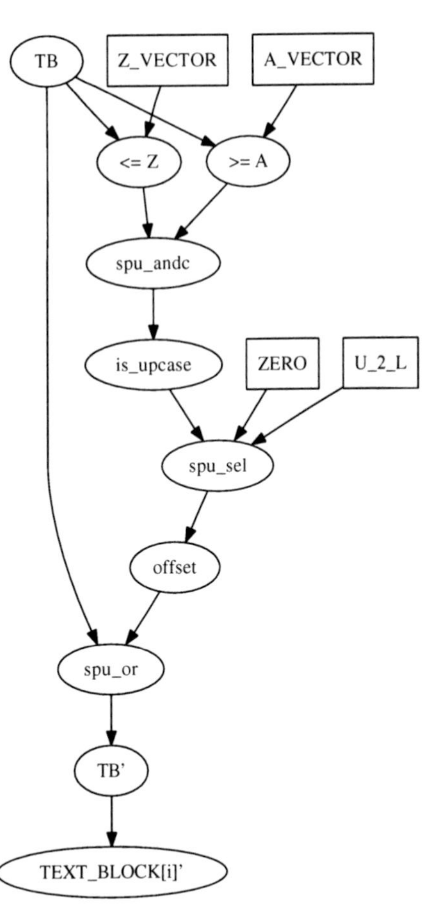

Fig. 8.2 Dependency diagram of the SIMD lowercase converter SPU code.

```
20  volatile unsigned char TEXT_BLOCK[ TEXT_SIZE*16 ]
    __attribute__ ((aligned (128)));
    #endif
    #define CHUNK_SIZE 16384

25  #ifndef USE_SIMD
    static void ProcessText( ControlBlock* cb, int tag_id ) {
      int i,j;
      for(j=0;j< (cb->M/CHUNK_SIZE)-1;++j) {
        mfc_get(TEXT_BLOCK,cb->addr+(j*CHUNK_SIZE),
30        sizeof(unsigned char)*CHUNK_SIZE,
                tag_id, 0, 0);
        mfc_write_tag_mask(1<< tag_id);
        mfc_read_tag_status_all();
        for(i=0;i<CHUNK_SIZE;++i) {
35        TEXT_BLOCK[i] = ( (TEXT_BLOCK[i] >= (int)'A') &&
          (TEXT_BLOCK[i] <= (int)'Z') )
            ?  TEXT_BLOCK[i]+32 : TEXT_BLOCK[i];
        }
        mfc_put(TEXT_BLOCK,cb->addr+(j*CHUNK_SIZE),
```

```
      sizeof(char)*CHUNK_SIZE,tag_id,0,0);
    mfc_write_tag_mask(1<<tag_id);
    mfc_read_tag_status_all();
  }
}
#else

static void print_vector_char(const char* s1, const vec_uchar16* V ) {
  printf( "%s = [%c %c %c %c %c %c %c %c %c %c %c %c %c %c %c %c]",
          s1, spu_extract( *V, 0 ), spu_extract( *V, 1),
          spu_extract( *V, 2), spu_extract(*V,3), spu_extract( *V, 4),
          spu_extract( *V, 5), spu_extract(*V,6), spu_extract(*V,7),
          spu_extract( *V, 8), spu_extract( *V, 9), spu_extract(*V,10),
          spu_extract(*V,11),spu_extract( *V, 12), spu_extract( *V, 13),
          spu_extract(*V,14), spu_extract(*V,15));
}
#define NO_UNROLL
#ifdef NO_UNROLL
static void ProcessText( ControlBlock* cb, int tag_id ) {
  int i,j;
  vec_uchar16 A_VECTOR = {'@','@','@','@','@','@','@','@',
                          '@','@','@','@','@','@','@','@'};
  vec_uchar16 Z_VECTOR = {'Z','Z','Z','Z','Z','Z','Z','Z',
                          'Z','Z','Z','Z','Z','Z','Z','Z'};
  vec_uchar16 ZERO = (vec_uchar16){0,0,0,0,0,0,0,0,
                                   0,0,0,0,0,0,0,0};
  vec_uchar16 U_2_L = (vec_uchar16){32,32,32,32,32,32,32,32,
                                    32,32,32,32,32,32,32,32};
  for(j=0;j< (cb->M/CHUNK_SIZE)-2;++j) {
    mfc_get(TEXT_BLOCK,cb->addr+(j*CHUNK_SIZE),
      sizeof(unsigned char)*CHUNK_SIZE,
            tag_id, 0, 0);
    mfc_write_tag_mask(1<< tag_id);
    mfc_read_tag_status_all();
    for(i=0;i<=CHUNK_SIZE/16;++i) {
      vec_uchar16 TB = TEXT_BLOCK[i];
      vec_uchar16 is_gte_A = spu_cmpgt(TB,A_VECTOR);
      vec_uchar16 is_lte_Z = spu_cmpgt(TB,Z_VECTOR);
      vec_uchar16 is_upc = (vec_uchar16)
        spu_andc( is_gte_A, is_lte_Z );
      vec_uchar16 offset = (vec_uchar16)
        spu_sel( ZERO, U_2_L,  is_upc );
#ifdef DEBUG_VERBOSE
      print_vector_char("TB_orig",&TB);
      printf("\n");
      print_vector_char("GTA =",&is_gte_A);
      printf("\n");
      print_vector_char("LTZ =",&is_lte_Z);
      printf("\n");
      print_vector_char("UPC =",&is_upc);
      printf("\n");
      print_vector_char("OFFSET =",&offset);
#endif
      TB = spu_or( TB, offset );
      TEXT_BLOCK[i] = TB;
#ifdef DEBUG_VERBOSE
      printf("\n");
      print_vector_char("TB'",&TB);
#endif
    }
    mfc_put(TEXT_BLOCK,cb->addr+(j*CHUNK_SIZE),
      sizeof(char)*CHUNK_SIZE,tag_id,0,0);
    mfc_write_tag_mask(1<<tag_id);
    mfc_read_tag_status_all();
  }
}
#else
```

```c
// 4-Unrolled version
#endif // loop unrolling
#endif

int main(unsigned long long speid,
         unsigned long long argp,
         unsigned long long envp __attribute__ ((__unused__))) {
  unsigned int ball=0;
  int tag_id = mfc_tag_reserve();
  ball = spu_read_in_mbox( ); // first read mbox

  mfc_get(&cb, argp, sizeof(cb), tag_id, 0, 0);
  mfc_write_tag_mask(1<< tag_id);
  mfc_read_tag_status_all();

  printf("\nSPE LOWECASE (0x%llx) says Hello to %d.\n",speid, cb.M);
  ball++;
  ProcessText( &cb,tag_id );
  spu_write_out_mbox( ball ); // back to PPE
  return 217;
}
```

Listing 8.11 File transformation example spe_lowercase.c

```
SPE (0x1002f160) says Pong to 2.
orig= [I N   T H I S   f i l e   i s  ]
GTA = [1 1 0 1 1 1 1 0 1 1 1 1 0 1 1 0]
LTZ = [0 0 0 0 0 0 0 0 1 1 1 1 0 1 1 0]
UPC = [1 1 0 1 1 1 1 0 0 0 0 0 0 0 0 0]
TB' = [i n   t h i s   f i l e   i s  ]

real    0m24.73 WITH SIMD
real    0m33.42 WITHOUT SIMD
```

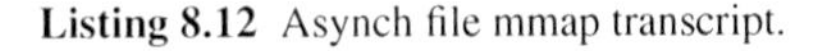

Listing 8.12 Asynch file mmap transcript.

8.5 Measuring Performance using High Resolution Timers

In this section we describe the use of the SPU Decrementer register to calculate number of processor ticks which occur in a giver interval. By dividing the number of ticks by the *timebase* of the processor, you can calculate elapsed time for a calculation with (a) high-resolution and (b) minimal intrusion. This can be used to develop profiling tools, but in this book we shall use high-resolution timing around performance critical sections of the code to guide our optimization decisions.

Consider the following code as shown in Listing 8.13.

```c
////////////////////////////////////////
// Program: Collatz Conjecture
// File   : collatz_spule.c
// Author : Sandeep Koranne
////////////////////////////////////////
#include <stdio.h>
#include <stdlib.h>
#include <spu_mfcio.h>
#include <spu_timer.h>

typedef unsigned long long int ULL;
```

```
ULL Collatz(ULL a, ULL b) {
  ULL i;
  ULL sum=0;
15  for(i=a;i<b;++i) {
    ULL ct=a;
    while(ct!=1)
      ct = (ct&1) ? 3*ct+1 : ct/2;
    sum += ct;
20  }
  return sum;
}
#define MEASURE_TIME
unsigned int total_tags = 0;
25 void ev_handler( int time_tag ) {
  total_tags++;
}
int main( unsigned long long spuid,
    unsigned long long argp ) {
30  // sweep from 1 to 100
  ULL retval=0;
  ULL a = 1;
  ULL b = a<<28;
#ifdef MEASURE_TIME
35  int id;
  uint64_t work_time;
  spu_slih_register( MFC_DECREMENTER_EVENT, spu_clock_slih);
  id = spu_timer_alloc( 10000, ev_handler );
  if(id<0) {
40    return 1;
  }
  spu_clock_start();
#endif
  printf("\nSweeping Collatz from 1 to 1<<28\n",a,b);
45 #ifdef MEASURE_TIME
  spu_timer_start( id );
#endif
  retval+=Collatz(a,b);
#ifdef MEASURE_TIME
50  spu_timer_stop(id);
  spu_timer_free(id);
  work_time = spu_clock_read();
  spu_clock_stop();
  printf("\n Got %d total interrupts @10000.", total_tags);
55 #else
  printf("\n Got %d in %d\n", (retval&0xFFFF),0);
#endif
  return (0);
}
```

Listing 8.13 Using SPU Decrementer register for high-resolution timer.

This program (Listing 8.13) needs to be compiled with including the `spu_timer.h` file and the `sputimer.a` library. The output of running this program is shown below; we sweep all natural numbers from 1 to $1 << 28$ to check for the Collatz conjecture (cf. Section 10.1.1). We allocate a timer with a resolution of 1000 clock ticks (on a virtual clock). The callback function we associate with this timer just increments a global variable called `total_tags`, and thus the total number of calls made to the callback function are recorded and reported to the user at the end of the program. As you can see below in the transcript, a total of 41648 calls were made during this sweep. We also ran the command using `elfspe` under `time` command to measure the wall clock time of the program. This corresponds to 5.233 seconds. We invoke the `bc` calculator to check if the number of clock ticks in our timer divided by the *timebase* of the PS3 Cell (79800000) is the same as the wall clock time, and we see that the time reported by the timer is 5.219 seconds. We know that SPU context setting time is approximately 400 mili-seconds, and the printf call just above the critical second, as well as the overhead (albeit outside the measured code) of setting up the timer also costs us some time.

```
[skoranne@localhost tmp]%time ./collatz
Sweeping Collatz from 1 to 1<<28
Got 41648 total interrupts @10000.
real     0m5.233s
user     0m0.001s
sys      0m0.006s
[skoranne@localhost tmp]% grep timebase /proc/cpuinfo
timebase         : 79800000
[skoranne@localhost tmp]% bc
bc 1.06
Copyright 1991-1994, 1997, 1998, 2000 Free Software Foundation, Inc.
This is free software with ABSOLUTELY NO WARRANTY.
For details type 'warranty'.
4164800000/798
5219047
```

Observe, that the high-resolution timing agrees rather well with the *real* time reported by the `time` command. Thus, we can use the high-resolution timer to measure and record timing information as we make changes to the code.

8.6 Synchronization

Since the Cell is all about effective use of SPUs running in parallel it is very important to understand the facilities for providing mutual exclusion to critical sections of the code. Exclusions may pertain to shared data-structures, file-access, I/O statements and extern function calls. The SPU MFC provide low-level methods for atomic operations on variables present in effective address space, or the global memory space, which is visible to all processors coherently. The low-level routines are based on 4 entries in the MFC cache table which can be loaded from the L2 cache. The MFC function which provides this facility is the `mfc_gerllar` function, for MFC Get Lock Line and Reserve. This is documented in the C/C++ Language Extensions for CBEA [33]. Other related functions for lock line reservation are given below:

```
typedef unsigned long long atomic_ea_t;
extern void atomic_set(atomic_ea_t v, int val);
extern void atomic_add(int a, atomic_ea_t v);
extern void atomic_sub(int a, atomic_ea_t v);
extern void atomic_inc(atomic_ea_t v);
extern void atomic_dec(atomic_ea_t v);
extern int atomic_read(atomic_ea_t v);
extern int atomic_add_return(int a, atomic_ea_t v);
extern int atomic_sub_return(int a, atomic_ea_t v);
extern int atomic_inc_return(atomic_ea_t v);
extern int atomic_dec_return(atomic_ea_t v);
extern int atomic_sub_and_test(int a, atomic_ea_t v);
extern int atomic_dec_and_test(atomic_ea_t v);
extern int atomic_dec_if_positive(atomic_ea_t v);
typedef unsigned long long mutex_ea_t;
```

Table 8.6 SPE MFC Atomic Update Command Summary

Function	Name	Description
mfc_getllar	Get Lock Line and Reserve	Lock line is obtained and reservation is created
mfc_putllc	Put Lock Line if Reservation for EA Exists	Lock line is put if reservation exists.
mfc_putlluc	Put Lock Line Unconditional	Lock line is put regardless of reservation
mfc_read_atomic_status	Read Atomic Command Status	Stalls till status is available,

```
extern void mutex_init(mutex_ea_t lock);
extern void mutex_lock(mutex_ea_t lock);
extern int mutex_trylock(mutex_ea_t ea);
extern void mutex_unlock(mutex_ea_t lock);
typedef struct
{
  int num_threads_signal;
  int num_threads_waiting;
} condition_variable_t __attribute__ ((aligned (128)));
typedef eaddr_t cond_ea_t;
extern void cond_init (cond_ea_t  cond);
extern void cond_signal (cond_ea_t cond);
extern void cond_broadcast (cond_ea_t cond);
extern void cond_wait (cond_ea_t cond, mutex_ea_t  mutex);
typedef unsigned long long completion_ea_t;
extern void init_completion(completion_ea_t completion);
extern void wait_for_completion(completion_ea_t completion);
extern void complete(completion_ea_t completion);
extern void complete_all(completion_ea_t completion);
```

Listing 8.14 libsync functions for synchronization.

```
extern void read_lock(eaddr_t ea);
extern void read_unlock(eaddr_t ea);
extern int  read_trylock(eaddr_t ea);
extern void write_lock(eaddr_t ea);
extern void write_unlock(eaddr_t ea);
extern int  write_trylock(eaddr_t ea);
```

Listing 8.15 Synchronization functions for SPU only.

The semantics of the mutex functionality provided by the libsync library is similar to the pthread mutexes. The advantage is of-course transparent access to the mutex in the SPU using the cache line locking as the underlying mechanism. As the SPE does not implement any L1 cache[1], the `putlluc` command results in a direct store to the referenced real memory and the invalidation of all other caches that contain the affected line. The `getllar`, `putllc`, and `putllc` commands support high-performance

[1] This statement should probably read the SPE of the current Cell implementation do not implement any L1 cache. The CBEA architecture standard has defined a L1 cache between the LS and DMA.

lock acquisition and release between SPEs and the PPE by performing direct cache-to-cache transfers. This is in contrast to pthread mutex which reside in the main memory, and can cause severe cache-line invalidation problems.

Consider the following example which demonstrates the use of GNU/Linux system information calls with synchronization for IO locking in the SPU code, so that output from any SPU is not mixed with other output. Listing 8.16 shows the PPE code for creating a mutex and its associated EA address is passed to the SPU which creates its own mutex based on the EA of the PPE mutex. Using `mutex_lock`, every SPU in turn locks (or waits for the lock) the mutex and prints its output. The SPU code is shown in Listing 8.17, and the output transcript is shown in Listing 8.18.

```
////////////////////////////////////////
// Program: Info
// Author : Sandeep Koranne
////////////////////////////////////////
#include <iostream>
#include <fstream>
#include <cassert>
#include <libspe2.h>
#include <pthread.h>
#include <string.h>
#include <getopt.h>
#include <sys/utsname.h>
#include <sys/time.h>
#include <sys/resource.h>
#include <stdlib.h>
#include <unistd.h>
#include <ppu_intrinsics.h>
#include <libsync.h>
#include "../align.h"


#define MAX_SPES 16

typedef struct _control_block {
  unsigned int N;
  unsigned int M;
  unsigned long long addr;
} CB;
CB *cb[MAX_SPES];

extern spe_program_handle_t info_function;
typedef struct ppu_pthread_data {
  void *argp;
  spe_context_ptr_t speid;
  pthread_t pthread;
} ppu_pthread_data_t;
ppu_pthread_data_t *datas;

void *ppu_pthread_function(void *arg) {
  ppu_pthread_data_t *datap = (ppu_pthread_data_t *)arg;
  unsigned int entry = SPE_DEFAULT_ENTRY;
  int rc = spe_context_run(datap->speid, &entry,
         0, datap->argp, NULL, NULL);
  pthread_exit(NULL);
}

void PreProcessorDefines( std::ostream& os ) {
  #ifdef __PPC64__
  os << std::endl << "Running PPC64 mode.";
  #else
  os << std::endl << "Running PPC32 mode.";
  #endif
```

```
    #ifdef __BIG_ENDIAN__
55  os << std::endl << "Running in BIG ENDIAN mode.";
    #else
    os << std::endl << "Running in LITTLE ENDIAN mode.";
    #endif

60  #ifdef __powerpc__
    os << std::endl << "Running in PowerPC mode.";
    #endif
  }

65 void PrintSizeOf( std::ostream& os ) {
    os << std::endl << "------------------------------"
       << std::endl << " sizeof(int)         = " << sizeof(int)
       << std::endl << " sizeof(long)        = " << sizeof(long)
       << std::endl << " sizeof(long long)   = " << sizeof(long long)
70     << std::endl << " sizeof(float)       = " << sizeof(float)
       << std::endl << " sizeof(double)      = " << sizeof(double)
       << std::endl << " sizeof(long double) = " << sizeof(long double)
       << std::endl << "------------------------------\n";
  }
75
  void PrintOS( std::ostream& os ) {
    struct utsname buf;
    int rc = uname( &buf );
    os << std::endl << "------------------------------"
80     << std::endl << "SYSNAME = " << buf.sysname
       << std::endl << "NODENAME= " << buf.nodename
       << std::endl << "RELEASE = " << buf.release
       << std::endl << "VERSION = " << buf.version
       << std::endl << "MACHINE = " << buf.machine
85 #ifdef _GNU_SOURCE
       << std::endl << "DOMAIN  = " << buf.domainname
  #endif
       << std::endl << "------------------------------";
  }
90
  void CopyFile( std::ostream& os, std::ifstream& ifs ) {
    os << std::endl << "------------------------------";
    while( ifs ) {
      std::string st;
95    getline( ifs, st );
      os << std::endl << st;
    }
    os << std::endl << "------------------------------";
  }
100
  void PrintCpuInfo( std::ostream& os ) {
    os << std::endl << "CPUInfo";
    std::ifstream ifs( "/proc/cpuinfo");
    CopyFile( os, ifs );
105 }

  void PrintMemInfo( std::ostream& os ) {
    os << std::endl << "MemInfo";
    std::ifstream ifs( "/proc/meminfo");
110 CopyFile( os, ifs );
  }

  void PrintSysInfo( std::ostream& os ) {
    os << std::endl << "Sys Info"
115    << std::endl << "------------------------------"
       << std::endl << "ARG MAX       = " << sysconf( _SC_ARG_MAX )
       << std::endl << "CHILD MAX     = " << sysconf( _SC_CHILD_MAX)
       << std::endl << "OPEN MAX      = " << sysconf( _SC_OPEN_MAX)
       << std::endl << "PAGESIZE      = " << sysconf( _SC_PAGESIZE)
```

```
120         << std::endl << "POSIX VERSION = " << sysconf( _SC_VERSION)
            << std::endl << "------------------------------\n";
    }
    unsigned int lock_value;
    mutex_ea_t  LOCKER_EA;
125 void RunPPESPETests( std::ostream& os ) {
      LOCKER_EA = (mutex_ea_t)(uintptr_t)(&lock_value);
      mutex_init( LOCKER_EA );
      int num_cpu = spe_cpu_info_get( SPE_COUNT_PHYSICAL_CPU_NODES, -1 );
      os << "There are " << num_cpu << " CPUs.\n";
130   int physical_spes,usable_spes,total_spes;
      for(int i=0; i<num_cpu; ++i) {
        physical_spes = spe_cpu_info_get( SPE_COUNT_PHYSICAL_SPES, i );
        os <<"There are " <<physical_spes<<" physical SPUs on CPU "<<i<<".\n";
        usable_spes = spe_cpu_info_get( SPE_COUNT_USABLE_SPES, i );
135     os <<"There are "<<usable_spes<<" usable SPUs on CPU "<<i<<".\n";
      }
      total_spes = spe_cpu_info_get( SPE_COUNT_PHYSICAL_SPES, -1 );
      os << "There are " << total_spes << " total SPUs.\n";
      if( total_spes > MAX_SPES ) {
140     std::cerr << std::endl << "Please recompile: " << __FILE__
            << " with MAX_SPES set to " << total_spes << std::endl;
        exit(1);
      }
      for(int i=0; i < total_spes; ++i ) {
145     cb[i] = new CB();
        cb[i]->addr = (unsigned long long int)&LOCKER_EA;
        cb[i]->N=i;
      }
      datas = new ppu_pthread_data_t[ total_spes ];
150   for(int i=0; i < total_spes; ++i) {
        datas[i].speid = spe_context_create (0, NULL);
        spe_program_load (datas[i].speid, &info_function);
        datas[i].argp = (unsigned long long*) cb[i];
        pthread_create(&datas[i].pthread,NULL,&ppu_pthread_function,&datas[i]);
155   }
      for (int i = 0; i < total_spes ; i++) {
        pthread_join (datas[i].pthread, NULL);
      }
      delete [] datas;
160   for(int i=0; i < total_spes; ++i) delete cb[i];
    }

    void PrintBanner( std::ostream& os ) {
      os << std::endl << "Cell Information Display Tool. Ver 1.0";
165   os << std::endl << "Author: Sandeep Koranne\n";
    }

    int main( int argc, char *argv [] ) {
      std::ostream& os( std::cout );
170   PrintBanner( os );
      PrintOS( os );
      PreProcessorDefines( os );
      PrintSizeOf( os );
      PrintCpuInfo( os );
175   PrintMemInfo( os );
      PrintSysInfo( os );
      os << std::endl << "Running Cell PPE/SPE Tests...\n";
      RunPPESPETests( os );
      return (EXIT_SUCCESS);
180 }
```

Listing 8.16 Sample program for printing Cell information info_ppe.C.

```
//////////////////////////////////////////////////////////////////
// Program: info_spu.c
```

```
// Author : Sandeep Koranne
//
////////////////////////////////////////////////////////////////////
#include <stdio.h>
#include <spu_mfcio.h>
#include <spu_intrinsics.h>
#include <stdlib.h>
#include <libsync.h>

typedef struct _control_block {
  unsigned int N;
  unsigned int M;
  unsigned long long addr;
} CB;
CB cb __attribute__ ((aligned(128)));

volatile vec_uint4 DATA_LS[1024<<2] __attribute__ ((aligned(128)));

extern int _etext;
extern int _edata;
extern int _end;
mutex_ea_t mutex;
register volatile unsigned int *RETADDR asm("0");
register volatile unsigned int *SP asm("1");
register volatile unsigned int *SSIZE asm("2");
static void MemoryInfo(void) {
    printf("     &_etext= %p\n", &_etext);
    printf("     &_edata= %p\n", &_edata);
    printf("       &_end= %p\n", &_end);
    printf("RETADDR     = %p\n", SP);
    printf("SPTR        = %p\n", SP);
    printf("SSIZE       = %p\n", SSIZE);
}

int main( unsigned long long speid,
    unsigned long long argp,
    unsigned long long envp ) {
  int tag_id = mfc_tag_reserve();
  int lock_code = 0;
  mfc_get(&cb, argp, sizeof(cb), tag_id,0,0);
  mfc_write_tag_mask( 1<<tag_id);
  mfc_read_tag_status_all();
  mutex = cb.addr;
  #ifdef USE_IO_LOCK
  mutex_lock( mutex );
  if( mutex_trylock( mutex ) ) {
    printf("\n This is incorrect.\n");
    return (-1);
  }
  #endif
  printf("\n------------------------------\n");
  printf("Hello, World! from %d\n", cb.N );
  MemoryInfo();
  printf("\n------------------------------\n");
  #ifdef USE_IO_LOCK
  mutex_unlock( mutex );
  #endif
  return (0);
}
```

Listing 8.17 Sample SPU program demonstrating IO Locking info_spu.c.

```
Running on Cell

Cell Information Display Tool. Ver 1.0
Author: Sandeep Koranne
```

```
5

------------------------------
SYSNAME = Linux
NODENAME= localhost.localdomain
RELEASE = 2.6.21-1.3194.fc7
10 VERSION = #1 SMP Wed May 23 22:13:52 EDT 2007
MACHINE = ppc64
DOMAIN  = (none)
------------------------------
Running PPC64 mode.
15 Running in BIG ENDIAN mode.
Running in PowerPC mode.
------------------------------
 sizeof(int)         = 4
 sizeof(long)        = 8
20 sizeof(long long)   = 8
 sizeof(float)       = 4
 sizeof(double)      = 8
 sizeof(long double) = 16
------------------------------
25
CPUInfo
------------------------------
processor : 0
cpu     : Cell Broadband Engine, altivec supported
30 clock    : 3192.000000MHz
revision  : 5.1 (pvr 0070 0501)

processor : 1
cpu     : Cell Broadband Engine, altivec supported
35 clock    : 3192.000000MHz
revision  : 5.1 (pvr 0070 0501)

timebase  : 79800000
platform  : PS3
40
------------------------------
MemInfo
------------------------------
MemTotal:        217068 kB
45 MemFree:           2736 kB
Buffers:           5288 kB
Cached:           54932 kB
SwapCached:      122388 kB
Active:          154088 kB
50 Inactive:         31824 kB
SwapTotal:       524280 kB
SwapFree:        385664 kB
Dirty:              264 kB
Writeback:            0 kB
55 AnonPages:       125540 kB
Mapped:            8688 kB
Slab:             18220 kB
SReclaimable:      6184 kB
SUnreclaim:       12036 kB
60 PageTables:        2128 kB
NFS_Unstable:         0 kB
Bounce:               0 kB
CommitLimit:     632812 kB
Committed_AS:    237780 kB
65 VmallocTotal: 8589934592 kB
VmallocUsed:       4648 kB
VmallocChunk: 8589929120 kB
HugePages_Total:      0
HugePages_Free:       0
70 HugePages_Rsvd:       0
Hugepagesize:     16384 kB
```

```
   ------------------------------
   Sys Info
75 ------------------------------
   ARG MAX        = 131072
   CHILD MAX      = 999
   OPEN MAX       = 1024
   PAGESIZE       = 4096
80 POSIX VERSION = 200112
   ------------------------------
   Running Cell PPE/SPE Tests...
   There are 1 CPUs.
   There are 6 physical SPUs on CPU 0.
85 There are 6 usable SPUs on CPU 0.
   There are 6 total SPUs.
   ------------------------------
   Hello, World! from 0
        &_etext= 0xd9c
90      &_edata= 0x1260
          &_end= 0x11380
   RETADDR      = 0x3ffa0
   SPTR         = 0x3ffa0
   SSIZE        = 0x4c0
95 ------------------------------
   ------------------------------
   Hello, World! from 2
        &_etext= 0xd9c
        &_edata= 0x1260
100       &_end= 0x11380
   RETADDR      = 0x3ffa0
   SPTR         = 0x3ffa0
   SSIZE        = 0x4c0
   ------------------------------
105 ------------------------------
   Hello, World! from 3
        &_etext= 0xd9c
        &_edata= 0x1260
          &_end= 0x11380
110 RETADDR     = 0x3ffa0
   SPTR         = 0x3ffa0
   SSIZE        = 0x4c0
   ------------------------------
   ------------------------------
115 Hello, World! from 1
        &_etext= 0xd9c
        &_edata= 0x1260
          &_end= 0x11380
   RETADDR      = 0x3ffa0
120 SPTR        = 0x3ffa0
   SSIZE        = 0x4c0
   ------------------------------
   ------------------------------
   Hello, World! from 5
125     &_etext= 0xd9c
        &_edata= 0x1260
          &_end= 0x11380
   RETADDR      = 0x3ffa0
   SPTR         = 0x3ffa0
130 SSIZE       = 0x4c0
   ------------------------------
   ------------------------------
   Hello, World! from 4
        &_etext= 0xd9c
135     &_edata= 0x1260
          &_end= 0x11380
   RETADDR      = 0x3ffa0
   SPTR         = 0x3ffa0
```

```
    SSIZE          = 0x4c0
140 ------------------------------
```

Listing 8.18 Transcript of Info program.

8.7 Conclusions

In this chapter we presented our first compiled code executable for the Cell platform. Since the PS3 runs standard GNU/Linux, many of the commands and utilities of GNU/Linux are available on the PS3, and we used some of the commands to gather system information. The use of libspe2 for run-time management of the SPEs is shown and explained with small examples. The use of hardware based high resolution timers for fine grained performance measurement was also shown. Synchronization between PPE and multiple SPEs is performed using the Synchronization facilities and an example was demonstrated. Code similar to the one presented in this chapter forms the basic structure of many examples in the remainder of this text; thus you should try compiling and running the small examples present in this chapter first. A *ping-pong* style communication example, with the SPE kernel performing file transformation is explained in detail.

Chapter 9
An Overview of the SDK

Abstract A brief overview of the major components of the SDK are given. No attempt is made to recreate the documentation, except, to list the major functionality provided by the SDK as a whole in one place. Our collection is organized by functional use and not just by API library

9.1 Major components of the Cell Broadband Engine SDK 3.0

At the time of this writing, the Cell Broadband Engine SDK version has been bumped up to 3.1, but none of the changes are dramatic, and many of them refer to the increased performance of the double-precision floating point unit in the SPU. Since in this book we rarely use double-precision, we have concentrated on the major components of the SDK, the APIs, which provide additional facilities which will aid us in writing efficient programs for the Cell.

In the coming chapters we shall also develop a number of data-structures, implement algorithms and perform compute intensive tasks on the Cell architecture. As you shall discover, the code and algorithms presented in this book are orthogonal, even complimentary to the SDK. We use the SDK to provide facilities for DMA, SPU intrinsics, synchronization, run-time control of SPEs, and more. But when it comes to writing a Graph library, spanning-tree, shortest-path, even sorting, we have presented our own solutions. This is not out of (complete) ignorance or overt enthusiasm. Our code serves at least two purpose, (i) I have presented the thinking which went with the code and also the pitfalls which lie on the way, (ii) the code is written in a type generic manner, and once you understand it, it should be fairly easy for you to change it to suit your particular need.

Nevertheless, for many applications, the out-of-the-box SDK functionality does the job, and does it extremely well. Consider a cut-and-dry application like matrix transpose, even 4x4 matrix transpose. We will recreate it as we use it to sort `unsigned int`, whereas the library version only supports matrices of floating point. For large matrices (where our data is already in floating point), random number

S. Koranne, *Practical Computing on the Cell Broadband Engine*,
DOI: 10.1007/978-1-4419-0308-2_9, © Springer Science + Business Media, LLC 2009

generation, SIMD math functions, and Fast Fourier Transform, we use the SDK libraries gleefully.

In this chapter we provide a list of major functionality which is part of the Cell SDK 3.0, and refer back to this list to point out some related functionality during the sequel of this text.

9.2 Application programming libraries in the SDK

The following libraries and frameworks are available with the SDK 3.0.

SPE Runtime Library : [28],
SPU Timer Library : [31],
Monte-Carlo Library : [24],
BLAS Library : [16],
LAPACK Library : [23],
SIMD Math API and Library : [25],
SPU Software Managed Cache : [22],
FFT Library : [22],
Game Math Library : [22],
Image Library : [22],
Large Matrix Library : [22],
Matrix Library : [22],
Misc Library : [22],
Sync Library : [22],
Vector Library : [22],
Multiprecision Math library : [22],
ALF Library and Framework : [26],
DACS Library and Framework : [21]

9.3 SPE Runtime Library

From the SPE Runtime Library manual [28] we infer the SPE Runtime Management Library to be the standardized low-level API that enables application level access to the SPEs. SPE Runtime library is organized around the concept of SPE contexts. SPE context is described in [28] as one of the base data structures for the library implementation, and it holds all persistent information about a logical SPE. We have described the main APIs of the SPE Runtime Management library in Section 8.4.1, pp. 109, hence we shall not repeat it here.

9.4 SPU Timer Library

From the SPU Timer Library documentation [31] we note that the SPU timer library was designed to provide virtual clock and interval timer services to programs executing on the SPU. The virtual clock is a 64-bit software managed, monotonic increasing *time-base* counter. The interval timers provide the ability to register a handler which is called on specified intervals. An example of the SPU timer library is shown in this book in Section 8.5, pp. 121. Code is shown in Listing 8.13, which implements an SPU timer and handler for event counting.

In order to understand the SPU Timer code and documentation the following terminology is essential:

1. FLIH: first-level interrupt handler, handler code must be placed at address 0x0 and SPU branches to this address whenever an interrupt occurs,
2. SLIH: second-level interrupt handler, code that serves a specific interrupt type,
3. Time base: a hardware register defined by PowerPC architecture which represents an elapsed time. It is a monotonically increasing counter that ticks at an implementation specific *time-base frequency*,
4. Decrementer: another hardware register defined by PowerPC which counts down from its programmed value at time-base frequency and generates an interrupt when the count has expired.

In addition to these hardware specific features performance measurement based on hardware counters is also explained by PAPI library[1]. PAPI stands for Performance Application Programming Interface.

9.5 Monte-Carlo Library

The Monte-Carlo Library [24] provides random number generators (RNGs) and distribution transformations. We have used the RNG in this book in a number of programs for generating random moves of objects in the simulated annealing framework, or permuting a sequence for random perturbations. The Monte-Carlo library provides the following RNGs:

1. Hardware based: not usable on the PS3,
2. Kirkpatrick-Stoll:
3. Mersenne Twister (MT): used by programs in this book,
4. Sobol

Example code using the MT RNG is shown in Section 10.4.2, pp. 161 and in Listing 10.9 where we use MT RNG to generate pseudo-random numbers prior to sorting. As mentioned above we have also used MT RNG in Simulated Annealing 11.11 and Sequence Pair 21.2.

---

[1] See `http://icl.cs.utk.edu/papi/` for more information.

9.6 BLAS Library

Basic Linear Algebra Subprograms (BLAS) library is based upon the BLAS report available from http://www.netlib.org. BLAS is used as the low-level commonly used linear algebra operations in high-performance computing and as the foundations on which LAPACK is built. BLAS is divided into three parts:

1. Level 1: scalar and vector operations,
2. Level 2: matrix-vector operations,
3. Level 3: matrix-matrix operations.

Each routine has 4 versions for (a) real single-precision, (b) real double-precision, (c) complex single-precision, and (d) complex double-precision, represented by prefixing S, D, C and Z respectively to the function name. The Cell BLAS in SDK 3.0 supports only the S and D forms. Many of the BLAS functions have been optimized to use SPEs even when running on the PPE. The environment variable `BLAS_NUMSPES` specifies the number of SPEs to use. For the PS3 this variable should be set to 6. Moreover since BLAS for large matrices can be memory intensive memory allocation should be done from huge-pages.

Some of the BLAS SPU functions are given below:

sscal_spu: BLAS 1 routine to scale vector by constant,

$$x \leftarrow \alpha x$$

where x is a vector and α is a constant. For SPU, n should be a multiple of 16 for double-precision and 32 for single-precision. The x vector should be aligned on a 16-byte boundary.

```
void sscal_spu( float*  sx, float sa, int n);
void dscal_spu( double* dx, double da, int n);
```

Listing 9.1 sscal_spu and dscal_spu

scopy_spu: BLAS 1 routine to copy vector from source to destination

$$y \leftarrow x$$

```
void scopy_spu( float*  sx, float*  sy, int n);
void dcopy_spu( double* dx, double* dy, int n);
```

Listing 9.2 scopy_spu and dcopy_spu

saxpy_spu: BLAS 1 routine to scale vector and perform element-wise add

$$x \leftarrow \alpha x + y$$

where x is a vector and α is a constant. For SPU, n should be a multiple of 32 for double-precision and 64 for single-precision. The x vector should be aligned on a 16-byte boundary.

```
void saxpy_spu( float*  sx, float* sy, float sa, int n);
void daxpy_spu( double* dx, double* dy, double da, int n);
```

Listing 9.3 saxpy_spu and daxpy_spu

sdot_spu: BLAS 1 routine destination

$$result \leftarrow x \cdot y$$

```
void sdot_spu( float*  sx, float*  sy, int n);
void ddot_spu( double* dx, double* dy, int n);
```

Listing 9.4 sdot_spu and ddot_spu

sgemv_spu: BLAS 2 routine to multiply matrix and vector destination

$$y \leftarrow \alpha Ax + y$$

```
void sgemv_spu( int m, int n, float alpha,
                float *a, float *x, float *y);
void dgemv_spu( int m, int n, double alpha,
                double *a, double *x, double *y);
```

Listing 9.5 sgemv_spu and dgemv_spu

sgemm_spu: BLAS 3 routine to multiply matrix and matrix destination

$$C \leftarrow AB + C$$

```
void sgemm_spu( int m, int n, int k, float alpha,
                float *a, float *x, float *c);
void dgemm_spu( int m, int n, int k, double alpha,
                double *a, double *x, double *c);
```

Listing 9.6 sgemm_spu and dgemm_spu

9.7 LAPACK Library

The LAPACK standard is also based upon a published standard interface for commonly used linear algebra operations in high-performance computing. The LAPACK standard is available from http://www.netlib.org. As with BLAS LAPACK routines are prefixed with S, D, C and Z depending upon the type of data. The following LAPACK functions have been optimized for SPE,

1. DGETRF: LU factorization of general matrix,
2. DGETRI: computes inverse of general matrix,
3. DGEQRF: compute QR factorization of general matrix,
4. DPOTRF: compute Cholesky factorization of general matrix,

5. DBDSQR: compute SVD (singular value decomposition) of real bi-diagonal matrix using implicit zero-shift QR algorithm,
6. DSTEQR: compute SVD of real symmetric tridiagonal matrix using implicit QR algoritm.

9.8 SIMD Math API and Library

The SIMD Math Library [25] provides SIMD optimized math functions. The functions are grouped into the following sections:

1. Absolute value and sign functions,
2. Classification and comparison functions,
3. Divide, multiply, modulus, remainder and reciprocal functions,
4. Exponentiation, root, and logarithmic functions,
5. Gamma and Error functions,
6. Minimum and maximum functions,
7. Rounding and next functions,
8. Trigonometric functions,
9. Hyperbolic functions

Please see the SIMD Math Library API Reference [25] for the full list of functions.

9.9 SPU Software Managed Cache

A brief discussion of SPU Software Managed Cache is given in Section 6.9, pp. 88.

9.10 FFT Library

The FFT library provides the following functions:

1. fft_1d_r2: single precision complex, FFT using DFT (discrete fourier transform) with radix-2 decimation in time,
2. fft_2d: transforms 4 rows of complex 2d data from time domain to frequency domain.

9.11 Game Math Library

The Game math library provides the following functions:

1. cos8, cos14, cos18: accurate upto atleast 8, 14 and 18 bits respectively for all angles in $-\pi$ to 2π for the cosine function,
2. sin8, sin14, sin18: accurate upto atleast 8, 14 and 18 bits respectively for all angles in $1/2\pi$ to $3/2\pi$ for the sine function,
3. pack_color8, pack_normal16, pack_rgba8: color clamping functions.

9.12 Image Library

The Image library provides the following functions:

1. Convolution: compute output pixels as the weighted sum of the input images's 3x3, 5x5, 7x7 or 9x9 neighborhood and the mask m,
2. Histogram: generates histogram of unsigned bytes.

9.13 Large Matrix Library

The Large Matrix library provides the following functions:

1. Calculate index of maximum absolute value in column,
2. lu2_decomp: compute LU decomposition of a dense general m by n matrix,
3. madd_matrix_matrix: performs matrix-matrix operation $C = AB + C$,
4. nmsub_matrix_matrix: performs matrix-matrix operation $C = C - AB$,
5. solve_unit_lower: solves matrix equation,
6. solve_upper_1:
7. solve_linear_system_1: computes solution to real system of linear equations,
8. transpose_matrix:

9.14 Matrix Library

The Matrix library operates on 4x4 matrices. The Matrix library provides the following functions:

1. identity_matrix4x4: constructs a 4x4 identity matrix,
2. inverse_matrix4x4: calculates the inverse of the 4x4 matrix,
3. mult_matrix4x4:
4. rotate_matrix4x4: performs rotation of angle (in radians) about the normalized vector,
5. scale_matrix4x4:
6. splat_matrix4x4:

7. transpose_matrix4x4:

9.15 Synchronization Library

The Synchronization library provides the following functions:

1. atomic_add: atomically adds integer x to the 32-bit integer pointed by the EA. The function returns the pre-added integer value pointed by EA; this function is roughly equivalent to:

```
int atomic_add(int x, int* ea) {
  int retval = *ea;
  *ea += x;
  return retval;
5 }
```

2. atomic_dec: atomically subtract 1 from the integer pointed to by EA, return the pre-decremented value as the return value of this function,
3. atomic_inc: atomically add 1 to the integer pointed to by EA, return the pre-decremented value as the return value of this function,
4. atomic_read: equal to a volatile load, atomically returns the 32-bit integer pointed by EA,
5. atomic_set: atomically writes the given 32-bit value to the location pointed by EA. The EA should be correctly aligned (32-bit alignment),
6. atomic_sub: atomically subtracts integer x from the 32-bit integer pointed by the EA. The function returns the pre-added integer value pointed by EA,
7. mutex_init: Mutex are used around critical sections of the code to protect shared data. This function initializes the specified mutex by setting its value to 0,
8. mutex_lock: acquires a lock on the specified mutex by spinning. The mutex must be 32-bit aligned address,
9. mutex_trylock: tries to acquire the lock, if mutex is already locked then returns 0 (specifying failed), else if the lock is available, lock is acquired and 1 is returned,
10. mutex_unlock: releases the mutex,
11. cond_broadcast: unblocks all threads waiting on the conditional variable specified in the argument,
12. cond_init:
13. cond_signal: used to unlock a single thread waiting on a condition,
14. cond_wait: see pthread function,
15. complete_all: see section on *Completion Variables*
16. init_completion:
17. wait_for_completion:
18. read_lock: see section in API on Reader/Writer locks
19. read_trylock:
20. read_unlock:
21. write_lock:

22. write_trylock:
23. write_unlock:

9.16 Vector Library

The Vector library provides the following functions:

1. clipcode_ndc: computes set of bit-flags indicating if the specified vertex is outside the halfspace defined by -1.0 to 1.0 volume,
2. clip_ray: computes the linear interpolation factor for the ray passing through the vertices specified as arguments to this function,
3. cross_product: computes cross-product of vectors,
4. dot_product: computes dot-product of vectors,
5. intersect_ray_triangle:triangle is 3d, function calculates point of intersection,
6. inv_length_vec: computes reciprocal magnitude of the 3d vector specified by v,
7. length_vec: computes length of vector,
8. normalize: normalizes the given vector,
9. reflect_vec: for lighting style calculation,
10. sum_across_float: returns the sum of the 4 components of the 4-way float vector,
11. xform_vec

9.17 Multiprecision Math Library

The Multiprecision Math library provides the following functions; the purpose and behavior of the function should be clear by function name:

1. mpm_abs:
2. mpm_add:
3. mpm_add_partial:
4. mpm_cmpeq:
5. mpm_cmpge:
6. mpm_cmpgt:
7. mpm_div:
8. mpm_fixed_mod_reduction:
9. mpm_gcd:
10. mpm_madd:
11. mpm_mod:
12. mpm_mod_exp:
13. mpm_mul:
14. mpm_mul_inv:
15. mpm_neg:

16. mpm_sizeof:
17. mpm_square:
18. mpm_swap_endian:

9.18 ALF Library and Framework

ALF stands for Accelerated Library Framework and ALF for Cell Broadband Engine is included as a framework in the SDK. From the ALF Programmer's Guide and API Reference [26], we see that ALF provides a programming environment for data and task parallel applications and libraries. The ALF API provides a set of interfaces to simplify library development on heterogeneous multi-core systems. We have used ALF in Chapter 13 when we solve equations using parallel search implemented on the SPUs.

ALF functionality includes:

1. Data transfer management
2. Parallel task management
3. Double buffering
4. Dynamic load balancing for data parallel tasks

From ALF's point-of-view there are three distinct types of tasks within an application:

1. Application: this is the program at the host (or main PPE server) only. The provided API libraries provided by acceleration engines can be used without knowing their inner workings. This is the most abstract level and requires least intrusion of application code,
2. Accelerated library: the library writer uses the ALF API to write the computation kernels on the accelerators which are most often the SPUs of the Cell Broadband Engine. Input partitioning and output partitioning has to be done, as well as the division of work between the server process running on the host and the accelerators running on SPUs,
3. Computational Kernel: this is where the bulk of the work is performed. The computational kernel is written for efficient execution on the accelerators and thus uses SIMD and dual-issue optimization. Work partitioning happens at a higher level, so the compute kernel should be written assuming a work load and data-packet which has already been assembled on the accelerator.

The ALF runtime framework implements the underlying data transport mechanism which is responsible for data movement from host to accelerator local store. ALF runtime also implements task management and provides dynamic load balancing which frees up the compute kernel writer to concentrate on the kernel code. ALF internally manages the data movement by appropriately choosing double buffering and scheduling DMA based data transfers of the input, and the output produced on the accelerator. ALF runtime for task management can support data dependencies, resource conflicts and task sequencing. For an example of an ALF application look at Listing 13.11 on pp. 279.

9.19 DACS Library and Framework

From the DACS manual [21] we infer Data Communication and Synchronization (DACS) library to provide a set of services which ease the development of applications and application frameworks in a heterogeneous multi-tiered system. DACS services can be divided into the following categories:

1. Resource reservation
2. Process management
3. Group management
4. Remote memory
5. Message passing
6. Mailboxes
7. Process Synchronization
8. Data Synchronization
9. Error Handling

9.20 Conclusions

In this chapter we covered the outline of SDK API Libraries available on the Cell. Please refer to the individual API documentation for further details on any specific function call.

Chapter 10
Basic Algorithms

Abstract We implement several data-structures and algorithms to solve programming problems which we shall use in the later chapters. We begin with writing integer programs to search for counterexamples of the Collatz conjecture. We analyze Barker sequences, calculate their merit factor. We implement factorial, gcd functions while discussing the properties of tail-recursive functions and how to implement them on the SPU which has limited stack space. We then implement searching and sorting data-structures and algorithms. Finally some advanced data-structures such as hash, perfect-hash, union-find, and min-heap and max-heap are implemented.

10.1 Some number theoretic problems

Number theory has been called the *queen of mathematics*[1], and thus it is fitting that we being our mathematical and algorithmic excursions in the Cell space by implementing some number theoretic programs. Moreover, this exercise will also let us introduce some of the special features of the SPU (such as the 128-bit vector), and demonstrate common programming paradigms which we shall refer to in the sequel of this book.

10.1.1 Collatz conjecture

Consider the function defined by:

$$Collatz(n) = \begin{cases} 3n+1 & : \text{ if } n \text{ odd} \\ n/2 & : \text{ if } n \text{ even} \end{cases}$$

[1] By C. F. Gauss.

S. Koranne, *Practical Computing on the Cell Broadband Engine*,
DOI: 10.1007/978-1-4419-0308-2_10,

The still open Collatz conjecture hypothesizes that repeated application of `Collatz` to its input will eventually saturate to 1, or equivalently $Collatz^+(Collatz(n)) == 1 \ \forall n \in Z^+$. Despite significant computer analysis (including by the author) no breakthrough has been achieved. Numbers in the range of $1 \ldots 2^{58}$ have been analyzed. A related problem is finding the number of recursive Collatz calls for a given n, and this is implemented in Algorithm 1, and plotted in Figure 10.1.

```
  Data: Unsigned integer N
  Result: Returns Collatz termination count
1 int CollatzTermination(int N) begin
2     int T=0;
3     while N>1 do
4         T++;
5         N = (N%2) ? (3*N+1) : (N/2);
6     end
7     return T;
8 end
```

Algorithm 1: Collatz function and its implementation to calculate termination count.

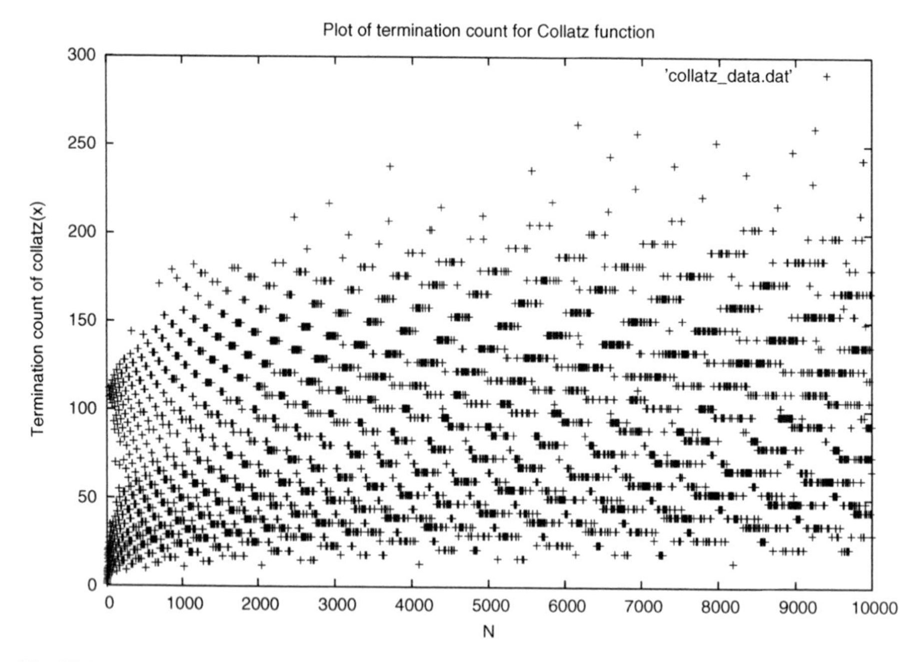

Fig. 10.1 Plot of n vs length of collatz sequence of n.

10.1.2 The Takeuchi Benchmark

A similar problem which has long been considered a benchmark problem for Common Lisp, is the *takeuchi*-benchmark defined as below in Listing 10.1.

```
;; program : takeuchi
;; author  : Sandeep Koranne

(defun takeuchi (x y z)
  "Takeuchi function"
  ;(format t "Called with ~D ~D ~D~%" x y z)
  (if (not (< y x)) z
      (takeuchi (takeuchi (1- x) y z)
    (takeuchi (1- y) z x)
    (takeuchi (1- z) x y))))
(defun takeuchi-orig (x y z)
  "Original Takeuchi benchmark"
  ;(format t "Called with ~D ~D ~D~%" x y z)
  (cond ((> x y)
   (takeuchi-orig (takeuchi-orig (1- x) y z)
      (takeuchi-orig (1- y) z x)
      (takeuchi-orig (1- z) x y)))
  (t y)))
(defconstant +ITER+ 20)

(defun takeuchi-sum(iter)
  (let ((id 0))
    (dotimes (i iter)
      (dotimes (j iter)
  (dotimes (k iter)
    (incf id (takeuchi i j k)))))
    id))
;; (defun main()
;;    (takeuchi-sum +ITER+))
;; (time (main))
;; (quit)
```

Listing 10.1 The Takeuchi benchmark for Common Lisp, takeuchi.lisp.

The same algorithm coded up in SPU C is shown in Listing 10.2.

```
/////////////////////////////////////////
// Program : Takeuchi
// Author  : Sandeep Koranne
/////////////////////////////////////////
#include <spu_mfcio.h>
#include <stdio.h>
#include <math.h>
#include <spu_intrinsics.h>
#include <simdmath.h>
#include <spu_timer.h>
unsigned long int total_tags = 0;
void ev_handler( int time_tag ) {
  total_tags++;
}

inline int takeuchi(int x, int y, int z) {
  if( y >= x ) return z;
  return takeuchi( takeuchi(x-1, y, z),
       takeuchi(y-1, z, x),
       takeuchi(z-1, x, y));
}
#define ITER 18
int main(unsigned long long speid,
         unsigned long long argp,
```

```
25          unsigned long long envp __attribute__ ((__unused__))) {
      int id;
      int i,j,k;
      spu_slih_register( MFC_DECREMENTER_EVENT, spu_clock_slih);
      id = spu_timer_alloc( 10000, ev_handler );
30    spu_clock_start();
      total_tags=0;
      id = 0;
      spu_timer_start( id );
      for(i=0;i<ITER;++i)
35      for(j=0;j<ITER;++j)
          for(k=0;k<ITER;++k)
      id += takeuchi( i, j, k );
      spu_timer_stop(id);
      printf("\n Takeuchi sum = %d in %d\n", id, total_tags);
40    return 0;
   }
```

Listing 10.2 Takeuchi benchmark in SPU, takeuchi_spu.c.

Table 10.1 Takeuchi benchmark timing comparison between Opteron Lisp and SPU.

Iteration	Lisp (Opteron)	C (SPU)
15	5.0	2.28
16	17.57	8.2
17	65	29.8
18	234	108.9

As can be seen from the above Table 10.1 the SPU code without parallelism or SIMD is still 2x faster than the Common Lisp code running on the Opteron. With parallelism this can be improved by a factor of 6x using multiple SPUs.

10.2 Merit Factor of Sequences and Auto-correlation

A Barker sequence is a finite sequence of integers, each ± 1, whose aperiodic auto-correlations are minimal for all sequences of that length. It is conjectured that there are only a finite number of Barker sequences. Analyzing auto-correlations of a certain period is also important in design of microwave antennas, since beam-spreading can cause the *side-lobes* to create auto-correlated spurious images which have to be filtered using auto-correlation techniques. Consider a aperiodic auto-correlation defined on a sequence $a_0, a_1, \ldots, a_n$ of integers as (in the microwave domain these will be complex numbers, as we shall see in Chapter 19):

$$c_k = \sum_{j=0}^{n-1-k} a_j \bar{a}_{j+k}$$

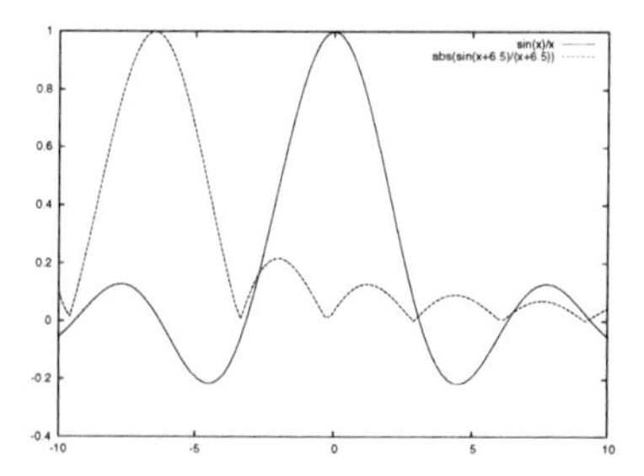

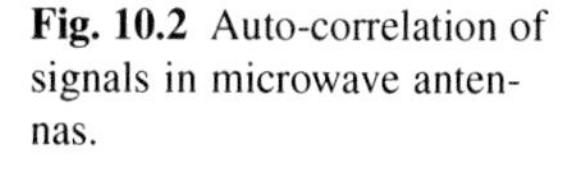
Fig. 10.2 Auto-correlation of signals in microwave antennas.

Consider a small example if $A_n = \{1101\}$ and we want to compute c_1, then $n = 4, k = 1$ therefore:

$$c_1 = \sum_{j=0}^{2} a_j \bar{a}_{j+1} = a_0\bar{a}_1 + a_1\bar{a}_2 + a_2\bar{a}_3 = 1\cdot 0 + 1\cdot 1 + 0\cdot 0 = 0 + 1 + 0 = 1$$

$$c_2 = \sum_{j=0}^{1} a_j \bar{a}_{j+2} = a_0\bar{a}_2 + a_1\bar{a}_3 = 1\cdot 1 + 1\cdot 0 = 1 + 0 = 1$$

thus this is a Barker sequence.

The auto-correlation itself is a sequence albeit of length $n-1-k$ for any given k. Depending on k, the periodicity, the value of c_k can achieve a high of n which is the *peak auto-correlation*, and this is the sequence which is of most importance, as it causes the maximum interference. A sequence in which $c_k = \pm 1$ is called a Barker sequence. In microwave engineering a common measure of c_k is the ratio of the square of the peak (n^2) to the sum of the squares of the off-peak values; this measure is called the *merit factor*, and it is simply

$$merit\_factor(A_n) = \frac{n^2}{2\sum_{k=1}^{n-1}|c_k|^2}$$

In the above example with $A_n = \{1101\}$ the merit factor is calculated as $\frac{4^2}{2\cdot(|1|^2+|1|^2)}$, which is 4.0.

Merit factor computations can also be performed for polynomials, although in this section we shall give an SPU function which computes the merit factor of a given sequence. The key idea behind presenting this example is the 128-bit quad-word bit operations that are possible using the SPU. We will represent a sequence as a bit string, and perform the auto-correlations. Then we can use byte-wide sums and *popcount* functions to calculate number of set bits. Obviously, this method will not work directly for complex numbers, but for ± 1 sequences it is ideal.

We have computed merit-factors of all Barker sequences upto 2^{32}, and they are given below:

The algorithm to compute the Merit factor is shown in Algorithm 2.

Table 10.2 Merit-factor for Barker sequences.

Length	Vector	Merit Factor
2	+1-1	2.0
3	+1-1-1	4.5
5	+1-1+1+1+1	6.25
7	+1-1+1+1-1-1-1	8.166667
11	+1-1+1+1-1+1+1+1-1-1-1	12.1
13	+1-1+1-1+1+1-1-1+1+1+1+1+1	14.083333

Data: N length of maximum sequence
Result: Hash table of merit-factors

1 **HashTable** `MeritFactor`(int N) **begin**
2 HashTable MFHash;
3 **for** $int p = 0; p < 2^N; ++p$ **do**
4 Sequence A = `GenerateSequence`(p);
5 int denominator=0;
6 **for** $int i = 0; i < (N-1); ++i$ **do**
7 **for** $int k = i+1; k < (N-1); ++k$ **do**
8 int $c_k = 0$;
9 **for** $int j = 0; j < (N-k-1); ++j$ **do**
10 $c_k += a_j \bar{\oplus} \bar{a}_{j+k}$;
11 **end**
12 **end**
13 **end**
14 denominator += $|c_k|^2$;
15 mf = $\frac{|p| \cdot |p|}{2*denominator}$;
16 MFHash.insert(mf, $|p|$, p);
17 **end**
18 **return**(MFHash);
19 **end**

Algorithm 2: Tabulating merit-factor of Barker sequences.

```
(defun compute-function ( a b )
  (if (= 0 (logxor a b)) -1 1))
(defun ck (A n k)
  (let ((ans 0)(temp_j 0)
5       (n_temp_k 0)(ub (- n k )))
    (dotimes (j ub)
      (setf temp_j (aref A j))
      (setf n_temp_k (logxor (aref A (+ j k)) 1))
      (incf ans (compute-function temp_j n_temp_k)))
10    ans))
(defun cs (A n)
  (let ((sum 0))
  (dotimes (i (- n 1))
    (let ((val (ck A n (1+ i))))
```

```
15          (incf sum (* val val))
            (format t "~%c_~D = ~D" (1+ i) val)))
      sum))
    (cs #(1 0  1 0 1) 5)
    (quit)
20  (defun merit-factor (A)
      (let ((n (length A)))
        (* 1.0 (/ (* n n) (* 2 (cs A n))))))
    (defun log2 (n)
      (dotimes (j 64)
25      (when (> (ash 1 j) n) (return-from log2 j))))
    (defun eb (n pos)
      (let ((SHIFT (1- (log2 n))))
        (ash (logand
        n (ash 1 (- SHIFT pos)))
30     (- pos SHIFT))))
    (defun convtnum-to-array (n)
      (let* ((expected-len (log2 n))
             (retval (make-array expected-len :element-type 'bit)))
        (dotimes (i expected-len)
35        (setf (aref retval i)
                (eb n i)))
        retval))
    (defvar *best-mf-hash* (make-hash-table))
    (defun do-mf-searh (n)
40    (let ((min-factor 100)
            (max-factor 0))
        (dotimes (x n)
          (let*
        ((i (+ x 2))
45       (A (convtnum-to-array i))
         (f (merit-factor A)))
            (when (< f min-factor) (setf min-factor f))
            (when (> f max-factor)
        (progn
50        (setf (gethash f *best-mf-hash*)
          (list (length A) i f A))
          (setf max-factor f)))
            (when (or
             (> f 2.0)
55           (= 0 (rem i 1000))))))))
```

Listing 10.3 Merit-factor of Barker sequences.

```
  Length Merit-Factor Sequence
    2      2.0          10
    3      4.5          100
    5      6.25        10111
5   7      8.166667    1011000
    11     12.1        10110111000
    13     14.083333   1010110011111
```

The direct method of iterating over the bit-vector is inefficient, and cannot be directly converted to SIMD. An equivalent formulation is to consider the bit-vector as :$a_0a_1a_2\ldots a_n$ then $a_i\bar{a}_{i+k}$ can be written as $a_0a_1a_2\ldots a_n \oplus (a_0a_1a_2\ldots a_n) << k$. We shift the bit-vector left by k, and do an XOR. The `spu_cntb` function performs a population-count (or the number of 1s in the bit-vector). By setting lower k bits of the shifted vector to a copy of A, we know that their contribution after the XOR to the population-count will be zero. This is necessary, as the shift operator shifts in ZEROS. The code is shown below:

```
////////////////////////////////////////////
```

```
// Program : Barker Sequences
////////////////////////////////////////
#include <stdio.h>
#include <spu_intrinsics.h>

vec_uchar16 SHV[]
= {{0,0,0,0,0,0,0,0,0,0,0,0,0,0,0,1},
   {0,0,0,0,0,0,0,0,0,0,0,0,0,0,0,3},
   {0,0,0,0,0,0,0,0,0,0,0,0,0,0,0,7},
   {0,0,0,0,0,0,0,0,0,0,0,0,0,0,0,15},
   {0,0,0,0,0,0,0,0,0,0,0,0,0,0,0,31},
   {0,0,0,0,0,0,0,0,0,0,0,0,0,0,0,63},
   {0,0,0,0,0,0,0,0,0,0,0,0,0,0,0,127},
   {0,0,0,0,0,0,0,0,0,0,0,0,0,0,0,255},
   {0,0,0,0,0,0,0,0,0,0,0,0,0,0,1,0xFF},
   {0,0,0,0,0,0,0,0,0,0,0,0,0,0,3,0xFF},
   {0,0,0,0,0,0,0,0,0,0,0,0,0,0,7,0xFF},
   {0,0,0,0,0,0,0,0,0,0,0,0,0,0,15,0xFF},
   {0,0,0,0,0,0,0,0,0,0,0,0,0,0,31,0xFF},
   {0,0,0,0,0,0,0,0,0,0,0,0,0,0,63,0xFF},
   {0,0,0,0,0,0,0,0,0,0,0,0,0,0,127,0xFF},
   {0,0,0,0,0,0,0,0,0,0,0,0,0,0,255,0xFF},
   {0,0,0,0,0,0,0,0,0,0,0,0,0,1,0xFF,0xFF},
   {0,0,0,0,0,0,0,0,0,0,0,0,0,3,0xFF,0xFF},
   {0,0,0,0,0,0,0,0,0,0,0,0,0,7,0xFF,0xFF},
   {0,0,0,0,0,0,0,0,0,0,0,0,0,15,0xFF,0xFF},
   {0,0,0,0,0,0,0,0,0,0,0,0,0,31,0xFF,0xFF},
   {0,0,0,0,0,0,0,0,0,0,0,0,0,63,0xFF,0xFF},
   {0,0,0,0,0,0,0,0,0,0,0,0,0,127,0xFF,0xFF},
   {0,0,0,0,0,0,0,0,0,0,0,0,0,0xFF,0xFF,0xFF}};

static void print_binary( vec_ullong2 A ) {
  unsigned long int x = spu_extract(A,0);
  unsigned long int y = spu_extract(A,1);
  printf("\n%d\n%d",x,y);
}
static void print_vector_u16( vec_uchar16* p) {
  printf("\n %d %d %d %d %d %d %d %d %d %d %d %d %d %d %d %d",
   spu_extract(*p,15),spu_extract(*p,14),spu_extract(*p,13),
   spu_extract(*p,12),spu_extract(*p,11),spu_extract(*p,10),
   spu_extract(*p,9),spu_extract(*p,8),spu_extract(*p,7),
   spu_extract(*p,6),spu_extract(*p,5),spu_extract(*p,4),
   spu_extract(*p,3),spu_extract(*p,2),spu_extract(*p,1),
   spu_extract(*p,0));
}
int CalculateFunction(int n, vec_ullong2 A, int L, int R,int k) {
  vec_uchar16 coA = (vec_uchar16)A;
  vec_uchar16 co  = spu_cntb(coA);
  int sum = spu_extract(co,0);
  sum += spu_extract(co,1);sum+=spu_extract(co,2);
  sum += spu_extract(co,3);sum+=spu_extract(co,4);
  sum += spu_extract(co,5);sum+=spu_extract(co,6);
  sum += spu_extract(co,7);sum+=spu_extract(co,8);
  sum += spu_extract(co,9);sum+=spu_extract(co,10);
  sum += spu_extract(co,11);sum+=spu_extract(co,12);
  sum += spu_extract(co,13);sum+=spu_extract(co,14);
  sum += spu_extract(co,15);
  //printf("\n Total bits = %d", sum);
  // +1 are sum-k
  // -1 are n-sum
  //printf("\n +1 = %d, -1 = %d", (sum-k), (n-sum));
  sum = (sum-k) - (n-sum);
  return sum;
}
int CK(vec_ullong2 A, int L, int R,int k) {
  vec_uchar16 t1 = SHV[R-1];
  A = (vec_ullong2)spu_and((vec_uchar16)A,t1);
```

```
      vec_ullong2 B = spu_slqw(A, k);
70    vec_ullong2 temp = (vec_ullong2)spu_and((vec_uchar16)A,SHV[k-1]);
      B = spu_or( B, temp );
      vec_ullong2 p = spu_eqv(A,B);
      p = (vec_ullong2)spu_and((vec_uchar16)p,t1);
      int sum = 0;
75    sum = CalculateFunction(R,p,L,R,k);
      printf("\n [%d] = %d", k, sum);
    }
    int CS(vec_ullong2 n, int L, int R) {
      int i=0;
80    for(i=1;i<R;++i) {
      CK(n,L,R,i);
      printf("\n");
      }
      return 0;
85  }

    #if 0
    int main( unsigned long long spuid,
              unsigned long long argp )
90  {
      int rc;
      printf(" Hello, World! from %d with %d \n", spuid, argp );
      vec_ullong2 A = (vec_ullong2){0,13};
      int cs = CS(A,124,4);
95    printf("\n");
      return (0);
    }
    #endif

100 int main( unsigned long long spuid,
              unsigned long long argp )
    {
      int rc;
      // 995 =1111100011
105   printf(" Hello, World! from %d with %d \n", spuid, argp );
      //vec_ullong2 A = (vec_ullong2){0,4}; // 21 for 10101
      //int cs = CS(A,125,3);
      vec_ullong2 A = (vec_ullong2){0,995}; // 21 for 10101
      int cs = CS(A,118,10);
110   printf("\n");
      return (0);
    }
```

Listing 10.4 Computing Merit factors of Barker sequences on the SPU, barker_spu.c.

Data: Unsigned integer n
Result: $n!$

1 **int** `factorial`(int n) **begin**
2   **if** $n < 2$ **then**
3   return 1;
4   **end**
5   return n * `factorial`($n-1$);
6 **end**

Algorithm 3: Recursive definition of factorial algorithm, is not tail recursive.

Data: Unsigned integer $n, acc = 1$
Result: $n!$
1 **int** factorial(int n, int $acc = 1$) **begin**
2   **if** $n < 2$ **then**
3   return acc;
4   **end**
5   return factorial($n - 1, n * acc$);
6 **end**

Algorithm 4: Tail recursive definition of factorial algorithm.

When we formulate the tail recursive algorithm to return a pre-computed result at the end of the recursion, we can also use this method to compute values across a vector using SIMD. The algorithm presented in Algorithm 4 when applied to a floating point vector of $\{4.0, 5.0, 6.0, 7.0\}$ with the appropriate function calls (using SPU intrinsics) for multiplication and comparison with 1.0, are inserted, correctly returns a vector of $\{24.0, 120.0, 720.0, 5040.0\}$. As you can see I have substituted floating-point arithmetic for factorial as we know integer multiplies are expensive on the SPE, and also since we use factorial values when expanding polynomial or transcendental series, where we are using the return value in a floating point context. For example consider the following fictional function:

$$f(x,a,b,n) = \sum_{i=0}^{n} f(x-a)^{n/b} \frac{\binom{2n}{3}}{n!} \tag{10.1}$$

The evaluation of the combinatoric quantity in this context does not use integer division, so the conversion of factorial to floating point has to be coerced.

```
static const vector float CONST_ONE = {1.0,1.0,1.0,1.0};
static const vector unsigned int CONST_ONE_INT = {1,1,1,1};
static vector float factorial_accumulate( vector float N,
                                          vector float acc ) {
  vector unsigned int GT_2 = (vector unsigned int)spu_cmpgt( N, CONST_ONE );
  vector unsigned int is_zero = spu_gather( GT_2 );
  if( spu_extract( is_zero, 0) == 0 ) return acc;
  else {
    vector unsigned int dm = spu_and( GT_2, CONST_ONE_INT );
    vector float dmf = spu_convtf( dm, 0 );
    vector float next_n = spu_sub( N, dmf );
    acc = spu_mul( N, acc );
    return factorial_accumulate( next_n, acc ); }
  return N;
}
#define MAX_COUNT 32
int main( unsigned long long spuid __attribute__ ((unused)),
          unsigned long long argp ) {
  vector float CONST_ZERO_VECTOR =
    (vector float) spu_splats( 0.0 );
  vector float CONST_FOUR = {4,4,4,4};
  vector float A = { 1.0, 2.0, 3.0, 4.0 };
  vector float B = A,C;
  int i=0;
  for(i=0; i < MAX_COUNT/4; ++i) {
    C = factorial_accumulate( A, CONST_ONE );
    print_vector( &C );
    A = spu_add( A, CONST_FOUR );
```

```
    printf("\n"); }
  return 0;
}
```

Listing 10.5 SIMD Factorial

The output of this program is shown below:

```
VF = [1 2 6 24]
VF = [120 720 5040 40320]
VF = [362880 3.6288e+06 3.99168e+07 4.79002e+08]
VF = [6.22702e+09 8.71783e+10 1.30767e+12 2.09228e+13]
VF = [3.55687e+14 6.40237e+15 1.21645e+17 2.4329e+18]
VF = [5.10909e+19 1.124e+21 2.5852e+22 6.20448e+23]
VF = [1.55112e+25 4.03291e+26 1.08889e+28 3.04888e+29]
VF = [8.84176e+30 2.65253e+32 8.22283e+33 2.63131e+35]
```

Listing 10.6 SIMD factorial computation

Data: Unsigned integer n
Result: $n!$
1 **int** factorial(int n) **begin**
2   int facn=1;
3   **for** $n; n > 1; --n$ **do**
4   facn *= i;
5   **end**
6   return facn;
7 **end**

Algorithm 5: Iterative algorithm for calculating factorial.

10.3 Computing Greatest Common Divisor (GCD)

For every non-negative numbers x and positive y we know: $gcd(x,y) = gcd(y, x\%y)$ Using the Multiprecision Library `libmpm.h` we can code this as:

```
/////////////////////////////////////////
// Program: GCD
// Author : Sandeep Koranne
// Uses Multi-precision library
/////////////////////////////////////////
#include <stdio.h>
#include <spu_intrinsics.h>
#include <libmpm.h>

void GCD(vector unsigned int* A,
   vector unsigned int* B,
   vector unsigned int *C) {
  const vector unsigned int ZERO=(vec_uint4){0,0,0,0};
  if(mpm_cmpeq(A,&ZERO,4))
    *C = *A;
  else {
    vector unsigned int temp;
    vector unsigned int temp2;
    mpm_fixed_mod_reduction(&temp,B,A,&temp2,4);
```

```
20        GCD(B, &temp,C);
      }
    }
```

Listing 10.7 Computing GCD using MPM Library.

The MPM provides GCD as one of the functions `mpm_gcd` so the above example should be seen as an example of MPM functionality.

```
  Data: Unsigned integer x, int y
  Result: gcd(x,y)
1 int gcd(unsigned int x, int y) begin
2     if y == 0 then
3         return x;
4     end
5     return gcd(y,x%y);
6 end
```

Algorithm 6: Recursive algorithm for calculating GCD.

```
  Data: Unsigned integer x, int y
  Result: extended_gcd(x,y)
1 int extended_gcd(unsigned int x, int y) begin
2     if (x%y) == 0 then
3         return [0, 1];
4     end
5     [x',y'] = extended_gcd(y, x%y);
6     return [y',x' - y' * (x/y)];
7 end
```

Algorithm 7: Recursive algorithm for calculating extended_gcd.

Extended GCD is used for large number reconstruction using prime modulo arithmetic.

```
  Data: Unsigned integer x, int y
  Result: gcd(x,y)
1 int gcd(unsigned int x, int y) begin
2     while true do
3         int y'=x%y;
4         if y' == 0 then
5             return x;
6         end
7         x = y,y = y';
8     end
9 end
```

Algorithm 8: Iterative algorithm for calculating GCD.

10.3.1 Suggested enhancements

1. Implement the Frame-Stewart algorithm for k-way Tower's of Hanoi and calculate number of moves w.r.t k,
2. Implement probabilistic primality checking algorithms,
3. Implement Chinese remainder based large number representation.

```
   Data: Unsigned integer x
   Result: 1 if x is prime, 0 otherwise
1  int IsPrime(unsigned int x) begin
2      if !(x&1) then
3          return 0;
4      end
5      int sqx = sqrt(x);
6      for i = 3; i < sqx; i+ = 2 do
7          if (x%i) == 0 then
8              return 0;
9          end
10     end
11     return 1;
12 end
```

Algorithm 9: Primality test for n.

10.4 Sorting Algorithms

Sorting is one of the most important operations for many commercial applications, especially database management systems. Hence many sequential and parallel sorting algorithms have been studied in the past [71]. Sanders and Winkel [107] noted that the performance of sorting is mostly dominated by pipeline stalls caused by branch mispredictions. Their proposed algorithm sss-sort (super-scalar sample sort) [107] eliminates conditional branches. However popular sorting algorithms, such as quicksort, are not suitable for exploiting SIMD instructions [5]. Previous sorting algorithms for the Cell have been proposed in [52, 62].

We have implemented a simple merge-sort algorithm which does out-of-core merges using a vector shuffle SIMD instruction. The in-core sort (upto 16K integers) is done using a variant of insertion sort and and non-recursive inplace sorting. The general sorting algorithm first divides all of the data into blocks that fit in the local store of the SPE. Then it sorts each block with quick-sort sorting algorithm. Finally it merges the sorted blocks with SIMD shuffle moves in a Batcher style network, implemented using SPU intrinsics.

The vector compare instruction reads from two input registers and writes to one output register. It compares each value in the first input register to the corresponding

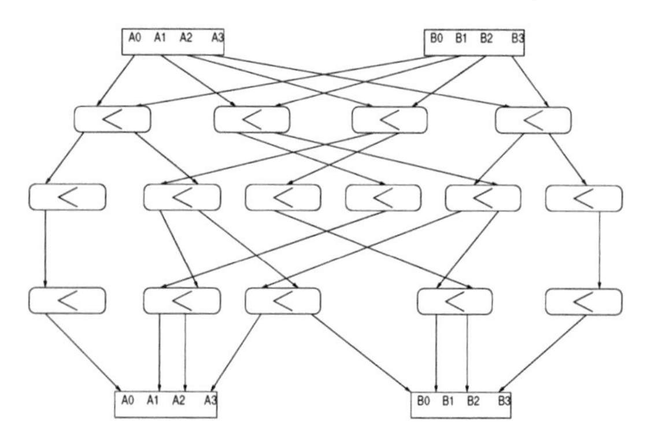

Fig. 10.3 Sorting network implemented using SPU intrinsics (from [62]).

value in the second input register and returns the result of comparisons as a mask in the output register. The SPU intrinsic is `spu_cmpgt`.

The vector select instruction takes three registers as the inputs and one for the output. It selects a value for each bit from the first or second input registers by using the contents of the third input register as a mask for the selection. The SPU intrinsic is `spu_selb`.

The vector permutation instruction also takes three registers as the inputs and one for the output. The instruction can reorder the single-byte values of the input arbitrarily. The first two registers are treated as an array of 32 single-byte values, and the third register is used as an array of indexes to pick 16 arbitrary bytes from the input register. The SPU intrinsic is `spu_shuffleb`. The complete sorting network is shown in Figure 10.3.

10.4.1 Implementation of SIMD Sort

Our implementation uses `spu_shuffle` and `spu_rlqwbyte`. These functions are heavily used in SIMD optimization and it is important to understand what they do.

10.4.2 The `spu_shuffle` function

The SPU intrinsic function `spu_shuffle` takes 3 arguments, `vector A`, `vector B` and the shuffle pattern `vector P`. If the value of the byte in the shuffle pattern is one of `10xxxxxx`, `110xxxxx` or `111xxxxx`, the predefined values of 0x00, 0xFF, and 0x80, respectively, are returned from the `spu_shuffle` function. Otherwise, the value in pattern is interpreted as a 5 bit value indexing into a byte array which contains a total of 32 bytes. The first 16 bytes of this indexed array are filled with values from vector A, and the remaining 16 bytes come from B. Thus is the pattern is

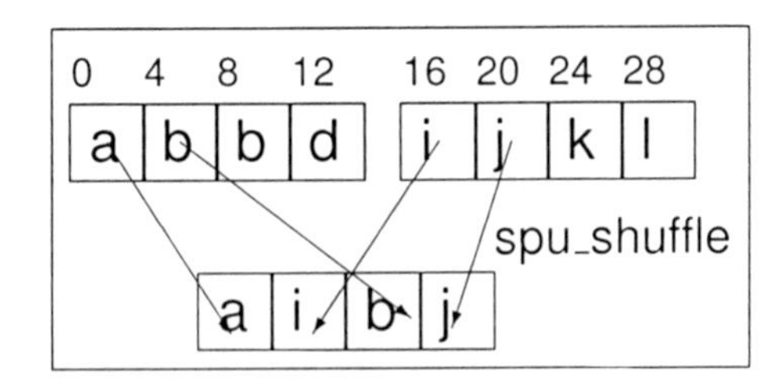

Fig. 10.4 SPU shuffle instruction example.

`0,1,2,3,4,5,6,7,8,9,10,11,12,14,15`, then calling `spu_shuffle(A,B,pattern)` will return A. To get every alternate byte from A interleaved with B, use the following pattern `0,16,1,17,2,18,3,19,4,20,5,21,6,22,7,23`. To do a matrix transpose of a 4x4 matrix we can use the pattern as follows; consider the matrix shown in Eqn. 10.2. Assume this matrix is a `vector unsigned int` starting at memory location T.

$$M = \begin{bmatrix} a & b & c & d \\ e & f & g & h \\ i & j & k & l \\ m & n & o & p \end{bmatrix} \tag{10.2}$$

We know $T = [abcd]$ and $T + 1 = [efgh]$ and so on. The transpose of M, written as M' is defined as:

$$M' = \begin{bmatrix} a & e & i & m \\ b & f & j & n \\ c & g & k & o \\ d & h & l & p \end{bmatrix} \tag{10.3}$$

Consider the first two rows of the matrices given in Eqn. 10.2 and Eqn. 10.3, the byte patterns can be calculated by observing, that in M, $T = [abcd, efgh]$, while in M' we have $T' = [aeim, bfjn]$. We construct the transpose in two parts by first computing intermediate values of $[aibj, ckdl, emfn, gohp]$. These can be easily computed using the following diagram shown in Figure 10.4.

To select $M7 = aibj$ from the combined vector we need to select bytes 0-3,16-19,4-7,20-23 from the combined byte arrays of $M1 = abcd$ and $M3 = ijkl$. Similarly construct temporaries $M9$, $M10$ and $M11$ as shown in Listing 10.8.

```
const vec_uchar16 SHA={0,1,2,3,16,17,18,19,
                       4,5,6,7,20,21,22,23};
const vec_uchar16 SHB={8,9,10,11,24,25,26,27,
                       12,13,14,15,28,29,30,31};

static inline
void TransposeMatrix(vector unsigned int *O,
                     vector unsigned int *I)
{
  vector unsigned int M1,M2,M3,M4,M5,M6,M7,M8,M9,M10,M11,M12;
  M1 = *(I+0);M3 = *(I+2); M7 = spu_shuffle(M1, M3, SHA);
  M8 = spu_shuffle(M1, M3, SHB); M4 = *(I+3);M2 = *(I+1);
  M9 = spu_shuffle(M2, M4, SHA);
  M5 = spu_shuffle(M7, M9, SHA);*(O+0) = M5;
```

```
15    M6 = spu_shuffle(M7, M9, SHB);*(O+1) = M6;
      M10 = spu_shuffle(M2, M4, SHB);
      M11 = spu_shuffle(M8, M10, SHA);*(O+2) = M11;
      M12 = spu_shuffle(M8, M10, SHB);*(O+3) = M12;
    }
```

Listing 10.8 SPU Implementation of 4x4 Matrix Transpose.

We use a nifty technique to sort individual vectors in non-decreasing order. We calculate the transpose of a 4-vector block, sort the 4 vectors using SIMD Bubblesort, and then retranspose them. The listing is shown in Listing 10.9. Consider 4 vectors arranged as follows:

$$A = \begin{bmatrix} 172 & 47 & 117 & 192 \\ 67 & 251 & 195 & 103 \\ 9 & 211 & 21 & 242 \\ 36 & 87 & 70 & 216 \end{bmatrix}$$

We run a pass of a sort on the vector A using `spu_cmpgt` based `MinMax` as the comparison and swap function. At the end of this sort the minimum vector is at location $A[0]$. This can be seen below:

$$A = \begin{bmatrix} 21 & 36 & 9 & 47 \\ 103 & 67 & 87 & 70 \\ 195 & 192 & 117 & 172 \\ 211 & 251 & 216 & 242 \end{bmatrix}$$

Now we perform a transpose of A, to get:

$$A = \begin{bmatrix} 21 & 103 & 195 & 211 \\ 36 & 67 & 192 & 251 \\ 9 & 87 & 117 & 216 \\ 47 & 70 & 172 & 242 \end{bmatrix}$$

Followed by a 4-element Bubblesort, to get:

$$A = \begin{bmatrix} 9 & 67 & 117 & 211 \\ 21 & 70 & 172 & 216 \\ 36 & 87 & 192 & 242 \\ 47 & 103 & 195 & 251 \end{bmatrix}$$

Followed by a retranspose to A, yielding the final sorted vector:

$$A = \begin{bmatrix} 9 & 21 & 36 & 47 \\ 67 & 70 & 87 & 103 \\ 117 & 172 & 192 & 195 \\ 211 & 216 & 242 & 251 \end{bmatrix}$$

```
static void BubbleSort( vector unsigned int *A ) {
```

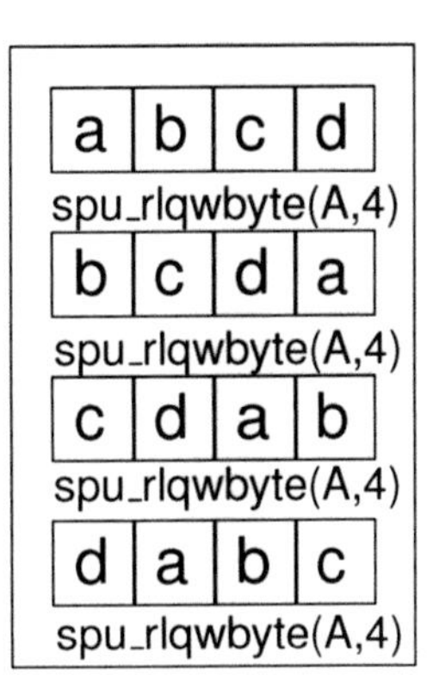

Fig. 10.5 SPU rlqwbyte example.

```
    vector unsigned int C0 = spu_cmpgt(*A,*(A+1));
    vector unsigned int T=spu_sel(*A,*(A+1),C0);
    *(A+1)=spu_sel(*(A+1),*A,C0);   *A = T;
5   C0=spu_cmpgt(*A,*(A+2)); T=spu_sel(*A,*(A+2),C0);
    *(A+2)=spu_sel(*(A+2),*A,C0);   *A =T;
    C0=spu_cmpgt(*A,*(A+3)); T=spu_sel(*A,*(A+3),C0);
    *(A+3)=spu_sel(*(A+3),*A,C0); *A  = T;
    C0=spu_cmpgt(*(A+1),*(A+2)); T=spu_sel(*(A+1),*(A+2),C0);
10  *(A+2)=spu_sel(*(A+2),*(A+1),C0); *(A+1)=T;
    C0=spu_cmpgt(*(A+1), *(A+3)); T=spu_sel(*(A+1),*(A+3),C0);
    *(A+3)=spu_sel(*(A+3),*(A+1),C0); *(A+1)=T;
    C0=spu_cmpgt(*(A+2),*(A+3));  T=spu_sel(*(A+2),*(A+3),C0);
    *(A+3)=spu_sel(*(A+3),*(A+2),C0); *(A+2)=T;
15 }
```

Listing 10.9 Ordering a SIMD vector using Transpose-Bubble-sort-Transpose

10.4.3 The `spu_rlqwbyte` function

Consider the example shown in Figure 10.5.

```
   static void MinMaxSmall( vector unsigned int *A,
                            vector unsigned int *B,
                            vector unsigned int *MAX,
                            vector unsigned int *MIN) {
5    vector unsigned int C0 = spu_cmpgt( *A, *B );
     vector unsigned int T = spu_sel( *B, *A, C0 );
     *MIN = spu_sel( *A, *B, C0 );
     *MAX = T;
   }
10
   static inline void
   MinMax( vector unsigned int *A,
           vector unsigned int *B,
           vector unsigned int *MAX,
15         vector unsigned int *MIN) {
     vector unsigned int vmin,vmax;
     MinMaxSmall( A, B, &vmax, &vmin );
     vmin = spu_rlqwbyte( vmin, 4 );
```

```
    MinMaxSmall( &vmin, &vmax, &vmax, &vmin );
20  vmin = spu_rlqwbyte( vmin, 4 );
    MinMaxSmall( &vmin, &vmax, &vmax, &vmin );
    vmin = spu_rlqwbyte( vmin, 4 );
    MinMaxSmall( &vmin, &vmax, &vmax, &vmin );
    *MAX = vmax; *MIN = vmin;
25 }
```

Listing 10.10 SPU Implementation of Min-Max using SIMD.

The SPU program shown in Listing 10.11 shows our implementation of SIMD sort.

```
static void
SortIndividualVector(vector unsigned int* DATA,int n ) {
  int i;
  for(i=0;i<n;i+=4) {
5   TransposeInPlace( &DATA[i] );
    BubbleSort( &DATA_LS[i] );
    TransposeInPlace( &DATA[i] ); }
}
static void Sort(vector unsigned int *T,int n ) {
10 int i,j;
  for(i=0;i<n-1;++i)
    for(j=i+1;j<n;++j)
      MinMax( &DATA_LS[i], &DATA_LS[j], &DATA_LS[j], &DATA_LS[i] );
  ShowData();
15 SortIndividualVector( T, n );
}
```

Listing 10.11 4 iteration Bubble-sort.

```
Hello, World! from 1 with 262112
[172 47 117 192] [67 251 195 103] [9 211 21 242] [36 87 70 216]
[21 36 9 47] [103 67 87 70] [195 192 117 172] [211 251 216 242]
[9 21 36 47] [67 70 87 103] [117 172 192 195] [211 216 242 251]
```

Listing 10.12 Transcript of running SIMD sort on 16 random integers.

Data: Array A
Result: Randomly permute A *in-place*
1 **int** `RandomlyPermute`(Array A) **begin**
2   int n = `Length`(A);
3   **for** $i = 0; i < n; ++i$ **do**
4   `Swap`($A[i], A[random(i,n)]$);
5   **end**
6 **end**

Algorithm 10: In-place permutation of an array

I am not going to write out the algorithms for finding the minimum, maximum elements for a given sequence as they are simple to implement. Both are $O(n)$ algorithms. Median estimation, k-th median estimation and other statistical functions on sequences are presented later in this book.

10.5 Table lookup

A common use of sorting is to arrange data in a table or search structure for $O(\lg n)$ lookup. Such table lookup is common in high performance computing for atmospheric sciences as many physical quantities (like the variation in the speed of sound with altitude) are non-trivial functions which are best represented as table lookup followed by interpolation. If the data is not uniformly placed then a search needs to be performed in the ordered table. The problem can be define as:

Given an ordered table of $x_0, x_1, \ldots, x_n$ data points find a integer k such that $x_k \leq y \leq x_{k+1}$, given a scalar value y comparable to x_i. When the scalar y is replaced by a vector of values, the the result is a vector k which meets the search criterion.

```
vec_uint4 SearchKeyHeader( vec_uint4 key ) {
  // p,q is given
  vec_uint4 CH[P];
  vec_uint4 header_key, next_key;
  vec_uint4 header_offset=(vec_uint4)spu_splats(N/P);
  int p;
  unsigned int x = 0;
  print_vector("SearchKey",&key);
  printf("\n Headers = \n");
  for(p=0;p < P; ++p) {
    print_vector("",&DATA_LS[x]);
    printf("\n");
    x += (N/P);
  }
  x=0;
  header_key = (vec_uint4)spu_splats(0);
  next_key = spu_add(header_key,header_offset);
  for(p=0;p < P; ++p) {
    CH[p] = spu_cmpgt( key, DATA_LS[x] );
    header_key = spu_sel( header_key, next_key, CH[p] );
    x += (N/P);
    next_key = spu_add(next_key, header_offset);
  }
  for(p=0;p < P; ++p) {
    print_vector("\n",&CH[p]);
  }
  printf("\n");
  header_key = spu_sub(header_key,header_offset);
  print_vector("HeaderKey",&header_key);
  printf("\n");
  return header_key;
}
```

Listing 10.13 Table lookup of 4 keys using SIMD.

```
Searching for key
[234 2567 1890 1024]
Headers =
[0 0 3 3]
[1102 1104 1107 1107]
[2083 2084 2086 2086]
[3091 3092 3094 3096]
KeyHeaders =
[1 1 1 1] [0 1 1 0] [0 1 0 0] [0 0 0 0]
KeyOffset =
[0 256 128 0]
```

Listing 10.14 Transcript of Table-lookup.

10.6 (Perfect) Hash Function

For simplicity we consider the case where key-value pairs are integers. A perfect hash function, is a function that maps integers from the set of keys into a table of values, one-to-one, without collisions, such that a table look can be performed in $O(1)$ time. Note that the set of keys can comprise large values with gaps, such as the set $\{1, 10, 100, 321, 1013\}$. Such functions are very useful in implementing fast table lookups. Obviously, a perfect hash function is key dependent, and therefore only makes sense when the set of keys is known in advance (such as the set of keywords in a language implementation for lexical analysis, where perfect functions are heavily used). We present a parallel search algorithm for finding a perfect hash function given the key-value pair. The meta-function we have considered is of the form:

$$H(k) = (k+s) \div M$$

where s is the step, and M is the table size. In a sequential algorithm we would check the search space of $s \times M$ (using the set k for guidance) and calculate the smallest table size which results in a collision free hash function. This can be done by the following (simple) algorithm:

```
static int FindPerfectHash( MBC* cb ) {
  int i;
  int N=10;
  int s=0,M=N-1,tableloc=0;
  int success;
  #if 1
  N=4;
  DATA_LS[0]=1,DATA_LS[1]=4,DATA_LS[2]=9,DATA_LS[3]=16;
  #else
  for(i=0;i<N;++i) DATA_LS[i] = ( rand() % 100 );
  #endif
  PrintData(N);
  for(s;s<N;++s) {
    for(M=2; M<(N*2);++M) {
      for(i=0;i<N;++i) UNIVERSE[i] = -1;
      success=1;
      for(i=0;i<N;++i) {
        tableloc = (DATA_LS[i]+s) / M;
        printf("\n\t[%d,%d]\t Trying (%3d + %d) div %d = %d",
                s,M, DATA_LS[i], s, M,tableloc);
        if( (UNIVERSE[tableloc] != -1) && (DATA_LS[i] != UNIVERSE[tableloc] )) {
          printf("\n COLLISION with %d", UNIVERSE[tableloc] );
          success=0;
          break;
        }
        else {
          UNIVERSE[tableloc] = DATA_LS[i];
        }
      }
      if( success ) {
        printf("\n***** SUCCESS with s=%d M=%d **** ",s,M);
        return 0;
      }
    }
  }
}
```

Listing 10.15 PerfectHashFunction search on SPU.

```
Data: Set k of keys
Result: Smallest s,M table size for perfect hash function
1  int,int FPH(Set k) begin
2      int i,s,M,success;
3      int n = |k|;
4      int C[n];
5      for s = 0; s < n; ++s do
6          for M = 2; M < (n*2); ++M do
7              for i = 0; i < n; ++i do
8                  C[i] = -1;
9              end
10             success=1;
11             for i = 0; i < n; ++i do
12                 int index = k[i] + s ÷ M;
13                 if C[index] ≠ -1 && C[index] ≠ k[i] then
14                     success=0;
15                     break;
16                 end
17                 else
18                     C[index] = k[i];
19                 end
20             end
21             if success then
22                 return (s,M);
23             end
24         end
25     end
26     return φ;
27 end
```

Algorithm 11: Searching s,M for a perfect hash function on k,v.

A search on the keys $\{1,4,9,16\}$ is shown in the transcript which find a hash function with $s = 1, M = 5$.

```
Running on Cell
sizeof(MBC) = 16
Hello from SPE (0x1001c120)
k=[ 1 4 9 16 ]
        [s,M]

  [0,2]  Trying (  1 + 0) div 2 = 0
  [0,2]  Trying (  4 + 0) div 2 = 2
  [0,2]  Trying (  9 + 0) div 2 = 4
 COLLISION with 0
  [0,3]  Trying (  1 + 0) div 3 = 0
  [0,3]  Trying (  4 + 0) div 3 = 1
  [0,3]  Trying (  9 + 0) div 3 = 3
  [0,3]  Trying ( 16 + 0) div 3 = 5
 COLLISION with 0
  [0,4]  Trying (  1 + 0) div 4 = 0
  [0,4]  Trying (  4 + 0) div 4 = 1
```

```
     [0,4]  Trying (  9 + 0) div 4 = 2
     [0,4]  Trying ( 16 + 0) div 4 = 4
20  COLLISION with 0
     [0,5]  Trying (  1 + 0) div 5 = 0
     [0,5]  Trying (  4 + 0) div 5 = 0
   .....
     [1,4]  Trying ( 16 + 1) div 4 = 4
25  COLLISION with 0
     [1,5]  Trying (  1 + 1) div 5 = 0
     [1,5]  Trying (  4 + 1) div 5 = 1
     [1,5]  Trying (  9 + 1) div 5 = 2
     [1,5]  Trying ( 16 + 1) div 5 = 3
30  ***** SUCCESS with s=1 M=5 ****
```

Listing 10.16 Transcript of perfect-hash computation on SPE.

Of-course, the hash function provided above is not *minimal*, the table size needed to store the keys can be arbitrary. In the next section we provide a 2-step perfect hash function which produces minimal hashes with very little overhead.

10.7 Minimal Perfect Hashing

The linear hash function we have seen above suffers from a sub-optimality in that it can produce larger tables than necessary. In this section we shall implement a probabilistic algorithm which can map integers in the range $1 \ldots M$ into another table of size k much smaller size than M. This algorithm is based on the paper by Czech et. al. [35]. The algorithm generates minimal perfect hash functions of the form:

$$H(x) = (g(f_1(x)) + g(f_2(x))) \bmod k$$

We have chosen f_1 as $f_1(x) = (3x+5) \bmod 13$ and $f_2(x) = (7x+1) \bmod 13$, with a table size $k = 13$. We present a simple example before describing the implementation on the SPU. Consider the set of data given to the algorithm as $\{3, 4, 7, 11, 12, 16, 18, 21\}$. We construct two tables T_1 and T_2 of $f_1(x)$ and $f_2(x)$ respectively:

```
Data values     = 3   4   7   11 12   16   18 21
-----------------------------------------------
T1  =  (3x+5)%13 = 1   4   0   12   2    1    7  3
T2  =  (7x+1)%13 = 9   3  11    0   7   12   10  5
```

Now consider a graph of the 13 nodes labelled $0 \ldots 12$ as shown in Figure 10.6.

Now order the edges of the graph in random order and label them $1 \ldots m$, this is shown in Figure 10.7. The roots of connected components are colored blue (nodes 1,3 and 7).

This graph should be acyclic; if the graph has a cycle then we need to find a different set of f_1 and f_2, and/or increase the table size (thus increasing the mod from 13 to some larger number), until the graph has no cycles. Now define a g label for all nodes of the graph inductively as follows: w.l.o.g label a starting vertex 0,

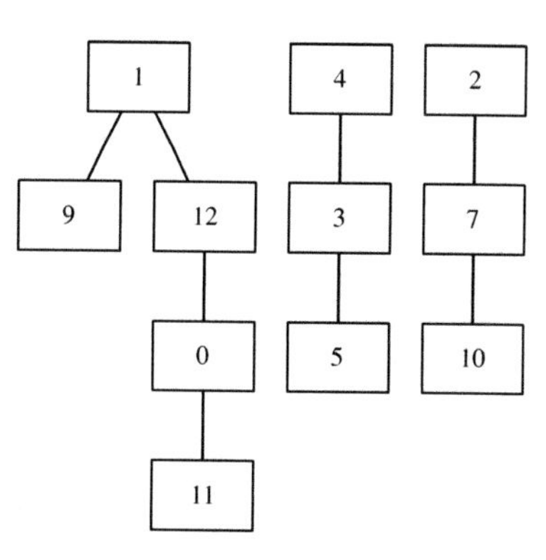

Fig. 10.6 Graph of nodes from T_1 and T_2

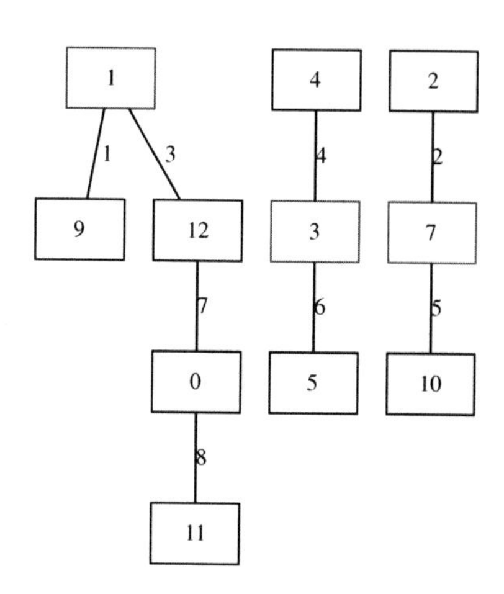

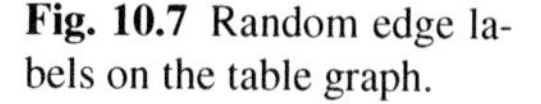

Fig. 10.7 Random edge labels on the table graph.

$g(9) = 0$, define $g(v) : u \rightarrow v$ as $l(e) - g(u)$, continuing from node 9, we go to node 1, and thus $g(1) = l(1) - g(9)$, thus $g(1) = 1$, $g(12) = 2$, $g(0) = 5$ and $g(11) = 3$. Then one connected component of the graph is complete. We repeat this process for the next component, but we start of the g label from 0, this gives us $g(7) = 0$, $g(2) = 2$, and $g(10) = 5$. And finally, $g(3) = 0$, $g(5) = 6$, and $g(4) = 4$. The g table is shown below:

```
label  = 9 1 12 0 11 3 5 4 7 2 10
g(l)   = 0 1  2 5  3 0 6 4 0 2  5
```

The table of g values (of 13 entries) needs to stored alongside the hash table. Now we can calculate the perfect hash for each of the given data: consider $H(3)$:

$$H(3) = (g(f_1(3)) + g(f_2(3))) \bmod 13 = (g(1) + g(9)) = 0 + 1 = 1$$

$$H(4) = (g(f_1(4)) + g(f_2(4))) \bmod 13 = (g(4) + g(3)) = 4 + 0 = 4$$

$$H(7) = (g(f_1(7)) + g(f_2(7))) \bmod 13 = (g(0) + g(11)) = 5 + 3 = 8$$

The complete list of $H(n)$ is given below, this is the perfect hash table:

```
D    = 3 4 7 11 12 16 18 21
H()  = 1 4 0  7  2  3  5  6
```

We can see that the hash function is correct, as $H(16)$ is

$$\begin{aligned} H(16) &= (g(f_1(16)) + g_2(f_2(16))) \bmod 13 & (10.4) \\ &= (g(1) + g(12)) \bmod 13 & (10.5) \\ &= (1 + 2) \bmod 13 = 3 & (10.6) \end{aligned}$$

Using this algorithm we have mapped integers in the range $1 \ldots 21$ into a table of 13 entries (although we have to also keep around the g table and perform two lookups in that for every hash).

10.8 Implementing minimal perfect hash in the SPU

The basic idea is to send the value data array to multiple SPUs in parallel and let them compete to find the smallest mod number (13 in our example) such that there is no cycle in the graph generated from the functions f_1 and f_2. Each SPU is free to choose any table function as long as it is of the form $ax + b$. We use the UNION-FIND data-structure (cf. 10.11) to find cycles, and if so abort the current search and restart with a different function, or a larger table size (both methods are tried alternating). At the end of each search iteration, the SPE send a Mailbox message to the PPE to make sure no other SPE has won the competition; if so, then the losing SPE can abort the search, if not, the competition goes on. Winners are decided on a first come first won basis, table size comparison can also be done.

The input to the SPE is the number of original data, and their address as effective address (EA) using a control block. The SPUs DMA the original data, this is constant during the search, thus this problem has low communication overhead, and effectively uses parallel search in the 3-dimensional search of f_1, f_2, and mod T table size. The function is communicated as the coefficient a of $ax + b$ and the constant term. The size of the table of g labels and its contents are also returned by the SPU which wins.

10.9 Using perfect hash functions with large graph clusters

One of the potential use of perfect hash algorithms is in encoding graph vertices to reduce the size of the node table in the local store of the SPE in solving graph problems. In the sequel of this book we will assume that *all* graph (or graph clusters) in the SPE are consecutively labelled from $0 \ldots n$, and there is a perfect hash which maps this re-labelled vertices to the global vertex numbers which are valid outside of the local store. Each edge has a bit which is set when one of the node numbers of the edge is not-local, and then it is not looked up, but an approximation using terminal propagation is used.

To repeat, all nodes in the local store graph are of two distinct types, either local or foreign, identified by a bit in the edge. Furthermore, all local node numbers are compressed using a perfect hash. Using terminal propagation, we create extra local nodes and replace all foreign nodes in the graph cluster by an equivalent terminal node. This is used in graph partitioning, and also in graph based placement in the chapter on VLSI placement.

10.10 Heap Data Structure

The following sequence implements a max-heap-order:

```
A[] = {8,7,5,4,3,4,1,1,2,1,1}
```

The root element in the max-heap must be greater than both of its children. A common method of implementing heaps is using an array with the following convention implicit in the indexing: `get_lhs(r) = A(r<<1)`,`get_rhs(r)=A(r<<1 +1)`, and `parent(r)=(r>>1)`. This indexing scheme is only valid when the tree is a complete binary tree.

```
static inline int HeapLength(void) { return N; }
static inline int HeapSize(void) { return gHeapSize; }
static inline unsigned int Parent(unsigned int i) { return (i >> 1); }
static inline unsigned int Left(unsigned int i) { return (i << 1); }
static inline unsigned int Right(unsigned int i) { return ((i << 1)+1);}
static inline void InitUniverse(void) { gHeapSize=0;}
static inline void Swap( unsigned int a, unsigned int b ) {
    int temp = UNIVERSE[ a ];UNIVERSE[ a ] = UNIVERSE[ b ];
    UNIVERSE[ b ] = temp;
}
#define MIN_HEAP
#ifdef MIN_HEAP
#define CMP_FUNC <
#define CMP_FUNC_INV >
#define HVALUE INT_MAX
#else
#define CMP_FUNC >
#define CMP_FUNC_INV <
#define HVALUE INT_MIN
#endif
static void MaxHeapify( unsigned int i ) {
    unsigned int L = Left(i);
    unsigned int R = Right(i);
```

```
        unsigned int largest =0;
25      largest=((L<=HeapSize())&&(UNIVERSE[L] CMP_FUNC UNIVERSE[i]))?L:i;
        if( (R<=HeapSize())&&(UNIVERSE[R] CMP_FUNC UNIVERSE[largest]))
          largest = R;
        if( largest != i ) { Swap(i,largest); MaxHeapify( largest );}
    }
30  static void BuildMaxHeap(void) {
        int i;
        gHeapSize = HeapLength();
        for(i=N/2; i >= 1; --i )  MaxHeapify(i);
    }
35  static void HeapSort(void) {
        int i;
        BuildMaxHeap();
        for(i=N;i>=2;--i) {
            Swap(1, i);gHeapSize--;MaxHeapify(1);
40      }
    }
    static inline int HeapMaximum(void) {return UNIVERSE[1];}
    static int ExtractMax(void) {
        int max_v = HeapMaximum();
45      UNIVERSE[1] = UNIVERSE[ gHeapSize ];
        gHeapSize--;  MaxHeapify(1);
        return max_v;
    }
    static void HeapIncreaseKey(unsigned int i, int key) {
50      UNIVERSE[i] = key;
        while( (i>1) && (UNIVERSE[ Parent(i) ] CMP_FUNC_INV UNIVERSE[i]) ) {
            Swap(i, Parent(i) ); // sift-up
            i = Parent(i);
        }
55  }
    static void MaxHeapInsert( int key ) {
        gHeapSize++;
        UNIVERSE[ gHeapSize ] = HVALUE;
        HeapIncreaseKey(gHeapSize, key);
60  }
```

Listing 10.17 Heap implementation in SPU

10.11 Disjoint Set Union-Find Algorithms

A disjoint set data structure supports the following operations. Given set S and elements x and y,

1. FINDSET(X): returns the set x is a member of, or x if it is a singleton,
2. UNION(X,Y): combine the sets of x and y to be the same set,
3. CARD(): returns cardinality of universe in number of sets,

The union-find data structure can be used in many graph algorithms. As a simple example consider the connected-components algorithms to find the connected components of an undirected graph. More detailed examples are given in the next chapter on Graph algorithms on the SPE.

Data: Undirected graph G
Result: Number of connected components
1 **int** `ConnectedComponents`(Graph G) **begin**
2 Set $S(|G|)$;
3 **for** *each vertex* $v \in G$ **do**
4 S.UNION(v, v)
5 **end**
6 **for** *each edge* $e \in G$ **do**
7 S.UNION(E.U,E.V)
8 **end**
9 return S.Card();
10 **end**

Algorithm 12: Connected components of an undirected graph.

```
// Disjoint Set-Union
static void InitializeUniverse( GraphControlBlock cb ) {
  int i;
  for(i=0; i < cb.N; ++i ) UNIVERSE[ i ] = -1;
}
static void ClearUniverse( GraphControlBlock cb ) {
  int i;
  for(i=0; i < cb.N; ++i ) UNIVERSE[ i ] = 0;
}
static int FindSet( int n ) {
  do {
    if( UNIVERSE[ n ] < 0 ) return n;
    else n = UNIVERSE[ n ];
  } while( 1 );
}
#define SAME_SET(i) ((FindSet(N1E(i))) == (FindSet(N2E(i))))
static int Union( int i, int j ) {
  int root_i = FindSet( i );
  int root_j = FindSet( j );
  if( root_i == root_j ) return;
  if( UNIVERSE[ root_i ] < UNIVERSE[ root_j ] )
    UNIVERSE[ root_j ] = root_i;
  else
    UNIVERSE[ root_i ] = root_j;
}
```

Listing 10.18 Disjoint set union-find implementation in SPU.

Listing 10.18 shows our implementation of disjoint set union-find using the algorithm discussed above, in the SPU. We use this data-structure in a number of programs in the sequel of this chapter and in the remainder of this book.

In the SPE Local Store we can implement the universe as an array A of N length, expecting the total number of entries not to exceed N for a single SPE. We initialize A:

```
vector int C_I = (vector int){-1,-1,-1,-1};
for(int i=0; i < N/4;++i) N[i] = C_I;
```

At the end of this initialization every element of A is -1. A negative value in the array implies that the element at that index is the root element of the set, and the absolute value of $A[i]$ (1 at initialization) is the depth of the set-tree formed at that index. To implement FINDSET(X) we implement:

```
int FindSet(int x) {
  if( A[x] < 0 )
        return x;
  else
        return FindSet( A[x] );
}
```

To implement the UNION(X,Y) function we do:

```
void Union(int x, int y) {
  int rx = FindSet(x);
  int ry = FindSet(y);
  if( rx == ry ) return; // they are in same set
  if( A[rx] < A[ry] ) // Note: A[rx] is negative for a root
    A[ry] = rx;
  else
    A[rx] = ry;
}
```

With these functions our disjoint set can perform union-find operations, but it is not optimal. Neither in its coding, which suffers from non-SIMD friendly indexing, and quadword misaligned loads of integer values, nor in its algorithmic behavior which does not implement *path compression* to link up intermediate values in the set-tree directly to its root in the FINDSET function. The latter is easy to fix. The implementation of the disjoint-union(3 on the SPE is given in Listing 11.3.

10.12 Computing Zech logarithm using SPE in batch mode

With the increasing importance of online commerce and financial transactions, cryptographic systems have become very important in current generation computer systems. Modern processors include hardware based cryptographic functions on chip to efficiently compute discrete mathematical functions. Most of these functions rely on properties of arithmetic on Galois fields. Decoding algorithms for error-correcting codes cyclic codes also use similar computations. For a standard representation of the Galois field elements addition is a simple modulo operation where as multiplication in the field is compute intensive. Multiplication is similar to the operations of inversion and division, and these too are costly in the standard representation. Recently logarithmic representation has also been investigated as a substitute for Galois field representation.

In alternative representation, the non-zero elements of $GF(2^m)$ are represented as the powers of a primitive element α , thus making multiplication trivial, at the cost of introducing additional complexity in performing addition. For any element $\gamma \in GF(2^m)$, the *discrete logarithm* of γ to the base α is the unique integer c with $0 \leq c \leq 2^m - 2$ satisfying $\alpha^c = \gamma$; equivalently $c = \log_\alpha \gamma$. Since $\alpha^x = 0$ will not have a solution for any x, we denote a special symbol ϕ for this purpose. Given the primitive element of a Galois Field $GF(2^m)$, we can represent any element $u \in$

$GF(2^m)$ by the exponent $z(u) \in N_q$ where the set N_q is the set of logarithms of the elements of $GF(2^m)$, plus an unique symbol denoting 0. Thus we have:

$$u = \alpha^{z(u)} \tag{10.7}$$

and $z(u)$ is called the logarithm. Additionally, if we define $Z : N_q \rightarrow N_q$ to satisfy:

$$\alpha^{Z(u)} = 1 + \alpha^u \ \forall u \in N_q \tag{10.8}$$

then $Z(u)$ is called the Zech logarithm. The key property of the Zech logarithm (in addition to the fact that it reduces multiplication to simple addition) is that addition of the field elements can be calculated without returning to the polynomial expansion (which is expensive). The catch however is that unlike simple addition, now we are required to calculate the Zech logarithm of any element in N_q before we can process it for addition:

$$\alpha^x + \alpha^y = \alpha^z \tag{10.9}$$

Using Zech logarithm we can write addition as:

$$\alpha^x + \alpha^y = \alpha^{x+Z(y-x)} = \alpha^{y+Z(x-y)} \tag{10.10}$$

Towards this end we describe an efficient method to calculate the Zech logarithm table given $GF(2^m)$ in the SPE. This can substitute for table lookup methods for small fields, and can also be used when batched logarithmic conversions are needed (in the case of constructing a lookup table). Our method to compute the Zech logarithm is based upon the original method proposed by K. Imamura [61]. Given a primitive element α in $GF(2^m)$ a field of order 2^m and characteristic 2, we know that all elements of $GF(2^m)$ can be written as: $0, 1, \alpha, \alpha^2, \alpha^3, \ldots, \alpha^{2m-2}$, since $\alpha^{2m-1} = 1$, and we denote a special symbol ϕ to denote $\alpha^\phi = 0$, so that we can express all elements of $GF(2^m)$ as α^x, even though without the special symbol this relation would not make sense for $\alpha^x = 0$.

10.12.1 How to calculate $Z(x)$

Given that α is a root of a primitive polynomial:

$$P(X) = X^n + P_{n-1}X^{n-1} + P_{n-2}X^{n-2} + \cdots + P_1X + P_0 \tag{10.11}$$

since $P(\alpha) = 0$, substituting in the above equation gives us

$$\alpha^n = -(P_{n-1}X^{n-1} + P_{n-2}X^{n-2} + \cdots + P_1X + P_0) \tag{10.12}$$

Thus we get

$$\alpha^x = x_0 + x_1\alpha + x_2\alpha^2 + \cdots + x_{n-1}\alpha^{n-1} \tag{10.13}$$

for $x = \phi, 0, 1, \ldots, 2^m - 2$. Next we define $N(x)$ as:

$$N(x) = x_0 + 2x_1 + 4x_2 + \cdots + 2^{n-1}x_{n-1} \tag{10.14}$$

Using the above equations we get:

$$N(Z(x)) = \begin{cases} N(x) - p + 1 & : \text{ if } N(x) \equiv p-1 (\bmod\ p) \\ N(x) + 1 & : \text{ if } N(x) \not\equiv p-1 (\bmod\ p) \end{cases} \tag{10.15}$$

As an example consider, $p = 2$, $m = 4$ and the primitive polynomial as $1 + X + X^4$,

$$P(X) = X^4 + X + 1 \tag{10.16}$$

solving for α^n gives us $\alpha^4 = \alpha + 1$. We apply this equality to calculate α^x for all $x \in N_q$. Since $p = 2$, Equation 10.15 reduces to:

$$N(Z(x)) = \begin{cases} N(x) - 1 & : \text{ if } N(x) \text{ odd} \\ N(x) + 1 & : \text{ if } N(x) \text{ even} \end{cases} \tag{10.17}$$

Combining the powers of α with the result of Equation 10.17 is sufficient to calculate $N(Z(x))$. This is shown below.

Table 10.3 Table of Zech logarithms for $p = 2, m = 4$ and primitive polynomial $1 + X + X^4$

x	α^x	$N(x)$	$N(Z(x))$	$Z(x)$
ϕ	0	0	1	0
0	1	1	0	ϕ
1	α	2	3	4
2	α^2	4	5	8
3	α^3	8	9	14
4	α^4	3	2	1
5	α^5	6	7	10
6	α^6	12	13	13
7	α^7	11	10	9
8	α^8	5	4	2
9	α^9	10	11	7
10	α^{10}	7	6	5
11	α^{11}	14	15	12
12	α^{12}	15	14	11
13	α^{13}	13	12	6
14	α^{14}	9	8	3

This table can be read as follows:

$$\alpha^{Z(2)} = \alpha^2 + 1 \quad (10.18)$$
$$\alpha^8 = \alpha^2 + 1 \quad (10.19)$$
$$N(8) = N(2) + 1 \quad (10.20)$$
$$5 = 4 + 1 \quad (10.21)$$

Data: p, m, Primitive polynomial
Result: Table of Zech logarithms in $GF(2^m)$ with that polynomial

```
1  void CalculateZechTable(m,n,Polynomial P) begin
2  |  for (x = φ,0,1,...,2^m − 2) do
3  |  |  Tabulate(N(x))
4  |  end
5  |  for (x = φ,0,1,...,2^m − 2) do
6  |  |  Tabulate(N(Z(x)))
7  |  end
8  |  for (x = φ,0,1,...,2^m − 2) do
9  |  |  ReverseLookup(x,N(Z(x)))
10 |  end
11 end
```

Algorithm 13: Imamura's method to calculate Zech logarithm given primitive polynomial.

10.13 Conclusions

In this chapter we have presented several algorithms and their implementation on the SPU. We implemented number theoretic functions to calculate the Collatz sequence, the Takeuchi benchmark, merit factor of binary sequences and the Barker sequence. We also presented tail-recursive and iterative algorithms for the factorial $n!$, and implemented it on the SPU. Using the multi-precision library (MPM) we implemented the GCD (greatest common divider) algorithm. A SIMD optimized version of sorting was discussed and its relation to matrix transpose was used to introduce the *spu_rlqwbyte* SPU intrinsic which is used in many of the programs in the later chapters. We presented algorithms and SPU code for Table lookup on sorted data, perfect hash functions, disjoint-union and Zech logarithms. We will use these algorithms in many of the projects later on in this book.

Chapter 11
Graph Theory on the CBEA

Abstract Graph theory is a cornerstone of computer science and recently has also become very useful in high-performance computing. In this chapter we present several graph theory data-structures and algorithms to solve basic graph problems. These algorithms are implemented on the SPU using 4-way SIMD in the general case. The problems considered include (a) min-cut, (b) minimum spanning tree, (c) comb-sort algorithm, (d) shortest-path algorithm, (e) Floyd-Warshall all-pairs shortest-path algorithm, (f) maximal independent set, (g) graph partitioning, (h) spectral methods for partitioning, and (i) on-line algorithms for streaming properties.

11.1 Introduction

In this chapter we discuss some basic concepts of graph theory, which is an important part of discrete mathematics and computer science. Many of the later examples in this book make extensive use of graph theoretical concepts, and the ability of compute graph properties or computations based on structures which are modeled as graphs. In particular the sections on polytopes, VLSI placement, channel-routing, micro-word scheduling, are presented and implemented as graph algorithms. Towards this end, an efficient graph solver in the SPE was key to our plan of solving many compute intensive problems on the Cell. But first some definitions of graph terms.

11.2 Definitions

A graph $G = (V, E)$ consists of two-sets V, the vertex set, and E, the edge set. Each edge is defined as a relation on two vertices (in hypergraphs an edge may have more that 2 vertices), and we denote the two vertices of an edge e by $e.u$ and $e.v$. There is

S. Koranne, *Practical Computing on the Cell Broadband Engine*,
DOI: 10.1007/978-1-4419-0308-2_11,

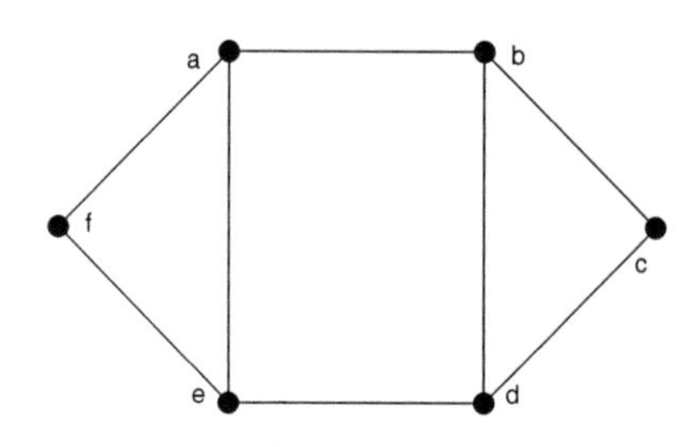

Fig. 11.1 Simple graph $G = (V, E)$ with $n = 6, m = 8$.

no implicit order between u and v, we can imply and order such that the edge now has a *direction* from $u \longrightarrow v$, rendering the graph a *directed graph*. A graph is other *undirected*. We seldom deal with cases where $u = v$ (in this case the edge is called a *loop*), and we can assume that both V and E are finite as well as non-null. It is customary to denote $|V|$ as n, and $|E|$ as m, the number of vertices and edges of the graph respectively. Vertices and edges can have properties attached to them, common vertex properties are node-names, node-numbers, and relative ordering functions. Edge properties are often referred to as edge-weights and can represent distance, unit-cost, flow-demand, flow-capacity, and a total order.

11.3 Graph representation

Depending on the size of n, m and comparing m to $O(n^2)$ we can classify the graphs we deal with it as:

1. LS SMALL: if n and m are small enough to fit in local store of an SPE,
2. PPE SMALL: if n and m are larger than LS size, but can fit in main memory of the system,
3. PPE LARGE: if n and m are such that the graph has to be cached on disk,
4. WEB GRAPH: when n and m are of the order of millions of nodes and billions of edges then we operate on them as an online streaming model,
5. SPARSE: if $m \ll O(n^2)$, if the number of edges is small compared to the $n(n-1)/2$ total edges there can be in a simple graph, then we say that the graph is sparse.

Basic textbook graph representation follow either a linked-list or adjacency matrix representation. For LS SMALL graphs both are fine, except that, every adjacency matrix algorithm comes with a $O(n^2)$ tag, which may be theoretical sub-optimal (for example the graph traversal algorithms are order $O(m)$ which could be much less than the $O(n^2)$ demanded by adjacency matrix representation). But the adjacency matrix representation simplifies the implementation of many graph algorithms, and in many cases allows the use of SIMD optimizations which should override the algorithmic sub-optimality.

11.3.1 Min-cut

Given $G = (V,E)$ we can define a CUT as set of edges (C_e) whose removal disconnects the graphs into atleast 2 connected components, provided no proper subset of (C_e) disconnects G. In the graph example we have seen above, deletion of edges connecting $v1v3, v3v5, v4v5, v4v2$ is a Cut, but adding any extra edge (eg., $v1v2$) is not, neither is deleting any edge from (C_e). The edge set can also be defined as a minimal set of edges whose removal reduces the rank of the graph by one. Cut sets are important in models of electrical, transportation and road-networks where disruption of any node or edge could cause outages in the disconnected parts. Cut sets can be used in this context to increase the redundancies and fail-safe operation of these models. We can define edge connectivity of a graph as the number of edges in the smallest cut set. Similarly we can define vertex connectivity of a graph as the minimal number of vertices which must be deleted to disconnect the graph (the assumption of-course being that deletion of a vertex is equivalent to the simultaneous deletion of *all* edges incident on that vertex). Graphs with vertex connectivity equal to 1 are called separable, and their properties will be used to design an efficient encoding scheme for large Web-graphs in a later chapter. Obviously the edge connectivity of G cannot exceed the minimum degree of G (minimum degree of G is the degree of the vertex with the smallest degree).

We would like to design an efficient SPE resident algorithm for LS Small class graphs which finds minimal or approximately minimal cut sets given the graph as an adjacency list or adjacency matrix formulation. We describe the algorithm which is based upon Karger's randomized cut method with optimizations for SPE coding. We first give an outline of Karger's algorithm.

```
   Data: Undirected graph G
   Result: Edges in min-cut of G
1  int MinCut(Graph G) begin
2      while GetNumberNodes(G) > 2 do
3          Edge e = RandomEdge(G);
4          ContractEdge(e);
5      end
6      return G.Edges();
7  end
```

Algorithm 14: Karger's randomized min-cut algorithm

David Karger gave an elegant randomized algorithm to probabilistically calculate min-cut of a graph [68]. His method was based on random sampling, and the main result he showed was that *skeleton* of a graph, which is the residue of a graph when random edges are contracted, will accurately approximate the value of *all* cuts in the original graph. Additional research into whether other properties also retain their bounds under random sampling was sparked by this discovery and we shall revisit that when we discuss property checking on graphs.

Karger's randomized min-cut can be classified as a Monte Carlo (since it is randomized) and Las Vegas (as it is probabilistic) algorithm for approximating and exactly finding cuts and flows in undirected graphs. For graphs which have edge weights, a form of scaling is proposed to achieve a sub-linear dependence on maximum edge weight. We focus on the unweighted case first. A *cut* as we stated above is a partition of the graph's vertices into two non-empty sets. The number of edges cut is the *value* of the cut. Karger's algorithm is shown in Algorithm 14.

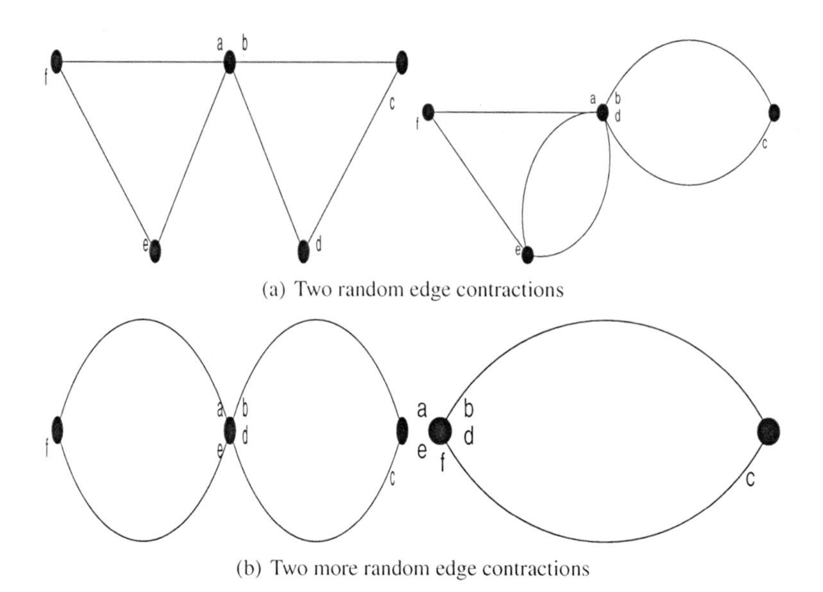

Fig. 11.2 Illustration of Karger's randomized min-cut method.

Consider the graph shown in Figure 11.1; let e the random edge be ab, when we contract nodes a and b, one self-loop is formed of the single edge which was contracted. The graph formed by this contraction (with the self-loop removed) is shown in Figure 11.2 Step 1. Next we contract edge bd at random, forming the graph shown in Figure 11.2 Step 1. Next is edge ae, and then finally af. The final graph remaining with only 2 nodes is shown in Figure 11.2 Step 4. The edges of this final graph are the min-cut computed by Karger's algorithm, the edges bc and bd, which are a min-cut. This is a probabilistic algorithm (in addition to being a randomized one), so we need to run it $O(n^2 \lg n)$ times to get the correct answer (a minimal cut) with high probability. There are simple optimizations, for examples when the number of nodes as given by *G*.GetNumberNodes(), returns a number which is less than $n/\sqrt{2}$ we can repeat the algorithm ($2 \lg n$ times) and for each independent run record the minimal cut. An equivalent formulation of Karger's min-cut using minimum spanning tree is also given in the next section.

11.4 Minimum Spanning Tree using Kruskal's Method

Let $G = (V,E)$ be a connected undirected graph, with a weight function on E mapping every edge to a real number (single-precision floating point in our implementation),

$$w : E \rightarrow R$$

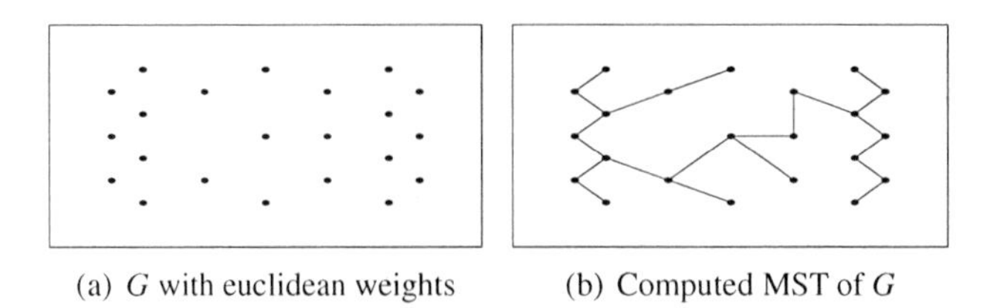

(a) G with euclidean weights (b) Computed MST of G

Fig. 11.3 Illustration of Minimum Spanning Tree (MST) construction on a graph.

A *forest* in G is a subgraph $F = (V,E')$ with no cycles; note that F has the same vertex set as G. A *spanning tree* in G is a forest with exactly one connected component. Given the weight function, a *minimum spanning tree (MST)* in G is a spanning tree whose total edge weight (sum of the weights of the edges in the tree) is minimum over all spanning trees. Observe that the MST condition involves minimum over all spanning trees, but due to a special property of this problem (for more details search for *matroids*), the problem can be solved by a *greedy* algorithm. The algorithm (called Kruskal's algorithm) is given below:

```
   Data: Undirected graph G with edge weights
   Result: MST of G
 1 int MST(Graph G) begin
 2     Edges sorted_edges = sort(G.Edges());
 3     Edges mst;
 4     int n = GetNumberNodes(G);
 5     Set S(n);
 6     for (i = 0; i < (n - 1); ++i) do
 7         Edge e = GetNextEdge((sorted_edges));
 8         if S.FindSet(e.u) ≠ S.FindSet(e.v) then
 9             mst.Add( e ); S.Union(e.u, e.v);
10         end
11     end
12     return mst;
13 end
```

Algorithm 15: Kruskal's greedy algorithm to compute MST of G.

The sorting function used in Algorithm 15 is the SPE local store sort function we have discussed in Chapter 10. The implementation of Algorithm 15 is given in Listing 11.5.

```
    ////////////////////////////////////////////////////////////////
    // Program   : Graph algorithms for CBE/SPE
    // Author    : Sandeep Koranne
    // Purpose   : Show function dispatch based on DMAed parameter
 5  #ifndef __GRAPH_H__
    #define __GRAPH_H__

    #include <stdlib.h>
    #define GRAPH_SIZE (1 << 10)
10  typedef struct _graph_control_block {
      unsigned int N;
      unsigned int M;
      unsigned long long addr;
    } GraphControlBlock;
15  #define N1_MASK 0xFFF00000
    #define N1_SHIFT 20
    #define N2_MASK 0xFFF00
    #define N2_SHIFT 8
    #define WT_MASK 0xFF
20  #define WT_SHIFT 0
    #define EDGE_WT(i) ( GRAPH[ (i) ] & WT_MASK )
    #define N1E(i) (( GRAPH[(i)] & N1_MASK ) >> N1_SHIFT )
    #define N2E(i) (( GRAPH[(i)] & N2_MASK ) >> N2_SHIFT )

25  #ifndef __SPU__
    #include "../align.h"

    class Graph {
    public:
30    Graph( const char* fileName,bool is_binary=false );
      ~Graph();
      unsigned int* get_graph() { return m_graph; }
      unsigned int get_number_edges() const { return M; }
      unsigned int get_number_nodes() const { return N; }
35    void write_binary( const char* file_name ) const;
      void read_binary( const char* file_name );
      void assign_random_weights();
    private:
      unsigned int* m_graph;
40    unsigned int N,M;
    };

    Graph::Graph( const char* fileName, bool is_binary ) {
      m_graph = (unsigned int*)
45      _malloc_align( (GRAPH_SIZE)*sizeof(int), 7 );
      if( is_binary ) {
        read_binary( fileName );
        return;
      }
50    std::ifstream ifs( fileName );
      if( !ifs ) {
        std::cerr << "Cannot open : " << fileName;
        exit( -1 );
      }
55    ifs >> N >> M; // check LS overflow
      for(unsigned int i=0; i < M; ++i ) {
        unsigned int edge = 0x0;
        unsigned int n1,n2,wt;
        ifs >> n1 >> n2 >> wt;
60      edge = (n1 << 20) | (n2 << 8) | wt;
        #ifdef DEBUG
        std::cout << std::endl << "(" << n1 << "," << n2
            << "," << wt << ") := " << edge;
        #endif
```

```
65        m_graph[i] = edge;
        }
      }
      Graph::~Graph( ) {
        _free_align( m_graph );
70    }

      void Graph::assign_random_weights( ) {
        for(unsigned int i=0;i<M;++i) {
          m_graph[i] &= ~(WT_MASK);
75        m_graph[i] |= ( rand() % 256 );
        }
      }

      void Graph::write_binary( const char* file_name ) const {
80      FILE* fp = fopen( file_name, "wb" );
        fwrite( &M, sizeof(unsigned int),1, fp );
        fwrite( &N, sizeof(unsigned int),1, fp );
        for(size_t n=0; n <= M; ) {
          n += fwrite( m_graph, sizeof( unsigned int), M, fp );
85      }
        fclose( fp );
      }

      void Graph::read_binary( const char* file_name ) {
90      FILE* fp = fopen( file_name, "rb" );
        fread( &M, sizeof(unsigned int),1, fp );
        fread( &N, sizeof(unsigned int),1, fp );
        for(size_t n=0; n <= M; ) {
          n += fread( m_graph, sizeof( unsigned int), M, fp );
95      }
        fclose( fp );
      }
      #endif // PPU only code
      #endif
```

Listing 11.1 Graph algorithm graph.h header file for Graph algorithms

The main graph class for our algorithms is presented in Listing 11.1. The macros to extract the node and edge weights are written using a 12-bit vertex code. If the number of nodes in one SPU is expected to be larger, then these macros can be changed.

```
      ///////////////////////////////////////////////////////////////////
      // Program    : Graph
      // Author     : Sandeep Koranne
      // Purpose    : Implement graph algorithms on the Cell Broadband
5     ///////////////////////////////////////////////////////////////////
      #include <iostream>
      #include <fstream>
      #include <libspe2.h>
      #include <stdlib.h>
10    #include <cassert>
      #include <pthread.h>
      #include <string.h>
      #include "graph.h"
      #define DEBUG // switch off in production
15    #define MAX_SPES 4
      static bool min_cut_mode=true;
      GraphControlBlock *cb[MAX_SPES];
      extern spe_program_handle_t graph_function;
      typedef struct ppu_pthread_data {
20      void *argp;
        spe_context_ptr_t speid;
        pthread_t pthread;
```

```
} ppu_pthread_data_t;
ppu_pthread_data_t *datas;
25
void *ppu_pthread_function(void *arg) {
  ppu_pthread_data_t *datap = (ppu_pthread_data_t *)arg;
  unsigned int entry = SPE_DEFAULT_ENTRY;
  int rc = spe_context_run(datap->speid, &entry,
30        0, datap->argp, NULL, NULL);
  pthread_exit(NULL);
}

int main( int argc, char* argv [] ) {
35  std::cout << "sizeof(CB) = "
      << sizeof( GraphControlBlock) << std::endl;
  assert( (sizeof( GraphControlBlock ) % 16) == 0);
  for(int i=0;i<MAX_SPES;++i) {
    cb[i] = (GraphControlBlock*)
40    _malloc_align( sizeof(GraphControlBlock), 5 );
  }
  Graph G( argv[1], true ); // enable binary mode
  // for Karger's cut, randomly change weights in SPU
  std::cout << "Read G=(V,E) N=" << G.get_number_nodes()
45      << " M= " << G.get_number_edges() << "\n";
  datas = (ppu_pthread_data_t*)
    malloc( sizeof(ppu_pthread_data_t)*MAX_SPES);
  for (int i = 0; i < MAX_SPES ; i++) {
    datas[i].speid = spe_context_create (0, NULL);
50    spe_program_load (datas[i].speid, &graph_function);
    cb[i]->addr = (unsigned long long int) G.get_graph();
    cb[i]->M = G.get_number_edges();
    cb[i]->N = G.get_number_nodes();
    datas[i].argp = (unsigned long long*) cb[i];
55    pthread_create (&datas[i].pthread, NULL,
        &ppu_pthread_function, &datas[i]);
  }
  for (int i = 0; i < MAX_SPES ; i++)
    pthread_join (datas[i].pthread, NULL);
60  // now we can read the results
  for (int i = 0; i < MAX_SPES ; i++) {
    unsigned int mbret;
    while( spe_out_mbox_status( datas[i].speid ) <= 0 ) ;
    spe_out_mbox_read( datas[i].speid, &mbret, 1 );
65    printf("\n On %x I read %d", datas[i].speid, mbret );
    spe_context_destroy (datas[i].speid);
  }
  free( datas );
  _free_align( cb );
70  return (EXIT_SUCCESS);
}
```

Listing 11.2 PPE code for graph algorithms

The PPE code for graph algorithms is responsible for loading the graph from the file system, either in text format or binary and transferring the data to SPE using SPE initiated DMA. Now we can present the individual graph algorithms implemented on the SPU.

```
static void InitializeUniverse( GraphControlBlock cb ) {
  int i;
  for(i=0; i < cb.N; ++i ) UNIVERSE[ i ] = -1;
}
5 static void ClearUniverse( GraphControlBlock cb ) {
  int i;
  for(i=0; i < cb.N; ++i ) UNIVERSE[ i ] = 0;
}
```

```
    static int FindSet( int n ) {
10    do {
        if( UNIVERSE[ n ] < 0 ) return n;
        else n = UNIVERSE[ n ];
      } while( 1 );
    }
15  #define SAME_SET(i) ((FindSet(N1E(i))) == (FindSet(N2E(i))))
    static int Union( int i, int j ) {
      int root_i = FindSet( i );
      int root_j = FindSet( j );
      if( root_i == root_j ) return;
20    if( UNIVERSE[ root_i ] < UNIVERSE[ root_j ] )
        UNIVERSE[ root_j ] = root_i;
      else
        UNIVERSE[ root_i ] = root_j;
    }
```

Listing 11.3 SPE file for Graph algorithms, disjoint-union implementation.

The code in Listing 11.3 implements the disjoint-set union-find algorithms presented in Section 10.11.

```
22  static void SortEdges( GraphControlBlock cb ) {
      unsigned int gap = cb.M;
      unsigned int progress = 1;
      int i;
      do {
27      if( gap > 1 )
          gap = gap / 1.3;
        i=0;progress = 0;
        #ifdef DEBUG_SORT
        printf("\n Gap = %d", gap);
32      #endif
        for(i=0;i+gap<cb.M;++i) {
          #ifdef DEBUG_SORT
          printf("\n Comparing %d (%d) : %d (%d)",
                 i,EDGE_WT(i),i+gap,EDGE_WT(i+gap));
37        #endif
          if( EDGE_WT(i) > EDGE_WT(i+gap) ) {
            unsigned temp = GRAPH[i];
            GRAPH[i] = GRAPH[i+gap];
            GRAPH[i+gap] = temp;
42          progress = 1;
          }
        }
      } while( (gap > 1) || (progress) );
    }
```

Listing 11.4 SPU Comb-sort implementation.

An SPU implementation of the comb-sort algorithm for sorting edges is presented in Listing 11.4. This sorting is used for minimum spanning tree construction using Kruskal's algorithm as presented below in Listing 11.5.

```
    static unsigned int MST( GraphControlBlock cb, int* how_many ) {
      unsigned int M = cb.M;
      unsigned int N = cb.N;
47    unsigned int i=0;
      unsigned int mst_cost = 0;
      unsigned int added_edges = 0;
      for(i=0; ( (i < M) && (added_edges < (N-1))); ++i ) {
    #ifdef DEBUG_MST
52      printf("\n Analyzing edge %d -> %d : %d",
               N1E(i), N2E(i), EDGE_WT(i) );
```

```
#endif
    if( !SAME_SET(i) ) {
      #ifdef DEBUG_MST
57    printf("\n Adding edge %d -> %d : %d to MST",
             N1E(i), N2E(i), EDGE_WT(i) );
      #endif
      mst_cost += EDGE_WT(i);
      Union( N1E(i), N2E(i) );
62    added_edges++;
    }
  } *how_many = M-i;
  return mst_cost;
}
```

Listing 11.5 SPU MST implementation.

```
    //////////////////////////////////////////////////////////////////////
    // Graph Properties Checking
    // Note: Uses UNIVERSE for node storage
67  //////////////////////////////////////////////////////////////////////

    static void ElementaryGraphProperties( GraphControlBlock cb,
                                           int* maxd,
                                           int* mind) {
72    int i;
      int max_degree = 0, min_degree = INT_MAX;
      ClearUniverse( cb );
      for(i=0; i < cb.M; ++i ) {
        UNIVERSE[ N1E(i) ]++;
77      UNIVERSE[ N2E(i) ]++;
      }
      for(i=0; i < cb.N; ++i ) {
        max_degree = ( max_degree > UNIVERSE[ i ] ) ? max_degree : UNIVERSE[i];
        min_degree = ( min_degree < UNIVERSE[ i ] ) ? min_degree : UNIVERSE[i];
82    }
      *maxd = max_degree, *mind = min_degree;
    }

    static void RandomEdgeAssignment( GraphControlBlock cb ) {
87    int i=0;
      for(i=0; i < cb.M; ++i ) {
        GRAPH[i] &= ~(WT_MASK);
        GRAPH[i] |= ( rand() % 255 );
      }
92  }


    static void ConstructCostMatrix( GraphControlBlock cb ) {
      int i,j;
97    memset( (void*)&COST_MATRIX,INFTY,(GRAPH_SIZE*GRAPH_SIZE)*sizeof(int));
      for(i=0; i<cb.M;++i) {
        int n1 = N1E(i), n2 = N2E(i);
        COST_MATRIX[n1][n2] = COST_MATRIX[n2][n1] = EDGE_WT(i);
      }
102   for(i=0;i<cb.N;++i) COST_MATRIX[i][i] = 0;
    }

    static void FloydWarshallAPSP( GraphControlBlock cb ) {
      int i,j,k,sum;
107   for(k=0;k<cb.N;++k)
        for(i=0;i<cb.N;++i)
          for(j=0;j<cb.N;++j) {
            sum = COST_MATRIX[i][k] + COST_MATRIX[k][j];
            COST_MATRIX[i][j]=(COST_MATRIX[i][j] < sum ) ?
112           COST_MATRIX[i][j] : sum;
          }
```

```
}

int main(unsigned long long speid,
117         unsigned long long argp,
            unsigned long long envp __attribute__ ((__unused__))) {

  int ack = 0;
  int tag_id = mfc_tag_reserve();
122  int i,j,edge,weight;
  unsigned int mst_cost = -1;
  srand( speid );
  mfc_get(&cb, argp, sizeof(cb), tag_id, 0, 0);
  mfc_write_tag_mask(1<< tag_id);
127  mfc_read_tag_status_all();
  printf("Hello from SPE (0x%llx) \n", speid );
  // now read the data as well but get it in 4096 * 4 bytes
#define CHUNK_OFFSET ((sizeof(unsigned int))*4096)
  for(i=0;i<GRAPH_SIZE/4096;++i) {
132    mfc_get(GRAPH+(i*4096), cb.addr+(i*CHUNK_OFFSET),
            sizeof(int)*4096, tag_id, 0, 0);
    mfc_write_tag_mask(1<< tag_id);
  }
  mfc_read_tag_status_all();
137  ElementaryGraphProperties( cb, &i, &j );
  printf("\n N=%d M=%d MaxDeg=%d MinDeg=%d",cb.N,cb.M,i,j);
  RandomEdgeAssignment( cb );
  InitializeUniverse( cb );
  PrintUniverse( cb );
142  PrintEdges( cb );
  printf("\n Sort Start");
  SortEdges( cb );
  printf("\n Sort Complete");
  PrintEdges( cb );
147  mst_cost = MST( cb, &i );
  PrintUniverse( cb );
  spu_write_out_mbox( i ); // put mst_cost
  return 0;
}
```

Listing 11.6 Graph analysis and DMA functions.

```
   (defun distance (n i j)
     "Return euclidean distance on grid of sqrt(n)"
     (let* ((s (floor (sqrt n)))
            (ix (floor (/ i s)))(iy (floor (mod i s)))
5           (jx (floor (/ j s)))(jy (floor (mod j s)))
            (dx (- ix jx))(dy (- iy jy))
            (d  (floor (+ (* dy dy) (* dx dx))))) d))
   (defun make-euclidean-graph (n)
     "Make euclidean graph of n nodes"
10   (format t "~D ~D" n (floor (/ (* n (1- n)) 2)))
     (dotimes (i (1- n) )
       (dotimes (j (- n i 1) )
         (let ((cj (+ i j 1)))
           (format t "~%~D ~D ~D" i cj (distance n i cj))))))
15 (make-euclidean-graph 100)
   (format t "~%")(quit)
```

Listing 11.7 Euclidean distance graphs on a grid (Common Lisp).

```
  Read G=(V,E) N=5 M= 5
  Hello from SPE (0x1002e120)
  N=5 M=5 MaxDeg=3 MinDeg=1
  ----------------------------------------
5   -1  -1  -1  -1  -1
```

```
     0    1    2    3    4
   ----------------------------------------
    Edge 0 = (0 -> 1) : 1
    Edge 1 = (1 -> 2) : 2
10  Edge 2 = (2 -> 3) : 3
    Edge 3 = (4 -> 3) : 4
    Edge 4 = (0 -> 2) : 3
    Sort Start
    Sort Complete
15  Edge 0 = (0 -> 1) : 1
    Edge 1 = (1 -> 2) : 2
    Edge 2 = (2 -> 3) : 3
    Edge 3 = (0 -> 2) : 3
    Edge 4 = (4 -> 3) : 4
20  Analyzing edge 0 -> 1 : 1
    Adding edge 0 -> 1 : 1 to MST
    Analyzing edge 1 -> 2 : 2
    Adding edge 1 -> 2 : 2 to MST
    Analyzing edge 2 -> 3 : 3
25  Adding edge 2 -> 3 : 3 to MST
    Analyzing edge 0 -> 2 : 3
    Analyzing edge 4 -> 3 : 4
    Adding edge 4 -> 3 : 4 to MST
   ----------------------------------------
30   1    2    3   -1    3
     0    1    2    3    4
```

Listing 11.8 Transcript of MST session running on SPE.

```
   #include "graph.h"
   static void Usage() {
     std::cerr << std::endl << "gt2bin <txt-file> <bin-file>" << std::endl;
     exit( -1 );
5  }
   int main( int argc, char* argv [] ) {
     if( argc != 3 ) { Usage(); }

     const char* txt_file_name = argv[1];
10   const char* bin_file_name = argv[2];

     Graph G( txt_file_name );
     G.write_binary( bin_file_name );
     return( EXIT_SUCCESS );
15 }
```

Listing 11.9 Text to binary converter.

11.5 Revisiting min-cut using MST and Karger's Method

An equivalent formulation of Karger's min-cut can be made using the MST algorithm. The algorithm is given below:

```
   Data: Undirected graph G
   Result: Edges in min-cut of G
 1 int MinCutMST(Graph G) begin
 2    Edges random_edges = RandomlyWeight(G.Edges());
 3    Edges sorted_edges = sort(random_edges);
 4    Edges mst;
 5    int n = GetNumberNodes(G);
 6    Set S(n);
 7    for (i = 0; i < (n - 2); ++i) do
 8       Edge e = GetNextEdge((sorted_edges));
 9       if S.FindSet(e.u) ≠ S.FindSet(e.v) then
10          S.Union(e.u, e.v);
11       end
12    end
13    return random_edges;
14 end
```

Algorithm 16: Karger's algorithm, MST variant.

The MST variant of Karger's algorithm works by randomly assigning weights to edges. Then the edges are sorted by weight, and we begin a regular MST algorithm by adding edges in increasing (non-decreasing) cost of weight while preventing cycles. When $n-2$ edges have been added to the MST (which we are not computing, we want to keep track of connected components only), the next edge to be added is on the min-cut. So are all the edges which remain in the random_edges edge container (remember a connected graph with more than $n-1$ edges must have a cycle). This set is returned as the min-cut. If you already have MST code lying around then this variant is easier to get off the ground. We have implemented this method as well, by adding 2 lines of code in Listing 11.5, to calculate the index of the last added edge, we easily find out the *remaining edges* which form the cut.

11.6 Dijkstra's Algorithms for Shortest Paths

Let $G = (V, E)$ be an undirected graph and w be a function assigning *non-negative* weight to each edge. This weight function could be a measure of distance or cost of transportation on this edge. Extend w to the domain $V \times V$ by defining $w(v, v) = 0$ and $w(u, v) = \infty$, if $u, v \notin E$ (u is not connected to v). Define the *length* of a path from $p = e_1 e_2 \ldots e_n$ to be

$$w(p) = \sum_{i=1}^{n} w(e_i)$$

For a vertex pair u, v, define the *distance* $d(u, v)$ from u to v to be the length of the *shortest path* from u to v. We denote a path of length ∞ if there is no path. The single-source shortest path problem (SSSP) (which is what Dijkstra's algorithm solves) is to find, given a source vertex $s \in V$, the value of $d(s,u)$ for every vertex u in G.

If the graph is unweighted (all edge weights are 1) we can solve SSSP in $O(n)$ time using breadth-first search. Here we discuss the more general case when edges do have weights. We give Dijkstra's algorithm below.

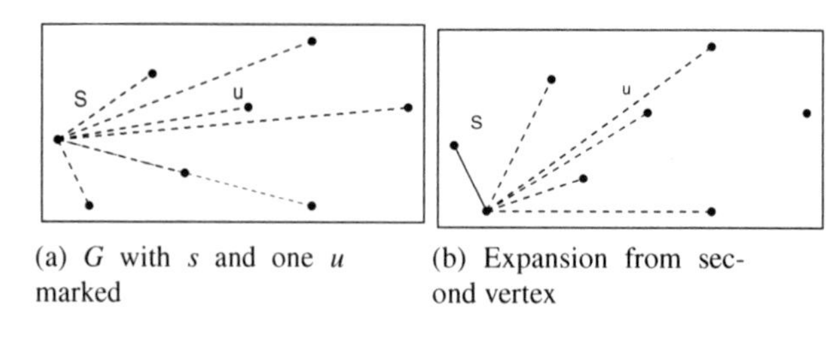

(a) G with s and one u marked

(b) Expansion from second vertex

Fig. 11.4 Illustration of Dijkstra's algorithm, 2 levels of *expansions* are shown.

```
   Data: Undirected graph G, s
   Result: Single-source shortest path from s to all u ∈ V
1  int DijkstaSSSP(Graph G) begin
2      int n = GetNumberNodes(G);
3      VertexSet X = s;
4      DistanceArray D(n);
5      D[s] = 0;
6      for u ∈ (V − {s}) do
7          D[u] = w(s,u);
8      end
9      while X ≠ V do
10         Vertex u = FindMinimumLength(D,V − X);
11         X = X ∪ {u};
12         for e = (u,v) : u ∈ (V − X) do
13             D[u] = min(D[v],D[u] + w(u,v));
14         end
15     end
16     return D;
17 end
```

Algorithm 17: Dijkstra's algorithm to solve single-source shortest path.

```
volatile NodeType GV[MAXNODES][MAXNODES] __attribute__ ((aligned(128)));
#define MAX_EDGES (MAXNODES<<4)
volatile Edge EDGE_MATRIX[MAX_EDGES] __attribute__ ((aligned(128)));
volatile int DISTANCE[MAXNODES] __attribute__ ((aligned(128)));

short int BF(volatile Edge edges[], int nE, int n, int source)
{
  int i, j;
  for (i=0; i < n; ++i) {
```

```
10        DISTANCE[i] = INFTY;
        }
        DISTANCE[source] = 0;
        for (i=0; i < n; ++i) {
          int progress = 0;
15        for (j=0; j < nE; ++j) {
            if (DISTANCE[edges[j].source] != INFTY) {
              int nd = DISTANCE[edges[j].source] + edges[j].weight;
              if (nd < DISTANCE[edges[j].dest]) {
                DISTANCE[edges[j].dest] = nd;
20              progress=1;
              }
            }
          }
          if (!progress) break;
25      }
        return -( DISTANCE[n-1] );
      }

      static short int LongestPath( volatile NodeType G[][MAXNODES], int n ) {
30      int edge_count=0;
        int i,j;
        for(i=0;i<(n-1);++i)
          for(j=0;j<(n);++j)
            if( G[i][j] ) {
35            EDGE_MATRIX[edge_count].source = i;
              EDGE_MATRIX[edge_count].dest = j;
              EDGE_MATRIX[edge_count].weight = -G[i][j];
              edge_count++;
            }
40      if( edge_count >= MAX_EDGES ) {
          printf("\n Increase EDGE_MATRIX table size to %d", edge_count );
          exit(-1);
        }
        return BF( EDGE_MATRIX, edge_count, n, 0 );
45    }

      static void ReadNodeInformation( const char* fileName ) {
        FILE* fp = fopen( fileName, "rt" );
        int i=0;
50      N = 0;
        fscanf( fp, "%d", &N );
        for(i=0; i < N; ++i) {
          int w,h;
          fscanf(fp, "%d %d",&w,&h);
55        W[i]=w, H[i]=h;
        }
        fclose( fp );
      }

60    static void PrintNodeInformation( ) {
        int i;
        printf("\n Total %d nodes.\n",N);
        for(i=0;i<N;++i) {
          printf("\n Node %d : %d %d",i,W[i],H[i]);
65      }
        printf("\n");
      }

      static inline void InitializeSequence( volatile NodeType *A, int n ) {
70      int i;
        for(i=0;i<n;++i) A[i]=i;
      }
```

Listing 11.10 SPU implementation of Dijsktra's shortest path

We have also implemented Bellman-Ford's shortest path algorithm on the SPU. The implementation is shown in Listing 11.10. We use this implementation to compute the *longest-path* in *adirected acyclic graph (DAG)* for scheduling problems and our floorplanning implementation (cf. Section 21.4, pp. 404).

11.7 Floyd-Warshall All-pairs shortest-path

In this section we describe an algorithm to compute the all-pairs shortest paths for all source-destination pairs in in a positively weighted graph G. It is well known that almost all dynamic programming problems can be equivalently viewed as problems seeking the shortest path in a directed graph. Thus, there are many practical applications of shortest path algorithms in a diverse fields such as geographical information systems, VLSI circuit routing, robotics and design of communication networks, This calculation is also used to compute graph properties such as graph diameter, which is important in polytopal calculations (see Chapter 18).

The APSP is solved by the well-known Floyd-Warshall (FW) algorithm [74], which computes the solution inplace from the weight matrix of the graph using a triple loop similar to MMM (matrix-multiply), but involving only additions and minimum operations, and with dependencies that restrict the ordering of the three loops (the k-loop has to be the outermost one). The problem of APSP is similar to computing reflexive transitive closures of G and indeed the same algorithm can be used (if the graph is given in the form of a matrix) to calculate closure. A code generator for DSP architectures (which include the Cell) specifically for APSP has been developed by Han, Franchetti and Püschel [60].

11.8 Formal definition of APSP (all-pairs shortest-path)

Let $G = (V, E)$ be a graph, with an appropriate weighting function $w : V \times V \rightarrow R^+$, with positive weights. As before we denote V as the vertex set with $n = |V|$. The APSP problem is to compute the minimum distance

$$d(u,v) = \min_{p \in \forall paths(u,v)} \sum_{(i,j) \in p} w(i,j)$$

for all pairs of vertices u, v in G. We assume the graph G is given by a $n \times n$ *cost matrix* (which is small enough to fit in local store of SPE). To solve problem where the matrix does not fit in the local store we adopt an approximation scheme based on clustering. This is discussed at the end of this chapter.

The cost matrix C is given as

$$C[i][j] = \begin{cases} 0 & : \text{ if } i = j \\ w(i,j) & : \text{ if } (i,j) \in E \\ \infty & : \text{ otherwise} \end{cases}$$

We compute the APSP in place by overwriting $C[i][j]$ with the shortest path from vertex i to j.

```
Data: Cost matrix C of G
Result: Modifies C in place
1  int FWAPSP(Matrix C) begin
2      int n = C.size();
3      for k = 1; k < n; ++k do
4          for i = 1; i < n; ++i do
5              for j = 1; j < n; ++j do
6                  C[i][j] = min((C[i][j]), C[i][k] + C[k][j]);
7              end
8          end
9      end
10 end
```

Algorithm 18: Floyd-Warshall algorithm to calculate all-pairs shortest paths.

```lisp
(declaim (optimize (speed 3) (safety 0)))
(defconstant +INFY+ (* 1000 1000))
(declaim (inline fw))
(defun fw (m n)
  (dotimes (k n)
    (dotimes (i n)
      (dotimes (j n)
        (setf (aref m i j)
              (min (aref m i j)
                   (+ (aref m i k) (aref m k j))))))))

(declaim (inline diameter))
(defun diameter (m n)
  "Calculate the maximum value in the matrix"
  (let ((ret-val 0))
    (dotimes (i n)
      (dotimes (j n)
        (setf ret-val (max ret-val (aref m i j)))))
    ret-val))
```

Listing 11.11 Common Lisp implementation of FW algorithm

The simple SPU implementation is shown in Listing 11.12. This code can be optimized using block, and loop unrolling as shown below.

```c
static void FloydWarshallAPSP( GraphControlBlock cb ) {
  int i,j,k,sum;
  for(k=0;k<cb.N;++k)
    for(i=0;i<cb.N;++i)
      for(j=0;j<cb.N;++j) {
        sum = COST_MATRIX[i][k] + COST_MATRIX[k][j];
        COST_MATRIX[i][j]=(COST_MATRIX[i][j] < sum ) ?
          COST_MATRIX[i][j] : sum;
      }
}
```

```
int main(unsigned long long speid,
         unsigned long long argp,
         unsigned long long envp __attribute__ ((__unused__))) {

  int ack = 0;
  int tag_id = mfc_tag_reserve();
  int i,j,edge,weight;
  unsigned int mst_cost = -1;
  srand( speid );
  mfc_get(&cb, argp, sizeof(cb), tag_id, 0, 0);
  mfc_write_tag_mask(1<< tag_id);
  mfc_read_tag_status_all();
  printf("Hello from SPE (0x%llx) \n", speid );
  // now read the data as well but get it in 4096 * 4 bytes
#define CHUNK_OFFSET ((sizeof(unsigned int))*4096)
  for(i=0;i<GRAPH_SIZE/4096;++i) {
    mfc_get(GRAPH+(i*4096), cb.addr+(i*CHUNK_OFFSET),
          sizeof(int)*4096, tag_id, 0, 0);
    mfc_write_tag_mask(1<< tag_id);
  }
  mfc_read_tag_status_all();
  ElementaryGraphProperties( cb, &i, &j );
  printf("\n N=%d M=%d MaxDeg=%d MinDeg=%d",cb.N,cb.M,i,j);
  RandomEdgeAssignment( cb );
  InitializeUniverse( cb );
  PrintUniverse( cb );
  PrintEdges( cb );
  printf("\n Sort Start");
  SortEdges( cb );
  printf("\n Sort Complete");
  PrintEdges( cb );
  mst_cost = MST( cb, &i );
  PrintUniverse( cb );
  spu_write_out_mbox( i ); // put mst_cost
  return 0;
}
```

Listing 11.12 APSP implementation using SPU

11.9 Maximal Independent Set

Maximal independent set of a graph is defined, and Luby's algorithm [83] to compute maximal independent set using parallel processors is implemented using SPU. We have followed the explanation given by Kozen in [74]. Luby's algorithm was originally designed in PRAM setting, but we can easily adopt it for a system such as Cell Broadband Engine which has distributed local-store. The algorithm runs in stages and in each stage the current independent set is removed from the graph.

11.10 Approximation algorithm for maximal matching

A matching M of graph $G = (V, E)$ is a subset of edges E of G such that no two edges in M have a common vertex. The maximum cardinality matching problem is to find a matching of maximum cardinality in the graph. If the graph is weighted,

$G = (V, E)$ and $w : E \rightarrow R^+$ be the edge weight function which assigns a positive weight to each of the edges of G. Then the weight $w(M)$ of the matching is defined as:

$$w(M) = \sum_{e \in M} w(e)$$

The weighted matching problem is to find a matching M of maximum weight. A simple greedy algorithm is shown in Algorithm 19. The greedy algorithm is an 1/2 approximation algorithm with $O(|E| \log |E|)$ time complexity due to the requirement of sorting edges by weight.

```
1 Set GreedyMatching(G = (V,E), w : E → R+) begin
2     Set M;
3     while E ≠ 0 do
4         let e be the heaviest edge in E;
5         add e to M;
6         remove e all edges incident to e from E;
7     end
8     return(M);
9 end
```

Algorithm 19: Greedy algorithm for finding maximum weight matchings.

The problem of maximum matching has been extensively studied and parallel algorithms have also been proposed. The key problem in distributed computation of matching has been getting the processors to compute towards the same matching as there are exponential number of matchings possible in a graph. Mulmuley, Vazirani and Vazirani had proposed a parallel algorithm for maximal matching where they showed that the complexity of matching is same as matrix inversion [90]. They used a clever randomization scheme to uniquify the matching across processors.

We implement a simple local edge-domination based approximate maximal matching algorithm based on Preis's algorithm [101]. The presented algorithm is a 1/2 approximation algorithm with a linear time complexity in the number of edges, i.e., its an $O(|E|)$ algorithm. The algorithm is depicted in Figure 11.5. For every vertex the locally maximal weighted edge and the associated vertex it is pointing to is identified. For every edge which is maximal for both source and destination vertices (u, v) is added to the current matching under construction and every other edge for (u, v) is deleted. This process is repeated till no edge remains. This algorithm has been implemented in the SPU using our graph representation. Another method could be to use the Tutte matrix method to find out if a graph admits a maximum matching of $|V|/2$. We have implemented Preis's approximate maximal-matching algorithm on the SPE; the code is shown in Listing 11.13.

```
  static int IsTainted(int e) {
    unsigned int dom_n1 = DOMINANT[ N1E(e) ];
    unsigned int dom_n2 = DOMINANT[ N2E(e) ];
    return (dom_n1 == -1 || dom_n2 == -1);
5 }
  static int UpdateEdgeStatus(int e) {
    unsigned int dom_n1 = DOMINANT[ N1E(e) ];
    unsigned int dom_n2 = DOMINANT[ N2E(e) ];
```

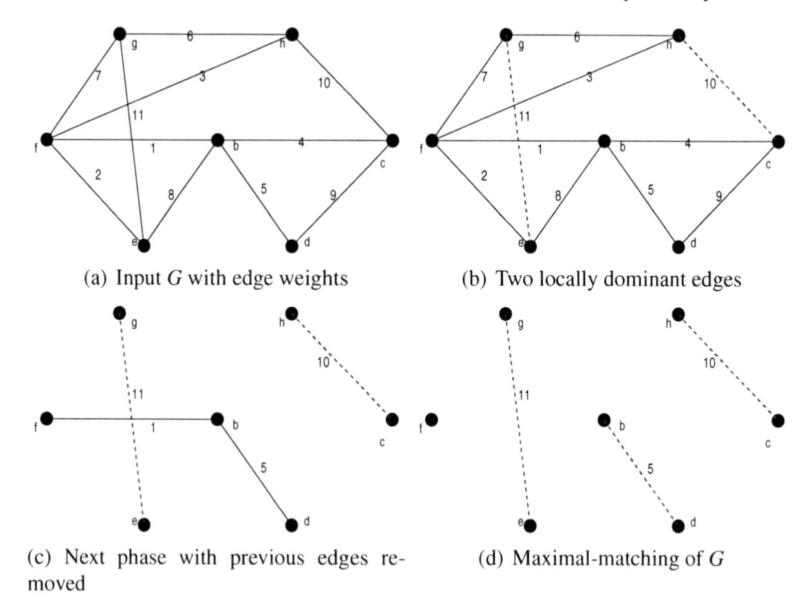

(a) Input G with edge weights

(b) Two locally dominant edges

(c) Next phase with previous edges removed

(d) Maximal-matching of G

Fig. 11.5 Illustration of Preis's algorithm for local edge domination.

```
  int et = EDGE_WT(e);
  int override=0;
#ifdef DEBUG_FULL
  printf("\n Analyzing edge %d (%d)-> %d(%d) of wt %d",
         N1E(e),dom_n1,N2E(e),dom_n2,et);
#endif
  unsigned int e1 = (dom_n1&0xFF000000)>>16;
  unsigned int e2 = (dom_n2&0xFF000000)>>16;
  // if either edge is tainted then reset wt
  if(IsTainted(e1) || IsTainted(e2)) override=1;
  if(override||((et > (dom_n1&0xFF)) && (et > (dom_n2&0xFF)))) {
    dom_n1 = dom_n2 = ((e&0xFF)<<16)|(et&0xFF);
    DOMINANT[ N1E(e) ] = DOMINANT[ N2E(e) ] = dom_n1;
  }
}
static int AnalyzeDomination(GraphControlBlock cb, int* wt) {
  int i=0,j=0,progress=0;
  for(i=0;i<cb.N-1;++i)
    for(j=i+1;j<cb.N;++j)
      if(DOMINANT[i] && (DOMINANT[i]!=-1) && (DOMINANT[i] == DOMINANT[j])) {
        printf("\nAdding pair (%d->%d) to M",i,j);
        progress++;
        *wt += DOMINANT[i]&0xFF;
        DOMINANT[i] = DOMINANT[j] = -1;
      }
  return progress;
}

static int CalculateMaximalMatching( GraphControlBlock cb ) {
  int i;
  int progress=1;
  int wt=0;
  while(progress) {
    for(i=0;i<cb.M;++i) {
```

```
45        if(DOMINANT[N1E(i)]==-1) continue;
          if(DOMINANT[N2E(i)]==-1) continue;
          UpdateEdgeStatus(i);
        }
        PrintDominant(cb);
        progress=AnalyzeDomination(cb,&wt);
      }
50    printf("\nDone %d.\n",wt);
      return wt;
    }
```

Listing 11.13 SPU Implementation of Preis's maximal-matching algorithm.

```
    sizeof(CB) = 16
    Read G=(V,E) N=7 M= 11
    Hello from SPE (0x1001d120)

5    N=7 M=11 MaxDeg=4 MinDeg=2
    Adding pair (1->6) to M
    Adding pair (3->5) to M
    Adding pair (0->2) to M
    Done 26.
10
     On 1001d120 I read 26
```

Listing 11.14 Transcript of maximal-matching on the graph shown in Figure 11.5.

11.11 Partitioning using Simulated Annealing

Graph partitioning is one of the fundamental problem in many diverse areas of science and technology. It is used extensively in VLSI CAD, ranging from system-level partitioning to low level physical place and route methods [72]. It is of paramount importance in FPGA's where it is a key sub-routine in *technology mapping*. Graph partitioning is also used in parallel computing in the course of the problem of mapping irregular and unstructured computations onto a distributed memory parallel machine to achieve load balance, and to reduce communication cost. Graph partitioning is NP-hard [49] and hence many heuristics and approximation algorithms have appeared in literature to give approximate solutions to the problem.

The most commonly used approach for a k-way partitioning is to recursively bisect the graph[1], i.e., it first divides the graph into two equal sized pieces and then recursively divides the two pieces. Some extended heuristics have been proposed that apply quadrisection in place of bisection.

Quadrisection, though more expensive than bisection finds better k-way partitions. Due to its greedy nature and the lack of global information, recursive bisection, in the worst case, may produce a partition that is very far away from optimal. This result is true even for sparse graphs and planar graphs. In our implementation we make use of an result by Simon, to give a method that produces approximately balanced partitions that are within $O(\log k)$ factor of the optimal k-way partition.

[1] When k is not a power of 2, some simple variants of recursive bisection is used.

The relaxation is in the form of the balancing condition, in our method each block in the partition is bounded by $(1 + \varepsilon)n/k$. We also make use of the computational power of a SIMD on SPU to implement *graph partitioning engine*.

11.11.1 k-way Partitioning : Problem definition

A *k-way partition* of a graph $G = (V, E)$ is a division of its vertex set V, into k subsets $(\pi_1 \pi_2 \ldots \pi_k)$ each of size n/k, where $n = |V|$, the total number of vertices in the graph. The *cost* of a k-way partition is the number of edges whose endpoints are in different subsets. The k-way partitioning for weighted graph is similar, except that the cost is the *sum of the weights* of edges whose end points are in different subsets. When k is two, we usually call graph partitioning as *bisection*. See Figure 11.6 for an example of graph bisection.

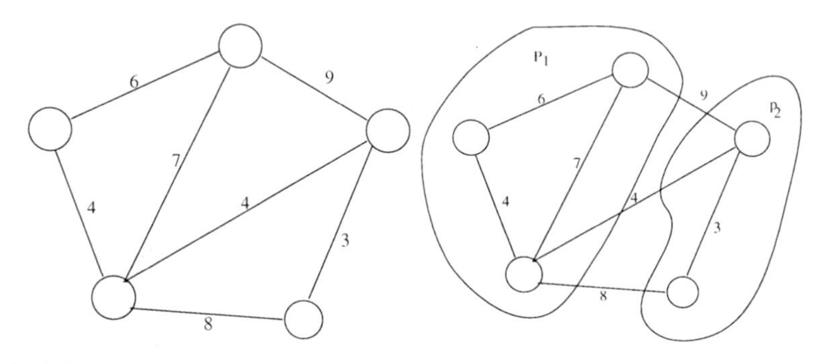

Fig. 11.6 Graph Bisection. Cost is 21

The k-way partitioning problem is to find a partition with minimum cost. Note that the problem of finding an optimal bisection itself is NP-hard [49]. We can write the k-way partitioning problem as follows :

$$V = \pi_1 \cup \pi_2 \cup \cdots \cup \pi_k$$

Where,

$$\pi_1 \cap \pi_2 \cap \cdots \cap \pi_k = 0 \tag{11.1}$$

The subsets are collectively exhaustive and mutually exclusive. The cost of the partition is defined to be :

$$C = \sum_{i,j \in V \;\; \pi(i) \neq \pi(j)} c(i,j)$$

The k-way graph partitioning problem is to minimize the cost of the partition.

11.11.2 The cost function

Hence, we can write the problem as :

$$\min C : \forall\,(i,j) \quad \text{abs}\,(\,|\pi_i| - |\pi_j|\,) \leq 1 \tag{11.2}$$

This is an example of an constrained minimization problem. In the present form Eq. 11.2 is not amenable to parallel distributed computation, due to the large cost associated with the verification of the constraint in Eq. 11.2. It is apparent that if we could write it in an unconstrained form, then it would give better performance in a distributed computing environment. To do this we use *SUMS*, or Sequential Unconstrained Minimization. We can write the cost as :

$$C' \quad = \quad C * (\text{Constraint factor})$$

To find the form of the *Constraint factor* we refer to Eq. 11.1. We are looking for a function on π_i that reaches maxima or minima at the condition represented by Eq. 11.2. To encode the balancing constraint as a cost function, we investigate the following form on π_i. Let a and b be two variable, with the proviso that $a + b = c$, then we know that ab is maximum at $a = b = c/2$. Generalizing, this to more than 2 variables, we have :

$$\max M = \prod_{i \in S}^{n} x_i \text{ is at}$$

$$x_i = \frac{|S|}{n}$$

Using this we can rewrite Eq. 11.1 and Eq. 11.2 as :

$$C \quad = \quad \sum_{i,j \in V \;\; \pi(i) \neq \pi(j)} \frac{\text{cost}(i,j)}{|\pi_i|\;|\pi_j|}$$

Now, we can remove the constraint from Eq. 11.2. The resulting cost function is also called the *ratio-cut*.

To partition graphs which are larger than the capacity allowed by the size of the local store, we have to communicate partition status of nodes across SPUs. Since communication costs can reduce the efficiency of the code, we have adopted a method from VLSI placement called *terminal propagation* which is conceptually equivalent to *fast multi pole* methods in n-body simulation. The idea is that, once clusters have been globally refined, it is sufficient to know the cost to some approximate representation of a node, not necessarily to the exact location of the node. In k-way partitioning when $k >> 4$, we use approximate methods to update the node positions. Terminal propagation is explained in detail below.

11.11.3 Terminal Propagation

All distributed partitioning approaches suffer with the problem of lose of accuracy due to bad initial partitions. The general idea is to divide the graph into two subsets with a good partitions and then solve the two sub graphs independently. A result by Simon shows that due to lack of global information the partitions resulting from this method may be far from optimal. We have used the method of *terminal propagation* to overcome this problem. Terminal propagation is used to share global information between co-operating processes. Consider the following example. Let $\pi_1 = \{a,b,c,d,e\}$ and let $\pi_2 = \{i,j,k,l,m\}$.

Then by terminal propagation we introduce a dummy vertex x in *both* the sub sets. x in π_1 is marked to represent some vertex in π_2 say i, similarly x in π_2 is supposed to represent a. Note that x can only be used to *compute* the cost, but cannot be moved by the SA code. This method has been very effective in broadcasting global information about partition assignment. See Figure 11.7 for an example of terminal-propagation.

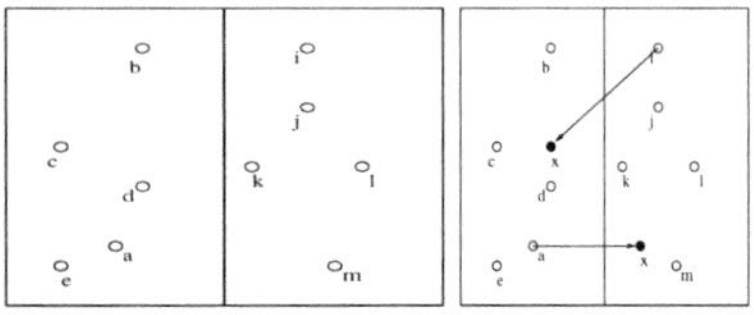

Fig. 11.7 Example of Terminal Propagation.

11.12 Simulated Annealing

Simulated Annealing is a general combinatorial optimization technique that is used for *hill-climbing* [69]. It works on the principle of thermodynamic cooling. In greedy approaches it is easy to reject some path from the current node due to the fact that that the greedy approach will reject *all* paths that result in a cost increase. It is quite possible that the optimal solution may lie on such a path. Simulated annealing overcomes this limitation by combining a randomization scheme with a cooling schedule that rejects cost increasing paths with a probability inversely proportional to an internal parameter, called the *temperature*.[2] The disadvantage of simulated annealing is the large amount of time it takes to compute the partition. To speed up this part we have written the code using SPU intrinsics. We shall also show how to use instruction scheduling and loop unrolling to optimize SPU code. Simulated annealing may be modeled as a random walk on a search graph, whose vertices are all possible

[2] SA is modeled after the thermodynamic cooling of crystals.

states, and whose edges are the candidate moves. An essential requirement for the next-state function is that it must provide a sufficiently short path on this graph from the initial state to any state which may be the global optimum. The maximum of the shortest-path on this state graph is called the *diameter* of the search space. We investigate the diameter of polytopal graphs in detail in our chapter on Polytopes 18. The general algorithm for simulated annealing is shown below:

```
Data: CostFunction, NextStateFunction
Result: State with lowest cost
1  int SimulatedAnnealing(CostFunction,NextStateFunction) begin
2      State s = InitialState();
3      float temp = InitialTemp();
4      int time=0;
5      int trials=1000;
6      float lowcost = ∞;
7      for time = 0;time < LIMIT;++time do
8          for i = 0;i < trials;++i do
9              temp = UpdateTemp(temp);
10             State ns = NextState(s);
11             float cost = CostFunction(ns);
12             if cost < lowcost then
13                 s = ns,lowcost = cost;
14             end
15             else
16                 Random coin = CoinToss();
17                 if coin > Probability(temp) then
18                     s = ns,lowcost = cost;
19                     // accept move
20                 end
21             end
22         end
23     end
24     return(s,lowcost);
25 end
```

Algorithm 20: General outline of Simulated Annealing.

The runtime of Simulated Annealing is controlled by the following *knobs*, which give a trade-off between speed and quality of solutions:

1. Temperature Cooling Schedule: the idea of simulated annealing is inspired by physical processes, where the temperature of the system is gradually reduced, the system approaches a state of minimum energy. Cooling schedule refers to the function which calculates the next temperature value. Popular choices are *exponential*, which starts off with temperature set to a high value like ∞, where all moves are allowed, and decreases using an exponential function,

15-bit n_1	15-bit n_2	2-bit e

Fig. 11.8 Edge data-structure in SPU Graph representation.

2. Cost function evaluation: this is the most time consuming part of the process in most applications since this step must have global knowledge about the system. This is the problem solved by terminal propagation where the cost function makes some approximations to avoid calculating global costs,
3. Trials per temperature: this is a clear run-time to quality trade-off, the more trials we allow, the more chance that a lower cost state will be searched,
4. Next state function: avoid cycling, spurious evaluations,
5. Backtracking, single-level, multi-level: the general simulated annealing algorithm does not need to keep track of low cost states, but as a practical matter most implementations (including ours) atleast maintains the best known solution thus far (and this is the solution returned at the end of annealing), but some implementations keep track of last n-best solutions,
6. Restarts: if the algorithm does not make progress in the last x-trials, the some implementations abandon the current state, and restart the simulation from a previously known best state. This can help escaping high-cost local-minima zones in the search space.

11.12.1 Implementation of Simulated Annealing on the SPU

We model the graph G as table of 32-bit unsigned integers, one for each edge:

Corresponding to every node in the graph we have a 32-bit record in a global table, called U, which is stored on the PPE main-memory, but parts of it can be sent to any SPU when it is needed. Using a perfect hash-function common properties of a cluster are stored in the local store of the SPU where they are needed. For example, consider a graph cluster of 7 nodes, $\{1,2,16,29,98,187,154\}$, and the SPU is going to be doing calculation on these, so we setup a DMA which brings in each of the quad-lines where these nodes are located in U, we will DMA 5 128-bit values. This gives us 20 nodes in the local store. Since we need to only use 7 of them (in this example), we create a perfect hash-function (see Chapter 10.6), to record information about the 7 nodes in a small table. But now we can index them directly using a function on the node number. This is very efficient compared to performing searches or even table lookups in the main memory. These 32-bit edges are stored as quad-words in the SIMD registers 4 at-a-time as shown here:

32-bit e_0	32-bit e_1	32-bit e_2	32-bit e_3
32-bit e_4	32-bit e_5	32-bit e_6	32-bit e_7
32-bit e_8	32-bit e_9	32-bit e_{10}	32-bit e_{11}
32-bit e_{12}	32-bit e_{13}	32-bit e_{14}	32-bit e_{15}

The cost function for partitioning can be formulated as:

$$\Pi_{cost} = \sum_{e_i=1}^{M} (\pi(e_i, n_1) \oplus \pi(e_i, n_2)) \cdot wt(e_i)$$

the function $\pi(e_i, n_{1,2})$ can be computed using the U table, and its perfect hashed counterpart in the local store of the SPU containing this cluster. The node numbers can be computed from the 32-bit edge numbers by bitwise AND and shift. This can be done using SIMD to compute 4 node number for n_1 and n_2 at the same time. Consider the following:

```
vector unsigned int N1M = {0xFE00,0xFE00,0xFE00,0xFE00};
vector unsigned int N2M = {0xFE,0xFE,0xFE,0xFE};
int cost=0;
for(int i=0; i < M; ++i ) {
  vector unsigned int e_1= EDGE_TABLE[i];
  vector unsigned int N1 = spu_sr( spu_and(N1M, e_1), 17);
  vector unsigned int N2 = spu_sr( spu_and(N2M, e_1), 1);
  vector unsigned int N1P = LoadPartInfo(N1);
  vector unsigned int N2P = LoadPartInfo(N2);
  vector unsigned int SAM = spu_xor(N1P,N2P);
  cost += spu_extract( spu_orx(SAM), 0 );
}
```

Listing 11.15 SPU code for *ComputeCost* function

As coded in Listing 11.15 the quadword load of the e_i from `EDGE_TABLE` creates a dependency chain in which the computation has to wait for the data to be loaded; similarly, the `LoadPartInfo` function which loads node information also has the same issue. By unrolling the loop for edges, we can overlap some of the computation for the node-partition-xor analysis with the edge table access. A four-way unrolling of this loop produces negligible expansion in clock cycle count for the for-loop, but now we have decreased the iteration count of this loop by a factor of 4. The savings in clock cycles come from the dual-issue nature of the SPU pipeline. Using the *asmvis* assembly language visualization tool (from IBM) we can save a lot of pen-paper (or our own automation work) by compiling the code to assembly language (use `-Wa,-adhl -O2 -g`) file. We then analyze this file using the visualizer, the resulting timing diagram with clock cycles, function names, branch labels and instruction mnemonics is shown in Figure 11.9.

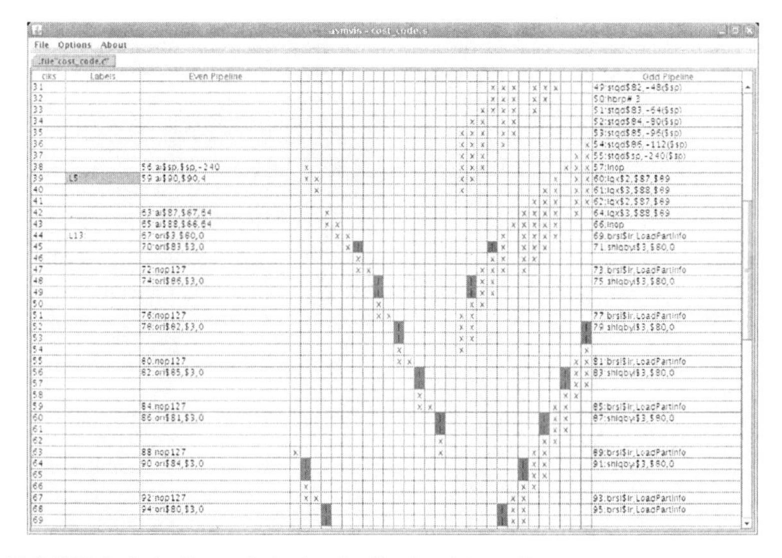

Fig. 11.9 SPU dual-pipeline optimization for Simulated Annealing.

From line 39–42, on branch L5: (which is highlighted) the for-loop on the edges begins. We can see the 4-loads (look in the rightmost column for the mnemonic `lqx`) which form the first part of the edge reading. Then we proceed to compute the node-numbers on these 2 edges, read the partition information, xor-it, and add it to the cumulative cost using `spu_orx`. This dependency graph is shown in Figure 11.10; inverting this dependency graph gives us a schedule of operations (which is how we arrived at the code in Listing 11.15). Unrolling the edge loop is equivalent to placing another copy of this graph next to the first one and scheduling instructions (taking care of dependencies and hazards).

Now that we have defined and implemented the simulated annealing solver we can use it to solve a number of problems. In this section we shall use it to compute a graph partition. In Chapter 17 we use simulated annealing to compute 3d nano-structures of molecules given their inter-atomic distance distributions, in Chapter 20 we shall see that a number of problems in VLSI CAD can be solved using annealing, eg., placement, micro-word minimization, floorplanning and scheduling. The graph partitioning algorithm is simple as it uses simulated annealing to do the actual work:

```
static void RunTrials( GraphControlBlock cb,int n,
           vector float temp ) {
  int i,j;
  unsigned int cost = INT_MAX,ncost;
  for(i=0;i<n;++i) {
    unsigned int n1,n2;
    n1 = rand() % cb.N;
    n2 = rand() % cb.N;
    SwapNodes( n1,n2 );
    ncost = ComputeCost(cb);
    if(__builtin_expect((ncost>cost),1)) {
      //printf("\n %d %d %d",i,n1,n2);
      SwapNodes( n1,n2 );
```

```
    Data: Graph G = (V,E)
    Result: Bi-partition cost with node-partition
 1  int GraphPartSA(Graph G, ControlBlock cb) begin
 2      for i = 0; i < n; ++i do
 3          if cb.EXACT then
 4              NODE_TABLE[i] = i & 1;
 5          end
 6          else
 7              NODE_TABLE[i] = rand() & 1;
 8          end
 9      end
10      int bestcost = ∞;
11      float temp = INITIAL_TEMP;
12      for i = 1; i < cb.TOPRUN; ++i do
13          int cost = 0, ncost = 0;
14          for trial = 0; trial < cb.TRIALS; ++trial do
15              int (n1,n2) = RandomNodePair();
16              SwapNodes(n1,n2);
17              ncost=ComputeCost(G);
18              if ncost < cost then
19                  cost = ncost;
20                  bestcost = (cost < bestcost) ? cost : bestcost;
21              end
22              else if CoinToss(temp) then
23                  cost = ncost;
24              end
25              else
26                  SwapNodes(n1,n2);
27              end
28          end
29          temp = UpdateTemp(temp,cooling);
30      end
31      return(bestcost);
32  end
```

Algorithm 21: Graph partitioning using Simulated Annealing.

Fig. 11.10 Dependency analysis of ComputeCost

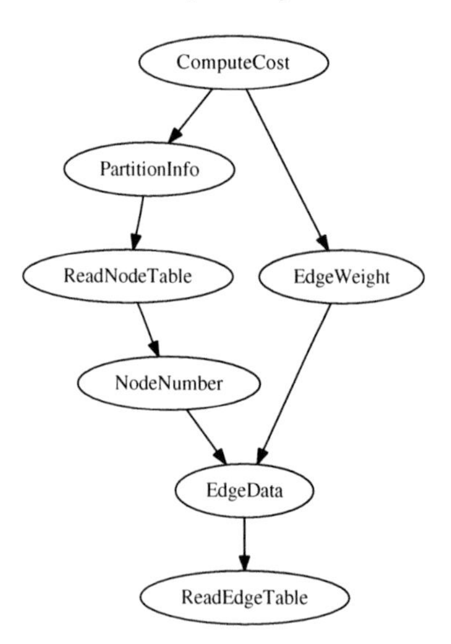


```
          continue; }
15      cost=ncost;
        if (__builtin_expect((ncost<best_cost),0)) {
          best_cost=ncost;
          if(cb.SAVE)
      for(j=0;j<cb.N;++j) { BEST_NODE_TABLE[j]=NODE_TABLE[j]; }
20      }
      }
    }
```

Listing 11.16 Implementation of simulated annealing session running on SPE.

```
   sizeof(CB) = 32
   Read G=(V,E) N=500 M= 1868
   Hello from SPE (0x1002e130)
   N=500 M=1868
5  Cost = 1390
   |P1| = 250 |P2| = 250
   [  0]  1 0 1 0 0 0 0 1 0 0 0 0 1 0 0 0 0 0 1 1
   [ 21]  1 0 0 0 1 0 0 0 0 0 0 0 0 0 0 0 0 0 0 1
   [ 42]  1 1 0 0 0 0 0 0 0 0 0 0 0 0 0 0 0 0 0 0
10 .....
   [420]  1 1 1 1 1 1 1 1 1 1 1 1 0 1 1 1 1 1 1 0
   [441]  1 0 1 1 1 1 1 1 1 1 1 1 1 1 1 1 1 1 1 1
   [462]  1 1 1 1 1 1 1 1 0 1 1 1 1 1 1 1 1 1 1 1
   [483]  0 1 1 1 0 1 1 1 1 1 1 1 1 1 1 1 1 0
15 SAnnealing on (1002e130) done,cost = 168
```

Listing 11.17 Transcript of simulated annealing session running on SPE.

The code for the `RunTrials` function is shown in Listing 11.16. We use the `__builtin_expect` function to provide a hint to the SPU instruction pipeline that it is unlikely that we will frequently derive lower `bestcost`. The `SwapNodes` function just swaps partition info data amongst `NODE_TABLE[n1]` and `NODE_TABLE[n2]`. We use

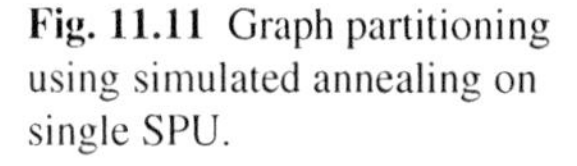

Fig. 11.11 Graph partitioning using simulated annealing on single SPU.

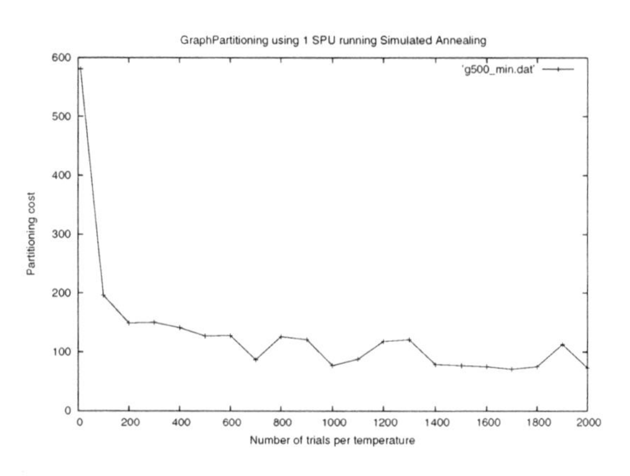

the same function to *undo* the swap when the swap does not result in either a cost reduction (greedy move), or the `coinToss` probability does not exceed the current temperature setting.

A sample output of running this program on a single SPU is shown in Listing 11.17. The initial cost of the *exact* binary partition is 1390, and it remains exact during the computation since we are only doing pairwise moves. The matrix of 0s and 1s denotes the partition assignment function; we can read the matrix to see that node 21 is assigned partition 1. The final cost of the graph partition is 168.

For randomized algorithms such as simulated annealing the multiple SPUs of the Cell offer an interesting paradigm for parallel simulations on the same data. In Figure 11.11 we show the run-time versus partitioning cost trade-off; as the number of trials per temperature iteration (shown on the X-axis) increases, we get a reduction in partitioning cost. In Figure 11.12 we have invoked the same SPU program on 4 SPUs, each SPU running with a local copy of the graph cluster in its local store. We seed the random number generator of each SPU with its `speid`. We see that although the partitioning cost curve is similar to the single SPU case, there are points where choosing the best out of 4 results is almost 30% better (look at the cost for iteration 1400 in both figures). With this experiment we have verified that our simulated annealing method is able to climb out of hills, and at the same time, now we have a working DMA based graph partitioning and clustering tool. We shall use both these facilities in the coming chapters.

11.13 Partitioning using Spectral Methods

Spectral partitioning has become one of the most successful heuristics for partitioning graphs and matrices [100]. It is also used in VLSI circuit design and simulation. Experimental work has demonstrated that spectral methods find good partitions of

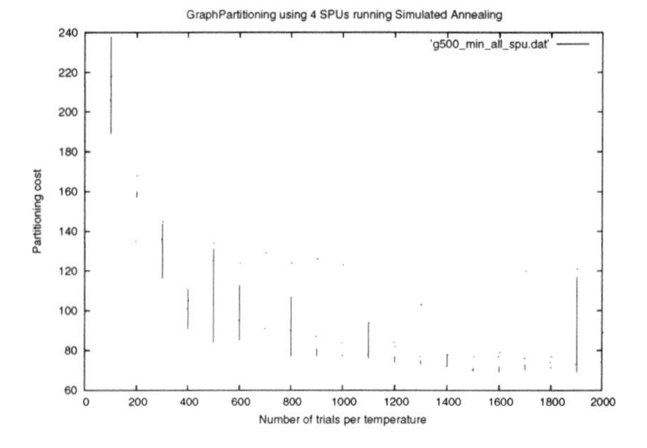

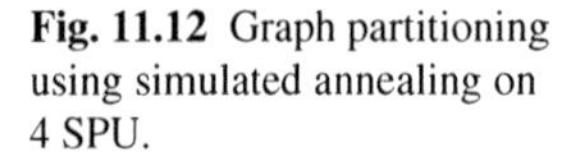
Fig. 11.12 Graph partitioning using simulated annealing on 4 SPU.

the graphs and matrices that arise in many applications. In this section we discuss the theory behind spectral partitioning and show an implementation on the SPU using SIMD. A good introduction to the theory of spectral partitioning in given in [118], which also contains excellent references for this important problem.

Spectral partitioning methods use the Fiedler vector, the eigenvector of the second-smallest eigenvalue of the Laplacian matrix—to find a small separator of a graph. These methods are important components of many scientific numerical algorithms and have been demonstrated by experiment to work extremely well. In this paper, we show that spectral partitioning methods work well on bounded-degree planar graphs and finite element meshes— the classes of graphs to which they are usually applied. While naive spectral bisection does not necessarily work, Spielman and Teng [118] prove that spectral partitioning techniques can be used to produce separators whose ratio of vertices removed to edges cut is $O(\sqrt{n})$ for bounded degree planar graphs and two-dimensional meshes. Since these are very frequently the graphs that occur in VLSI design, it is useful to have an SPU implementation of a spectral partitioner.

Donath and Hoffman [39] first suggested using the eigenvectors of adjacency matrices of graphs to find partitions. Fiedler associated the second-smallest eigenvalue of the Laplacian of a graph with its connectivity and suggested partitioning by splitting vertices according to their value in the corresponding eigenvector.

11.13.1 Theoretical formulation of spectral partitioning

As before we write a graph as $G = (V,E)$ representing a connected graph on n vertices. As shown above in Eqn 11.2 on pp. 199 the cost function for a partition is the *cut size*, or the number of edges with one vertex in one partition, and the other vertex in the different partition. The *cut ratio* is also denoted as above:

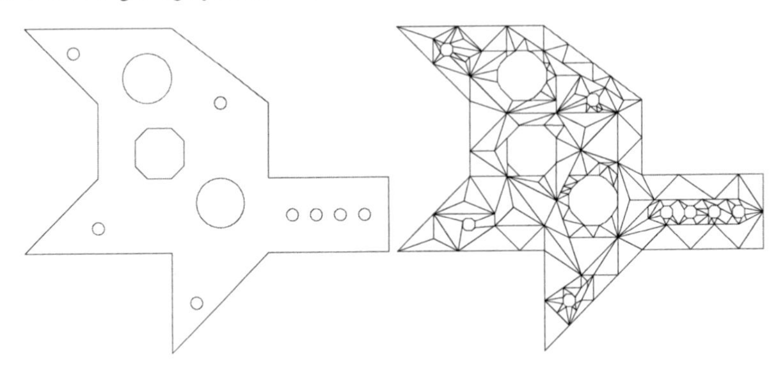

Fig. 11.13 CAD drawing of machine part and its triangulated mesh.

$$\phi(A,\bar{A}) = \frac{|E(A,\bar{A})|}{\min(|A|,|\bar{A}|)}$$

The adjacency matrix of graph G written as $A(G)$ is the $n \times n$ matrix whole (i,j) entry is 1 iff $(i,j) \in E$, 0 otherwise. The diagonal entries of the form (i,i) are defined to be 0. Let matrix D be a diagonal matrix with $D_{i,i} = deg(v_i)$, the degree of vertex i of G. The *Laplacian* $L(G)$ of G is then defined as the matrix $D - A$. This is a $n \times n$ matrix with positive values on the diagonal and negative elsewhere. It is also symmetric and positive semi-definite.

11.13.2 Application to graph partitioning and finite-element meshes

Consider the CAD drawing of a machine part as shown in Figure 11.13 and its associated finite-element mesh division. The FEM can be represented as a graph as shown in Figure 11.14.

The Laplacian of the graph shown in Figure 11.14 is shown below (note that the data in incomplete as size has been truncated to fit the margins of the text). The actual graph has 190 vertices. The *band* structure of the planar graph can be shown below in Figure 11.15.

This graph is symmetric, positive definite and hence the eigenvalues are real. The second smallest eigenvector is computed using SPU code and the vertices are partitioned based on the median of this eigenvector. All vertices whose eigenvector component is less than the median are colored green and remaining are colored red. The resulting graph partition is shown in Figure 11.16.

Thus Spectral partitioning can be an effective method for graph partitioning especially with low degree graphs with strong separators (using $\sqrt{n}$ vertices). Planar mesh graphs, which are used frequently in FEM are thus good candidates for this method.

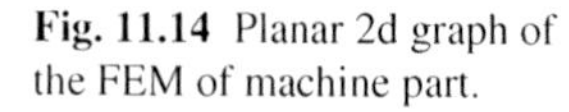

Fig. 11.14 Planar 2d graph of the FEM of machine part.

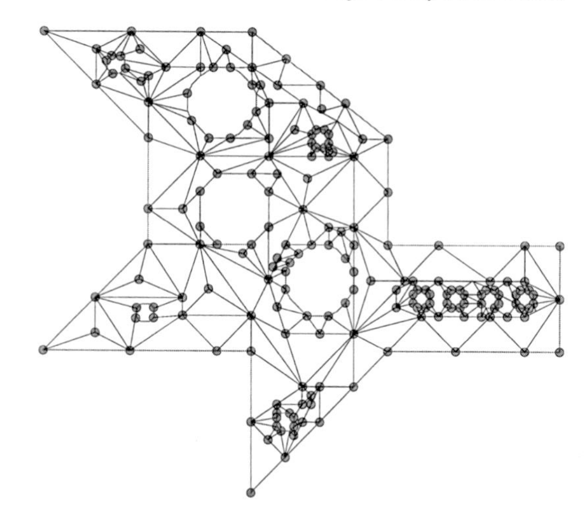

```
3 -1 -1 -1 0 0 0 0 0 0 0 0 0 0 0 0 0 0 0 0 0 0 0 0 0 0 0 0 0 0 0 0 0 0 0 0 0 0 0 0 0 0 0 0 0 0 0
-1 5 -1 -1 -1 -1 0 0 0 0 0 0 0 0 0 0 0 0 0 0 0 0 0 0 0 0 0 0 0 0 0 0 0 0 0 0 0 0 0 0 0 0 0 0 0 0
-1 -1 7 0 -1 0 -1 0 0 -1 -1 0 0 0 0 0 0 0 0 -1 0 0 0 0 0 0 0 0 0 0 0 0 0 0 0 0 0 0 0 0 0 0 0 0 0
-1 -1 0 6 0 -1 0 -1 0 0 0 0 0 -1 -1 0 0 0 0 0 0 0 0 0 0 0 0 0 0 0 0 0 0 0 0 0 0 0 0 0 0 0 0 0 0 0
0 -1 -1 0 4 -1 -1 0 0 0 0 0 0 0 0 0 0 0 0 0 0 0 0 0 0 0 0 0 0 0 0 0 0 0 0 0 0 0 0 0 0 0 0 0 0 0 0
0 -1 0 -1 -1 4 0 -1 0 0 0 0 0 0 0 0 0 0 0 0 0 0 0 0 0 0 0 0 0 0 0 0 0 0 0 0 0 0 0 0 0 0 0 0 0 0 0
0 0 -1 0 -1 0 3 -1 0 0 0 0 0 0 0 0 0 0 0 0 0 0 0 0 0 0 0 0 0 0 0 0 0 0 0 0 0 0 0 0 0 0 0 0 0 0 0
0 0 0 -1 0 -1 -1 5 0 0 0 0 -1 -1 0 0 0 0 0 0 0 0 0 0 0 0 0 0 0 0 0 0 0 0 0 0 0 0 0 0 0 0 0 0 0 0
0 0 0 0 0 0 0 0 3 0 0 -1 -1 -1 0 0 0 0 0 0 0 0 0 0 0 0 0 0 0 0 0 0 0 0 0 0 0 0 0 0 0 0 0 0 0 0 0
0 0 -1 0 0 0 0 0 0 2 -1 0 0 0 0 0 0 0 0 0 0 0 0 0 0 0 0 0 0 0 0 0 0 0 0 0 0 0 0 0 0 0 0 0 0 0 0 0
0 0 -1 0 0 0 0 0 0 -1 6 -1 0 -1 0 0 0 0 -1 -1 0 0 0 0 0 0 0 0 0 0 0 0 0 0 0 0 0 0 0 0 0 0 0 0 0 0
0 0 0 0 0 0 0 0 -1 0 -1 4 -1 -1 0 0 0 0 0 0 0 0 0 0 0 0 0 0 0 0 0 0 0 0 0 0 0 0 0 0 0 0 0 0 0 0 0
0 0 0 0 0 0 0 -1 -1 0 0 -1 4 -1 0 0 0 0 0 0 0 0 0 0 0 0 0 0 0 0 0 0 0 0 0 0 0 0 0 0 0 0 0 0 0 0 0
0 0 0 -1 0 0 0 -1 -1 0 -1 -1 -1 11 -1 -1 -1 -1 -1 0 0 0 0 0 0 0 0 0 0 0 0 0 0 0 0 0 0 0 0 0 0 0 0
0 0 0 -1 0 0 0 0 0 0 0 0 -1 4 -1 0 0 0 0 0 0 0 0 0 0 0 0 0 0 0 0 0 0 0 0 0 0 0 0 0 0 0 0 0 0 0
0 0 0 0 0 0 0 0 0 0 0 0 0 -1 -1 10 -1 0 0 0 0 0 0 0 0 0 0 0 0 0 0 0 0 0 0 0 0 0 0 0 0 0 0 0 0 0 0
0 0 0 0 0 0 0 0 0 0 0 0 0 -1 0 -1 4 -1 0 0 0 0 0 0 0 0 0 0 0 0 0 0 0 0 0 0 0 0 0 0 0 0 0 0 0 0 0
0 0 0 0 0 0 0 0 0 0 0 0 0 -1 0 0 -1 3 -1 0 0 0 0 0 0 0 0 0 0 0 0 0 0 0 0 0 0 0 0 0 0 0 0 0 0 0 0 0
0 0 0 0 0 0 0 0 0 0 -1 0 0 -1 0 0 0 -1 5 -1 -1 0 0 0 0 0 0 0 0 0 0 0 0 0 0 0 0 0 0 0 0 0 0 0 0 0
0 0 -1 0 0 0 0 0 0 0 -1 0 0 0 0 0 0 0 -1 6 -1 -1 0 -1 0 0 0 0 0 0 0 0 0 0 0 0 0 0 0 0 0 0 0 0 0 0
0 0 0 0 0 0 0 0 0 0 0 0 0 0 0 0 0 0 -1 -1 4 -1 0 0 0 0 0 0 0 0 0 0 0 0 0 0 0 0 0 0 0 0 0 0 0 0 0
0 0 0 0 0 0 0 0 0 0 0 0 0 0 0 0 0 0 0 -1 -1 4 -1 0 0 0 0 0 0 0 0 0 0 0 0 0 0 0 0 0 0 0 0 0 0 0 0
0 0 0 0 0 0 0 0 0 0 0 0 0 0 0 0 0 0 0 0 0 -1 7 -1 -1 -1 -1 0 0 0 0 0 0 0 0 0 0 0 0 0 0 0 0 0 0 0
0 0 0 0 0 0 0 0 0 0 0 0 0 0 0 0 0 0 0 -1 0 0 -1 3 -1 0 0 0 0 0 0 0 0 0 0 0 0 0 0 0 0 0 0 0 0 0
0 0 0 0 0 0 0 0 0 0 0 0 0 0 0 0 0 0 0 0 0 0 -1 -1 4 -1 0 0 0 -1 0 0 0 0 0 0 0 0 0 0 0 0 0 0 0 0 0
0 0 0 0 0 0 0 0 0 0 0 0 0 0 0 0 0 0 0 0 0 0 -1 0 -1 5 0 -1 -1 -1 0 0 0 0 0 0 0 0 0 0 0 0 0 0 0 0
0 0 0 0 0 0 0 0 0 0 0 0 0 0 0 0 0 0 0 0 0 0 -1 0 0 0 4 -1 0 0 0 0 0 0 0 0 0 0 0 0 0 0 0 0 0 0 0 0
0 0 0 0 0 0 0 0 0 0 0 0 0 0 0 0 0 0 0 0 0 0 0 0 0 -1 -1 5 -1 0 0 0 0 0 0 0 0 0 0 0 0 0 0 0 0 0 0
0 0 0 0 0 0 0 0 0 0 0 0 0 0 0 0 0 0 0 0 0 0 0 0 0 -1 0 -1 7 -1 -1 0 0 0 0 0 0 0 0 0 0 -1 0 -1 0 0
0 0 0 0 0 0 0 0 0 0 0 0 0 0 0 0 0 0 0 0 0 0 0 0 -1 -1 0 0 -1 5 -1 -1 0 0 0 0 0 0 0 0 0 0 0 0 0 0
0 0 0 0 0 0 0 0 0 0 0 0 0 0 0 0 0 0 0 0 0 0 0 0 0 0 0 0 -1 -1 4 -1 0 0 0 0 0 0 0 0 -1 0 0 0 0 0 0
0 0 0 0 0 0 0 0 0 0 0 0 0 0 0 0 0 0 0 0 0 0 0 0 0 0 0 0 0 -1 -1 5 -1 -1 0 0 0 0 0 0 -1 0 0 0 0 0
0 0 0 0 0 0 0 0 0 0 0 0 0 0 0 0 0 0 0 0 0 0 0 0 0 0 0 0 0 0 0 -1 4 -1 -1 0 0 0 0 0 -1 0 0 0 0 0 0
0 0 0 0 0 0 0 0 0 0 0 0 0 0 0 0 0 0 0 0 0 0 0 0 0 0 0 0 0 0 0 -1 -1 4 -1 -1 0 0 0 0 0 0 0 0 0 0 0
0 0 0 0 0 0 0 0 0 0 0 0 0 0 0 0 0 0 0 0 0 0 0 0 0 0 0 0 0 0 0 0 -1 -1 8 -1 -1 -1 0 0 0 0 0 0 0 0
0 0 0 0 0 0 0 0 0 0 0 0 0 0 0 0 0 0 0 0 0 0 0 0 0 0 0 0 0 0 0 0 0 -1 -1 4 0 0 0 0 0 0 0 0 0 0 0 0
0 0 0 0 0 0 0 0 0 0 0 0 0 0 0 0 0 0 0 0 0 0 0 0 0 0 0 0 0 0 0 0 0 0 -1 0 3 0 0 0 0 0 0 0 0 -1 -1
0 0 0 0 0 0 0 0 0 0 0 0 0 0 0 0 0 0 0 0 0 0 0 0 0 0 0 0 0 0 0 0 0 0 -1 0 0 2 -1 0 0 0 0 0 0 0 0 0
```

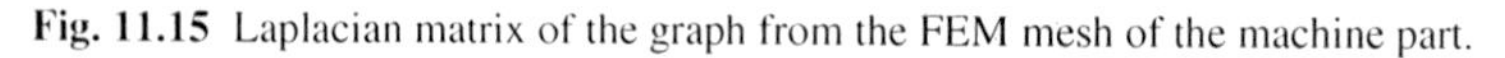

Fig. 11.15 Laplacian matrix of the graph from the FEM mesh of the machine part.

```
/////////////////////////////////////////
// Program: Spectral
// Author : Sandeep Koranne
/////////////////////////////////////////
float DotProd(const V& a, const V& b) {
  float sum=0;
```

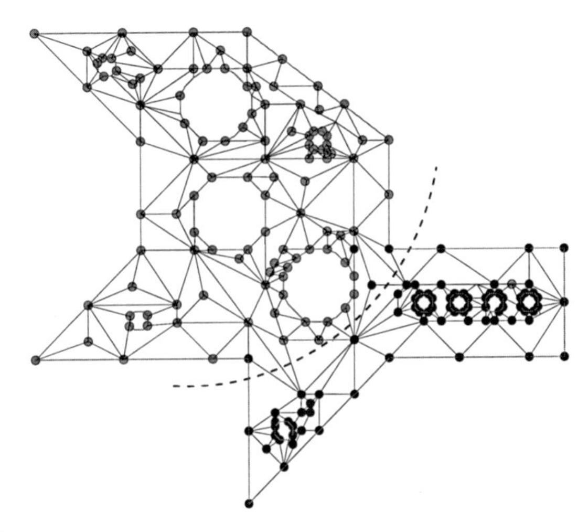

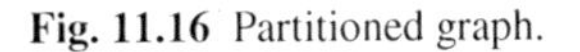

Fig. 11.16 Partitioned graph.

```
  assert( a.size() == b.size() );
  for(int i=0;i<a.size();++i) sum += a[i]*b[i];
  return sum;
}
void ScalarProd( V& v, float x ) {
  for(int i=0;i<v.size();++i) v[i] *= x;
}
float Max(const V& v) {
  float m=v[0];
  for(int i=0;i<v.size();++i) m=std::max(m,v[i]);
  return m;
}
V MV(const M& m, const V& x) {
  V ans(x.size(),0);
  assert(m.size() == x.size() );
  for(int i=0;i<m.size();++i) ans[i] = DotProd(m[i],x);
  return ans;
}
void Normalize( V& a ) { // modify in place
  float sum=0;
  for(int i=0;i<a.size();++i) sum += a[i]*a[i];
  sum = std::sqrt( sum );
  for(int i=0;i<a.size();++i) a[i] = a[i]/sum;
}
float Error(const M& m, const V& x) {
  float err=0;
  V mx = MV( m, x); //std::cout << std::endl << "mx = " << mx;
  float f = 1.0/mx[0];
  V vx = x;
  ScalarProd(vx,f); // in place
  float mm = Max(vx);
  float sum=0.0;
  for(int i=0;i<x.size();++i) {
    float t = (mx[i] - x[i])/mm;
    sum += (t*t);
  }
  return sum/(float)x.size();
}
#define MAX_ITER 100
void FindEigenVector(const M& m, V& guess, const V& v1) {
```

```
    for(int iter=0; iter < MAX_ITER; ++iter) {
      float dp = DotProd(guess,v1);
      V v2(v1.size(),0.0);
50    for(int i=0; i<v1.size(); ++i) {
        v2[i] = guess[i] - (dp*v1[i]);
      }
      v2 = MV( m,v2 );
      float e0 = 1.0/v2[0];
55    for(int i=0;i<v2.size();++i) v2[i] *= e0;
      guess = v2;
    }
  }
```

Listing 11.18 Spectral partitioning.

11.14 On-line streaming algorithm for graph properties

In this section we describe a method to count the number of distinct elements in a data stream. This algorithm can be used to count graph properties, such as triangles in a graph, in an online algorithm. The algorithm is based on the research paper by Flajolet and Martin [46].

Let $a_1, a_2, \ldots, a_n$ be a sequence of n elements from the domain $[m] = \{1, \ldots, m\}$. The zeroth-frequency moment of this sequence is the number of distinct elements that occur in the sequence and is denoted F_0.

The problem of estimating F_0 in an online stream has become very important with the advent of the Internet and associated routing protocols. Counting the number of distinct elements in a (column of a relational) table of data is a fairly fundamental problem in databases. This has applications to estimating the selectivity of queries, designing good plans for executing a query, etc.

Another application of counting distinct elements is in routing of Internet traffic. The router usually has very limited memory, but it is desirable to have the router gather various statistical properties (say, the number of distinct destination addresses) of the traffic flow. The number of distinct elements is also a natural quantity of interest in several large data set applications (eg., the number of distinct queries made to a search engine over a week). Distinct-values estimates are also useful for network resource monitoring, in order to estimate the number of distinct destination IP addresses, source-destination pairs, requested urls, etc. Distinct-values estimation can also be used as a general tool for duplicate- insensitive counting: Each item to be counted views its unique id as its value, so that the number of distinct values equals the number of items to be counted. Duplicate-insensitive counting is useful in mobile computing to avoid double- counting nodes that are in motion.

Obviously F_0 can be computed exactly in one pass through the entire data set, by keeping track of all the unique values observed in the stream. In the data stream model, an algorithm is considered efficient if it makes one (or a small number of) passes over the input sequence, uses very little space, and processes each element of the input very quickly. Since we are designing our implementation to use the local store of the SPE we have reserved 64 KB block of memory for our temporary buffer to process the stream which is brought into the local store using DMA. Algorithmically, the workspace and time-complexity of the data-stream algorithm are of importance. For implementation, we have to design for uniform register file, efficient use of dual-pipeline and 4-way SIMD optimization. Since data-stream is single-pass or multipass, depending on the algorithm we can choose to use single-buffer or double-buffering (if we use dual-pass mode to increase the accuracy, then double buffering is not possible, we would have to use a variant of triple buffering). We have been able to achieve more than 10x run-time reduction using the SPEs. Our implementation can handle TEXT, GZIPPED and binary data and uses upto 6 SPEs in parallel to compute F_0. We have also implemented a non-parallel PPE only version and run-time comparisons are presented in Figure 11.18.

Our implementation is a straightforward implementation of Flajolet and Martin's original paper [46] with a hash table built using SPU intrinsics. The basic method is as follows: the algorithm first picks a random hash function $h : [m] \rightarrow [0,1]$. It then applies h to all the elements in a and maintains the value $v = min_{j=1}^{n} h(a_j)$. The algorithm has the right approximation (in the expectation sense) because if there are F_0 independent and uniform values in [0; 1], then their expected minimum is around $1/F_0$.

The intuition behind the FM algorithm is as follows. Using a hash function ensures that all items with the same value will make the same selection; thus the final bit vector M is independent of any duplications among the item values. For each distinct value, the bit b is selected with probability $2^{(b+1)}$. Accordingly, we expect $M[b]$ to be set if there are atleast 2^{b+1} distinct values. We have used multiple hash functions where the same data is sent to the 6 SPEs in parallel and each SPE uses a different hash-function, and the value of F_0 is the average value of the result returned by each SPE. In this problem the SPEs send the result back using MailBox. A standalone PPE only code has also been implemented for performing run-time benchmarking, the results are shown in Figure 11.18. We get almost 100x speedup using the SPEs, as not only is there an 6x parallelism advantage, the SPE code also uses 4-way integer SIMD.

The implementation of the FM algorithm is shown below in Listing 11.20 (for the SPE code), and Listing 11.21 for the PPE part. A Lisp based exact F_0 calculating generator is shown first in Listing 11.19.

```
(defun generate-random-numbers (file-name n m)
  "Generate numbers, optionally keep track of properties"
  (let ((h (make-hash-table))
  (fs (open file-name :direction :output :if-exists :supersede)))
5   (dotimes (i n)
      (let ((r (random m)))
  (setf (gethash r h) r)
  (format fs "~D~%" r)))
    (format fs "~D~%" (hash-table-count h))
10  (close fs)
    (hash-table-count h)))
(defun main ()
  (generate-random-numbers "lnum.txt" 100000000 100000))
(format t "~%Generated ~D numbers " (main))
15 (quit)
```

Listing 11.19 Lisp based test case generator for F_0 moment.

```
/////////////////////////////////////////////////////////////////
// Program   : MBox Reduce Paradigm
// Author    : Sandeep Koranne
// Purpose   : Implementation of Graph Algorithms on Cell Broadband

#include <spu_mfcio.h>
#include <stdio.h>
#include <math.h>
#include <simdmath.h>

//#define DEBUG_FULL
#ifdef DEBUG_FULL
  #define DEBUG
//#define DEBUG_SORT
  #define DEBUG_MST
#endif
typedef struct _mbox_control_block {
  unsigned int N;
  unsigned int M;
  unsigned long long addr;
  unsigned int N2;
  unsigned int M2;
  unsigned long long addr2;
} MBC;
#define NUM_ELEMENTS (1024*4)

MBC cb __attribute__ ((aligned (128)));
volatile vector unsigned int DATA_LS [ NUM_ELEMENTS/4 ]
         __attribute__ ((aligned (128)));
static volatile int COMMAND;
static void BufferGet(MBC *cb, int c, int tag_id) {
  mfc_get(DATA_LS, c?cb->addr:cb->addr2,
          sizeof(int)*NUM_ELEMENTS, tag_id, 0, 0);
  mfc_write_tag_mask(1<< tag_id);
  mfc_read_tag_status_all();
}

int SOMER;
int ScalarHFunc(int key) {
  int c2=0x27d4eb2d; // a prime or an odd constant
  key = (key ^ 61) ^ (key >> 16);
  key = key + (key << 3);
  key = key ^ (key >> 4);
  key = key * c2;
  key = key ^ SOMER; // some randomness
  key = key ^ (key >> 15);
  return key;
}

vector unsigned int HFunc(vector unsigned int key) {
  int k0 = spu_extract(key,0);
  int k1 = spu_extract(key,1);
  int k2 = spu_extract(key,2);
  int k3 = spu_extract(key,3);
  k0 = ScalarHFunc(k0);
  k1 = ScalarHFunc(k1);
  k2 = ScalarHFunc(k2);
  k3 = ScalarHFunc(k3);
  key=spu_insert(k0,key,0);
  key=spu_insert(k1,key,1);
  key=spu_insert(k2,key,2);
  key=spu_insert(k3,key,3);
  return key;
}

static vector unsigned int hash_code;
```

```
   static void CalculateF0Moment( MBC cb ) {
     int i,j,k;
70   vector unsigned int hash_function;
     vector unsigned int ttz;
     unsigned int trailing_zeros[4];
     for(i=0;i < NUM_ELEMENTS/4; ++i ) {
       vector unsigned int a =  DATA_LS[i];
75     hash_function = HFunc( a );
       for(k=0;k<4;++k) {
         unsigned int value = spu_extract( hash_function, k);
         trailing_zeros[k]=0;
         for(j=0;j<32;++j,++trailing_zeros[k]) {
80         if( value & ( 1 << trailing_zeros[k] ) ) break;
         }
         trailing_zeros[k] = (1 << (trailing_zeros[k]-1));
         ttz=spu_insert( trailing_zeros[k],ttz,k );
       }
85     hash_code = spu_or(hash_code, ttz);
     }
   }

   static vector unsigned int ReturnF0Moment(void) {
90   vector unsigned int ans;
     vector float val = (vector float)spu_splats( 1/0.77351);
     val = spu_mul( spu_convtf(hash_code,0), val);
     ans = (vector unsigned int)spu_convts(val,1);
     return ans;
95 }

   int main(unsigned long long speid,
            unsigned long long argp,
            unsigned long long envp __attribute__ ((__unused__))) {
100  int tag_id = mfc_tag_reserve();
     int i,j,maxv,minv;
     vector unsigned int result;
     int final_result=0;
     srand( speid );
105  hash_code = (vector unsigned int)spu_splats(0);
     SOMER = ScalarHFunc( speid );
     mfc_get(&cb, argp, sizeof(cb), tag_id, 0, 0);
     mfc_write_tag_mask(1<< tag_id);
     mfc_read_tag_status_all();
110  spu_write_out_mbox( 17 ); // am alive
     COMMAND = spu_read_in_mbox();
     while( COMMAND != 8 ) {
       BufferGet( &cb, COMMAND,tag_id );
       CalculateF0Moment( cb );
115    COMMAND = spu_read_in_mbox();
     }
     // got termination orders
     result = ReturnF0Moment();
     for(i=0;i<4;++i)
120    final_result += spu_extract( result, i);
     if(spu_stat_out_mbox())
       spu_write_out_mbox( final_result );
     return 0;
   }
```

Listing 11.20 SPE implementation of F_0 moment calculation on data streams.

```
   /////////////////////////////////////////////////////////////////
   // Program   : F0 Moment Calculation in Stream Data
   // Author    : Sandeep Koranne
   // Purpose   : Implement graph algorithms on the Cell Broadband
5  /////////////////////////////////////////////////////////////////
```

```
#include <iostream>
#include <libspe2.h>
#include <stdlib.h>
#include <string.h>
#include <errno.h>                                                          10
#include <sys/types.h>
#include <sys/stat.h>
#include <fcntl.h>
#include <unistd.h>
#include <sys/mman.h>                                                       15
#include <cassert>

#include "../align.h"

#define DEBUG // switch off in production                                   20
#define MAX_SPES 6

typedef struct _mbox_control_block {
  unsigned int N;
  unsigned int M;                                                           25
  unsigned long long addr;
  unsigned int N2;
  unsigned int M2;
  unsigned long long addr2;
} MBC;                                                                      30
#define NUM_ELEMENTS (1024*4)
MBC *cb[MAX_SPES];
int *DATA __attribute__ ((aligned(128)));
int *DATA2 __attribute__ ((aligned(128)));
                                                                            35
extern spe_program_handle_t moment_function;
typedef struct ppu_pthread_data {
  void *argp;
  spe_context_ptr_t speid;
  pthread_t pthread;                                                        40
} ppu_pthread_data_t;
ppu_pthread_data_t *datas;

void *ppu_pthread_function(void *arg) {
  ppu_pthread_data_t *datap = (ppu_pthread_data_t *)arg;                   45
  unsigned int entry = SPE_DEFAULT_ENTRY;
  int rc = spe_context_run(datap->speid, &entry,
                           0, datap->argp, NULL, NULL);
  pthread_exit(NULL);
}                                                                           50

static int PrepData( const char* fileName, int expected_size ) {
  int filedes = -1;
  if( !strcmp(fileName,"-") ) {
    filedes = 0;                                                            55
  } else {
    filedes = open( fileName, O_RDONLY | O_LARGEFILE );
  }
  if( filedes == -1 ) {
    perror("open: failed");                                                 60
    exit(1);
  }
  int br=1;
  unsigned int command_code = 1;
  unsigned int retvalue=0;                                                  65
  for(int spi=0;spi<MAX_SPES;++spi) {
    spe_out_mbox_read( datas[spi].speid, &retvalue, 1 );
  }
  for(int BR=0;BR<expected_size;++BR) {
    if(command_code==1) {                                                   70
      br = read( filedes, DATA, NUM_ELEMENTS*sizeof(int) );
    }
```

```
75      else {
          br = read( filedes, DATA2, NUM_ELEMENTS*sizeof(int) );
        }
        //std::cout << std::endl << "Read " << br << " bytes." << std::endl;
        if( br < 0 )
          perror("read");
80      for(int spi=0;spi<MAX_SPES;++spi) {
          MBC* cbp=(MBC*)(datas[spi].argp);
          cbp->M=br;
          spe_in_mbox_write( datas[spi].speid,
                             &command_code, 1 , SPE_MBOX_ANY_BLOCKING);
85      }
        if (command_code==1)
          command_code=2;
        else
          command_code=1;
90    }
      for(int spi=0;spi<MAX_SPES;++spi) {
        if(spe_out_mbox_status( datas[spi].speid ) ) {
          spe_out_mbox_read( datas[spi].speid, &retvalue, 1 );
        }
95    }
      command_code=8;
      for(int spi=0;spi<MAX_SPES;++spi) {
        spe_in_mbox_write( datas[spi].speid,
                           &command_code, 1 , SPE_MBOX_ALL_BLOCKING);
100   }
      std::cout << std::endl << "Processing complete." << std::endl;
      return (br/sizeof(int));
    }

105 static void ShowData() {
      for(int i=0; i < NUM_ELEMENTS; ++i )
        std::cout << " " << DATA[i];
      std::cout << std::endl;
    }
110
    static void Usage() {
      std::cerr << std::endl << "moment: <file-or-stream>\n";
    }

115 int main( int argc, char* argv [] ) {
      int expected_size = 16;// buffers of 16K
      if( argc < 1 ) { Usage(); return 1; }
      std::cout << "F0 Moment Calculation Program on Data Streams\n";
      std::cout << "sizeof(MBC) = "<<sizeof(MBC)<<std::endl;
120   DATA = (int*)_malloc_align(sizeof(int)*NUM_ELEMENTS,7);
      DATA2 = (int*)_malloc_align(sizeof(int)*NUM_ELEMENTS,7);

      //ShowData();
      for(int i=0;i<MAX_SPES;++i) {
125     cb[i] = (MBC*)_malloc_align(sizeof(MBC), 5);
      }
      datas = (ppu_pthread_data_t*)
        malloc( sizeof(ppu_pthread_data_t)*MAX_SPES);

130   for (int i = 0; i < MAX_SPES ; i++) {
        datas[i].speid = spe_context_create (0, NULL);
        spe_program_load (datas[i].speid, &moment_function);
        cb[i]->addr = (unsigned long long int) DATA;
        cb[i]->addr2= (unsigned long long int) DATA2;
135     cb[i]->M = NUM_ELEMENTS; // number of data
        cb[i]->N = 7; // number to divide
        datas[i].argp = (unsigned long long*) cb[i];
        pthread_create (&datas[i].pthread, NULL,
                        &ppu_pthread_function, &datas[i]);
      }
```

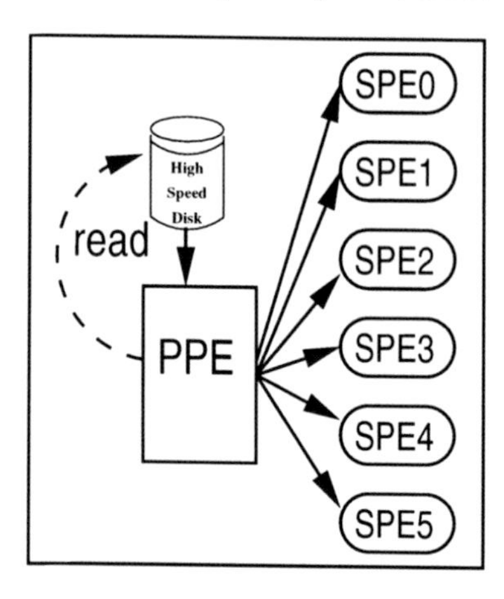

Fig. 11.17 Implementation of F_0 Moment calculator using SPEs, the PPE performs file read only. Each SPE initiates DMA and computes a unique hash function. When the PPE send a mailbox message indicating end-of-file or request for hash-code, each SPE sends the currently computed F_0 back to the PPE using 32-bit mailbox message. The PPE then performs a 6-way average and returns the result to the user.

```
140      sleep(2); // or put mutex
         if( argc > 2 ) expected_size = atoi( argv[2] );
         int num_elements = PrepData( argv[1], expected_size );

         for (int i = 0; i < MAX_SPES ; i++) {
145        pthread_join (datas[i].pthread, NULL);
         }
         std::cout << std::endl << "Processing complete..." << std::endl;
         float avg=0.0;
         for (int i = 0; i < MAX_SPES ; i++) {
150        unsigned int mbret=0;
           while( spe_out_mbox_status( datas[i].speid ) <= 0 ) ;
           spe_out_mbox_read( datas[i].speid, &mbret, 1 );
           avg += mbret;
           spe_context_destroy (datas[i].speid);
155      }
         avg /= (MAX_SPES*1.25);
         std::cout << std::endl << "Moment F0 Mean is " << (int)avg << std::endl;
         free( datas );
         _free_align( cb );
160      _free_align( DATA );
         return (EXIT_SUCCESS);
       }
```

Listing 11.21 PPE implementation of F_0 moment calculation on data streams.

11.14.1 Discussion of implementation

It is instructive to consider the implementation of the F_0 Moment calculator as many of the design choices in the program were made with deliberation. Consider the following:

1. How to use the F_0 Moment calculator on streaming files: we have written the interface to cater to 4-bit signed integer format binary files. In order to process text

data, we wrote a simple ASC2BIN converter which reads in data from `std::cin` and writes binary output to `std::cout`. This is used as a pipe on the command line; it can be used in conjunction with `gunzip -c` pipe to process gzipped data. An example is shown below:

```
time gunzip -c lnum.txt.gz | ./asc2bin | ./moment.cell.exe -
```

In the above command line, the gzipped file lnum.txt.gz is unzipped using gunzip, and converted to binary using asc2bin, it is then consumed by our binary `moment.cell.exe` which has a `-` on the commandline telling it to not open a disk file, but to read input from stdin (file descriptor 0). Our implementation also has a third argument which tells how many 16K blocks it should process. This is useful in benchmarking and performing scalability tests.

2. How to use SPE for parallel processing: in the Flajolet and Martin algorithm, the hashing computation is carried out several time using different hash functions and the results are averaged. This obviously is a place where SPEs can carry out execution of an individual hash calculation in parallel. Internally, each SPE also uses SIMD to process 4 integers at the same time, thus we expect a 6x4 or 24x speedup over conventional program which has to apply the 6 hash functions in serial.
3. Using `open` system call, instead of `fopen`, or `mmap`: as we discussed above, each SPE computes a different hash function on the same data. This situation is unlike the pipeline computation we have seen in Section 8.4.1, where data is moved from one SPE to another. In that example, the first SPE was given the EA (effective address) of the file in memory (using `mmap`) and as that was the *only* SPE reading from the file, the MMU of that SPE was causing page-faults. Now consider what would happen if the file's memory address was passed to each SPE, and each of them were to map the contents into EA, and issue DMA commands. Not only would the MMUs of each SPE page-fault incessantly, but since all of them would only be accessing the memory controller, we would get nowhere close to the theoretical performance of the EIB. In our current implementation we have also used many-to-one communication (as the EIB lacks a *broadcast* mode), but we have limited the impact using double-buffering. During one iteration the PPE reads the disk contents into a buffer (`DATA`), and subsequently informs the SPEs that data is ready in `cb.addr`, in the next iteration, which overlaps with the DMA of DATA, PPE is reading data into `DATA2`, and this file read is overlapped with the SPU computation. We expect the data to be in the L2 cache for all the SPEs which send a DMA request for the data. Moreover, since we are performing the asc2bin conversion outside, we do not need to perform formatted input, nor do we need the explicit buffering of the file contents. Thus we used the `open` command to read the contents of the disk.
4. Using double buffering on the PPE, but not on the SPE: this was discussed above,
5. Why is there no processing for the remainder of the file, as file sizes don't match DMA size exactly: in the examples provided with the SDK in the DMA code, a lot of code is devoted to handling the contents of the file which lie outside of multiples of the DMA transfer count. For example, if the DMA size is 4096 bytes, and the length of the file is 4100 bytes, additional code has to be written

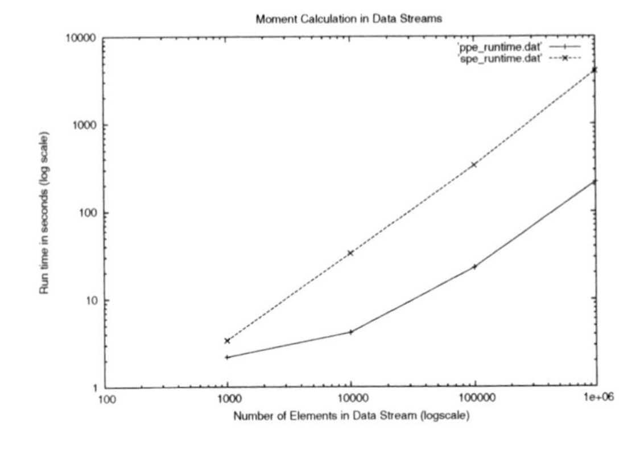

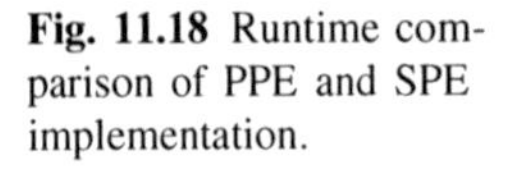

Fig. 11.18 Runtime comparison of PPE and SPE implementation.

to handle the last remaining 4 bytes. If the SPE cannot handle data sets which are multiples of the DMA count, then this processing has to be done on the PPE, which means replicating code, or perhaps even rewriting code. For the Moment calculator, we used an interesting property of the algorithm to avoid this problem. Consider a file size of 4100 bytes which is read using a DMA buffer of 4096, thus we need 2 DMA transfers, but only 4 bytes of the second transfer contribute to F_0. We use the property of the hash function to observe that the statement above also implies that the remaining 4092 bytes of the second transfer (which in our implementation will be stale data, or copies of the last transfer) do not contribute to the moment calculation. Thus we simply compute the moment for all data.

6. How to process textual data or gzipped data: we use a pipeline chain on the command line as shown above,
7. Comparing performance of our SPE accelerated solution: there is a significant performance gain when using the SPEs to compute the hash functions in parallel as opposed to computing them serially on the PPE. Since we assume 6 distinct hash functions, the PPE has to perform 6x more work than any single SPE. Moreover, since the gunzip and the asc2bin processes are also running, they consume PPE cycles, while in the case of the SPE, the PPE part of moment calculation is idle, leaving the dual-thread EMT PPE free to perform the gunzip and the asc2bin; both of these programs are heavy users of file I/O, hence good candidates for EMT.
8. Why do we get super-linear performance speedup: by combining all the aspects of parallelism we have described above, and note that all the reasons are orthogonal (i.e., they do not depend on each other), we get almost 100x speedup as shown in Figure 11.18. We can process a 1 Gb compressed data file in seconds.

11.15 Conclusion

In this chapter we have presented graph algorithms and their SPU implementation. Graph theory is a cornerstone of computer science and recently has also The problems considered included (a) min-cut, (b) minimum spanning tree, (c) comb-sort algorithm, (d) shortest-path algorithm, (e) Floyd-Warshall all-pairs shortest-path algorithm, (f) maximal independent set, (g) graph partitioning, (h) spectral methods for partitioning, and (i) on-line algorithms for streaming properties. Areas of further study can be:

1. Comprehensive graph library for the SPU: similar to the Boost Graph Library [63],
2. Shortest path computation using highway routing:
3. Graph property checking.

Chapter 12
Alternative methods for parallel programming on SPE

Abstract In this chapter we introduce alternative methods for parallel programming on the Cell Broadband Engine. These include NESL, the Nested data-parallel language. The Ct model/language proposed by Intel has striking similarities to the NESL language and we discuss Ct in this chapter as well. We also highlight other interesting parallel programming projects, including POWERLIST, ALF, Cilk++, Intel Ct and RapidMind.

12.1 Introduction

In this chapter we present 2 different models of parallel programming. The first is the nested data-parallel language NESL [6], and the second is a data-structure called Power List [87] which is designed for formal methods in parallel algorithms. We also briefly discuss Accelerated Library Framework (ALF), a parallel programming framework designed to abstract out the details of parallel programming, by letting the developer concentrate on writing *compute kernels* which run on *accelerators*. The ALF system provides facilities for task-scheduling, data-motion, automatic double-buffering, and even hybrid (a combination of Cell and Opteron) accelerators. We conclude this chapter by a brief overview of the Berkeley *dwarf* [43] paper as it is quite influential and brings the current challenges of parallel programming to the fore.

12.2 NESL: Nested Data Parallel Language

NESL is a data parallel language designed to support nested parallelism, even when the nesting occurs deep inside loops. It was designed by Blelloch et. al. at CMU. It supports the following parallel constructs:

1. Parallel sequences,

S. Koranne, *Practical Computing on the Cell Broadband Engine*,
DOI: 10.1007/978-1-4419-0308-2_12, © Springer Science + Business Media, LLC 2009

2. Parallel apply-to-each construct to apply any expression in parallel to each element of the sequence,
3. Set of parallel operations on sequences, such as summation.

NESL supports parallelism through operations on sequences. The sequences are similar to one-dimensional arrays in sequential languages but on a parallel machine are somehow spread across the processors. In NESL sequences are specified using square brackets $[2,1,9,-3]$,and indices are zero based, i.e. $a[0]$ extracts the first element of the sequence. Elements of a sequence can be of any type including other sequences for example, $[alpha, beta, gamma]$, but they must all be of the same type.

12.2.1 The Parallel Apply-to-each construct

This construct takes a set-like notation: $\{x*x : xin[1,2,3]\}$; produces another sequence, $[2,4,6]$, except that the semantics of the parallel apply are such that the multiplication operation is performed in parallel on $1,2,3$. Instead of a builtin operation like a multiplication, used defined functions can also be supplied as the operation in the parallel apply-to-each. Consider:

```
function factorial(n) =
  if (n < 2) then 1
  else n*factorial(n-1);

{factorial(x) : x in [1,2,3]};
```

As expected this produces another sequence $[1,2,6]$. On a parallel back-end for NESL all factorial calls are computed in parallel. The parallel apply-to-each construct can also be applied over multiple sequences, consider:

```
{a+b : a in [1,2,3]; b in [3,4,5]};
```

produces $[4,6,8]$ (again in parallel if the back-end supports it). The apply-to-each can also be combined with a *predicate* function to selectively pick elements from the sequences, consider:

```
{a+a : a in [-1,2,-3] | a > 0 };
```

produces a sequence $[4]$. Selected elements which pass the predicate filter are passed onto the computation in their relative order. In addition to apply-to-each, there are also parallel functions like *sum*, *take* which operate on sequences. There are functions to extract, replace, insert values in sequences.

12.2.2 Nested parallelism

In NESL the parallelism extends to all members of the sequence, horizontally (all elements of the sequence are operated in parallel) and vertically (if the element type is a sequence, then parallelism is recursively performed on that sequence as well). Consider the following example:

```
{sum(a) : a in [[1,2,3], [3,4,5]]};
```

Many parallel languages or systems would either allow parallelism to occur on the top of the sequence (on the 2 element sequence that make up the argument to apply-to-each), or they would be applied to *only* the leaf-level elements of the computation. NESL was designed to exploit parallelism all the time in the computation. This implies that not only do the 2 *sum* calls (on $[1,2,3]$ and $[3,4,5]$) issued in parallel, the system can use parallel adders to sum up the individual sequences themselves. Another example is sparse matrix multiplication. Given matrix M as follows:

$$M = \begin{bmatrix} 7.0 & 2.0 & 0 & 0 \\ 2.0 & 0.0 & 0.0 & 0 \\ 0.0 & 1.0 & 0.0 & 1.0 \\ 0.0 & 0 & 0.0 & 8.0 \end{bmatrix} \tag{12.1}$$

As a sequence this matrix M is represented as:

```
M = [[(0, 7.0), (1, 2.0)],
     [(0, 2.0)],
     [(1, 1.0), (3, 1.0)],
     [(3, 8.0)]]
```

A common operation on sparse matrices is to multiply them by a dense vector. In such an operation, the result is the dot-product of each sparse row of the matrix with the dense vector. The NESL code for taking the dot-product of a sparse row with a dense vector x is:

```
function sparse_matrix_mult(M,x) =
  {sum({v * x[i] : (i,v) in row})
   : row in M};
```

This example has nested parallelism since there is parallelism both across the rows and within each row for the dot products. The total depth of the code is the maximum of the depth of the dot products, which is the logarithm of the size of the largest row. The total work is proportional to the total number of nonzero elements.

So what, you may ask, is the connection between NESL and SPE programming (well apart from the obvious connection that NESL ran on a Connection machine) ? I have found that not only does coding an algorithm in NESL provide a working executable model to verify SPE results, it also forces you to think in terms of applying predicates (think of conditional selection) on nested sequences (think of vector float), and the parallel execution semantics are recreated using SIMD (atleast to get 4-way parallelism). NESL uses a machine specific back-end for remote parallel execution on diverse hardware (eg., Cray Y-MP, Connection Machine CM-2, and even PVM/MPI based network clusters). We have designed a back-end which is specific to SPE and have used similar nested application of functions to SIMD floats to optimize general codes on irregularly structured data. The details of our implementation are presented in the next section.

12.3 CVL and VCODE: Vector Libraries

CVL is a low-level vector library callable from the C-language. This library includes a wide variety of vector operations such as element-wise function applications, scans, reduces and permutations. It is also extremely amenable to an implementation on the SPE. In this section we describe some of the CVL design concepts and how they map on the SPE.

The key idea behind data representation in CVL is that of segments and descriptors:

```
s = [[1,2,3],[9,8,7,2]];
```

is a sequence and can be represented as an unsegmented vector of 7 elements:

```
v = [1,2,3,9,8,7,2];
```

and a segment descriptor of 2 elements:

```
d = [3,4];
```

Together v,d provide the necessary information to operate on the sequence. For the SPE we have designated local store areas where descriptors are located and a separate area for segments. We have implemented garbage collection on segment descriptors and segments which helps in mapping complex operations to individual SIMD operations on vectors.

There are six primitive data types in VCODE

1. Integer
2. Boolean
3. Float
4. Character
5. Segment descriptor

As can be seen these map very closely to the SPE data-types. In particular the single-precision floating point 4-way SIMD vector of the SPE makes the implementation of floating point VCODE operations simple. VCODE instructions are classified into five types: element-wise operations, vector instructions, segment descriptor instructions, control instructions, and I/O instructions. VCODE is a stack based language. All VCODE instructions take and remove their operands from the stack, and return their results to the stack. We have simulated a stack in the local store of the SPE. We give some examples of the vector instructions we have implemented.

1. Operation $+, -, *, /, \%$:supported types $int, float$ returns the result of performing the given operation on the top 2 elements on the stack, pushes result on the stack,

2. Operation $<,>,=,\leq,\geq,\neq$:supported types $int, float$,
3. Operation $LSHIFT, RSHIFT, NOT, AND, OR, XOR$: supported types $bool, int$,
4. Operation $SELECT$: supported types $bool, int, float$, returns the result of selecting from the 2 top stack arguments based on the 3rd argument. Equivalent to the conditional select operation of SPE (except SPE is register based),
5. Operation $RAND$: returns a random vector of integers based on the ranges specified in the data vector,
6. Operation $FLOOR, CEIL, ROUND$: returns the vector values truncated towards negative infinity (for FLOOR) and positive infinity (for CEIL),
7. Operation I_TO_F: converts integer vector to float, equivalent to SPU convtf instruction,
8. Operation $LOG, SQRT, EXP$: returns the corresponding transcendental function of the argument,
9. Trigonometric Operation: real values sin, cos, etc., functions are supported. Using the SIMD math library these can be efficiently mapped to SPU SIMD instructions,
10. Vector PLUS Scan function $+\_SCAN$: returns the segmented plus-scan of the vector. The segmented plus-scan of $1,2,3,4$ is $0,1,3,5$,
11. Vector MULT Scan function $*\_SCAN$: returns the segmented mult-scan of the vector. The segmented mul-scan of $1,2,3,4$ is $1,1,2,6$,
12. Vector MAX Scan function MAX_SCAN: returns the segmented max-scan of the vector. The segmented max-scan of $1,5,2,8,6$ is $-\infty,1,5,5,8$,
13. Vector MIN Scan function MIN_SCAN: returns the segmented max-scan of the vector. The segmented min-scan of $5,2,8,6$ is $\infty,5,2,2,2$,
14. Vector PLUS Reduce $+\_REDUCE$: returns plus-reduce of the top of stack, plus-reduce for $[5,2,8,6],[1,2]$ is $[21,3]$,
15. Vector MULT Reduce $*\_REDUCE$: returns mult-reduce of the top of stack, mult-reduce for $[5,2,8,6],[1,2]$ is $[480,2]$,
16. Vector MAX Reduce MAX_REDUCE: returns max-reduce of the top of stack, max-reduce for $[5,2,8,6],[1,2]$ is $[8,2]$,
17. Vector MIN Reduce MIN_REDUCE: returns min-reduce of the top of stack, min-reduce for $[5,2,8,6],[1,2]$ is $[2,1]$,

These operation give a flavor of the data-parallel operation capabilities of the VCODE model. In addition there are EXTRACT and REPLACE instructions along with PERMUTE instructions for rearranging vector entities. These are useful when implementing scalar access of vector operands. A simple function is LENGTH which returns the length of the vector argument.

In order to correctly implement control flow VCODE also defines control instructions which rearrange the stack. Since we are anyway simulating the stack in local store memory, our implementation of VCODE runtime implements these control instructions as index shuffles of the local store. For simplicity our first version of the VCODE library also does not support recursive functions, we implement tail recursive functions with accumulators as iterative instructions.

12.3.1 CBE/SPE specific changes to VCODE

For communicating with EA memory we have integrated DMA as a vector instruction to implement double buffered memory access. The vector instruction *DMA_GET* takes the 3 top of the stack sequences as the local store vector location, effective address index (all memory visible to VCODE is indirected through a coherent indexing scheme which is global across all SPEs), and the number of elements to copy from EA to LS. There is an equivalent instruction *DMA_PUT* which writes data to EA vectors. These calls can be made blocking or non-blocking depending upon a run-time switch.

12.4 A Simplified SPU ISA Simulator

```
(setf (gethash "spu\\_add" *GH*) (list #'+ "+" "a,fa"))
(setf (gethash "spu\\_sub" *GH*) (list #'- "-" "sf,fs"))
(setf (gethash "spu\\_mul" *GH*) (list #'* "*" "mpy,fm"))
(setf (gethash "spu\\_div" *GH*) (list #'/ "/" ""))
(setf (gethash "spu\\_rem" *GH*) (list #'mod "\\%" ""))
```

Listing 12.1 SPU ISA Simulator

```
(defun T2 (a b name)
  (let* ((hv (gethash name *GH*))
         (f (first hv))
         (sym (second hv))
         (isa (third hv))
         (res (map 'vector f a b)))
    (format t "~%%res of ~A on ~A ~A " name a b)
    (wvc t a b ". . . . . . ." name sym isa
         res ". . . . . . ." ". . . . . . ." )))
```

Listing 12.2 SPU ISA Simulator contd..

The following schematic of the VCODE stack was generated automatically using the program shown above in Listing 12.1. Using this program we have a full functionally equivalent model of the SPU and its registers and local store. We can write programs in NESL, compile them to VCODE and see the corresponding SPU assembly. We can simulate the program using our simulator, tune the assembly for dual-issue and SIMD.

1. `spu_add`: SPU intrinsic and instruction corresponding to `spu_add`

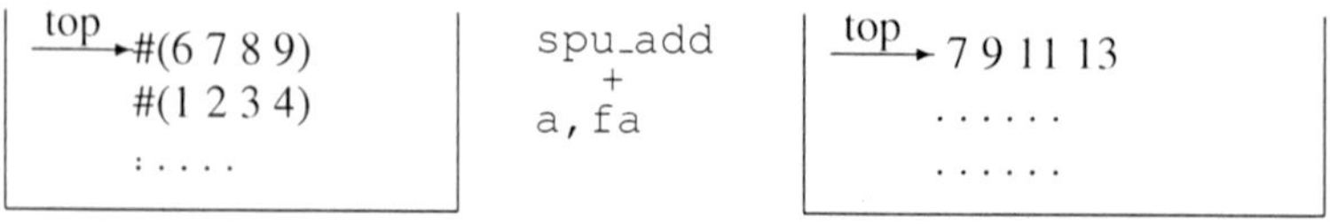

top → #(6 7 8 9)	`spu_add`	top → 7 9 11 13
#(1 2 3 4)	+	. . . . . .
:	`a,fa`	. . . . . .

2. spu_sub: SPU intrinsic and instruction corresponding to spu_sub

Stack before	Operation	Stack after
top → #(6 7 8 9)	spu_sub	top → 5 5 5 5
#(1 2 3 4)	−	......
:....	sf, fs	......

3. spu_mul: SPU intrinsic and instruction corresponding to spu_mul

Stack before	Operation	Stack after
top → #(6 7 8 9)	spu_mul	top → 6 14 24 36
#(1 2 3 4)	*	......
:....	mpy, fm	......

4. spu_div: SPU intrinsic and instruction corresponding to spu_div

Stack before	Operation	Stack after
top → #(6 7 8 9)	spu_div	top → 6 7/2 8/3 9/4
#(1 2 3 4)	/	......
:....		......

5. spu_rem: SPU intrinsic and instruction corresponding to spu_rem

Stack before	Operation	Stack after
top → #(6 7 8 9)	spu_rem	top → 0 1 2 1
#(1 2 3 4)	%	......
:....		......

6. spu_idiv: SPU intrinsic and instruction corresponding to spu_idiv

Stack before	Operation	Stack after
top → #(6 7 8 9)	spu_idiv	top → 0 1 2 1
#(1 2 3 4)	/	......
:....		......

7. spu_cmpgt: SPU intrinsic and instruction corresponding to spu_cmpgt

Stack before	Operation	Stack after
top → #(6 7 8 9)	spu_cmpgt	top → T T T T
#(1 2 3 4)	>	......
:....	cgt, fcgt	......

8. spu_cmpgte: SPU intrinsic and instruction corresponding to spu_cmpgte

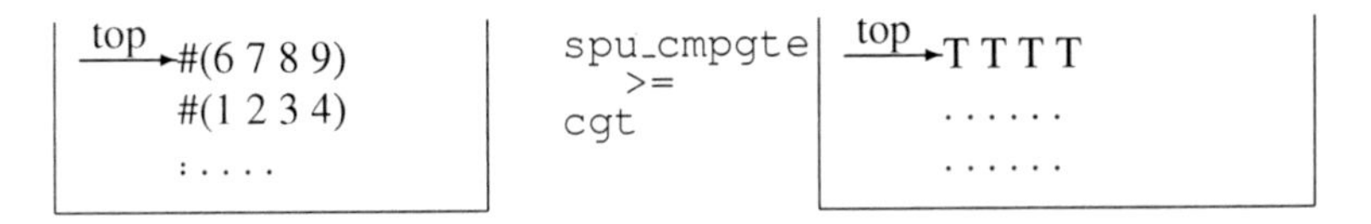

9. `spu_cmplt`: SPU intrinsic and instruction corresponding to `spu_cmplt`

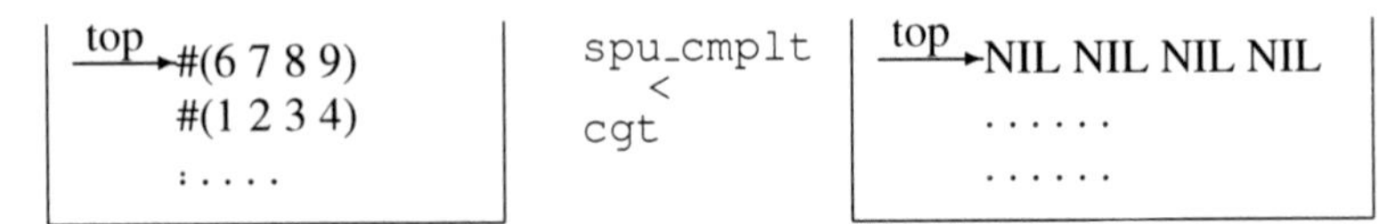

10. `spu_cmplte`: SPU intrinsic and instruction corresponding to `spu_cmplte`

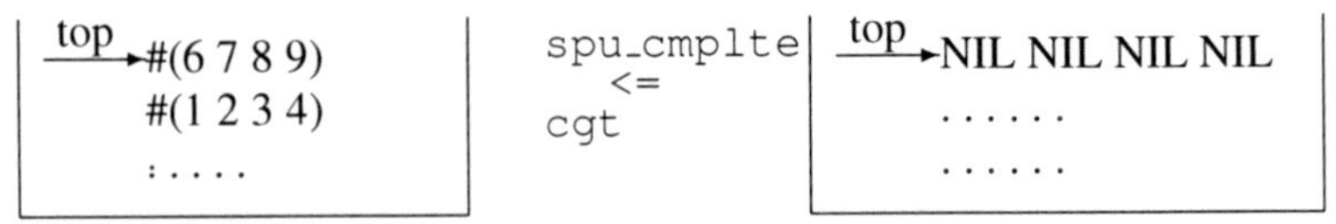

11. `spu_cmpeq`: SPU intrinsic and instruction corresponding to `spu_cmpeq`

top → #(6 7 8 9)
#(1 2 3 4)
: . . . .

spu_cmpeq
==
ceq, fceq

top → NIL NIL NIL NIL
. . . . . .
. . . . . .

12. `spu_lshift`: SPU intrinsic and instruction corresponding to `spu_lshift`

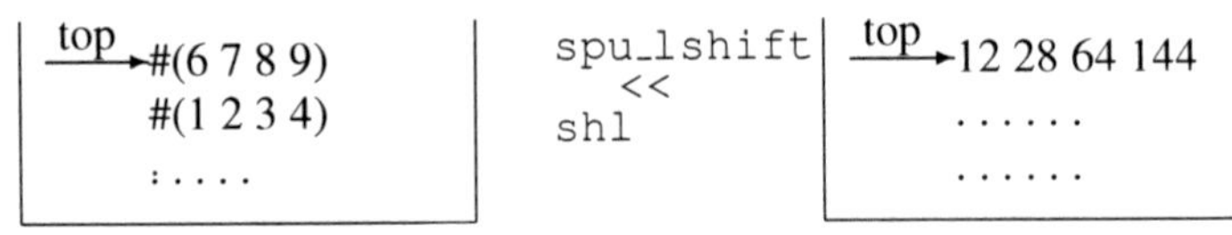

13. `spu_rshift`: SPU intrinsic and instruction corresponding to `spu_rshift`

top → #(6 7 8 9)
#(1 2 3 4)
: . . . .

spu_rshift
>>
shl

top → 3 1 1 0
. . . . . .
. . . . . .

14. `spu_rl`: SPU intrinsic and instruction corresponding to `spu_rl`

top → #(6 7 8 9)
#(1 2 3 4)
: . . . .

spu_rl
<<
rot

top → 12 28 64 144
. . . . . .
. . . . . .

15. `spu_rr`: SPU intrinsic and instruction corresponding to `spu_rr`

top → #(6 7 8 9)
#(1 2 3 4)
: . . . .

spu_rr
>>
rot

top → 3 3221225473 1 2415919104
. . . . . .
. . . . . .

16. `spu_and`: SPU intrinsic and instruction corresponding to `spu_and`

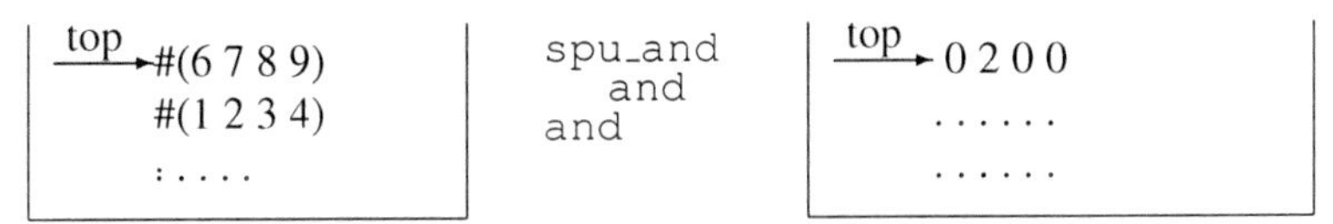

17. `spu_andc`: SPU intrinsic and instruction corresponding to `spu_andc`

18. `spu_eqv`: SPU intrinsic and instruction corresponding to `spu_eqv`

```
top → #(6 7 8 9)        spu_eqv        top → -8 -6 -12 -14
      #(1 2 3 4)          eq                  ......
      :....             eqv                   ......
```

19. `spu_nand`: SPU intrinsic and instruction corresponding to `spu_nand`

```
top → #(6 7 8 9)        spu_nand       top → -1 -3 -1 -1
      #(1 2 3 4)          nand                ......
      :....             nand                  ......
```

20. `spu_nor`: SPU intrinsic and instruction corresponding to `spu_nor`

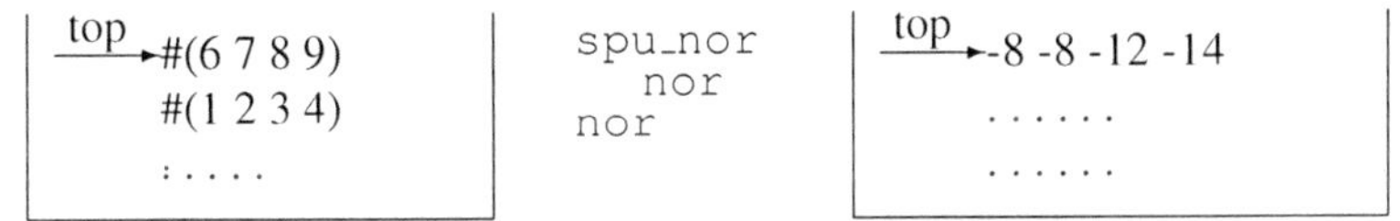

21. `spu_or`: SPU intrinsic and instruction corresponding to `spu_or`

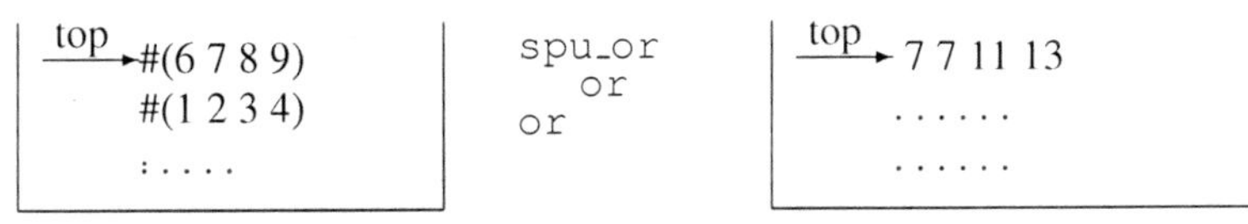

22. `spu_orc`: SPU intrinsic and instruction corresponding to `spu_orc`

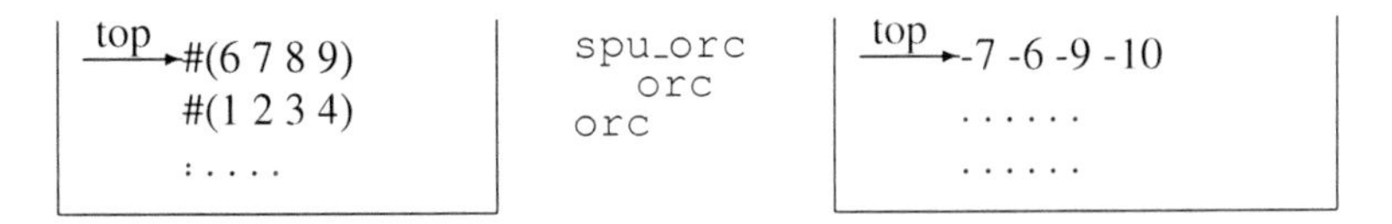

23. `spu_xor`: SPU intrinsic and instruction corresponding to `spu_xor`

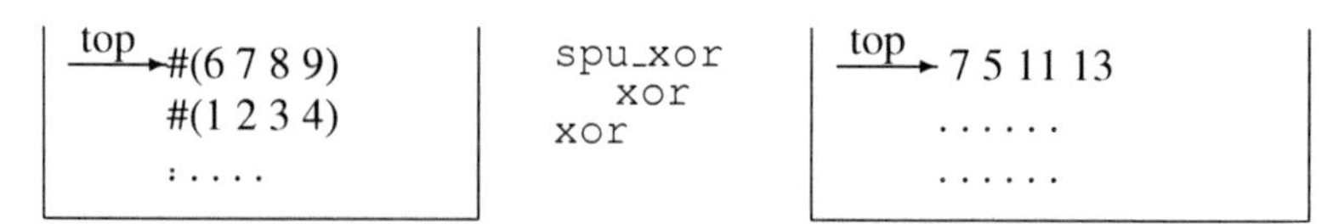

24. `spu_sel`: SPU intrinsic and instruction corresponding to `spu_sel`

top → #(6 7 8 9)
#(1 2 3 4)
#(T NIL NIL T)

spu_sel
select
selb

top → 6 2 3 9

25. `spu_rand`: SPU intrinsic and instruction corresponding to `spu_rand`

top → #(6 7 8 9)

spu_rand
rand
--

top → 1 5 6 1

26. `spu_floor`: SPU intrinsic and instruction corresponding to `spu_floor`

top → #(6.1 7.8 8.2 9.8)

spu_floor
$\lfloor x \rfloor$
--

top → 6 7 8 9

27. `spu_ceil`: SPU intrinsic and instruction corresponding to `spu_ceil`

top → #(6.1 7.8 8.2 9.8)

spu_ceil
$\lceil x \rceil$
--

top → 7 8 9 10

28. `spu_truncate`: SPU intrinsic and instruction corresponding to `spu_truncate`

top → #(6 7 8 9)

spu_truncate
truncate
--

top → 6 7 8 9

29. `spu_roundtf`: SPU intrinsic and instruction corresponding to `spu_roundtf`

top → #(6.1 7.8 8.2 9.8)

spu_roundtf
round
frds

top → 6 8 8 10

30. `spu_re`: SPU intrinsic and instruction corresponding to `spu_re`

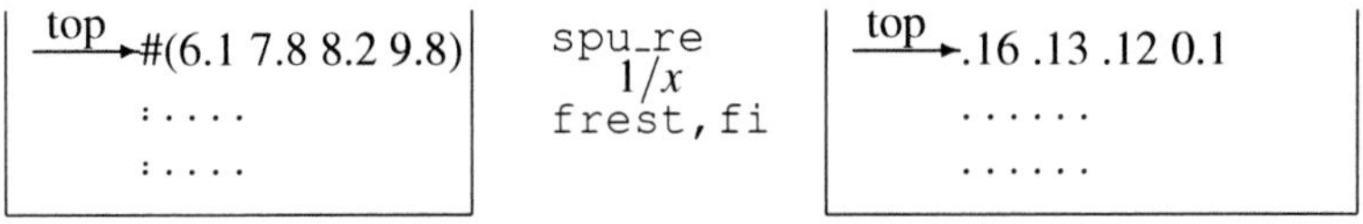

top → #(6.1 7.8 8.2 9.8)	spu_re $1/x$ frest,fi	top → .16 .13 .12 0.1
:....		......
:....		......

31. `spu_rsqrte`: SPU intrinsic and instruction corresponding to `spu_rsqrte`

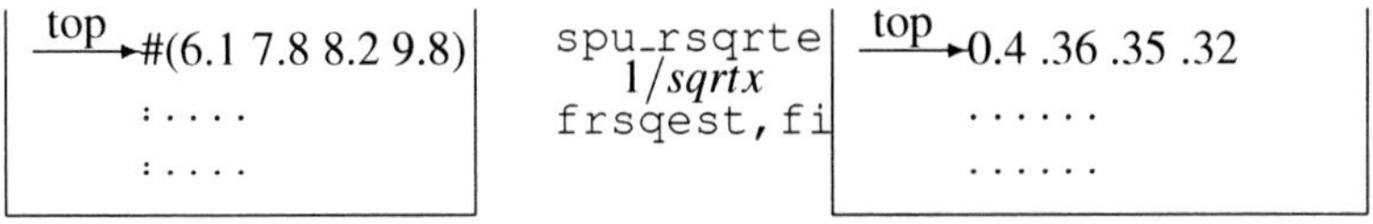

top → #(6.1 7.8 8.2 9.8)	spu_rsqrte $1/sqrtx$ frsqest,fi	top → 0.4 .36 .35 .32
:....		......
:....		......

32. `spu_convtf`: SPU intrinsic and instruction corresponding to `spu_convtf`

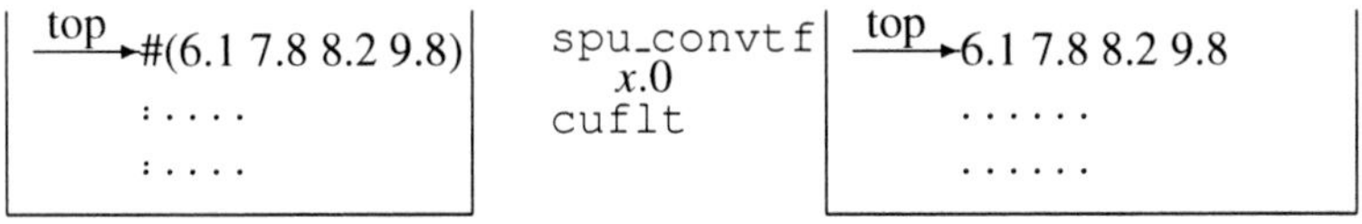

top → #(6.1 7.8 8.2 9.8)	spu_convtf $x.0$ cuflt	top → 6.1 7.8 8.2 9.8
:....		......
:....		......

33. `spu_convts`: SPU intrinsic and instruction corresponding to `spu_convts`

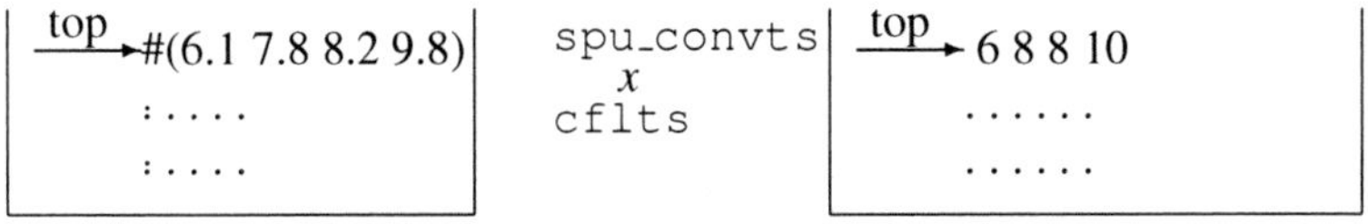

top → #(6.1 7.8 8.2 9.8)	spu_convts x cflts	top → 6 8 8 10
:....		......
:....		......

34. `spu_itob`: SPU intrinsic and instruction corresponding to `spu_itob`

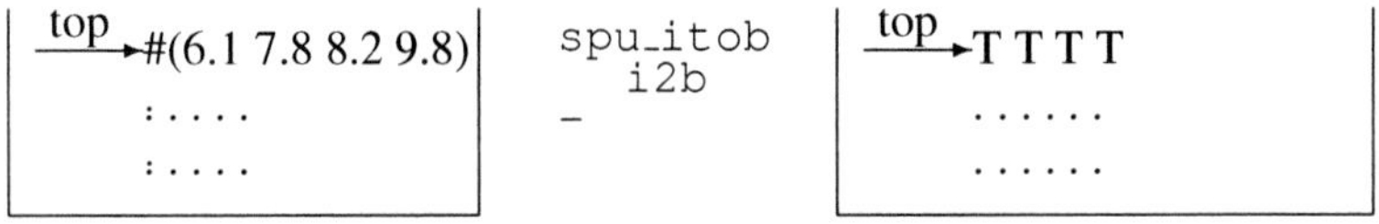

top → #(6.1 7.8 8.2 9.8)	spu_itob i2b –	top → T T T T
:....		......
:....		......

35. `spu_btoi`: SPU intrinsic and instruction corresponding to `spu_btoi`

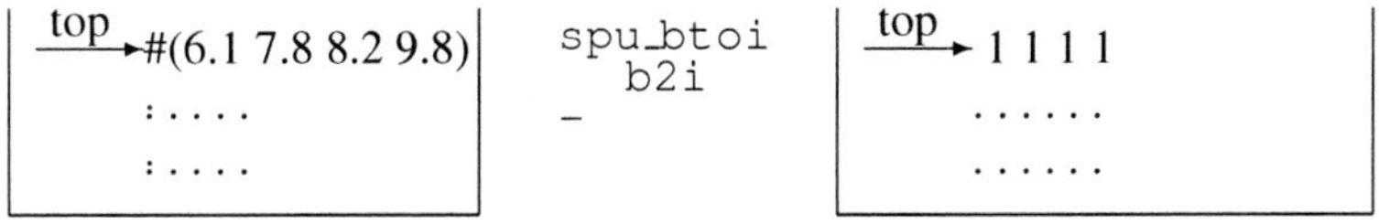

top → #(6.1 7.8 8.2 9.8)	spu_btoi b2i –	top → 1 1 1 1
:....		......
:....		......

36. `spu_log`: SPU intrinsic and instruction corresponding to `spu_log`

top → #(6.1 7.8 8.2 9.8)	spu_log log frest	top → 1.8 2.1 2.1 2.3
:....		......
:....		......

37. spu_sqrt: SPU intrinsic and instruction corresponding to spu_sqrt

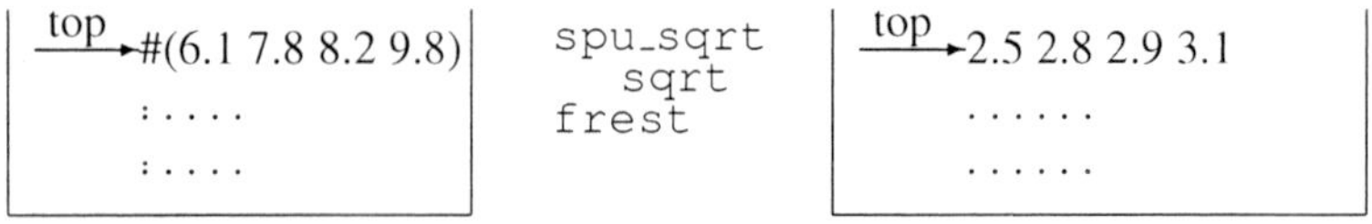

top → #(6.1 7.8 8.2 9.8) : . . . . :	spu_sqrt sqrt frest	top → 2.5 2.8 2.9 3.1 . . . .

38. spu_exp: SPU intrinsic and instruction corresponding to spu_exp

top → #(6.1 7.8 8.2 9.8) : . . . . :	spu_exp exp frest	top → 450. 2400. 3600. 18000. . . . .

39. spu_sin: SPU intrinsic and instruction corresponding to spu_sin

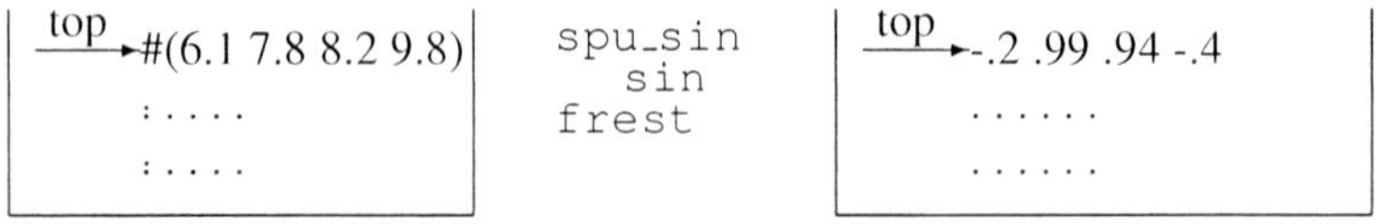

top → #(6.1 7.8 8.2 9.8) : . . . . :	spu_sin sin frest	top → -.2 .99 .94 -.4 . . . .

40. spu_cos: SPU intrinsic and instruction corresponding to spu_cos

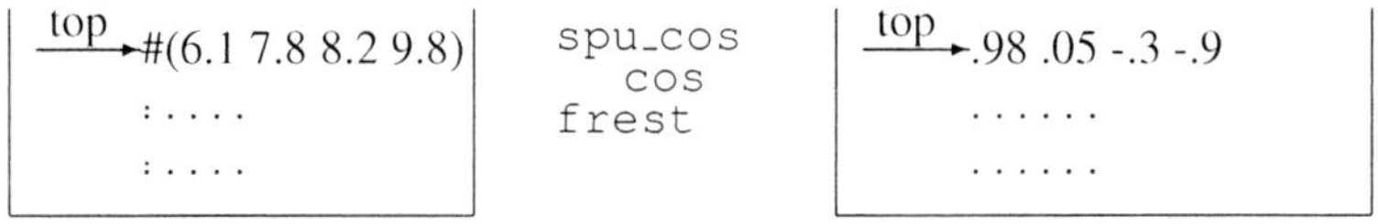

top → #(6.1 7.8 8.2 9.8) : . . . . :	spu_cos cos frest	top → .98 .05 -.3 -.9 . . . .

41. spu_tan: SPU intrinsic and instruction corresponding to spu_tan

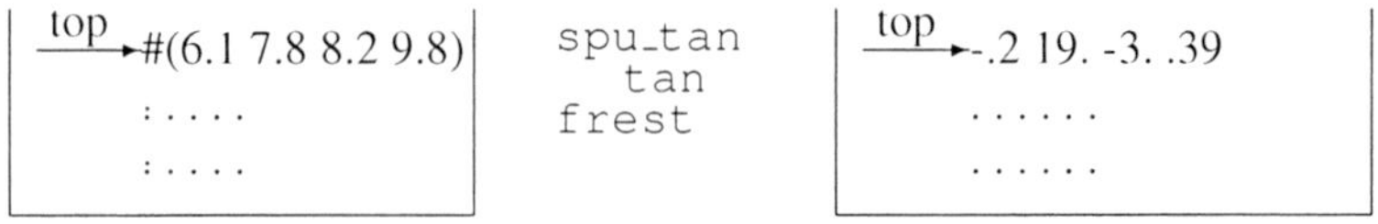

top → #(6.1 7.8 8.2 9.8) : . . . . :	spu_tan tan frest	top → -.2 19. -3. .39 . . . .

42. spu_asin: SPU intrinsic and instruction corresponding to spu_asin

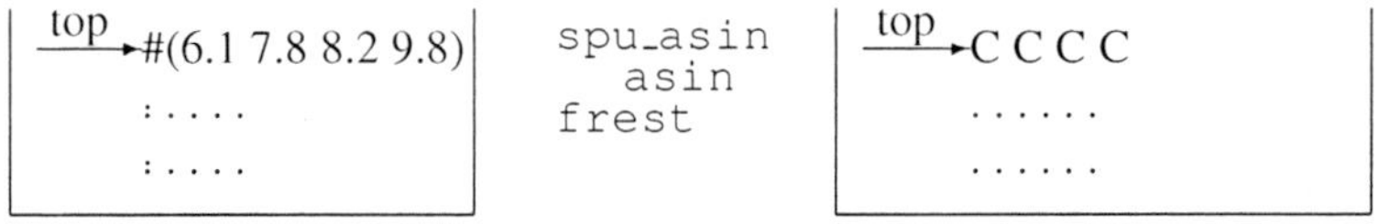

top → #(6.1 7.8 8.2 9.8) : . . . . :	spu_asin asin frest	top → C C C C . . . .

43. spu_acos: SPU intrinsic and instruction corresponding to spu_acos

top → #(6.1 7.8 8.2 9.8) : . . . . :	spu_acos acos frest	top → C C C C . . . .

44. `spu_atan`: SPU intrinsic and instruction corresponding to `spu_atan`

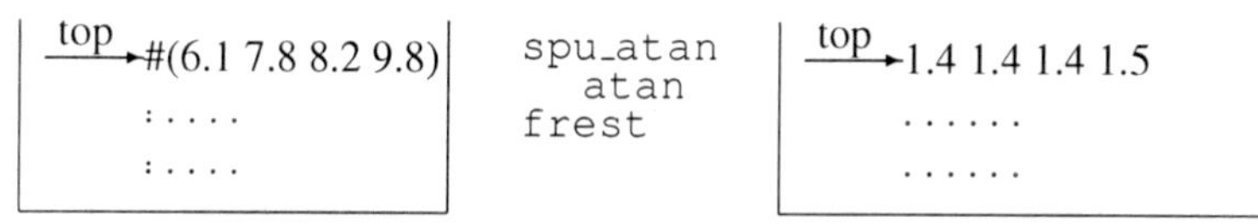

45. `spu_sinh`: SPU intrinsic and instruction corresponding to `spu_sinh`

top → #(6.1 7.8 8.2 9.8)

```
spu_sinh
  sinh
frest
```

top → 220. 1200. 1800. 9000.

46. `spu_cosh`: SPU intrinsic and instruction corresponding to `spu_cosh`

top → #(6.1 7.8 8.2 9.8)

```
spu_cosh
  cosh
frest
```

top → 220. 1200. 1800. 9000.

47. `spu_tanh`: SPU intrinsic and instruction corresponding to `spu_tanh`

top → #(6.1 7.8 8.2 9.8)

```
spu_tanh
  tanh
frest
```

top → 1.0 1.0 1.0 1.0

12.5 VCODE Compiler

I wrote a simple VCODE compiler which takes data from a text file and compiles it into a subset of SPE intrinsics. This instruction sequence is then simulated by our SSISA (Simplified SPU ISA Simulator) which we present in the next chapter, to execute SPE-VCODE examples for debugging. Unfortunately, not all of NESL produced VCODE is supported as the main intention of this program was not to write yet-another VCODE interpreter but to evaluate the feasibility and match of SPU intrinsics with the VCODE model. The compiler is written in Common Lisp, and does not use any of the NESL code. Some example code of the compiler is shown in Listing 12.3.

```
(defconstant +MAX-HANDLES+ 100)
(defvar *ARENA* nil "memory arena for vectors")
(defstruct vector-descriptor
  (length 0 :type fixnum)
5  (refcount 0 :type fixnum)
  (memory nil))
(defun pvd (n)
  (format t "~%VD[~D] = ~D" n (aref *ARENA* n)))
(defun vd-alloc (length)
```

```
10    "Allocate a new vector descriptor of this length"
      (dotimes (i +MAX-HANDLES+)
        (when (= 0 (VDR
        (aref *ARENA* i)))
          (setf (aref *ARENA* i)
15        (make-vector-descriptor
           :length length :refcount 1
           :memory (make-array length)))
          (return-from vd-alloc i)))
      (format t "~%ERROR: Arena low, rerun with more memory."))
20  (defun vd-length (n) (VDL (aref *ARENA* n)))
    (defun vd-refcount (n) (VDR (aref *ARENA* n)))
    (defun vd-memory (n) (VDM (aref *ARENA* n)))
    (defun initialize-arena ()
      "Initialize default vector descriptors"
25    (setf *ARENA* (make-array +MAX-HANDLES+))
      (dotimes (i +MAX-HANDLES+)
        (setf (aref *ARENA* i) (make-vector-descriptor))))
    (defun vd-alloci (a)
      (let ((vdn (vd-alloc (length a))))
30      (setf (VDM (aref *ARENA* vdn)) a)
        vdn))
```

Listing 12.3 VCODE compiler.

```
    (defun vd-scan-plus (a)
      "Return a new vd with scan plus of contents of
      a, eg [5 1 3 4 9 2] -> [0 5 6 9 13 22]"
32    (let ((vdn (vd-alloc (vd-length a)))
            (running-total 0))
        ;; identity of operation is 0
        (dotimes (i (vd-length a))
          (setf (aref (vd-memory vdn) i)
37        running-total)
          (incf running-total (aref (vd-memory a) i)))
        vdn))

    (defun vd-scan-mul (a)
42    "Return a new vd with scan plus of contents of
      a, eg [5 1 3 4 9 2] -> [0 5 6 9 13 22]"
      (let ((vdn (vd-alloc (vd-length a)))
            (running-total 1))
        ;; identity of operation is 0
47      (dotimes (i (vd-length a))
          (setf (aref (vd-memory vdn) i)
          running-total)
          (setf running-total
          (* running-total
52           (aref (vd-memory a) i))))
        vdn))

    (defun vd-scan-max (a)
      "Return a new vd with scan plus of contents of
57     a, eg [5 1 3 4 9 2] -> [0 5 6 9 13 22]"
      (let ((vdn (vd-alloc (vd-length a)))
            (running-total most-negative-fixnum))
        ;; identity of operation is most-negative-fixnum
        (dotimes (i (vd-length a))
62        (setf (aref (vd-memory vdn) i)
          running-total)
          (setf running-total
          (max (aref (vd-memory a) i)
         running-total)))
67      vdn))
```

Listing 12.4 VCODE compiler contd..

12.6 PowerList: Data structure for parallel programming

We have seen the use of Petri-nets to model concurrent execution of parallel programs. We also looked at Vector libraries which operate on segment descriptors of arbitrary length. We have implemented a VCODE compiler for the SPU, but as we will discuss in that chapter, tracking descriptors length, stack space and SIMD execution simultaneously is complicated.

In this section we present a data-structure explicitly designed to represent Parallel Recursion, called Powerlist (see the original paper on Powerlist by Jayadev Misra [87]). We introduce the Powerlist data-structure, and show how it can be implemented on the SPU. We then use this as an abstraction layer to present algorithms.

As many of the algorithms and implementation in this book have also been done using Common Lisp, it should not be a surprise, when we advocate that *list* is a basic data structure to model (not implement but, model) recursion. A list is either NIL (empty) or is a CONS (constructed by concatenating an element to a list). In the implementation on SPU all list code is implicitly translated to quad-word aligned vectors with an associated length for each vector.

When dealing with parallel algorithms, a simple list is inadequate as it provides no mechanism for parallel iteration on the successive elements of the list. Misra presented a modification to the simple list structure termed Powerlist [87], which is more suitable to model parallel recursion, and parallel programs operating on lists.

The smallest powerlist P_1 consists of a single element. A larger powerlist is constructed (similar to simple list) from the elements of two powerlists of the same length, using the following operations. If p, q are powerlists of the same length then:

1. $p \mid q$ is the powerlist formed by concatenating p and q,
2. $p \bowtie q$ is the powerlist formed by successively taking alternating items from p, and then q, starting from p.

Here are some example of using these operations:
$\langle 0 \rangle | \langle 1 \rangle = \langle 01 \rangle$
$\langle 0 \rangle \bowtie \langle 1 \rangle = \langle 01 \rangle$
$\langle 01 \rangle | \langle 23 \rangle = \langle 0123 \rangle$
$\langle 01 \rangle \bowtie \langle 23 \rangle = \langle 0213 \rangle$

Misra called the $\mid$ operator *tie* and $\bowtie$, *zip* respectively. From these construction we can see that powerlists always have 2^n elements in them, thats the reason why the base case for a powerlist is a list of a single (2^0) element. Using these operations powerlists can be constructed and used to represent data. Consider the following example:
$\langle 5 \rangle$ is a powerlist of length 1 containing a scalar, 5
$\langle \langle 2 \rangle \rangle$ is a powerlist of length 1 containing a powerlist of length 1 of containing a scalar 2,
$\langle \rangle$ is a not a powerlist, as length 0 is not allowed,

$\langle\langle 12\rangle\langle 34\rangle\langle 56\rangle$ is a powerlist representation of a matrix where each element of the outermost powerlist is a column of the 3x2 matrix.

12.6.1 Functions on Powerlists

Recursive function definitions in Lisp are usually defined by case analysis — a recursive function will have a base case for the empty list, and will recurse on a smaller sublist. Functions on powerlists are defined in the same manner. Consider the function $rev(p)$ defined as:
$rev(\langle x\rangle) = \langle x\rangle$
$rev(p|q) = rev(q)|rev(p)$
Function *rev* returns the element when given a powerlist of length 1, else it creates a new powerlist using the *tie* operation by deconstructing the argument into 2 equal length powerlists and then reversing the elements, and finally using tie to merge the results, but starting from q. It is instructive to carry out the operations for *rev* on a small example, and we do so below.
Let $A = \langle 12345678\rangle$ be a powerlist of log-length 3. *Log-length* is logarithmic length which is a more useful measure for powerlists as their length is always a power of 2. The *depth* of a powerlist is the number of *level* in it, or recursively count the depth, with the depth of a scalar defined to be 0, and depth of a powerlist of scalar to be 1. To reverse A we have:
$A = \langle p|q\rangle = \langle 1234\rangle \mid \langle 5678\rangle$
$rev(A) = rev(\langle 5678\rangle) \mid rev(\langle 1234\rangle)$
$rev(A) = \langle rev(\langle 78\rangle) \mid rev(\langle 56\rangle)\rangle \mid rev(\langle 34\rangle) \mid rev(\langle 12\rangle)\rangle$
$rev(A) = \langle\langle 87\rangle \mid \langle 65\rangle\rangle \mid \langle\langle 43\rangle \mid \langle 21\rangle\rangle$
$rev(A) = \langle 8765\rangle \mid \langle 4321\rangle$
$rev(A) = \langle 87654321\rangle$

This is also shown in Figure 12.1.

Using the tie and zip operations we can define other useful operations on powerlists; consider the right-rotation and left-rotation functions rr, rl:
$rr\langle x\rangle = \langle x\rangle$ for singleton
$rl\langle x\rangle = \langle x\rangle$ for singleton
$rr(p \bowtie q) = (rr(q)) \bowtie p$
$rl(p \bowtie q) = q \bowtie (rl(p))$
We use the $\bowtie$ as a deconstruction operator.

We conclude this section with an example algorithm where the powerlist description is succinct, and shows the advantages of modeling parallel recursive algorithms using powerlist.

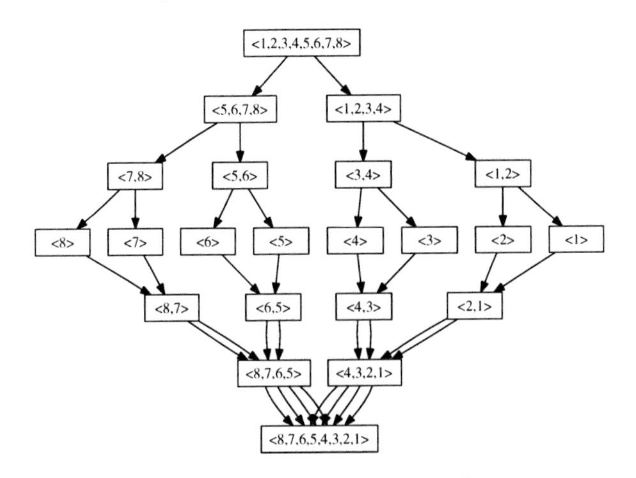

Fig. 12.1 Example of running *rev* on Powerlist $\langle 1,2,3,4,5,6,7,8\rangle$.

12.6.2 Using powerlists for polynomial evaluation

Let $f(x)$ be a polynomial with coefficients $p_j, 0 \le j < 2^n$, where $n \ge 0$. Then $f(x)$ can be represented as a powerlist A whose j^{th} element is p_j. The evaluation of the polynomial at a point w is

$$ep(f(x), x = w) = \sum_{0 \le j < 2^n} p_j \times w^j$$

By factoring the even powers of w we get:

$$ep(f(x), x = w) = \sum_{0 \le j < 2^{n-1}} p_{2j} \times w^{2j} + \sum_{0 \le j < 2^{n-1}} p_{2j+1} \times w^{2j+1}$$

We now deconstruct the coefficient powerlist as $A = \langle odd \mid even\rangle$ and evaluate $\langle even\rangle$ at $x = w^2$, we evaluate odd as $w \times ep(f(x), x = w^2)$. Formally:
$\langle x\rangle \; ep \; w = \langle x\rangle$, for singleton
$(p \bowtie q) \; ep \; w = [p \; ep \; w^2] + [w \times (q \; ep \; w^2)]$

Although we have not stated as much, it should be readily apparent, that after the list deconstruction, subsequent operations on p and q can be executed in parallel.

12.7 ALF: Accelerated Library Framework

The Accelerated Library Framework (ALF) [26] provides a programming environment for data and task parallel applications and libraries. The ALF API provides you with a set of interfaces to simplify library development on heterogeneous multi-core systems. You can use the provided framework to offload the computationally intensive work to the accelerators. More complex applications can be developed by combining the several function offload libraries. You can also choose to implement applications directly to the ALF interface. ALF supports the multiple-program-multiple-data (MPMD) programming model where multiple programs can be scheduled to run on multiple accelerator elements at the same time. The ALF functionality includes:

Data transfer management
Parallel task management
Double buffering
Dynamic load balancing for data parallel tasks.

With the provided API, you can also create descriptions for multiple compute tasks and define their execution orders by defining task dependency. Task parallelism is accomplished by having tasks without direct or indirect dependencies between them. The ALF runtime provides an optimal parallel scheduling scheme for the tasks based on given dependencies.

12.7.1 ALF workload division

From the application or library programmers point of view, ALF consists of the following two runtime components: a host runtime library, and an accelerator runtime library.

The host runtime library provides the host APIs to the application. The accelerator runtime library provides the APIs to the applications accelerator code, usually the computational kernel and helper routines. This division of labor enables programmers to specialize in different parts of a given parallel workload. ALF tasks The ALF design enables a separation of work. There are distinct types of task within a given application:

1. Application: you develop programs only at the host level. You can use the provided accelerated libraries without direct knowledge of the inner workings of the underlying system,
2. Accelerated library: you use the ALF APIs to provide the library interfaces to invoke the computational kernels on the accelerators.

You divide the problem into the control process, which runs on the host, and the computational kernel. Within the ALF framework, a computational kernel is defined as an accelerator routine that takes a given set of input data and returns the output data based on the given input. For more details on the ALF API and library, please read the ALF Programmers Guide and API Reference [26].

12.8 Tale of Dwarfs

In this section we discuss the UC Berkeley paper [43] on computational dwarfs (or patterns of high-performance computing) and how each dwarf would map to CBE. The material in this section is based on the paper "The Landscape of Parallel Computing Research: A View From Berkeley", [43]. The authors ask seven fundamental questions about parallel computing:

1. What are the applications?
2. What are common kernels of the applications?
3. What are the HW building blocks?
4. How to connect them?
5. How to describe applications and kernels?
6. How to program the hardware?
7. How to measure success?

The authors further define a *dwarf* as "an algorithmic method that captures a pattern of computation and communication". Previous studies have already identified common code and computational methods which are omnipresent in high-performance scientific computing, and indeed much of the linear-algebra and matrix solvers, FFT libraries are a realization of this stark manifestation of a 80-20 rule, but in the dwarf-paper, Berkeley researchers refine the list of compute kernels to a handful, and these are listed below before further discussion:

1. Dense Linear Algebra
2. Sparse Linear Algebra
3. Spectral Methods
4. n-body Methods
5. Structured Grids
6. Unstructured Grids
7. MapReduce
8. Combinational Logic
9. Graph Traversal
10. Dynamic Programming
11. Backtrack and Branch-and-Bound
12. Graphical Models
13. Finite State Machines

Many of the above problems have been addressed by commercial, off-the-shelf (COTS) software libraries, eg., BLAS, ATLAS, LAPACK, FFT. The dwarfs that we find of particular interest are graph traversal, which can be generalized to graph property checking, and finite-state machine property checking. In this book we have presented examples of graph algorithms which map efficiently onto the SPEs. We show in a later chapter (cf. Chapter 10) how graph data-structures can be combined with randomized algorithms and perfect hash-functions to solve important graph theoretic problems, while keeping the small size of SPU local-store in mind.

In this regards, the Cell Broadband Engine is a harbinger of things to come, small, fast local store, with large register files and an asynchronous but coherent memory. Rather than pontificate about how each dwarf could map onto the Cell Broadband Engine, we are going to keep the dwarf classification in mind, and refer back to them whenever we solve a particular compute problem, and we feel that it is representative of a general technique which can be used to map a dwarf on the SPE.

12.9 Introduction to Cilk++

Cilk++ [2] is a cross-platform solution which according to its authors offers "easiest, quickest and most reliable way to maximize application performance on multicore processors". Cilk++ provides very simple extensions to C++ and is based on a powerful runtime system for multicore application execution. Cilk++ extends the language to provide the following Cilk keywords:

1. cilk: identifies a function written in Cilk,
2. spawn: indicates that the procedure can safely operate in parallel with other threads of execution,
3. sync: an example of the *barrier* method and akin to pthread *join*, indicates that execution of current procedure cannot proceed until all previously spawned procedures have returned,

```
cilk int fib (int n)
{
  if (n < 2) return 1;
  else
  {
    int x, y;

    x = spawn fib (n-1);
    y = spawn fib (n-2);

    sync;

    return (x+y);
  }
}
```

Listing 12.5 Fibonacci number computation in Cilk

An example Cilk function is shown above. The Cilk scheduler uses a policy called *work stealing*. This divides procedure execution amongst processors by maintaining a stack of procedure frames suspended in wait for the scheduler. When a processor has executed all frames on its own stack it can *steal* work from another processor's stack (from the opposite end) and execute that frame. For more information on Cilk, see [2].

12.10 Introduction to Intel Ct

Intel Ct, or "C for throughput computing" is described as "a flexible parallel programming model for tera-scale architectures" in the white paper published recently by Intel [54]. Ct is a deterministic parallel programming model capable of leveraging the best features of GPU and CPUs for high performance computing. Similar to NESL, Ct also supports nested data parallelism with extended functionality to address irregularl algorithms. Examples presented in the initial documentation include sparse matrix algorithms using compressed sparse colunmn representation. The matrix representation is identical to the one presented earlier in this chapter.

Ct's implementation is based on "nested vectors" which are nested collections. Ct also has dynamic (run-time based) task scheduling to minimize threading overhead. Each task in Ct is represented by a *spawn* point, but the number of tasks created depends on the number of cores available and the size of the vector being processed. Thus, Ct is highly adaptive to varying data sizes and core loads, as claimed by its authors.

12.11 Introduction to RapidMind

From [103], RapidMind is a development and runtime platform that enables single threaded, manageable applications to fully access multi-core processors. With RapidMind, developers continue to write code in standard C++ and use their existing skills, tools and processes. The RapidMind platform then parallelizes the application across multiple cores and manages its execution. An article on RapidMind's API for the Cell Broadband Engine (available at IBM's alphaworks website) describes the work done by RapidMind for the Cell architecture. According to the article RapidMind interface is based upon three main types:

1. Value: `Value<N,T>` type represents a fixed-size container, where T can be any basic C++ numerical type and N is the number of elements of that type,
2. Array: the `Array<D,V>` type represents a variable-sized multidimensional container in which D is the dimensionality (akin to rank in Lisp) and V is the element type,
3. Program: this represents the actual computation. It is also represented as a container for program code and can be constructed dyamically. Program functions in RapidMind are first class objects in the sense they can be stored in an object.

Program objects can include arbitrary computations, including dynamic data-dependent control flow and random access reads from other arrays. For more information on RapidMind's API see [103].

12.11.1 Conclusion

We have seen above that powerlists combine the elegance and formalism of simple lists (which have a rich history of formal techniques to analyze program behavior), with the advantage of parallel recursion. We also saw that powerlist models of algorithms are powerful, and can be used to represent many common algorithms. We have implemented a complete powerlist library in the SPU with SPU intrinsics, and along-with our VCODE compiler, provides a simple and efficient manner of implementing parallel recursive algorithms on the SPU. We also discussed ALF, the Accelerated Library Framework, which provides facilities for rapid parallel programming, not only on the Cell but also in a hybrid model. We briefly discussed the Berkeley dwarf paper, and presented a summary of their main points. We presented an introduction to Cilk++, Intel Ct and RapidMind. There are also available Charm++ parallel objects, Barcelona Superscalar SMP (based on CellSS), which can be researched by the reader.

Chapter 13
Computational Mathematics on the CBEA

Abstract Although the purpose of computing is insight and not numbers[1], nevertheless, numerical computation forms the bulk of high-performance parallel scientific computing. In this chapter we introduce methods and examples of math programs on the Cell Broadband Engine. Some of the examples use API libraries already described in Chapter 9, while some other are built from scratch. We discuss the needle method of computing PI, (b) line fitting (or linear regression), (c) monomial and polynomial representation and computation, (d) SIMD polynomial evaluation with fixed interpolation, (e) parallel distributed local store polynomial evaluation, (f) evaluation of Boolean functions, (g) parallel matrix functions, (h) solving equations of a single complex variable using ALF.

13.1 Introduction

Most of the published work on Cell Broadband Engine has been on single-precision numerical computations in the context of image processing, matrix computations and digital signal processing. Numerical computation has long been the realm of expensive supercomputers grinding at teraflop and now petaflop speeds. The Cell Broadband Engine has been able to achieve impressive (close to 200 gigaflops) results on single-precision computation and the new PowerXCell 8i chip increases the double-precision performance as well.

13.2 Calculating PI

Consider a square of side $2r$. We generate n random points within this square. For every point we calculate whether it also lies within a circle of radius r centered in-

[1] Attributed to Richard W Hamming

S. Koranne, *Practical Computing on the Cell Broadband Engine*,
DOI: 10.1007/978-1-4419-0308-2_13,

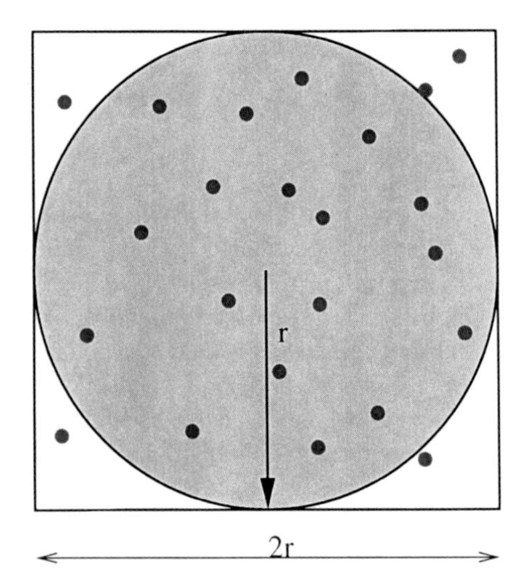

Fig. 13.1 Calculating PI as the ratio of number of points that lie within a circle of radius r, when they have been generated as random points lying in the square of radius $2r$.

scribed in the square. If the random number generator is good we expect the number of points which lie within the circle to be proportional to its area πr^2, yielding:

$$\frac{n}{4r^2} = \frac{n'}{\pi r^2}$$

or equivalently

$$\pi = \frac{4n'}{n}$$

We write a simple program to generate n points on each of the 6 SPUs, and ask the SPUs to return the number of points which lie within the circle. The code for single-precision is given in Listing 13.1.

```
   ////////////////////////////////////////////
   // Program: PI
   // Author : Sandeep Koranne
   ////////////////////////////////////////////
 5 #include <spu_mfcio.h>
   #include <stdio.h>
   #include <math.h>
   #include <mc_rand.h>
   #include <spu_intrinsics.h>
10 #include "../util.h"
   inline vec_float4 grx() {return mc_rand_mt_minus1_to_1_f4();}
   inline vec_float4 gry() {return mc_rand_mt_minus1_to_1_f4();}
   #define N 100000000
   vec_uint4 CalcPI(int n) {
15   int i;
     vec_uint4 Z=(vec_uint4){0,0,0,0};
     vec_uint4 O=(vec_uint4){1,1,1,1};
     vec_uint4 r=Z;
     vec_float4 C=(vec_float4){0.25,0.25,0.25,0.25};
20   for(i=0;i<n;++i) {
```

```
        vec_float4 x = grx(), y = gry();
        vec_float4 x2 = spu_mul( x,x ),y2 = spu_mul( y,y );
        vec_float4 d2 = spu_add( x2,y2);vec_uint4 igt= spu_cmpgt(d2,C);
        vec_uint4 off = spu_sel(O,Z,igt); r = spu_add(r, off);
25    }
      return r;
    }
    int main( unsigned long long spuid,
              unsigned long long argp ) {
30    int rc;
      printf(" Hello, World! from %d with %d \n", spuid, argp );
      mc_rand_mt_init( spuid );
      vec_uint4 r = CalcPI(N);
      print_vector("r",&r);
35    int ans=spu_extract(r,0); ans += spu_extract(r,1);
      ans += spu_extract(r,2);  ans += spu_extract(r,3);
      float apx_pi = 4*(float)ans/(float)(N);
      printf("\n PI %f = %d/%d\n",apx_pi,ans,4*N);
      return (0);
40  }
```

Listing 13.1 Calculating π using area of circle method.

Running this program on the PS3 we get:

```
 Hello, World! from 1 with 262112

[19637544 19635172 19629931 19641496]
 PI 3.141765 = 78544143/400000000
```

Listing 13.2 Transcript of PI calculator on the PS3.

You may find the following formula for π interesting as well:

$$\pi = \sum_{k=0}^{\infty} \frac{1}{16^k}\left(\frac{4}{8k+1} - \frac{2}{8k+4} - \frac{1}{8k+5} - \frac{1}{8k+6}\right)$$

This formula was discovered in 1996 by P. Borwein, S. Plouffe and D. H.Bailey and has the curious property that it can calculate the n-th binary or hexa-decimal digits of π without calculating the previous $n-1$ digits. This formula was discovered using the PSLQ algorithm.

13.3 Line fitting

Add math section here.

The Lisp code to perform line fitting is shown in Listing 13.3. The corresponding SPU code is shown in Listing 13.4.

```
  ;; program : line fitting
  ;; author  : sandeep koranne
  ;;
  (defun line-fit (x y)
5   "given array x/y compute best line fit"
    (let* ((n (* 1.0 (length x)))
           (xa (/ (reduce #'+ x) n))
           (ya (/ (reduce #'+ y) n))
```

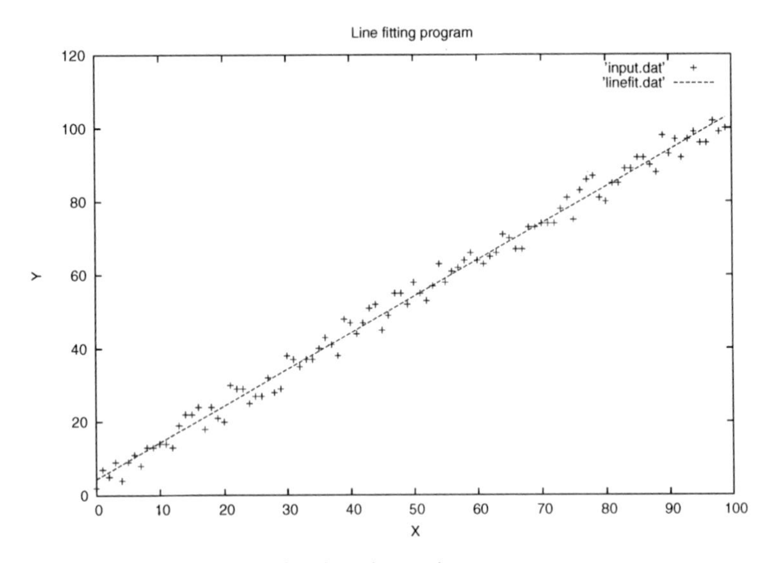

Fig. 13.2 Example of line fitting with points given as input.

```
            (stt (reduce #'+ (map 'vector
10               (lambda (p) (* (- p xa) (- p xa))) x)))
            (b  (/ (reduce #'+
         (map 'vector
              (lambda (px py)
                (* py (- px xa))) x y)) stt))
15          (a  (- ya (* b xa))))
       (values a b)))

  (defun linear-vector (n)
    (let ((a (make-array n :element-type 'float)))
20    (dotimes (i n)
        (setf (aref a i) i))
       a))

  (defun create-random-vector (n m)
25  (let ((a (make-array n :element-type 'float)))
      (dotimes (i n)
        (setf (aref a i) (+ i (random m))))
       a))
  (defvar x (linear-vector 100))
30 (defvar y (create-random-vector 100 10))
  (defun main()
    (let ((f1 (open "input.dat" :direction :output :if-exists :supersede))
          (f2 (open "linefit.dat" :direction :output :if-exists :supersede)))
      (map 'vector (lambda (px py) (format f1 "~D ~D~%" px py)) x y)
35    (multiple-value-bind (a b) (line-fit x y)
        (dotimes (i (length x))
          (format f2 "~D ~D~%" (aref x i) (+ a (* b (aref x i))))))
      (close f1)
      (close f2)))
40
  (main)
  (quit)
```

Listing 13.3 Common Lisp implementation of line-fitting.

Now we implement the algorithm on the SPU, as before consider the X and Y data to be input, and the function returns intercept *a* and the slope *b*.

```
 1  //////////////////////////////////////////
    // Program: Line Fitting
    // Author : Sandeep Koranne
    //////////////////////////////////////////
 5  #include <spu_mfcio.h>
    #include <stdio.h>
    #include <math.h>
    #include <mc_rand.h>
    #include <spu_intrinsics.h>
10  #include "../util.h"

    #define N 1024
    volatile vec_float4 X[N] __attribute__ ((aligned(128)));
    volatile vec_float4 Y[N] __attribute__ ((aligned(128)));
15
    static void CreateLinearVector(vec_float4 V[], int n) {
      int i;
      vec_float4 CT=(vec_float4){4,4,4,4};
      V[0] = (vec_float4){0,1,2,3};
20    for(i=1;i<n;++i) V[i] = spu_add( V[i-1],CT );
    }

    static void CreateRandomVector(vec_float4 V[], int n, vec_float4 m) {
      int i;
25    for(i=0;i<n;++i) {
        vec_float4 r = mc_rand_mt_minus1_to_1_f4();
        r = spu_mul(r,m); // spread around m
        r = spu_add(X[i],r);
        V[i] = r;
30    }
    }

    float SumArray( vec_float4 V[], int n ) {
      vec_float4 temp=(vec_float4){0,0,0,0};
35    float ans=0.0;
      int i;
      for(i=0;i<n;++i) temp = spu_add(temp,V[i]);
      ans=spu_extract(temp,0); ans+= spu_extract(temp,1);
      ans+=spu_extract(temp,2); ans+= spu_extract(temp,3);
40    ans /= (4*n);
      return ans;
    }

    float STT( vec_float4 V[], int n, float xa ) {
45    int i;
      vec_float4 temp=(vec_float4){0,0,0,0};
      vec_float4 vxa = spu_splats( xa );
      float ans=0.0;
      for(i=0;i<n;++i) {
50      vec_float4 t0 = spu_sub( V[i], vxa );
        vec_float4 t1 = spu_mul( t0,t0 );
        temp = spu_add( temp, t1 );
      }
      ans=spu_extract(temp,0); ans+= spu_extract(temp,1);
55    ans+=spu_extract(temp,2); ans+= spu_extract(temp,3);
      return ans;
    }

    float CalculateB( vec_float4 VX[], vec_float4 VY[],
60        int n, float xa, float stt ) {
      int i;
```

```
    vec_float4 temp=(vec_float4){0,0,0,0};
    vec_float4 vxa = spu_splats( xa );
    float ans=0.0;
65  for(i=0;i<n;++i) {
      vec_float4 t0 = spu_sub( VX[i], vxa );
      vec_float4 t1 = spu_mul( t0, VY[i] );
      temp = spu_add( temp, t1 );
    }
70  ans=spu_extract(temp,0); ans+= spu_extract(temp,1);
    ans+=spu_extract(temp,2); ans+= spu_extract(temp,3);
    ans /= stt;
    return ans;
  }
75
  float CalculateA( float ya, float xa, float b ) {
    return (ya - (xa * b));
  }

80 static void LineFit( vec_float4 VX[], vec_float4 VY[], int n) {
    float xa = SumArray( VX, n );
    float ya = SumArray( VY, n );
    float stt= STT( VX, n, xa );
    float B = CalculateB( VX, VY, n, xa, stt );
85  float A = CalculateA( ya, xa, B );
    //printf("\n Line = %f x + %f\n",A,B);
  }

  static void PrintData(vec_float4 V[], int n) {
90  int i;
    for(i=0;i<n;++i) {
      printf("\n");
      fprint_vector(&V[i]);
    }
95 }
  #define TRIALS 100000
  int main( unsigned long long spuid,
            unsigned long long argp )
  {
100 int rc,i;
    vec_float4 m=(vec_float4){0.4,0.4,0.4,0.4};
    printf(" LineFitting! from %d with %d \n", spuid, argp );
    mc_rand_mt_init( spuid );
    printf("\n");
105 CreateLinearVector(X,N);
    for(i=0;i<TRIALS;++i) {
      CreateRandomVector(Y,N,m);
  #ifdef DEBUG
      PrintData(X,N);
110   PrintData(Y,N);
  #endif
      LineFit(X,Y,N);
    }
    printf("\n\n");
115 return (0);
  }
```

Listing 13.4 SPU implementation of line-fitting.

13.4 Representing Monomials and Polynomials

A good reference for concepts about polynomial representation is the book Ideals, Varieties and Algorithms [34]. We assume the coefficients of the monomial and polynomials are defined over the field of real numbers. A *field* is a set where operations of addition, subtraction, multiplication and division are defined with the usual properties. Fields are important since linear algebra works over any field, not just R or C. Next we define a *monomial* in $x_1, x_2, \ldots, x_n$ as a product of the form

$$x_1^{\alpha_1} \cdot x_2^{\alpha_2} \ldots x_n^{\alpha_n}$$

where all the exponents α_i are non-negative integers. The *total-degree* of the monomial is the sum of the exponents.

A *polynomial* f in $x_1, x_2, \ldots, x_n$ with coefficients in field k is a finite linear combination of monomials in k, written as:

$$f = \sum_{\alpha} a_\alpha x^\alpha$$

13.4.1 Application of algebraic geometry in robotics

Consider a robotic arm as shown in Figure 13.3 (this example is based on the one given in [34]). The lengths of the links are l_1 and l_2 with $l_1 > l_2$ and l_1 tethered at the origin $x = 0, y = 0$. Let the coordinate of the joint at the end of link 1 be denoted as (x, y), and for the end of link 2 as (z, w), then the following hold:

$$x^2 + y^2 = l_1^2$$

$$(z - x)^2 + (w - y)^2 = l_2^2$$

These equations completely define the *state* of the arm and can be used to perform analysis of which positions are accessible to the arm and how much distance the joints need to travel to complete a task (which determines the time taken to complete the task which is the objective to be minimized in a shop floor).

13.4.2 Application of algebraic geometry in CAD

The next application is taken from computer-aided design. Computer representation of polynomials is of interest to all practitioners of numerical computing, and thus many papers and implementations have been proposed in literature. A new sparse multivariate polynomial representation was proposed by Monagan and Pearce in [99] which builds upon the 1974 work by Johnson on using *heaps* to mul-

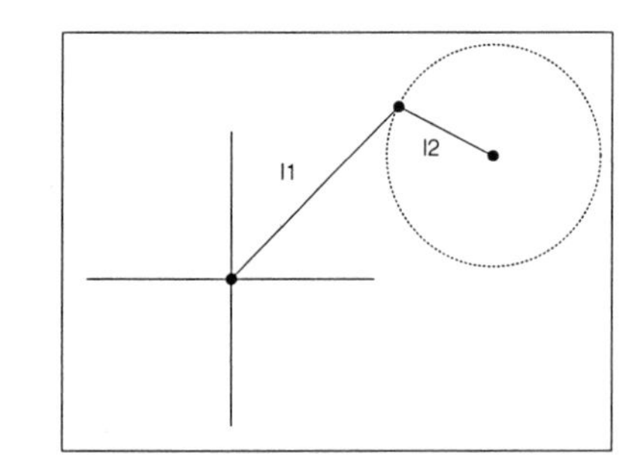

Fig. 13.3 Robotic arm with links of length l_1 and l_2.

tiply and divide sparse polynomials. In this chapter we concentrate on polynomial representations which enable efficient polynomial evaluation at multiple points. Our problem is motivated by the following discussion. Consider Figure 13.4 where a circle is translated on a parametric 2d curve $x = f(t), y = g(t)$. Given rectangles or other obstructions which are also present in the region the problem is to calculate which rectangles will be intersected by the circular shape. Given the parametric equation of the curve and equations for the circle this problem is equivalent to function evaluations.

13.4.3 Implicitization of Curves and Parametric Forms

Another area where symbolic analysis is needed is for implicitization of curves from their parametric representations and vice-versa. We have already seen the need to have the parametric representation of an ellipse in handy in the previous application. The equation of the circle in parametric form is ([34],pp. 15):

$$x = \frac{1-t^2}{1+t^2}$$

$$y = \frac{2t}{1+t^2}$$

The implicitization of this equation can be performed using an algebraic concept called the *resultant*. defined as the $(m+n) \times (m+n)$ matrix where m, n are the degrees of the polynomials for the parametric forms. Consider the parametric equations of the circle, a point (x, y) would lie on the circle iff the equations $x = \frac{1-t^2}{1+t^2}$ and $y = \frac{2t}{1+t^2}$ have a common root. The equations can be rewritten as:

$$(x+1)t^2 + x - 1 = 0$$

$$yt^2 - 2t + y = 0$$

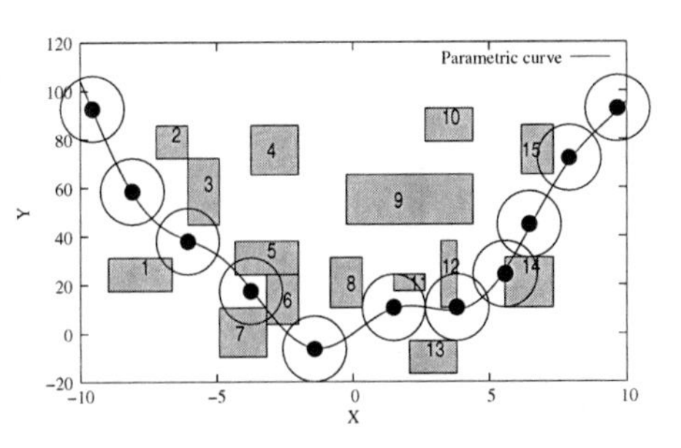

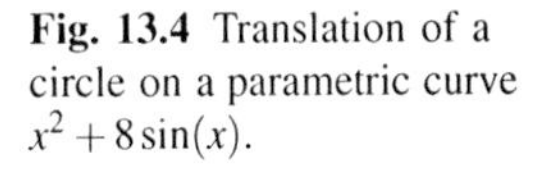

Fig. 13.4 Translation of a circle on a parametric curve $x^2 + 8\sin(x)$.

The resultant matrix can be written as:

$$Res = \begin{bmatrix} x+1 & 0 & x-1 & 0 \\ 0 & x+1 & 0 & x-1 \\ y & -2 & y & 0 \\ 0 & y & -2 & 0 \end{bmatrix}$$

Expanding $det(Res)$ gives us $4(x^2 + y^2 - 1)$ which is the required implicitization of the circle. The key idea in this example is the expansion of the determinant which has to be performed symbolically and thus requires polynomial operations. We discuss the implementation of polynomials in the next section.

In [99] the authors describe two types of polynomial representations, (a) the distributed representation in which terms are ordered in a monomial ordering and (b) the recursive representation where a multivariate polynomial is a polynomial in a single variable x whose coefficients are polynomials in y and so on. Examples of a distributed representation is

$$9xy^3z + 4y^3z^2 - 6xy^2z - 8x^3 - 5y^3$$

where the terms are sorted in *graded lexicographical order* with $x > y > z$. With this ordering xy^3z and y^3z^2 appear before other terms as their degree (5) is higher. The same polynomial in the recursive form is

$$-8x^3 + (9zy^3 - 6zy^2)x + (4z^2 - 5)y^3$$

It was believed that recursive dense representation was best overall even for sparse problems. Experimental results by Fateman [44], but Pearce and Monagan showed in [99] that infact distributed representation can be faster than the recursive representation.

```
vec_float4 EvaluateFunction(vec_float4 X) {
  vec_float4 MPI = (vec_float4){0.017,0.017,0.017,0.017};
  vec_float4 ans=spu_mul(X,X);
  vec_float4 xra=spu_mul(X,MPI);
5 vec_float4 s  = sinf4(xra);
  vec_float4 d8 = (vec_float4){8.0,8.0,8.0,8.0};
  s = spu_mul(s,d8);
```

```
  return spu_add(ans,s);
}
#define R 6.0
static vec_uint4 IntersectsRect(int r, vec_float4 X, vec_float4 Y) {
  // does Rect 'r' intersect circle of radius 6 @ X,Y
  vec_float4 CRAD  = (vec_float4){R,R,R,R};
  vec_float4 CMINX = spu_sub(X,CRAD), CMINY = spu_sub(Y,CRAD);
  vec_float4 CMAXX = spu_add(X,CRAD), CMAXY = spu_add(Y,CRAD);
  vec_uint4 pos = (vec_uint4)spu_splats(0);
  pos = spu_and( spu_cmpgt(RECT[r<<2+2],CMINX),
                 spu_cmpgt(RECT[r<<2+3],CMINY));
  pos = spu_and(pos,spu_and( spu_cmpgt(CMAXX,RECT[r<<2+0]),
                             spu_cmpgt(CMAXY,RECT[r<<2+1])));
  return pos;
}
```

Listing 13.5 SPU Implementation of parametric curve evaluation.

We experimented with monomial representation using word size packing using 32-bits.

```
(defun lex-order (a b)
  "Compare monomials a and b in lex order"
  (dotimes (i (length a))
    (let ((dega (aref a i))
          (degb (aref b i)))
      (if (< dega degb)
          (return-from lex-order t)
          (if (> dega degb) (return-from lex-order nil)))))
  t)

(defun mono-prod (a b)
  "Returns product of monomials"
  (let* ((len (length a))
         (out-mono (make-array len :element-type 'fixnum
           :initial-element 0)))
    (dotimes (i (1- len))
      (setf (aref out-mono i) (+ (aref a i) (aref b i))))
    (setf (aref out-mono (1- len)) (* (aref a (1- len)) (aref b (1- len))))
    out-mono))

(defun compatible? (a b)
  "Are the monomials a and b compatible in degrees"
  (dotimes (i (1- (length a)))
    (when (/= (aref a i) (aref b i)) (return-from compatible? nil)))
  t)

(defun mono-sum (a b)
  "Returns sum of monomials, assumes compatible"
  (let* ((len (length a))
         (out-mono (make-array len :element-type 'fixnum
           :initial-contents a)))
    (setf (aref out-mono (1- len))
    (+ (aref a (1- len)) (aref b (1- len))))
    out-mono))

(defun poly-prod (a b)
  "a and b are lists denoting polynomials"
  (let ((ans nil))
    (dolist (u a)
      (dolist (v b)
        (push (mono-prod u v) ans)))
    (sort ans #'lex-order)))

(defun simplify-poly (a)
  "Replace terms with same degree monomials"
```

x-degree	y-degree	z-degree	coeff

Fig. 13.5 Monomial representation using word size packing.

```
    (let ((ans nil)
          (u (car a)))
      (dolist (v (cdr a))
        (if (compatible? u v)
50          (setf u (mono-sum u v))
            (progn
              (push u ans)
              (setf u v))))
      (push u ans)
55    (reverse ans)))
```

Listing 13.6 Polynomial representation in Lisp.

13.5 Evaluating polynomials

Consider the example of an univariate polynomial:

$$f(x) = 3x^2 + 6x + 6$$

we would like to find the maxima of this function in the range $x \in [-4:4]$. This is can be done by evaluating the function at fixed points in the range. When evaluating a univariate polynomial of high-degree optimizations can be made by recognizing that powers of x can be evaluated using squaring rather than multiplication. A general form of this rule is the Horner Form:

$$(3x+6)x+6$$

When computing higher powers like x^{10}, we can use the fact that $x^{10} = (x^5)^2$. The case of univariate is important as even with multivariate polynomials, evaluations can sometimes reduce to univariate polynomial evaluation. Consider the multivariate polynomial:

$$f(x,y,z) = 12x^3y^2z^5 + 4xyz^2 + 6z$$

When evaluating $f(x,y,z)$ at $x = 0.5, y = 0.5, z = 0.5$, it is the same as

$$f(x) = 12x^{10} + 4x^4 + 6x$$

at $x = 0.5$. This case is particularly important for us when we evaluate probability polynomials with Boolean coefficients at a small number of fixed values.

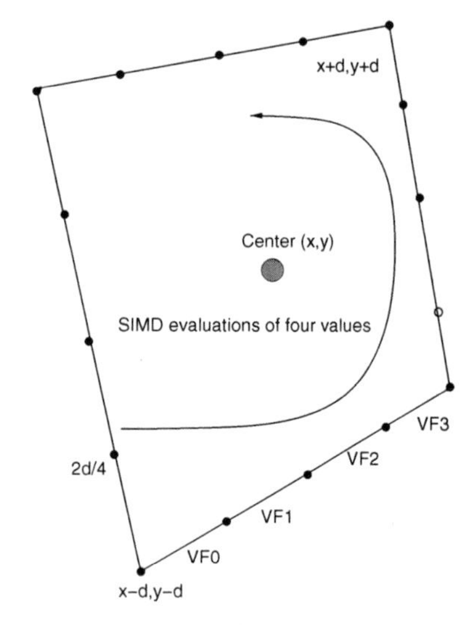

Fig. 13.6 SIMD Evaluation at fixed step of linearly bounded contour.

13.6 SIMD Polynomial evaluation with fixed step

In the section on solving complex equations of a single variable, we need to evaluate the complex polynomial $f(z)$ on the boundary of a contour sampled at equidistant points. If the contour boundary can be represented as a set of linear segments then the problem reduces to polynomial evaluation with fixed step. In Section 13.10, pp. 263 we use this scheme to evaluate a complex function (not necessarily restricted to a polynomial) at fixed step intervals using SIMD. Consider the example shown in Figure 13.6.

13.7 Parallel Polynomial Evaluation

An excellent example of parallel evaluation of polynomials is given in the following example of computing the Mandelbrot fractal, defined as:

$$f(z) = z^2 + c$$

region of the complex plane with $|f(z)| < 2$ are shaded as being inside the fractal. The boundary of the curve has fractal properties. The curve is shown in Figure 13.7 in the region $[-2,2], i[-1,1]$.

The Lisp code to calculate the fractal is shown in Listing 13.7 and the corresponding SPU code is shown in Listing 13.8.

```
;; Program: Mandelbrot for f(z) = z^2 + c
;; Author : Sandeep Koranne
```

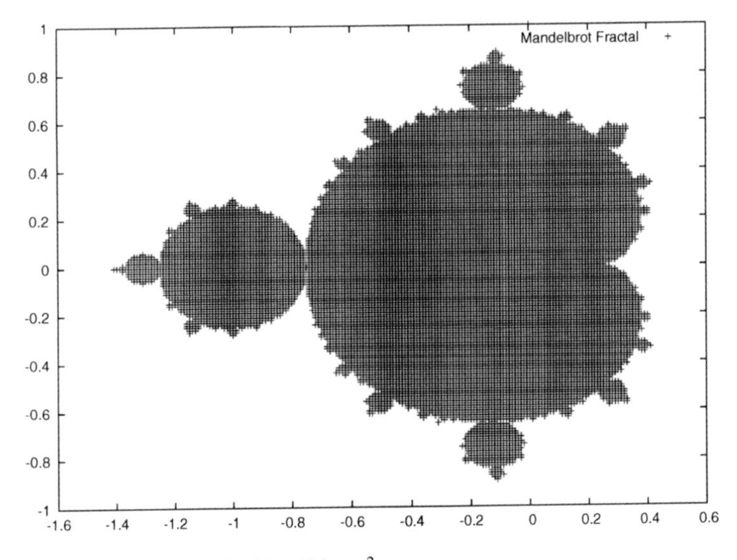

Fig. 13.7 Mandelbrot fractal defined by $f(z) = z^2 + c$.

```
(defconstant +EZ+ 2.0)
(defconstant +ITER+ 1024)
(defconstant +GRAIN+ 0.01)
(defun has-escaped(rez imz)
  (> (+ (* rez rez) (* imz imz)) +EZ+))
(defun next-value(cr ci por poi)
  (values (+ (* cr cr) (- (* ci ci)) por)
          (+ poi (* 2 cr ci))))
(defun check-point(rez imz)
  (let ((crez rez)(cimz imz))
    (dotimes (i +ITER+)
      (multiple-value-bind (a b)
          (next-value crez cimz rez imz)
        (setf crez a cimz b)
        (when (has-escaped crez cimz) (return-from check-point nil)))))
  t)
(defun main(cr ci ni nj grain)
  (dotimes (i ni)
    (dotimes (j nj)
      (when (check-point (+ cr (* grain j)) (- ci (* grain i)))
          (format t "~D ~D~%" (+ cr (* grain j)) (- ci (* grain i)))))))
(main -1.5 1.0 200 300 +GRAIN+)
(quit)
```

Listing 13.7 Lisp code to calculate Mandelbrot

```
////////////////////////////////////////
// Program: Mandelbrot Fractal
// Author : Sandeep Koranne
////////////////////////////////////////
#include <spu_mfcio.h>
#include <stdio.h>
#include <math.h>
#include "mdbl_ctrl.h"
#include "../util.h"
```

```
10  #define NUM_ELEMENTS 300
    #define USE_FLOAT
    #ifdef USE_FLOAT
    #define SFT float
    #define VFT vec_float4
15  #else
    #define SFT double
    #define VFT vec_double2
    #endif
    MDBL_CTRL cb __attribute__ ((aligned (128)));
20  volatile unsigned char DATA_LS [ NUM_ELEMENTS ][NUM_ELEMENTS]
    __attribute__ ((aligned (128)));
    static vec_float4 EZ = (vec_float4){2.0,2.0,2.0,2.0};
    static vec_uint4 has_escaped(VFT rez, VFT imz) {
      VFT rez2 = spu_mul( rez, rez ),imz2 = spu_mul( imz, imz );
25    VFT reimz= spu_add( rez2, imz2 );
      #ifdef USE_FLOAT
      vec_float4 ac_reimz = reimz;
      #else
      vec_float4 ac_reimz = spu_roundtf( reimz );
30    #endif
      vec_uint4  hesc = (vec_uint4)spu_cmpgt( ac_reimz, EZ );
      return (vec_uint4)spu_and(hesc,1);
    }
    static vec_uint4 check_point(VFT rez, VFT imz) {
35    int i,ITER=256;
      vec_uint4 res = (vec_uint4)spu_splats(0);
      VFT srez = rez,simz = imz;
    #ifdef USE_FLOAT
      VFT TWO = {2.0,2.0,2.0,2.0};
40  #else
      VFT TWO = {2.0,2.0};
    #endif
      for(i=0;i<ITER;++i) {
        res = spu_or(res, has_escaped(rez,imz));
45      rez = spu_add( srez, spu_sub( spu_mul(rez,rez), spu_mul( imz, imz)));
        imz = spu_add( simz, spu_mul( spu_mul(TWO,rez), imz));
      }
      return res;
    }
50  static int PerformComputation(MDBL_CTRL cb) {
      int result = 0,i,j,counter=0;
      #ifdef USE_FLOAT
      VFT GRAIN={0.01,0.01,0.01,0.01};
      #else
55    VFT GRAIN={0.01,0.01};
      #endif
      VFT cr=(VFT)spu_splats(cb.cr);
      VFT ci=(VFT)spu_splats(cb.ci);
      VFT cpr,cpi;
60    printf("\nSearching from %f,%f", cb.cr,cb.ci);
      for(i=0;i<cb.N;++i) {
        SFT di=(SFT)i;
        VFT iaf=spu_splats(di);
        cpi = spu_mul(iaf,GRAIN);
65      cpi = spu_sub(ci, cpi);
        for(j=0;j<cb.M;++j) {
          SFT dj=(SFT)j;
          VFT jaf=spu_splats(dj);
          cpr = spu_mul(jaf,GRAIN);
70        cpr = spu_add(cr, cpr);
          vec_uint4 res=check_point(cpr,cpi);
          DATA_LS[i][j] = spu_extract(res,0);
        }
      }
75    return result;
    }
```

Listing 13.8 SPU code to calculate Mandelbrot

13.8 Evaluating Boolean Functions

A special case of polynomial evaluation occurs when the field is drawn from Boolean algebra and is restricted to 0 or 1. In this case a monomial is simply a bit-vector and a polynomial can be represented as a collection of bit-vectors. Consider the example:

$$f(x_0, x_1, x_2) = x_0\bar{x_1}x_2 + \bar{x_0}x_1\bar{x_2}$$

We have implemented Boolean polynomials in the context of signal probability polynomials in Section 23.5, pp. 444.

13.9 Parallel Matrix Functions

For most matrix and linear algebra functions we are using standard SDK provided functionality. See Chapter 9 for an overview of the SDK functionality available on the Cell, especially in the context of linear algebra, see 9.6, pp. 136.

13.10 Solving Equations in a Single Complex Variable

This section presents the implementation of an algorithm to solve complex equations of a single variable. We shall cover the following topics in this chapter:

1. Numerical methods: overview on fundamental data types and their representation on the CBE
2. Using SIMD for numerical methods
3. Matrix methods, vectors and complex number libraries
4. Solving equations of holomorphic functions
5. General simulations on the SPU
6. Performance measurement, optimization using Self-Tuning

The functions we shall consider are analytic in the region under consideration, which for simplicity, we assume to be a rectangular region in the complex plane. Such functions have a convergent Taylor series at every point in this region. A definition of analytic functions is given below and can also be found in the standard text on complex function theory of a single variable, Conway [10].

Since solving for complex function zeros is important in engineering, where time-to-result is of great importance, parallel algorithms for solving this problem have also been previously described. Schaefer and Bubeck describe a distributed parallel complex zero finder based on a network-of-workstation (NoW) cluster in [111]. Parallelism in their approach is based on PVM and CThreads. Another (sequential) method for computing zeros is presented by Kravanja and van Barel [76]. Their approach is based on the evaluation of a symmetric bi-linear form via numerical integration, which involves computing the derivative of the function under investigation. A good mathematical description as well as numerical methods for handling clusters of zeros is presented by Kravanja, Sakurai and van Barel in [75].

A recent paper by Meylan and Gross [86] presents another parallel algorithm to find complex zeros of complex analytic functions; the focus of their paper is on solving non-linear eigenvalue problems. Their approach is very similar to the one presented in this paper, although the parallelism is based on NoW clusters of 16 Linux machines. Additionally, the paper does not strive for efficiency in solving for complex zero, rather, it focuses on showing distributed computing as an alternative method for achieving parallelism. Morgan and Shapiro [88] presented a very similar box-bisection algorithm, although they used the method for solving second-degree systems, not for solving complex equations. A very lucid description of the quadrant change phenomenon, as well as the concept of $u - change$ and $v - change$ is described by Cain in [53]. Finding approximate complex zero is also important as it can be used as an initial guess for Newton type methods. One method for computing approximate complex zeros is presented by Pan in [97]; his method is defined only for complex polynomials. Numerical software libraries typically implement complex equation solvers by using Newton's method seeded with random points in the complex plane.

Our choice of the quadrant labeling algorithm (as presented in [59]), combined with parallel domain decomposition is based on the following observations:

1. Admits general complex functions (not limited to polynomials),
2. No need to compute derivative $f'(z)$ as compared to the numerical integration method,
3. No need to give initial guess, real or complex, to start the method as compared to Newton method,
4. Exhibits multiple level of parallelism: we have successfully implemented, multi-core (using upto 6 SPUs), SIMD (computing four complex function evaluations at same time), dual instruction issue per cycle (implemented as part of SPU), distributed (by dividing the initial region onto a cluster farm of PS3), MPMD (by switching some SPUs to steepest descent search when approximate root is known),
5. Simple to program: implemented our algorithms in PPU/Opteron Standard C++, SPU C, Common Lisp
6. Simple to control parameters m, the step size, and d the precision required, influencing run-time and accuracy in a smooth trade off.

Problem Statement for solving $\mathbf{f(z) = 0}$

We give the definition of analytic functions, since properties of analytic functions are central to our algorithm:
Let G be an open set in C the complex plane, then a function $f(z) : G \rightarrow C$ is analytic if $f(z)$ is continuously differentiable on G. A function which is analytic in the whole complex plane is called an *entire* function. Examples of analytic functions include exponential, logarithmic (atleast in the branch that lies in G), trigonometric and combinations thereof. More details about the mathematical properties of analytic functions, which are central to our method, are given below in Section 13.10.

Analytic functions can be thought as *mappings* from connected sets G to Ω. We use this property when convincing ourselves about the existence of a zero of $f(z)$ in a rectangular region in G. We also state without proof the following theorem which is key to the idea of quadrant labeling:

Theorem 1. *Liouville's Theorem. (cf. Conway [10],pp. 77) If $f(z)$ is a bounded entire function curve then $f(z)$ is a constant.*

We now discuss the concept of *winding number* or *index* of a curve around a point. Let $f(z)$ be a continuous complex functions which maps the complex plane C into C, $f : C \rightarrow C$ within a bounded domain. Let γ be a Jordan curve that does not pass through any zeros of $f(z)$. We first consider the problem of computing all zeros of $f(z)$ that lie within γ. The *index* ([10] pp. 80) of $f(z)$ at 0 with respect to γ is called the *winding number*. The *argument principle* of complex analysis for holomorphic (or analytic functions) states that the change in argument of $f(z)$ as z moves around γ in a positive (counter clockwise) sense can be used to calculate the number of zeros of $f(z)$ lying within γ. Let $q_j, j = 1, \ldots, n$ denote the zeros of $f(z)$ inside γ, then

$$\int_{\gamma} \frac{f'(z)}{f(z)} = 2i\pi \sum_{k=1}^{n} \mu(q_j) \tag{13.1}$$

where $\mu(q_j)$ are the multiplicities of the zeros q_j. To simplify the algorithm we have used a square of size d centered around $a_0 + ib_0$ as the Jordan curve γ. By expanding the rectangle, or shrinking and sub-dividing the rectangle, we get another curve γ'. We can use the argument principle to count the number of complex zeros of $f(z)$ within γ'. By Liouville's theorem above, every non-constant complex function must have $u = 0$ and $v = 0$ curves which cross a large enough boundary; that is they cannot be bounded in the region, and thus our traversal around *any* Jordan curve is bound to locate these curves, provided the step size d is small enough. In the next section we describe methods to calculate the number of zeros within γ and how to refine γ will be described in detail.

Description of Quadrant Labeling Algorithm

Let S_0 be a square with center (x_0, y_0) and side length $2d$. We assume we have SPU functions to compute $u(x,y) = re(f(z))$ and $v(x,y) = im(f(z))$ (the implementations are shown in Section 13.10), we get

$$f(z) = f(x+iy) = u(x,y) + iv(x,y) \tag{13.2}$$

The two functions are also called the u-curve and the v-curve, respectively. To determine the number of zeros of $f(z)$ within S_0, we start evaluating $u(x,y)$ and $v(x,y)$ at $(x_0 - d, y_0 - d)$, stepping along the curve γ, in this case the square S_0, with a step size d/m, where m is an integer whose value decreases with the refinement depth converging to 1 when d is equal to the stopping precision of the system.

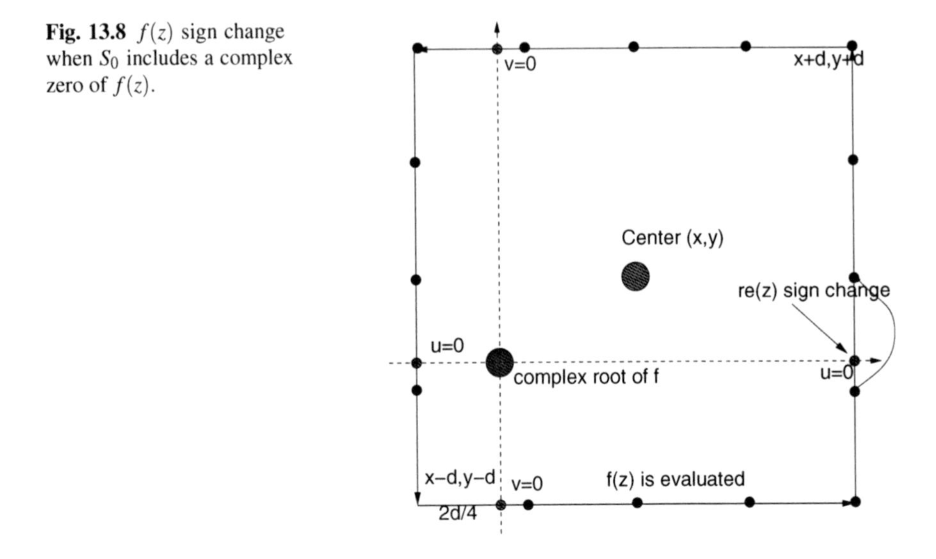

Fig. 13.8 $f(z)$ sign change when S_0 includes a complex zero of $f(z)$.

Let u_i be the value of $u(x,y) : x = x_i, y = y_i$, and similarly for u_{i+1}. If the sign of u_i is different from u_{i+1}, we know that we have crossed the imaginary axis between the two function evaluations. Similarly, if the sign of v_i is different from v_{i+1} we know we have crossed the real axis, where $u(x,y) = 0$, as is shown in Figure 13.8. The plot of this function where $u(x,y) = 0$ is called the u-curve, similarly defined for the v-curve. Obviously, a complex zero must lie on both u-curve, and v-curve. By keeping track of the sign change as we move around S_0 we can compute the number of complex zeros S_0 contains. If we detect both u and v sufficiently close to 0, we report a complex root at $x_i + iy_i$. Once we detect a zero in the square we begin the refinement process. Square S_0 is divided into 4 sub-squares, centered at $(x_0 - d/2, y_0 - d/2)$, $(x_0 + d/2, y_0 - d/2)$, $(x_0 + d/2, y_0 + d/2)$, and $(x_0 - d/2, y_0 + d/2)$,

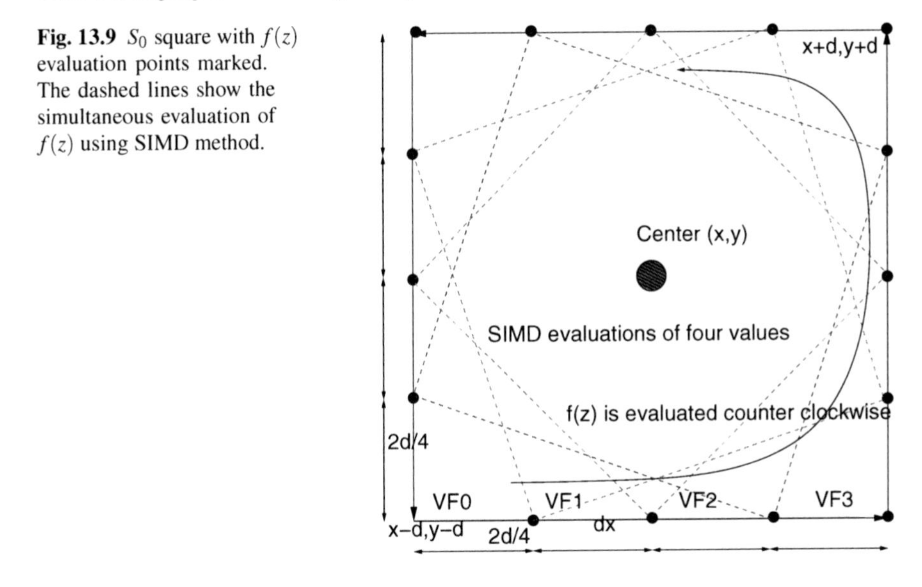

Fig. 13.9 S_0 square with $f(z)$ evaluation points marked. The dashed lines show the simultaneous evaluation of $f(z)$ using SIMD method.

and of size d. Since for every iteration we are dividing d by 2, this process is bound to terminate when $d < \varepsilon$, where ε is a floating point constant of the order of $10e-6$. The center of this square with side $d^* < \varepsilon$ is returned as a root of $f(z)$.

If both u and v functions change sign as we move from a point u_i to u_{i+1}, then we have to reduce the side $2d$ of the square (a process which we call *refinement*) and try again. Consider the square shown in Figure 13.9. The square is centered at the complex point $a+ib$ with initial side $2d$, giving the corners as $((a-d)+i(b-d))$, $((a+d)+i(b-d))$, $((a+d)+i(b+d))$, and $((a-d)+i(b+d))$. Each side of this square is then divided into m fragments of length $2d/m$. To compute the quadrant code of $f(z)$ as we traverse the square in counter-clockwise orientation, we have to evaluate $f(z)$ at the $4m$ locations. Using SIMD organization we reduce this by a factor of 4 as follows: we build a (vector float) containing the real components of the corners of the square, similarly we build a (vector float) containing the imaginary components of the corners. We build 2 constant vectors, CONST_DX and CONST_DY which contain the δ changes as we move from m_i to m_{i+1}. CONST_DX is simply $\{2d/m, 0, -2d/m, 0\}$, and CONST_DY is $\{0, 2d/m, 0, -2d/m\}$. These are the deltas added to the real and imaginary component as we move from one evaluation point to the next. Instead of $4m$ evaluations we only need to move along the single bottom edge of the square. When we evaluate $f((a-d)+i(b-d))$, the SIMD evaluation automatically evaluates the other 3 corners. We then add CONST_DX (CONST_DY) to the real (imaginary) component to get $f((a-d+2d/m)+i(b-d))$. The second and fourth element (for real component) does not change as these represent the two vertical sides of the square. Using this method we can reduce the computation by a factor of 4 as compared to a non-SIMD complex function evaluation. Once the real and imaginary component

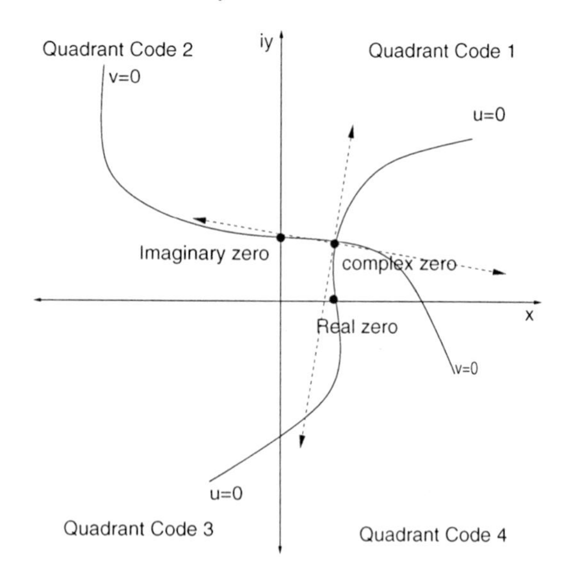

Fig. 13.10 Quadrant label and the $u = 0$ and $v = 0$ curves.

of $f(z)$ have been evaluated, their quadrant codes are evaluated, as described in the next section.

Quadrant Labeling

Given a square centered at x, y of side $2d$, and a function to evaluate $f(z)$, we define the quadrant code of the square as follows.

As we know, the complex plane can be defined by two orthogonal axis, namely, the real axis going from $-\infty$ to $+\infty$. The other axis is the imaginary axis, from $-i\infty$ to $+i\infty$. For every analytic region we consider, we can compute a quadrant code for the value of the function $f(z)$ for every z in the region. The quadrant code $qc(z)$ is a number defined as:

$$qc(z) = \begin{cases} \text{Text}-\text{book} \;\; \text{SPU} \;\; \text{C++/Lisp} & \\ \quad 1 \qquad 3 \qquad 2 & \text{if } re(f(z)) > 0 \text{ and } im(f(z)) > 0 \\ \quad 2 \qquad 1 \qquad 3 & \text{if } re(f(z)) < 0 \text{ and } im(f(z)) > 0 \\ \quad 3 \qquad 0 \qquad 5 & \text{if } re(f(z)) < 0 \text{ and } im(f(z)) < 0 \\ \quad 4 \qquad 2 \qquad 7 & \text{if } re(f(z)) > 0 \text{ and } im(f(z)) < 0 \\ \quad 0 \qquad 3 \qquad 11 & \text{if } re(f(z)) \approx 0 \text{ or } im(f(z)) \approx 0 \end{cases} \tag{13.3}$$

The quadrant codes are also shown in Figure 13.10.

Consider the function $f(z) = e^z + \sin(z)\cos(z)$. The resulting plot of the quadrant code is shown in Figure 13.11. For the human reader, the location of the zeros can be understood by looking for a point where all 4 quadrant labels, namely, 1,2,3 and 4 are present in the figure. Of-course, this is not the exact computer implementa-

```
-2.0  0.76  4444444444444333333333333333222222222222222211111111111111111111111111111111111111111
-2.0  0.72  4444444444444333333333333333222222222222222211111111111111111111111111111111111111111
-2.0  0.68  2444444444444333333333333332222222222222222111111111111111111111111111111111111111111
-2.0  0.64  4444444444444433333333333332222222222222221111111111111111111111111111111111111111111
-2.0  0.59  2444444444444433333333333332222222222222221111111111111111111111111111111111111111111
-2.0  0.55  2444444444444433333333333332222222222222211111111111111111111111111111111111111111111
-2.0  0.52  4444444444444433333333333332222222222222111111111111111111111111111111111111111111111
-2.0  0.48  4444444444444443333333333332222222222222111111111111111111111111111111111111111111111
-2.0  0.44  2444444444444443333333333332222222222221111111111111111111111111111111111111111111111
-2.0  0.40  4444444444444444333333333322222222222211111111111111111111111111111111111111111111111
-2.0  0.35  4444444444444444333333333322222222222111111111111111111111111111111111111111111111111
-2.0  0.31  4444444444444444333333333322222222221111111111111111111111111111111111111111111111111
-2.0  0.27  2444444444444444433333333322222222221111111111111111111111111111111111111111111111111
-2.0  0.24  2444444444444444433333333322222222211111111111111111111111111111111111111111111111111
-2.0  0.20  4444444444444444443333333322222222211111111111111111111111111111111111111111111111111
-2.0  0.16  1444444444444444443333333322222222111111111111111111111111111111111111111111111111111
-2.0  0.11  2444444444444444443333333322222222111111111111111111111111111111111111111111111111111
-2.0  0.07  4444444444444444443333333322222221111111111111111111111111111111111111111111111111111
-2.0  0.03  4444444444444444444333333322222221111111111111111111111111111111111111111111111111111
-2.0 -0.03  3111111111111111112222222333333444444444444444444444444444444444444444444444444444444
-2.0 -0.08  3111111111111111122222222333333444444444444444444444444444444444444444444444444444444
-2.0 -0.12  3111111111111111122222222333333344444444444444444444444444444444444444444444444444444
-2.0 -0.15  1111111111111111122222222333333344444444444444444444444444444444444444444444444444444
-2.0 -0.20  3111111111111111122222222333333334444444444444444444444444444444444444444444444444444
-2.0 -0.24  1111111111111111222222222333333333444444444444444444444444444444444444444444444444444
-2.0 -0.27  1111111111111111222222222333333333344444444444444444444444444444444444444444444444444
-2.0 -0.32  4111111111111112222222222233333333334444444444444444444444444444444444444444444444444
-2.0 -0.36  3111111111111112222222222233333333334444444444444444444444444444444444444444444444444
-2.0 -0.39  3111111111111112222222222233333333333444444444444444444444444444444444444444444444444
-2.0 -0.44  4111111111111122222222222223333333333344444444444444444444444444444444444444444444444
-2.0 -0.48  3111111111111122222222222223333333333334444444444444444444444444444444444444444444444
-2.0 -0.52  3111111111111222222222222223333333333334444444444444444444444444444444444444444444444
-2.0 -0.55  3111111111111222222222222223333333333333444444444444444444444444444444444444444444444
-2.0 -0.60  3111111111111222222222222223333333333333344444444444444444444444444444444444444444444
-2.0 -0.64  1111111111111222222222222223333333333333334444444444444444444444444444444444444444444
-2.0 -0.67  4111111111112222222222222223333333333333333444444444444444444444444444444444444444444
-2.0 -0.72  4111111111112222222222222222333333333333333444444444444444444444444444444444444444444
-2.0 -0.76  4111111111112222222222222222333333333333333344444444444444444444444444444444444444444
-2.0 -0.79  1111111111112222222222222222333333333333333334444444444444444444444444444444444444444
-2.0 -0.84  4111111111112222222222222222333333333333333334444444444444444444444444444444444444444
-2.0 -0.88  1111111111122222222222222222333333333333333333444444444444444444444444444444444444444
-2.0 -0.91  3111111111122222222222222222333333333333333333444444444444444444444444444444444444444
-2.0 -0.96  4111111111122222222222222222233333333333333333344444444444444444444444444444444444444
-2.0 -1.00  4111111111122222222222222222233333333333333333334444444444444444444444444444444444444
-2.0 -1.04  1111111111122222222222222222233333333333333333334444444444444444444444444444444444444
-2.0 -1.07  4111111111122222222222222222233333333333333333333444444444444444444444444444444444444
-2.0 -1.12  3111111111222222222222222222233333333333333333333444444444444444444444444444444444444
-2.0 -1.16  3111111111222222222222222222223333333333333333333344444444444444444444444444444444444
-2.0 -1.20  4111111111222222222222222222233333333333333333333334444444444444444444444444444444444
```

Fig. 13.11 Quadrant code plot for function $f(z) = e^z + \sin(z)\cos(z)$.

tion, but nevertheless it gives a pictorial depiction of the zeros. Equation 13.3 shows the various quadrant codes we have used throughout out our implementation. Since SIMD evaluation of quadrant codes is based on bit-wise operations, the first quadrant is labeled 3 (0b11) as both real and imaginary components are greater than 0. Similarly, the second quadrant has code 1 (0b01) as the real component is NOT greater than 0 (so the MSB of the quadrant code is 0), while the LSB (representing the sign of the imaginary component) is 1. Another variant of the quadrant code we have used in our implementation is using the first 5 prime numbers as the code; using prime numbers 2,3,5,7 and 11 (as shown in the third column of Eqn. 13.3), we can quickly check for sign changing properties as described below. We build a composite quadrant code by multiplying the individual quadrant codes of the 4 vertices. In Eqn. 13.3 the first column of codes is only for exposition, our computer implementations use prime number encoding or bit-encoding. Quadrant codes given real and imaginary SIMD (float vectors) on the SPUs can be done efficiently by comparing the real and imaginary value to a constant vector of 0.0; this gives us a bit-mask vector which contains the sign information of the real and imaginary components. By logical-AND of the real sign-mask with 0x10, and the imaginary

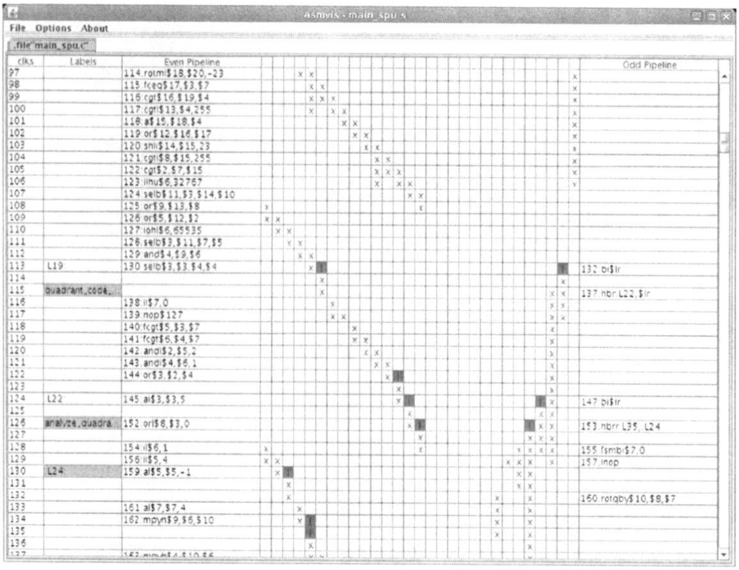

Fig. 13.12 8 cycles to compute quadrant code using dual-pipeline of the SPE.

sign-mask by 0x01, and then logical-OR of the two we get the quadrant code of 4 complex numbers simultaneously.

```
static vec_float4 CZ   = (vec_float4){0.0,0.0,0.0,0.0};
static vec_uint4 CO = (vec_uint4) {1,1,1,1};
static vec_uint4 CT = (vec_uint4) {2,2,2,2};

vec_uint4 calculate_quadrant_code( VF re_val, VF im_val ) {
    vec_uint4 REM   = spu_cmpgt( re_val, CZ ); // Re sign
    vec_uint4 RMA   = spu_and  ( CT, REM    ); // 2 LSB
    vec_uint4 IMM   = spu_cmpgt( im_val, CZ ); // Im sign
    vec_uint4 IMA   = spu_and  ( CO, IMM    ); // 1 LSB
    vec_uint4 qcode = spu_or   ( RMA, IMA   );
    return qcode;
```

Listing 13.9 Calculating quadrant code

We interleave quadrant code calculation (which is mostly executed on the even pipeline), with floating point code to calculate sinf4 and cosf4 functions of the next point. By executing SFS/SFX/SFP instructions together we are optimizing the quality of the SPU assembly code for dual instruction issue per cycle. The quadrant code function takes just 8 clock cycles (although an optimizing compiler could do this as well, we are consciously writing code knowing the instruction types of the SPU). This is shown in Figure 13.12. We tabulate $z, f(z), qc$ in Table 13.1.

We have also implemented a simple optimization when computing and accumulating the quadrant codes as follows: from Table 13.1 we can write the quadrant code as a list $(4^6, 0, 1^6, 2, 0, 3)$. We *compress* this list to $(4, 0, 1, 2, 0, 3)$, by removing consecutive duplicate entries. No information is lost by this compression as we

z	$f(z)$	qc		z	$f(z)$	qc
C(-1.0 -1.0)	C(0.19 -2.30)	4		C(-0.5 -1.0)	C(1.07 -1.51)	4
C(0.00 -1.0)	C(1.54 -0.84)	4		C(0.50 -1.0)	C(1.64 -0.38)	4
C(1.00 -1.0)	C(1.46 -0.28)	4		C(1.00 -0.5)	C(1.63 -0.30)	4
C(1.00 0.00)	C(1.73 0.00)	0		C(1.00 0.5)	C(1.63 0.30)	1
C(1.00 1.00)	C(1.46 0.28)	1		C(0.50 1.00)	C(1.64 0.38)	1
C(0.00 1.00)	C(1.54 0.84)	1		C(-0.5 1.00)	C(1.07 1.51)	1
C(-1.0 1.00)	C(0.19 2.30)	1		C(-1.0 0.50)	C(-0.42 1.17)	2
C(-1.0 0.00)	C(-0.63 0.00)	0		C(-1.0 -0.5)	C(-0.42 -1.17)	3

Table 13.1 Table of $z, f(z), qc$ for $e^z - z^2$ around a box centered at $0+i0$ of width 2. The evaluation is done in counter clockwise orientation. Observe that there is a cycle 1,2,3,4 when walking around this square, which implies that there is indeed a complex root of $e^z - z^2$ in this region. We subdivide the square further to refine the search. We can also see 2 $u = 0$ points where the quadrant code is 0 as the evaluation of $f(z)$ is real.

only need to count when we move from one quadrant to another. Secondly, we also remove 0 labeled entries when calculating the winding number, as $u = 0, v = 0$ is not used in that computation. This leaves us with a list $(4, 1, 2, 3)$ which has winding number 1, and hence shows that there is 1 root in this square for this function. The reader should be able to appreciate the simplicity of the quadrant labeling approach, as the only primitive computation being done is the function evaluation $f(z)$ and the sign comparison of $f(z)$ with Complex zero, which is numerically robust. Using prime numbers as labels we can optimize this even more as we traverse the list and maintain a running product which calculates the winding number, and also checks when we make a diagonal jump, in which case we have to decrease the step size for the square, as both $u = 0$ and $v = 0$ curves were crossed.

Domain Decomposition

We know that we can count the number of complex zeros of $f(z)$ inside a rectangular region defined by $a + ib \times A + iB$ by counting the quadrant label changes as we evaluate $f(z)$ along the boundary of the rectangle. We start off the general search around $C(0,0)$ with a square of side $2d$, we calculate the number of zeros found in this square. If we want to increase the number, or if there were no zeros found, we increase d by a suitable multiple, say $2d$. Thus we can assume that for the sequel of this paper, there is an analytic rectangular (square) region which has atleast one complex root.

Since we are designing our algorithm to execute on a PS3 which has 6 SPUs we do domain decomposition of this region into 6 smaller regions. To simplify implementation and increase robustness we overlap the computation of the regions by a small amount. This poses a problem if there were multiple zeros in the overlapped region, but we can identify these as a post-process. (Recently Sasaki has presented an interesting approach to resolving clusters of complex zeros in [110], and imple-

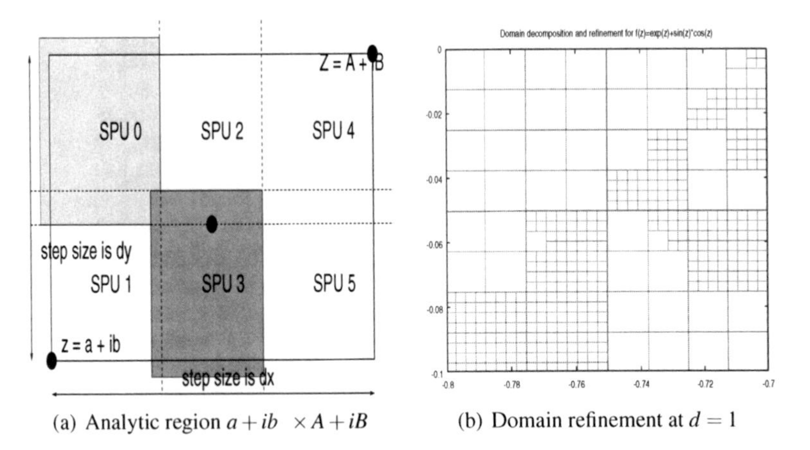

(a) Analytic region $a + ib \;\; \times A + iB$ (b) Domain refinement at $d = 1$

Fig. 13.13 Domain decomposition/refinement on SPU, $f(z) = e^z + \sin(z)\cos(z)$.

menting his method on the SPUs is planned for future versions of our system.) Most likely the application is interested in knowing any zeros there are in the given region. This division of the global analytic region into 6 smaller regions each covered by an SPU is shown in Figure 13.13(a).

We describe the process of domain refinement assuming our square (rectangle) is centered at a complex number $(c_0 + ic_0)$, and has a side $2d$. We divide $2d$ into m fragments where we will evaluate $f(z)$. The number m depends on the precision of the computation and is decreased as we gradually refine the domain, at the same time the size of each square becomes smaller. The function $f(z)$ is evaluated m times at $((c_0 - d) + i(c_0 - d))$ to $((c_0 + d) + i(c_0 - d))$; this is the bottom edge of the square. For every evaluation we keep track of the quadrant labels of $f(z)$. We know that when $qc(z)$ changes from 1 to 2 or from 3 to 4 we have encountered a $v = 0$ curve, similarly when $qc(z)$ changes from 2 to 3, or from 4 to 1 we have found a $u = 0$ curve. A plot of domain refinement is shown in Figure 13.13(b). If we detect that this square has non-zero roots then we proceed to recursively refine this square into 4 smaller sub-squares of side $2d/1.999$ to get 4 overlapping sub-squares. Since identical SPU code is invoked for each sub-square (only the work-block parameters containing the analytic region boundary are changed by recursion) no communication to the PPU is required, as the SPU can update the work-block parameter in its local-store and call the compute-kernel function recursively.

To the best of our knowledge complex number functions are not optimized using SIMD on the PPU or SPU by the provided simdmath library. Moreover, a significant part of the design of our algorithm separates the computation of the real and imaginary part of $f(z)$ so as to compute quadrant codes efficiently. Thus we have implemented our own complex number math library which is SIMD optimized, in the sense it operates on 4 complex numbers simultaneously.

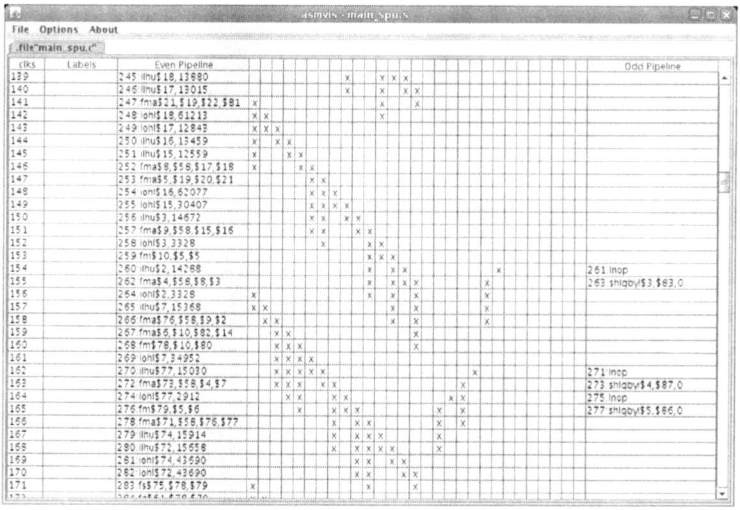

clks	Labels	Even Pipeline	Odd Pipeline
139		245 ilhu$18,13880	
140		246 ilhu$17,13015	
141		247 fma$21,$19,$22,$81	
142		248 iohl$18,61213	
143		249 iohl$17,12843	
144		250 ilhu$16,13459	
145		251 ilhu$15,12559	
146		252 fma$8,$58,$17,$18	
147		253 fma$5,$19,$20,$21	
148		254 iohl$16,62077	
149		255 iohl$15,30407	
150		256 ilhu$3,14672	
151		257 fma$9,$58,$15,$16	
152		258 iohl$3,3328	
153		259 fm$10,$5,$5	
154		260 ilhu$2,14288	261 lnop
155		262 fma$4,$58,$8,$3	263 shlqbyi$3,$83,0
156		264 iohl$2,3328	
157		265 ilhu$7,15368	
158		266 fma$76,$58,$9,$2	
159		267 fma$6,$10,$82,$14	
160		268 fm$78,$10,$80	
161		269 iohl$7,34952	
162		270 ilhu$77,15030	271 lnop
163		272 fma$73,$58,$4,$7	273 shlqbyi$4,$87,0
164		274 iohl$77,2912	275 lnop
165		276 fm$79,$5,$6	277 shlqbyi$5,$86,0
166		278 fma$71,$58,$76,$77	
167		279 ilhu$74,15914	
168		280 ilhu$72,15658	
169		281 iohl$74,43690	
170		282 iohl$72,43690	
171		283 fs$75,$78,$79	

Fig. 13.14 Even-issue floating point dominates the complex function evaluation.

We analyzed the generated code for our complex function evaluation SIMD library, and actively tried to increase the overlapping computation in these 4 execution units. On a conventional 64-bit microprocessor (like the Opteron 242), the floating-point unit has been optimized with multiple instructions per cycle, but the SPU is like an explicit schedule RISC processor in this regards. Additionally, the MADD and MSUB instructions also help in reducing the length of the inner loop of function evaluations. The SPU supports the following data types: (i) Byte (8 bits), (ii) Half-word (16 bits), (iii) Word (32 bits), (iv) Doubleword (64 bits), and Quadword (128 bits). For our implementation we have exclusively used the 32-bit word, in floating point and unsigned int modes. We have reduced the data transfers requirement of our implementation to a minimum. The PPU sends analytic regions to the SPUs, and the SPUs return either the number and value of computed roots, or 0 to signal no roots in the region. The evaluation of the complex value function $f(z)$ is the most important component of our system. The evaluation is dominated by the even-issue pipeline of the SPE, this is shown in Figure 13.14.

Design of a complex function SIMD library

Consider the complex $\sin(z)$ function. It is well known that

$$\sin(a+ib) = \sin(a)\cosh(b) + i\cos(a)\sinh(b) \quad (13.4)$$

Assume VFA and VFB to contain a (vector float) which represents the real and imaginary parts of 4 complex numbers, respectively. Then using SIMD we can say that $re(sin(VFA+iVFB)) = \sin(VFA)*\cosh(VFB)$ where this $\sin(x)$ is the sine function defined for real numbers as implemented in the simdmath library. Similarly, $im(\sin(VFA+iVFB))$ can be implemented. Some of the equations similar to the one in 13.4 for the functions we use in our system are described below. These functions and others are described in detail by Smith in [117].

Function	Real Arithmetic	Notes
$(a+ib)+(c+id)$	$(a+c)+i(b+d)$	-
$(a+ib)-(c+id)$	$(a-c)+i(b-d)$	-
$(a+ib)*(c+id)$	$(ac-bd)+i(ad+bc)$	-
$\frac{a+ib}{c+id}$	$\frac{aQ+b}{cQ+d}+i\frac{bQ-a}{cQ+d}$	$Q=\lvert c\rvert<\lvert d\rvert\ ?\ \frac{c}{d}:\frac{d}{c}$
e^{a+ib}	$e^a\cos(b)+ie^a\sin(b)$	$\sin,\cos$ computed together
$\sin(a+ib)$	$\sin(a)\cosh(b)+i\cos(a)\sinh(b)$	-
$\cos(a+ib)$	$\cos(a)\cosh(b)-i\sin(a)\sinh(b)$	-
$\tan(a+ib)$	$\frac{\sin(2a)}{\cos(2a)+\cosh(2b)}+i\frac{\sinh(2b)}{\cos(2a)+\cosh(2b)}$	Denom. common
$\sinh(a+ib)$	$\sinh(a)\cosh(b)+i\cosh(a)\sin(b)$	-
$\cosh(a+ib)$	$\cosh(a)\cos(b)+i\sinh(a)\sin(b)$	-
$\tanh(a+ib)$	$\frac{\sinh(2a)}{\cosh(2a)+\cos(2b)}+i\frac{\sin(2b)}{\cosh(2a)+\cos(2b)}$	Denom. common

Table 13.2 Complex function evaluation using real arithmetic functions [117].

Complex Function Evaluation by SPU

Since the function to be solved for is given by the user, we cannot assume that it is known at compile time. Thus, we have the following three scenarios for complex function evaluation:

1. Function known at compile time: the code of the function is already wrapped in a single function callable on the SPU and PPU,
2. Function input by user, compiled to SPU code by calling spu_gcc: for performance, and especially, if the function is going to be solved in many different complexes, it is more efficient to generate C or C++ code, dynamically, and to compile this resulting code into SPU and PPU callable functions,
3. Function input by user as symbolic sum-of-products form: in this case the user types in:
   `exp (z)-sin (z)-12*cos (8 * z)` to represent $e^z - sin(z) - 12cos(8z)$.

SIMD evaluation of complex numbers

Consider the example function $f(z) = e^z + \sin(z)\cos(z)$, we compute the real part of $f(z)$ as $e^a\cos(b) + \sin(a)\cos(a)\cosh(b)^2 + \sin(a)\cos(a)\sinh(b)^2$, where a and b are (vector float) type SIMD vectors representing real and imaginary part of four z, respectively. This expression can be easily evaluated on the SPU as follows (the (vector float) type casts have been removed for clarity):

```
vec_float4 cosa,sina;
vec_float4 ea        = expf4    ( a );
vec_float4 cosb      = cosf4    ( b );
vec_float4 cosh_b    = _coshf4  ( b );
vec_float4 sinh_b    = _sinhf4  ( b );
(void)         sincosf4( a, &sina, &cosa );
vec_float4 sc_prod   = spu_mul  ( sina, cosa );
vec_float4 csq       = spu_mul  ( cosh_b, cosh_b );
vec_float4 ssq       = spu_mul  ( sinh_b, sinh_b );
vec_float4 t0        = spu_mul  ( sc_prod, csq );
vec_float4 t2        = spu_madd( sc_prod, ssq, t0 );
vec_float4 ans       = spu_madd( ea, cosb,t2 );
```

Listing 13.10 Evaluating complex functions.

Quadrant labeling of the complex function on the SPU

We had earlier described the mathematical background of quadrant labeling in Section 13.10, in this section we describe the implementation of quadrant labeling as implemented on the SPU. We assume the function to be evaluated is available on the SPU either as a direct callable function, or as a token stream. The argument to this function is a 4-element vector of floats given to the SIMD evaluation phase.

Floating point performance measurement on the SPUs

By analyzing the generated assembly code for the complex function evaluation, measuring the total number of calls made to the functions, and knowing the real time elapsed for the computation we have arrived at approximately 140 gigaflops as a reasonable estimate of the floating point performance of our system. Since the PPU does very little compute intensive work, and we only have 6 SPUs active, we have effectively achieved 23.3 gigaflops per SPU from a theoretical capacity of 25.6 gigaflops.

Complete Algorithm Description

In this section we describe the full algorithm we have implemented for calculating complex roots of analytic functions. One optimization for load balancing on the SPUs (which is not shown in the following description) is the following change we have made to the recursive evaluation of sub-divided squares. After a certain recursive depth, instead of the SPU recursing on each of the 4 sub-squares, we send 1 square back to the PPU for adding it to the set of work-blocks. The SPU only recurses on 3 sub-squares. The rationale behind this scheme is the following observation: as the recursion depth becomes high many squares report zero roots in their interior. Thus, it may occur that one SPU can become idle when other SPUs have a lot of work ahead of them. By using the ALF work-block mechanism, we get automatic load balancing. In our current implementation this change is done after recursive depth of 8. The functions to compute the value of $f(z)$ at the m evaluation points, the calculation of the quadrant code $qc(f(z))$, and their SIMD implementation has already been discussed. The `calculate_corners` function takes the complex center of a square in real and imaginary component, the side of the square $2d$, and returns two (vector float) containing the four corners of the square, $((rc-d),i(ri-d))$, $((rc+d),i(ri-d))$, $((rc+d),i(ri+d))$, and $((rc-d),i(ri+d))$ in their real and imaginary components. The use of 4-element (vector float) to represent the corners of the square under investigation was a natural fit between this algorithm and the SPU.

```
Data: Function: f(z), Center :c, Initial Side: d, Step: m, Precision: ε
Result: Calculated roots of f(z)
1  void feval(VF rez, VF imz, VF *refz, VF *imfz); // evaluates f(rez,imz)
2  VI qcode(VF refz, VF imfz); // calculates qc for f(z)
3  void calculate_corners(float rec, float imc, VF *realcorners, VF *imagcorners);
4  void quadrisect(float rec, float imc, float d, VF *recorners, VF *imagcorners);
5  VI update_total(VIprev_qc, VIcurrent_qc);
6  complex_list num_roots(float rec, float imc, float d, int steps) begin
7     if d < ε then
8        return complex(rec,imc)
9     end
10    VF realcorners, imagcorners, const_dx = {d/2,0,-d/2,0}, const_dy = {0,d/2,0,-d/2};
11    VI running_total current_qcode, prev_qcode(0);
12    calculate_corners(rec, imc, &realcorners, &imagcorners);
13    for i = 0; i < m; ++i do
14       VI current_qcode = qcode(realcorners, imagcorners);
15       running_total = update_total(prev_qcode, current_qcode);
16       prev_qcode = current_qcode;
17       realcorners += cost_dx, imagcorners += const_dy;
18    end
19    if check_for_root( running_total) then
20       // there are some roots in this square, refine this square
21       quadrisect(rec, imc, d, &realcorners, &imagcorners);
22       complex_list root_list;
23       for i = 0; i < 4; ++i do
24          root_list.append( num_roots(realcorners[i], imagcorners[i],d/2,steps));
25       end
26       return root_list;
27    end
28 end
```

Algorithm 22: Algorithm for calculating complex roots of $f(z)$.

The function `quadrisect` takes the complex center of a square in real and imaginary component, the side of the square $2d$, and returns two (vector float) containing the new centers of the refined squares defined as $((rc-d), i(ri-d))$, $((rc+d/2), i(ri-d/2))$, $((rc+d/2), i(ri+d/2))$, and $(rc-d/2), i(ri+d/2))$. To make the sub-squares overlap, we can divide d by 1.999 instead of 2. Once the quadrant codes have been computed as (vector unsigned int), we use the function `update_total` which updates the count of number of roots in this square. Consider the quadrant codes given below:

	QC[0]	QC[1]	QC[2]	QC[3]
current_qcode iteration 0 =	7	5	5	2
current_qcode iteration 1 =	2	5	7	7
current_qcode iteration 2 =	2	7	7	2
current_qcode iteration 3 =	3	5	2	7

We maintain a running total for two rows at a time on this table; consider the first column, when the quadrant code changes from 7 to 2, it is a counter clockwise motion so we add +1. When we move from 2 to 7 we add -1. The direction change vector for the first 2 rows are $1, 0, 1, -1$, we then move down one row to get

0, 1, 0, 1. We get a running total vector of 2, 0, 2, 1. We then sum this vector to get a positive winding number of 1 in this square. The complete algorithm is shown in Algorithm 22.

Experiment Configuration Details

We have used the ALF framework for communicating between the PPU and the SPUs. Since the function to be evaluated stays constant, the function can be sent as a task context, and does not need to be present as a work-block parameter. But, there is nothing to prevent a function (as a tokenized stream of stack operands) to be included in the work-block parameters. The comp_kernel is implemented as a single function to search for complex roots in the analytic region included in the current work-block. Since the SPUs include all the code necessary to do domain decomposition and refinement, no further communication was necessary to/from the PPU, unless the SPU communicates the end of the work-block by sending either the computed complex roots, and its multiplicity, or 0 for number of roots found. The input_prep for this task setup is minimal, and so is the output_prep.

Aspect	CELL	Opteron
cpu	2 Cell Broadband Engine	2 AMD Opteron(tm) 242
clock	3192.0 MHz	1600 MHz
revision	5.1 (pvr 0070 0501)	Family 15 model 5
gcc	4.1.2	4.1.1

Table 13.3 Summary of machines used in performance comparison.

For debugging we used gdb for SPUs, spu-top, and by printf statements in SPU code. SPU code optimization was done using spu_timing. As we have mentioned previously, we were consciously trying to improve dual-issue rates during SPU executions. PPU domain decomposition code was not SIMD enabled and instead used Standard C++ std::<complex> class for complex function evaluation. Even though the SPUs have limited stack space, by carefully limiting the depth of recursion, we were able to write the square refinement code recursively; although the calls are purely tail-recursive, and can be removed easily (translated to iteration). All code was compiled with gcc 4.1.2 at highest optimization levels.

The other machine for the reference was a dual-Opteron 242 running Gentoo Linux 2.6.23 GNU gcc 4.1.1 was used as the compiler, and again all compiler optimizations were switched on (-O3 -mtune=opteron). Using pthreads, we attempted to optimize for multicore performance by dividing the initial analytic region into 4 equal parts. Again, no adaptive refinement was done for load balancing.

We have also experimented with connecting 2 PS3 (connected over the internet) and using top-level region bisection to divide the problem. Again the decomposition algorithm implemented was not adaptive. This implies that if 1 division has less or no zeros, then there is sub-optimal load balancing. Doing adaptive region sub-division over distributed PS3 is one of the planned enhancements in the near future. But in the case when zeros were uniformly distributed, we got almost linear speedup using distributed PS3s (as expected).

Using ALF

Setting up ALF for communicating with the SPUs was done using the following:

1. **alf_init**: initializes the ALF sub-system,
2. **alf_num_instances_set**: command line argument (default 6 SPUs),
3. **alf_task_desc_create**: task was set to SPE,
4. **alf_task_desc_set_int32**: setup STACK_SIZE, WB_PARAM_CTX_SIZE, and size of result,
5. **alf_task_create**: actually created the task,
6. **alf_wb_param_add**: to add parameters to the work-block,
7. **alf_wb_enqueue**: starts SPU function given as compute_kernel,
8. **alf_task_finalize**: finalizes the task parameters
9. **alf_task_wait**: same as MT *wait*, result available after this.

13.10.1 Performance Comparisons

To compare the performance of our SPU complex root finding algorithm we ran it on the following complex functions as shown in Table 13.4. These functions were extracted from previous studies on the problem of solving complex functions. Two of these functions (as $|f(z)|$) are plotted in Figure 13.15.

Function	Region	Reference
$e^z - z^2$	$[-\pi - 0.5i] \times [2\pi + i\pi]$	[59]
$e^z + \sin(z)\cos(z)$	$[-1.0 - i1.0] \times [1.0 + i1.0]$	
$e^{3z} + 2z\cos(z) - 1$	$[-\pi - i\pi] \times [2\pi + i2\pi]$	[76]
$\sin(\frac{z^2+\pi^2}{z+\pi(2i-3)})$	$[-10 - i5] \times [10 + i10]$	[111]

Table 13.4 Complex functions used for performance comparison

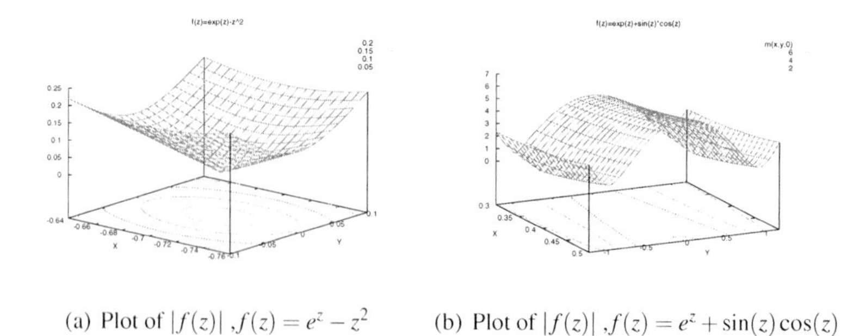

(a) Plot of $|f(z)|$,$f(z) = e^z - z^2$ (b) Plot of $|f(z)|$,$f(z) = e^z + \sin(z)\cos(z)$

Fig. 13.15 Graphical plots of the complex roots of the functions.

Baseline Floating Point Comparison

SPU	PPU	Opteron
3.0	132.8	33.176

Table 13.5 Baseline timing numbers (in seconds) for floating point evaluation of $e^z + \sin(z)\cos(z)$.

We first give baseline numbers which can be used to gauge relative floating point performance of the machines involved in the performance comparison. The baseline method involved computing $f(z) = e^z + \sin(z)\cos(z)$ around a square centered at $0 + i0$, a side of 2.0, and $m = 10000000$. This involves computing $f(z)$ 4x10e7 times. To keep the optimizing compilers from removing these evaluations, we compute maximum and minimum of $|f(z)|$ at these evaluation points. Even though the SPU calculations are performed only on a single SPU they are still an order of magnitude faster than the next fastest time. This is due in part to the SIMD organization of evaluating only 1 edge of the square m times; the other 3 edges are automatically evaluated using the 4-way SIMD organization of our complex function library. The dual issue intruction optimizations we have performed in the complex math library also help to reduce stalls, and reduce total clock cycle count.

Performance comparison of CELL/SPU with PPU and Opteron

We evaluate the performance of the SPU implementation with PPU and Opteron. The experiment was performed with 6 SPUs activated using ALF. For all experiments the initial starting square was $C(0,0)$, the initial d side was 2.0, m the number of steps was 100000, and was increased by a factor of 2 by the algorithm every time a diagonal quadrant jump was detected. The stopping precision was 1e-7, the recursion depth was limited to 12. We have investigated the performance of our implementation by analyzing the generated assembly code of the SPU C-code, and by running spu-top while the system was computing roots.

The Opteron experiment was run on a 1.6 GHz Opteron 242 SMP (multi-threaded with 4 threads on 2 processors). The Opteron has a high performance floating point unit, moreover the C++ libraries and GCC g++ compiler have been tuned to produce good code for Opteron when compiled with -mtune=opteron switch (which we did). This could explain the significant difference in numbers between Opteron and PPU execution. Nevertheless, when compared to CBE as a whole, the CELL with 6 SPUs in parallel outperforms the Opteron by a large margin. As the functions to be solved become more complex, the runtime gap continues to widen. Unfortunately, we have not been able to migrate our complex SIMD library to GCC intrinsic vector support. This is also planned for the future.

Discussion of results

Our SPU implementation is approximately 60 times faster than the multi-threaded 2-way SMP Opteron implementation. For brevity we have only shown the first computed root, all computations were timed for their entire duration. The weak performance of the PPU on this code can be explained by the lack of dual execution floating point unit. We analyzed the assembly code for Opteron and SPU in some detail, and observed that our dual-issue optimizations were reducing stalls, and enabling dual-issue on a large fraction of the code for evaluating common complex function groups such as $\sin(z)\cos(z)$.

The SIMD organization of the function evaluations on the contour clearly gives the CELL/SPU a 4x advantage over the Opteron. Moreover, the quadrant calculation code is also optimized by the SIMD architecture, optimizing the whole m-way loop. We have removed all branches and partially unrolled this m-way loop (for all three implementations). But it appears that the SPUs large register file for floating point computation is the only one to significantly benefit from this loop unrolling. For small complex functions such as $f(z) = e^z - z^2$, the loop unrolling is producing good distribution of even pipeline integer operations, floating point operations, and odd pipeline load-store. For more complex functions, the floating point unit is getting saturated. On the other hand, for the code generated on the Opteron, complex numbers are passed as 8-byte (2*sizeof(float)) stack offsets using the %rsp register. Even declaring the functions to accept complex numbers by value, adding register declarations does not alter this behavior. Since our complex math SIMD library is optimized for register use (especially when passing and returning real and imaginary components), we save clock cycles on manipulating stack frames. This difference is also amplified 4x since our register optimizations are done on (vector float)s of 4 complex numbers, while the Opteron's assembly code is doing stack frame operations for every single complex number.

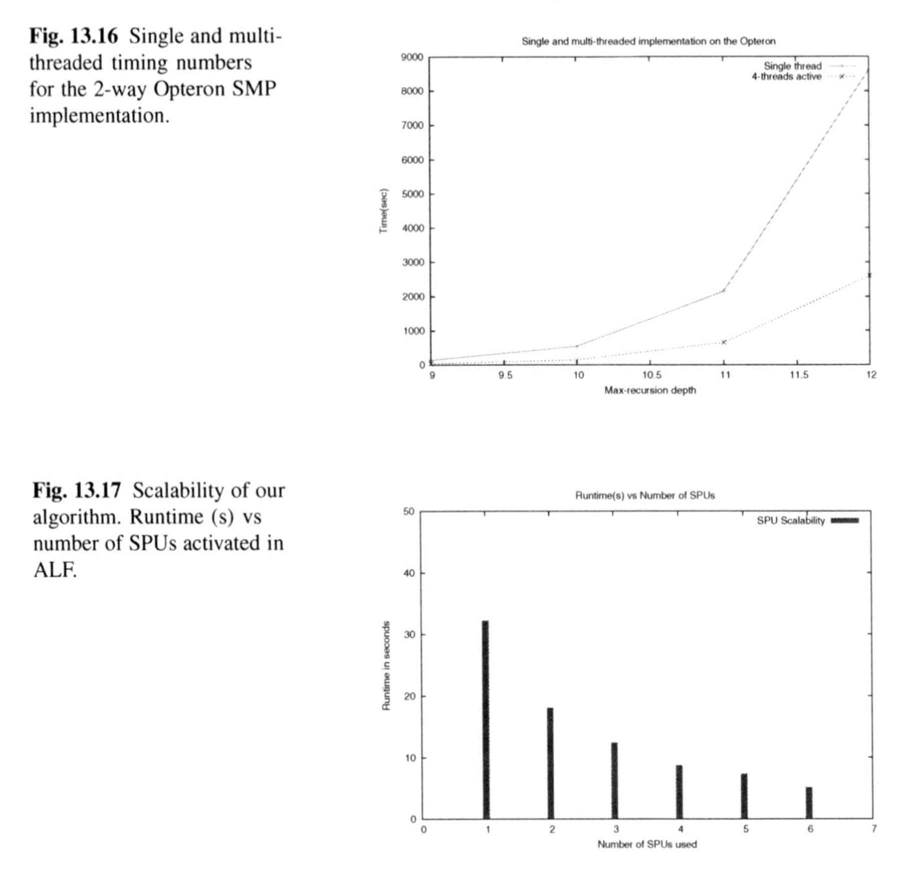

Fig. 13.16 Single and multi-threaded timing numbers for the 2-way Opteron SMP implementation.

Fig. 13.17 Scalability of our algorithm. Runtime (s) vs number of SPUs activated in ALF.

We calculated the runtime of the Opteron implementation by varying the maximum recursion depth from 9 to 12 (inclusive), step count m was set constant to 10000. We have plotted the runtime vs max-depth for the single threaded, and multi-threaded implementations in Figure 13.16. We see that the multi-threading gives close to 2x improvement, as we were running on a 2-way SMP system.

Scalability of algorithm w.r.t number of SPUs

We plot the runtime versus number of SPUs used to evaluate $f(z) = e^z + \sin(z)\cos(z)$ with $d = 2$ and precision set to 1e-9. We varied the number of active SPUs from 1 to 6 (the maximum usable in PS3), the resulting plot is shown in Figure 13.17. As can be seen we get good scalability as the number of SPUs are increased.

We have presented the design and implementation of a multicore SIMD parallel algorithm to compute complex zeros of analytic functions in a given region. As compared to other complex root finding methods our method proved to be amenable to parallel implementation on the SPUs exhibiting significant speedup over conventional processors. In our opinion the PS3 is an extremely versatile research-and-development vehicle for experimenting with the CELL Processor. The availability of a standard tool chain made development much easier as compared to developing scientific applications on the Emotion Engine chip in the PS2. The standardization of the hardware also means that it is easy to connect multiple PS3s (having a gigabit Ethernet also helps here) to form a cluster. The SIMD capabilities of the SPUs have often been neglected in other multicore experiments, which have solely concentrated on the multicore aspects of the 6 SPUs. We have been able to use all available parallelism of the CELL processor using multicore, SIMD and dual-instruction issue per cycle to get high performance. Previous implementations were either sequential, or utilized distributed computing, which has a significant penalty in data communication. The SPU compute times are close to 2 orders of magnitude faster than conventional microprocessors. As more SPUs become available for supercomputing applications, we expect this number to improve even more.

We first present the SPU based complex function evaluation system which computes the quadrant code. The code is presented in Listing 13.11. The `comp_kernel` implements the ALF kernel function.

```
   ////////////////////////////////////////
   // Program: Complex roots of f(z)
   // Author : Sandeep Koranne
   ////////////////////////////////////////
 5 #include <stdio.h>
   #include <alf_accel.h>
   #include <simdmath.h>
   #include <simdmath/coshf4.h>
   #include <simdmath/sinhf4.h>
10 #include <math.h>
   #include "../transfer.h"

   static int debug = 0; // set to 0 to turn-off debug

15 #define QUADRANT_ZERO 0
   #define QUADRANT_ONE 1
   #define QUADRANT_TWO 2
   #define QUADRANT_THREE 3
   #define QUADRANT_FOUR 4
20 #define QUADRANT_ERROR 5

   static int quadrant_code ( float re_val,
                              float im_val ) {
     /* when both are > 0 return 1 or 2 for prime code */
25   if((re_val==0.0)||(im_val==0.0)) return QUADRANT_ZERO;
     if((re_val>0)&&(im_val>0)) return QUADRANT_ONE;
     if((re_val<0)&&(im_val>0)) return QUADRANT_TWO;
     if((re_val<0)&&(im_val<0)) return QUADRANT_THREE;
     if((re_val>0)&&(im_val<0)) return QUADRANT_FOUR;
30   return QUADRANT_ERROR;
   }

   static vector unsigned int quadrant_code_simd( vector float re_val,
                                                  vector float im_val) {
35   vector float CONST_ZERO_VECT = (vector float) spu_splats(0.0);
```

```
    vec_uint4 RE_QMASK    = (vec_uint4) spu_splats(2);
    vec_uint4 IM_QMASK    = (vec_uint4) spu_splats(1);
    // compare re_val with 0 and set the mask
    vec_uint4 RE_MASK  = spu_cmpgt(re_val, CONST_ZERO_VECT);
40  vec_uint4 IM_MASK  = spu_cmpgt(im_val, CONST_ZERO_VECT);
    RE_MASK = spu_and( RE_MASK, RE_QMASK );
    IM_MASK = spu_and( IM_MASK, IM_QMASK ); // now only last 2 bits
    vec_uint4 quadrant_code = spu_or(RE_MASK, IM_MASK);
    quadrant_code = spu_add(quadrant_code,5);
45  return quadrant_code;
  }

  #if 1
  static vector float real_part_of_f ( vector float a,
50                                       vector float b ) {
    /* a function to compute real part of the complex function */
    /* for example e^z + sin(z)*cos(z) =
       e^a cos(b) + sin(a)cos(a)* cosh(b)^2 + sin(a)cos(a)*sinh(b)^2
    */
55  vec_float4 ea      = (vec_float4) expf4(a);
    vec_float4 cosb    = (vec_float4) cosf4(b);
    vec_float4 cosh_b    = (vec_float4) _coshf4(b);
    vec_float4 sinh_b    = (vec_float4) _sinhf4(b);
    vec_float4 cosa,sina;
60  (void) sincosf4(a,&sina,&cosa);
    vec_float4 sc_prod = (vec_float4) spu_mul(sina, cosa);
    vec_float4 csq     = (vec_float4) spu_mul(cosh_b, cosh_b);
    vec_float4 ssq     = (vec_float4) spu_mul(sinh_b, sinh_b);
    vec_float4 t0      = (vec_float4) spu_mul(sc_prod, csq);
65  vec_float4 t2      = (vec_float4) spu_madd(sc_prod, ssq,t0);
    vec_float4 ret_val = (vec_float4) spu_madd(ea,cosb,t2);
    return ret_val;
  }
  #endif
70

  static vector float imag_part_of_f ( vector float a,
                                       vector float b ) {
    /* a function to compute real part of the complex function */
75  /* for example e^z + sin(z)*cos(z) =
       e^a cos(b) + sin(a)cos(a)* cosh(b)^2 + sin(a)cos(a)*sinh(b)^2
    */
    vec_float4 ea      = (vec_float4) expf4(a);
    vec_float4 sinb    = (vec_float4) sinf4(b);
80  vec_float4 cosh_b    = (vec_float4) _coshf4(b);
    vec_float4 sinh_b    = (vec_float4) _sinhf4(b);
    vec_float4 cosa,sina;
    (void) sincosf4(a,&sina,&cosa);
    vec_float4 ccq     = (vec_float4) spu_mul(cosa,cosa);
85  vec_float4 ssq     = (vec_float4) spu_mul(sina,sina);

    vec_float4 sc_prod = (vec_float4) spu_mul(sinh_b, cosh_b);

    vec_float4 t0      = (vec_float4) spu_mul(sc_prod,ccq);
90  vec_float4 t1      = (vec_float4) spu_mul(sc_prod,ssq);
    vec_float4 t2      = (vec_float4) spu_sub(t0,t1);
    vec_float4 ret_val = (vec_float4) spu_madd(ea,sinb,t2);
    return ret_val;
  }
95
  struct complex_number {
    float re,im;
  };

100 static int
  analyze_quadrant_code_multiplicities( vec_uint4 qcode ) {
    unsigned int prod_code = 1;
```

```
    int i=0;
    for(i=0;i<4;++i)
105   prod_code *= spu_extract( qcode, i);

    // full house = 5*6*7*8
    if( prod_code == (5*6*7*8) )
      return 1;
110 else if( ( prod_code == 625) || (prod_code == 1296) ||
             (prod_code == 2401) || (prod_code == 4096) )
      return 0;
    else return 1;
  }
115 //#define DEBUG_VERBOSE
  static struct
  complex_number search_for_root( float cr,
                                  float ci,
                                  float d,
120                               int depth ) {
    int i=0;
    vector float a,b,real_corners,imag_corners;
    struct complex_number retval;
    a = (vector float)spu_splats(cr+d);
125 b = (vector float){2*d,0,0.0,2*d};
    real_corners = (vector float) spu_sub(a,b);

    a = (vector float)spu_splats(ci+d);
    b = (vector float){2*d,2*d,0.0,0.0};
130 imag_corners = (vector float) spu_sub(a,b);
  #ifdef DEBUG_VERBOSE
    printf("\n Corners of square are \n");
    for(i=0;i<4;++i)
      printf("\n %f %f ",
135          spu_extract(real_corners,i),
             spu_extract(imag_corners,i));
  #endif
    /* computing complex function at the 4 corners */
    a = real_part_of_f( real_corners, imag_corners );
140 b = imag_part_of_f( real_corners, imag_corners );
    vec_uint4 qcode = quadrant_code_simd(a,b);

  #ifdef DEBUG_VERBOSE
    printf("\n f(z) = \n");
145 for(i=0;i<4;++i)
      printf("\n %f,%f ", spu_extract(a,i),spu_extract(b,i));
  #endif
    // analyze the qcode for this square
  //#define DEBUG_QCODE
150 #ifdef DEBUG_QCODE
    printf("\n QCODE for (%f,%f,%f) = [", cr, ci,d);
    for(i=0;i<4;++i)
      printf(" %d ", spu_extract(qcode,i)-5);
    printf(" ]");
155 #endif
    #define EPSILON 0.000001
    #define ROOT_EPSILON 0.2
    #define MAX_DEPTH 20
    int qcode_analysis =
160   analyze_quadrant_code_multiplicities(qcode);
    float rez = fabs(spu_extract(a,0));
    float imz = fabs(spu_extract(b,0));
    int i_am_root = ((rez < ROOT_EPSILON) && (imz < ROOT_EPSILON) );
    if(i_am_root ||  (d < EPSILON) || (depth > MAX_DEPTH ) ) {
165   if( i_am_root )
        printf("\n Potential root @(%f,%f) f(z) = "\
               "(%f + i %f) when d = %f and depth = %d",
               cr,ci,rez,imz,d,depth);
      else
```

```
170         qcode_analysis = 0;
      }
      if(depth > MAX_DEPTH) qcode_analysis = 0;
      if( qcode_analysis ) {
        // search deeper
175     search_for_root(cr-d/2, cr-d/2,d/2,depth+1);
        search_for_root(cr+d/2, cr-d/2,d/2,depth+1);
        search_for_root(cr+d/2, cr+d/2,d/2,depth+1);
        search_for_root(cr-d/2, cr+d/2,d/2,depth+1);
      } else {
180     //printf("\n Square (%f,%f,%f) does not contain any root",cr,ci,d);
      }
      return retval;
    }

185 int comp_kernel(void *p_task_context,
                    void *p_parm_context,
                    void *p_input_buffer,
                    void *p_output_buffer,
                    void *p_inout_buffer,
190                 unsigned int current_count,
                    unsigned int total_count) {
      udata_t *args    = (udata_t *) p_parm_context;
      result_t *result = (result_t *) p_task_context;
      float cr = args->re_val, ci = args->im_val, d=args->d;
195   current_count = total_count = 0;
      p_input_buffer = p_input_buffer;
      p_output_buffer = p_output_buffer;
      /* print some information on this work block */
    #ifdef DEBUG_VERBOSE
200   printf("\n Computing on center (%f,%f) with d = %f",cr,ci,d);
    #endif
      search_for_root( cr, ci, d, 1 );
      result->quadrant_code = 2;
      //printf("Exiting alf_accel_comp_kernel\n");
205   return 0;
    }

    int input_prep(void *p_task_context,
                   void *p_parm_context,
210                void *p_dtl,
                   unsigned int current_count,
                   unsigned int total_count) {
      if (debug)
        printf("Exiting alf_accel_input_list_prepare\n");
215   return 0;
    }

    int output_prep(void *p_task_context,
                    void *p_parm_context,
220                 void *p_dtl,
                    unsigned int current_count,
                    unsigned int total_count) {
      if (debug)
        printf("Exiting alf_accel_output_list_prepare\n");
225   return 0;
    }
    ALF_ACCEL_EXPORT_API_LIST_BEGIN
       ALF_ACCEL_EXPORT_API("", comp_kernel);
       ALF_ACCEL_EXPORT_API("", input_prep);
230    ALF_ACCEL_EXPORT_API("", output_prep);
    ALF_ACCEL_EXPORT_API_LIST_END
```

Listing 13.11 Complex Function SPE code

Listing 13.12 implements the PPE side of things. The use of ALF for system initialization can be seen.

```
#include <stdio.h>
#include <alf.h>
#include <unistd.h>
#include <stdlib.h>
#include <string.h>
#include "transfer.h"

#define DEBUG

result_t result;

#ifdef DEBUG
#define debug_print(fmt, arg...) printf(fmt,##arg)
#else
#define debug_print(fmt, arg...) { }
#endif

#define IMAGE_PATH_BUF_SIZE 1024
char spu_image_path[IMAGE_PATH_BUF_SIZE];
char library_name[IMAGE_PATH_BUF_SIZE];
char spu_image_name[] = "alf_hello_world_spu";
char kernel_name[] = "comp_kernel";
char input_dtl_name[] = "input_prep";
char output_dtl_name[] = "output_prep";

#ifdef _ALF_PLATFORM_HYBRID_
char ppu_image_path[IMAGE_PATH_BUF_SIZE];
#endif

int main(int argc, char* argv [])
{
  int ret;
  alf_handle_t handle;
  alf_task_desc_handle_t task_desc_handle;
  alf_task_handle_t task_handle;
  alf_wb_handle_t wb_handle;
  void *config_parms = NULL;
  int i,j;
  udata_t my_data;
  int nodes = 1;
  float rcr = -0.73;
  float rci = -1.54;
  float rd  = 2.0;

  if(argc > 1) nodes = atoi( argv[1] );
  if(argc > 2) rcr   = atof( argv[2] );
  if(argc > 3) rci   = atof( argv[3] );
  if(argc > 4) rd   = atof( argv[4] );
  result.quadrant_code = 0;

#ifdef _ALF_PLATFORM_HYBRID_
  sprintf(library_name, "alf_hello_world_hybrid_spu64.so");
#else
  if (sizeof(void *) == 4)
    sprintf(library_name, "alf_hello_world_cell_spu.so");
  else
    sprintf(library_name, "alf_hello_world_cell_spu64.so");
#endif

  if ((ret = alf_init(config_parms, &handle)) < 0) {
    fprintf(stderr, "Error: alf_init failed, ret=%d\n", ret);
    return 1;
  }
  alf_num_instances_set(handle, nodes); /* set this to 6 later */
```

```
65  alf_task_desc_create(handle, ALF_ACCEL_TYPE_SPE, &task_desc_handle);
    alf_task_desc_set_int32(task_desc_handle,
                            ALF_TASK_DESC_MAX_STACK_SIZE, 4096);
    alf_task_desc_set_int32(task_desc_handle,
                            ALF_TASK_DESC_WB_PARM_CTX_BUF_SIZE,
70        sizeof(udata_t));
    alf_task_desc_set_int32(task_desc_handle,
                            ALF_TASK_DESC_WB_IN_BUF_SIZE, 0);
    alf_task_desc_set_int32(task_desc_handle,
                            ALF_TASK_DESC_WB_OUT_BUF_SIZE, 0);
75  alf_task_desc_set_int32(task_desc_handle,
                            ALF_TASK_DESC_WB_INOUT_BUF_SIZE, 0);
    alf_task_desc_set_int32(task_desc_handle,
                            ALF_TASK_DESC_TSK_CTX_SIZE,
          sizeof(result_t));
80  alf_task_desc_set_int64(task_desc_handle,
                            ALF_TASK_DESC_ACCEL_IMAGE_REF_L,
                            (unsigned long long)spu_image_name);
    alf_task_desc_set_int64(task_desc_handle,
                            ALF_TASK_DESC_ACCEL_LIBRARY_REF_L,
85                          (unsigned long long)library_name);
    alf_task_desc_set_int64(task_desc_handle,
                            ALF_TASK_DESC_ACCEL_KERNEL_REF_L,
                            (unsigned long long)kernel_name);
    alf_task_desc_set_int64(task_desc_handle,
90                          ALF_TASK_DESC_ACCEL_INPUT_DTL_REF_L,
                            (unsigned long long)input_dtl_name);
    alf_task_desc_set_int64(task_desc_handle,
                            ALF_TASK_DESC_ACCEL_OUTPUT_DTL_REF_L,
                            (unsigned long long)output_dtl_name);
95
    alf_task_desc_ctx_entry_add(task_desc_handle,
                                ALF_DATA_INT32, 1);
    alf_task_desc_ctx_entry_add(task_desc_handle,
                                ALF_DATA_INT32, 1);
100
    alf_task_create(task_desc_handle, &result, 6, 0, 0, &task_handle);

    result.quadrant_code = 0;
    alf_task_desc_destroy(task_desc_handle);
105

    /* now add some squares to the queue */
    #define I_COUNT 2
    #define J_COUNT 2
110
    for(i=1; i < nodes; ++i) {
      alf_wb_create(task_handle, ALF_WB_SINGLE, 1, &wb_handle);
      my_data.re_val = (rcr-rd) / (nodes/2) + (i*rd/2);
      my_data.im_val = rci - ((i%2)*rd/2);
115   my_data.d = rd;
      my_data.dp = rd;
      alf_wb_parm_add( wb_handle,
                       (void*)&(my_data.re_val),1,ALF_DATA_DOUBLE,2);
      alf_wb_parm_add( wb_handle,
120                    (void*)&(my_data.d),1,ALF_DATA_DOUBLE,1);
      alf_wb_enqueue(wb_handle);
    }
    alf_task_finalize(task_handle);
    alf_task_wait(task_handle, -1);
125 debug_print("In main: alf_task_wait done.\n");
    debug_print("Before alf_exit\n");
    if ((ret = alf_exit(handle, ALF_EXIT_POLICY_FORCE, 0)) < 0) {
      fprintf(stderr, "Error: alf_exit failed, ret=%d\n", ret);
      return 1;
130 }
    float icount = i, jcount = j;
```

```
    printf("\n Computed f(z) inside %f squares\n", icount*jcount);
    debug_print("Execution completed successfully, exiting.\n");
    return 0;
135 }
```

Listing 13.12 Complex Function PPE code

13.11 Conclusions

In this chapter we have introduced SPU implementations of common mathematical computations including polynomial representation, linear regression, interpolation, complex number representation and equation solving. We have also used the SDK for matrix operations. This chapter should be used in conjunction with the other math libraries for the Cell Broadband, e.g., LAPACK, BLAS and FFT for Cell. Our efforts at implementing mathematics algorithms on the SPU stems from the need to present concept which are obvious when presented in this context; as well as using many of the techniques developed here in the subsequent chapters on line-of-sight, and VLSI CAD to solve real world problems.

Chapter 14
Vector Graphics on SPU

Abstract In this chapter we design and implement several computational geometry kernels based on 4-way 32-bit SIMD functions. The geometrical kernels we describe include (a) manipulation of polygonal data, (b) checking polygon convexity, (c) convex-hull computation, (d) computing signed polygon area, and (e) line-clipping.

14.1 Introduction

In this chapter we discuss the design and implementation of algorithms for for computational geometry parallelism; in the form of a Single Instruction Multiple Data (SIMD) model. For more details on computational geometry refer to the following text books [36, 7, 102, 89]. High performance computational geometry applications using SIMD have been described by the author in [73] using the Emotion Engine chip present in the PlayStation2 game console. Other seminal references for computational geometry include [112, 124].

14.2 Basic Computational Geometry on SPU

Consider a simple rectilinear polygon as shown in Figure 14.1; such polygons are often seen in CAD applications for VLSI and indeed the figure shows two polygons overlapping. We shall represent the polygon using SIMD structures and perform various computational geometry operations on the polygon. The algorithms for these operations are described in detail in [36] thus we shall not repeat them here.

S. Koranne, *Practical Computing on the Cell Broadband Engine*,
DOI: 10.1007/978-1-4419-0308-2_14,

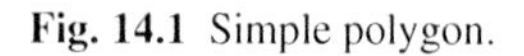

Fig. 14.1 Simple polygon.

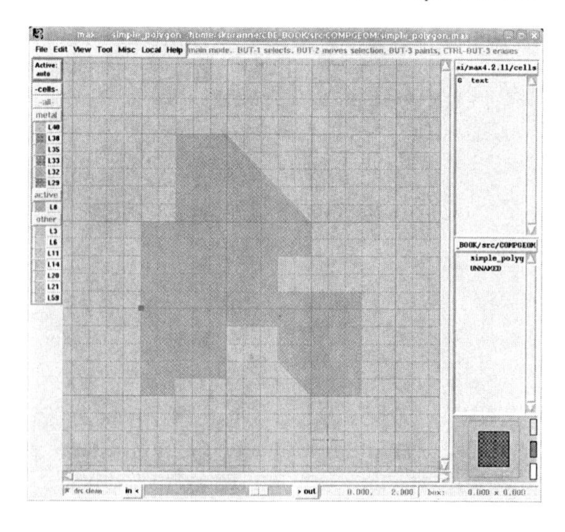

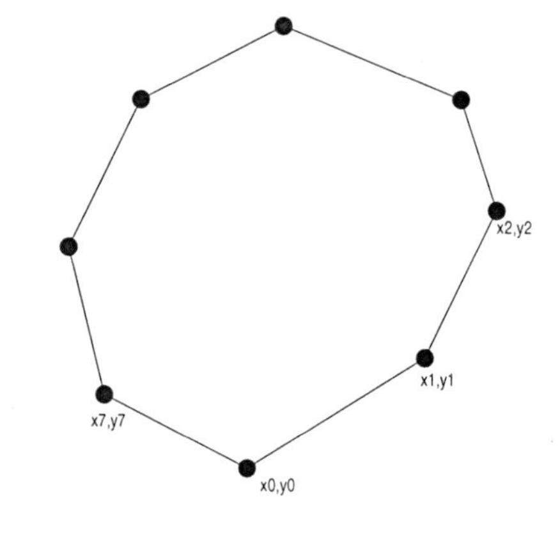

Fig. 14.2 Simple polygon represented as array of vertices in counter-clockwise orientation.

14.2.1 Reading Polygon Data

In order to efficiently support all-angle edges, we use 32-bit floating point representation of the edge co-ordinates. An edge pq can be represented as $(p_x, p_y), (q_x, q_y)$, with the convention that the interior of the polygon is to the left of the edge. Four pairs of edges $(0-0, 1-1, 2-2, 3-3)$ to be compared are thus:

$$
\begin{aligned}
M[0:3] &= [p0_x|\ p0_y|\ q0_x|\ q0_y] \\
M[4:7] &= [r0_x|\ r0_y|\ s0_x|\ s0_y] \\
M[8:11] &= [p1_x|\ p1_y|\ q1_x|\ q1_y] \\
M[12:15] &= [r1_x|\ r1_y|\ s1_x|\ s1_y]
\end{aligned}
$$

$$
\begin{aligned}
M[16:19] &= [p2_x|\ p2_y|\ q2_x|\ q2_y] \\
M[20:23] &= [r2_x|\ r2_y|\ s2_x|\ s2_y] \\
M[24:27] &= [p3_x|\ p3_y|\ q3_x|\ q3_y] \\
M[28:31] &= [r3_x|\ r3_y|\ s3_x|\ s3_y]
\end{aligned}
$$

where M denotes a memory location. A block of 4 edge pairs is located from $M[0:31]$. We can re-write this block as:

$$
\begin{aligned}
M[0:3] &= [p0_x|\ p1_x|\ p2_x|\ p3_x] \\
M[4:7] &= [p0_y|\ p1_y|\ p2_y|\ p3_y] \\
M[8:11] &= [q0_x|\ q1_x|\ q2_x|\ q3_x] \\
M[12:15] &= [q0_y|\ q1_y|\ q2_y|\ q3_y] \\
M[16:19] &= [r0_x|\ r1_x|\ r2_x|\ r3_x] \\
M[20:23] &= [r0_y|\ r1_y|\ r2_y|\ r3_y] \\
M[24:27] &= [s0_x|\ s1_x|\ s2_x|\ s3_x] \\
M[28:31] &= [s0_y|\ s1_y|\ s2_y|\ s3_y]
\end{aligned}
$$

This re-organization of memory is done while reading in the edge data.

```
void PrepData( int n ) {
  int i;
  for(i=0;i<MAX_E;++i)
    EDGES[i] = (vec_float4)spu_splats(0.0);

  EDGES[0] = (vec_float4){0,2000,1600,0}; // X
  EDGES[1] = (vec_float4){0,600,-600,-1000}; // Y
  EDGES[2] = (vec_float4){0,1600,1600,0}; // X
  EDGES[3] = (vec_float4){1000,600,-200,0}; // Y
  EDGES[4] = (vec_float4){400,1600,1000,0}; // X
  EDGES[5] = (vec_float4){1000,200,-200,0}; // Y
  EDGES[6] = (vec_float4){400,2600,1000,0}; // X
  EDGES[7] = (vec_float4){2000,200,-800,0}; // Y
  EDGES[8] = (vec_float4){1000,2600,400,0}; // X
  EDGES[9] = (vec_float4){2000,-1000,-800,0}; // Y
  EDGES[10] = (vec_float4){2000,2000,400,0}; // X
  EDGES[11] = (vec_float4){1000,-1000,-1000,0}; // Y
  EDGES[12] = (vec_float4){2000,1600,0,0}; // X
  EDGES[13] = (vec_float4){600,-600,-1000,0}; // Y
}
```

Listing 14.1 Polygon coordinates.

14.2.2 Checking Polygon Convexity

Before implementing the convexity predicate on the whole polygon we need to define an useful predicate on three points which calculate whether the points form a positive or negative signed area triangle. This can be used to calculate the *inside-outside* relation of a point with respect to an edge.

Consider the problem of calculating orientation of edge *pq* w.r.t. edge *rs*. For SPACING rule, both *p* and *q* must be on the EXTERIOR side of the edge *rs*; from elementary geometry (cf. [36], pp. 16), we know that the sign of the determinant *D* as calculated below in Equation 14.1 determines whether point *q* lies to the right or left of edge *rs*:

$$D = \begin{vmatrix} 1 & r_x & r_y \\ 1 & s_x & s_y \\ 1 & q_x & q_y \end{vmatrix} \tag{14.1}$$

By repeating this computation for *p*, we can determine if edge *pq* is either (i) completely to the exterior of edge *rs*, in which case it is an comparable edge, (ii) completely to the interior of *rs*, in which it can be discarded, or (iii) is in an intersecting orientation (ignoring projection constraints), in which case the decision for comparability can be decided based on the flag for INTERSECTION test. Similar tests can be constructed for checking of PERPENDICULAR and PARALLEL predicates.

The SPU architecture (described below) supports 4 way SIMD on 32-bit floating point numbers by operating on four fields of an 128-bit register. Let registers $A0, B0, C0, D0$ denote the p_x, p_y, q_x, q_y fields of the *pq* edge; and let registers $A1, B1, C1, D1$ denote the r_x, r_y, s_x, s_y fields of the *rs* edge in an edge-pair group; then we can re-write Equation 14.1 as:

$$D_{simd} = \begin{vmatrix} 1 & A0 & B0 \\ 1 & C0 & D0 \\ 1 & A1 & B1 \end{vmatrix} \tag{14.2}$$

Evaluation of Equation 14.2 will result in the calculation of orientation information for $r0, r1, r2, r3$ at the same time. This is the essence of the 4-way SIMD computational geometry method in our proposed method.

```
vec_uint4 Determinant(vec_float4 rx, vec_float4 ry,
                      vec_float4 sx, vec_float4 sy,
                      vec_float4 qx, vec_float4 qy) {
  vec_float4 ZERO = (vec_float4)spu_splats(0.0);
  vec_float4 syqx = spu_mul(sy,qx);
  vec_float4 d0   = spu_msub(sx,qy,syqx);
  vec_float4 d1   = spu_mul( rx, spu_sub(sy,qy) );
  vec_float4 d2   = spu_mul( ry, spu_sub(qx,sx) );
  vec_float4 d21  = spu_add(d1,d2);
  vec_float4 ans  = spu_add(d21,d0);
  return (vec_uint4)spu_and(spu_cmpgt(ans,ZERO),1);
}
```

Listing 14.2 Orientation checking for points w.r.t an edge.

Using the determinant function as shown in Listing 14.2, we can define the polygon convexity predicate as below. The figure in 14.3 shows the check being executed.

```
/////////////////////////////////////////
// Checking polygon convexity
/////////////////////////////////////////
int CheckPolygonConvexity(vec_float4* edges,
                          int vn) {
  int i;
```

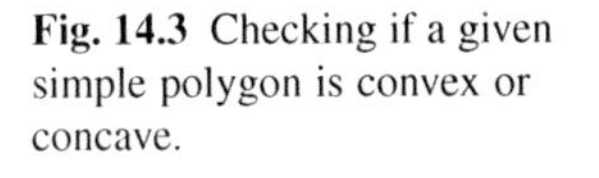

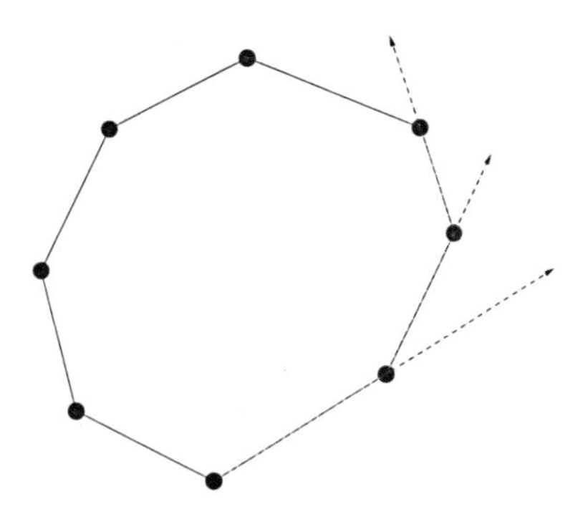

Fig. 14.3 Checking if a given simple polygon is convex or concave.

```
    int n = (((vn/4)+1)*2);
    vec_uint4 isconvex = (vec_uint4)spu_splats(1);
    for(i=0;i<=(n+2);i+=2) {
      vec_uint4 dircheck = Determinant(edges[i],edges[i+1],
                                        edges[i+2],edges[i+3],
                                        edges[i+4],edges[i+5]);
      isconvex = spu_and(isconvex,dircheck);
    }
    return (spu_extract(isconvex,0) && spu_extract(isconvex,1) &&
            spu_extract(isconvex,2) && spu_extract(isconvex,3));
}
```

Listing 14.3 Checking a polygon for convexity.

14.2.3 Computing Convex Hull of 2d Data

The problem of computing the convex hull of a point set is a fundamental problem in computational geometry which is extensively discussed in textbooks on that subject [36, 102]. The problem is best visualized with the following aid; imagine the vertices of the polygon to consist of nails and next a rubber-band is placed around these nails. The convex hull is the shape of the rubber band as it contracts around the polygon. See the figure 14.4 for an illustration of the input point set and its computed hull. Convex hull computations (especially in higher dimensions) are important as due to basic linear-programming, the objective function (either maximization or minimization) optimum is located at one of the vertices of the convex hull.

In this section we discuss the 2-dimensional convex hull problem. We have used a parallel, merge-scan based algorithm which uses Graham's scan to compute the upper and lower hull when the point set is small. These smaller convex hulls are then merged by the PPE. We present the SPU based point set convex hull below.

```
/////////////////////////////////////////
// ConvexHull computation
// Algorithm is based on upper chain and lower chain
// then accepting right turns only
/////////////////////////////////////////
```

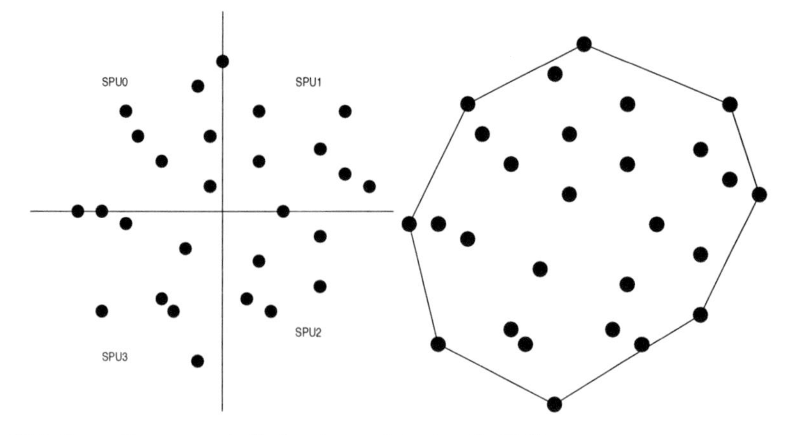

Fig. 14.4 Computing convex hull of a given point set on the SPU.

```
void ComputeConvexHull(vec_float4* edges,
                       int n,
                       vec_uint4* upper_hull,
                       vec_uint4* lower_hull) {
10
  int i;
  vec_float4 ppx,px,ppy,py;
  const vec_uint4 ONE=(vec_uint4)spu_splats(1);
  for(i=0; i < 2*n; i+=2) { // upper hull
15  if(i==0) { upper_hull[i]=ONE; ppx=edges[i];ppy=edges[i+1];continue;}
    if(i==1) { upper_hull[i]=ONE; px=edges[i];py=edges[i+1];continue;}
    vec_uint4 dir_check = Determinant(ppx,ppy,px,py,edges[i],edges[i+1]);
    ppx = spu_sel(ppx,px,dir_check);
    ppy = spu_sel(ppy,py,dir_check);
20  upper_hull[i] = dir_check;
    px = edges[i]; py = edges[i+1];
  }
  for(i=2*n-2; i >= 0; i-=2) { // lower hull
    if(i==(2*n-2)){lower_hull[i]=ONE; ppx=edges[i];ppy=edges[i+1];continue;}
25  if(i==(2*n-4)){lower_hull[i]=ONE; px=edges[i];py=edges[i+1];continue;}
    vec_uint4 dir_check = Determinant(ppx,ppy,px,py,edges[i],edges[i+1]);
    ppx = spu_sel(ppx,px,dir_check);
    ppy = spu_sel(ppy,py,dir_check);
    lower_hull[i] = dir_check;
30  px = edges[i]; py = edges[i+1];
  }
}
```

Listing 14.4 Computing the convex hull of 2d-point set on SPU.

The convex hulls computed by the SPU can be merged by the PPE.

14.2.4 Partitioning point set data to do parallel convex hull

We have adopted simple quadrisection based point partitioning for large point sets as shown in Figure 14.5.

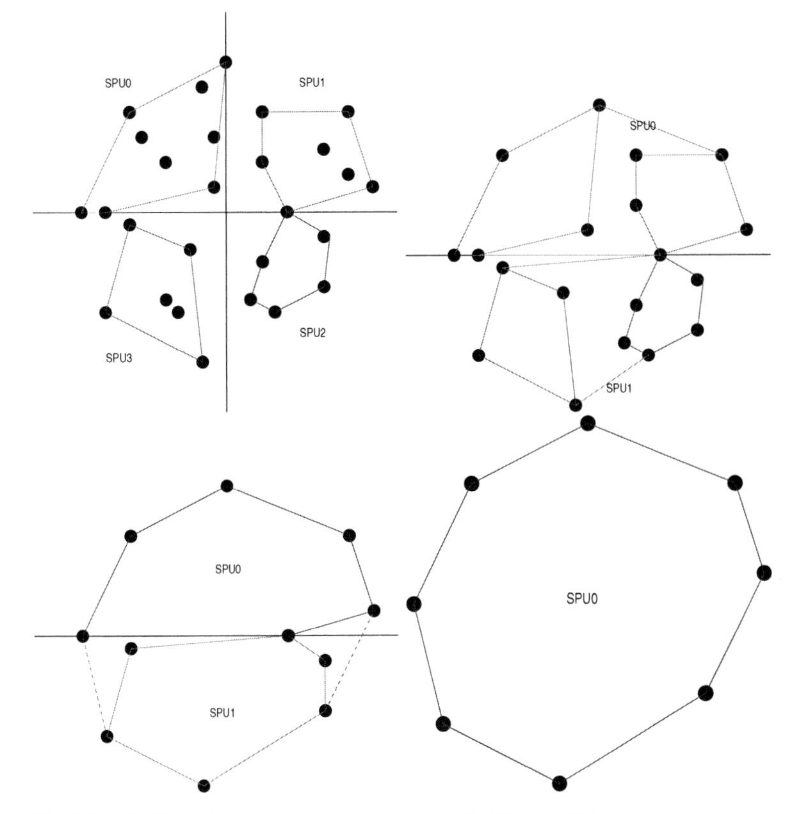

Fig. 14.5 Using divide-and-conquer to compute convex hull in parallel.

There are atleast two methods of achieving parallelism in computational geometry:

1. Solve for each component in parallel: given n different point sets, this would compute n convex hulls in parallel. This is coarse grained parallelism, but is very effective at a high level when the operations are complex, and do not require communication,
2. Use a parallel algorithm at the low level and distribute it on the multiple SPUs: this is more difficult as almost all non-trivial computational geometry algorithm have a lot of transient state which would need to be shared amongst the SPUs. We will look at some techniques and algorithms to reduce this problem.

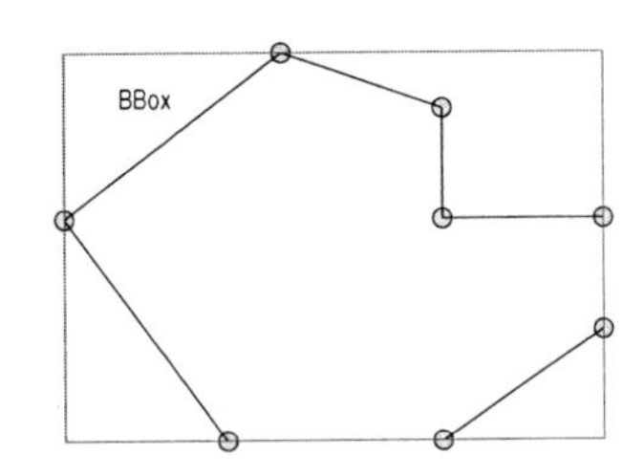

Fig. 14.6 Polygon with its computed bounding-box.

14.2.5 Computing the bounding box of a polygon

The *bounding-box* of a polygon is the 4-tuple *minx,miny* and *maxx,maxy* which represents the minimum box with sides parallel to the axes which will contain the polygon. See Figure 14.6. The code to compute the polygon bounding box is shown in Listing 14.5.

```
//////////////////////////////////////////
// Bounding Box Computation
//////////////////////////////////////////
#define MIN(x,y)  (((x) < (y)) ? (x) : (y))
#define MAX(x,y)  (((x) > (y)) ? (x) : (y))
vec_float4 ComputeBoundingBox(vec_float4* edges, int vn) {
  int i;
  int n = (((vn/4)+1)*2);
  vec_float4 min_x = edges[0];
  vec_float4 min_y = edges[1];
  vec_float4 max_x = edges[0];
  vec_float4 max_y = edges[1];
  for(i=0;i<=n;i+=2) {
    vec_uint4 sgtx = spu_cmpgt( edges[i], edges[i+2] ); // compare x
    vec_float4 gtx = spu_sel(edges[i+2],edges[i],sgtx); // greater
    vec_float4 ltx = spu_sel(edges[i],edges[i+2],sgtx); // lesser
    sgtx = spu_cmpgt( gtx, max_x );
    max_x = spu_sel( max_x, gtx, sgtx );
    sgtx = spu_cmpgt( min_x, ltx );
    min_x = spu_sel( min_x, ltx, sgtx ); // now max_x and min_x are correct
    vec_uint4 sgty = spu_cmpgt( edges[i+1], edges[i+3] ); // compare y
    vec_float4 gty = spu_sel(edges[i+3],edges[i+1],sgty); // greater
    vec_float4 lty = spu_sel(edges[i+1],edges[i+3],sgty); // lesser
    sgty = spu_cmpgt( max_y, gty );
    max_y = spu_sel( gty, max_y, sgty );
    sgty = spu_cmpgt( min_y, lty );
    min_y = spu_sel( min_y, lty, sgty ); // now max_y and min_y are correct
  }
  float smaxx,sminx,sminy,smaxy;
  sminx = MIN(MIN(spu_extract(min_x,0),spu_extract(min_x,1)),
              MIN(spu_extract(min_x,2),spu_extract(min_x,3)));
  sminy = MIN(MIN(spu_extract(min_y,0),spu_extract(min_y,1)),
              MIN(spu_extract(min_y,2),spu_extract(min_y,3)));
  smaxx = MAX(MAX(spu_extract(max_x,0),spu_extract(max_x,1)),
              MAX(spu_extract(max_x,2),spu_extract(max_x,3)));
  smaxy = MAX(MAX(spu_extract(max_y,0),spu_extract(max_y,1)),
              MAX(spu_extract(max_y,2),spu_extract(max_y,3)));
  vec_float4 ans = {sminx,sminy,smaxx,smaxy};
  return ans;
}
```

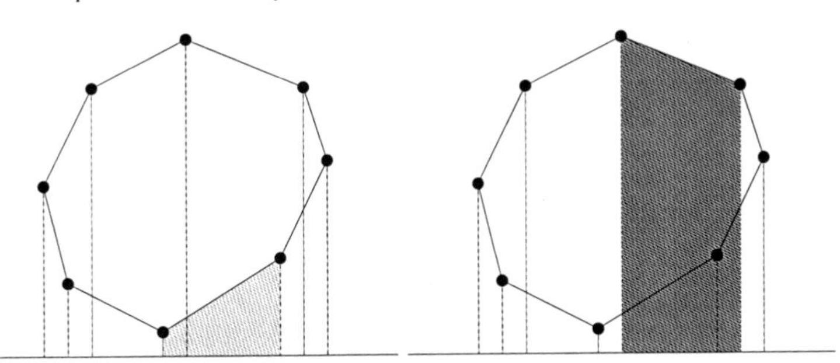

Fig. 14.7 Computing polygon signed area using SIMD.

Listing 14.5 Computing the bounding box of polygon on SPU.

14.2.6 Computing Polygon Signed Area using SIMD

The algorithm to compute the signed area of a polygon uses the trapezoid-cover algorithm. The area of a region can be calculated as the area difference of the trapezoid formed by the upper enveloping edge and the area of the trapezoid formed by the lower enveloping edge. See Figure 14.7. The signed area of a polygon is positive if the polygon is traversed in counter clockwise orientation, or equivalently if the polygon interior is to the *left* of the edge.

The SPU code to compute polygon area is given below in Listing 14.6.

```
/////////////////////////////////////////////
// Signed Area Computation
/////////////////////////////////////////////
float ComputeArea(vec_float4* edges, int vn) {
  int i;
  int n = (((vn/4)+1)*2);
  float ret=0.0;
  vec_float4 area = (vec_float4)spu_splats(0.0);
  for(i=0;i<n;i+=2) {
    vec_float4 dx = spu_sub( edges[i], edges[i+2] );
    vec_float4 dy = spu_add( edges[i+1], edges[i+3] );
    area = spu_add( area, spu_mul(dx, dy ) );
  }
  for(i=0;i<4;++i)
    ret += spu_extract(area,i);
  ret /= 2e6;
  return ret;
}
```

Listing 14.6 Computing the signed area of polygon on SPU.

The function to compute polygon perimeter is similar and is shown next.

```
/////////////////////////////////////////////
// Perimeter computation
/////////////////////////////////////////////
```

```
float ComputePerimeter(vec_float4* edges, int vn) {
  int i;
  int n = (((vn/4)+1)*2);
  float ret=0.0;
  vec_float4 perim = (vec_float4)spu_splats(0.0);
  for(i=0;i<=n;i+=2) {
    vec_float4 dx = spu_sub( edges[i], edges[i+2] );
    vec_float4 dy = spu_sub( edges[i+1], edges[i+3] );
    vec_float4 dx2 = spu_mul(dx,dx);
    vec_float4 dy2 = spu_mul(dy,dy);
    vec_float4 dis2 = spu_add(dx2,dy2);
    vec_float4 rd = spu_rsqrte(dis2);
    vec_float4 d = spu_re(rd);
    perim = spu_add(perim, d);
  }
  for(i=0;i<4;++i)
    ret += spu_extract(perim,i);
  ret /= 1e3;
  return ret;
}
```

Listing 14.7 Computing the perimeter of polygon on SPU.

14.2.7 Implementation of parallel line-clipper

The problem of line-clipping is important when *viewports* are implemented [93] to *clip* portions of the scene-graph which are rendered to coordinates outside of the currently viewable coordinates. The viewport is usually defined as a bounding-box and thus clipping a set of lines to a bounding-box is a common operation.

The algorithm we have implemented uses a clip code known as *Cohen-Sutherland* clipping which is defined as follows. See Figure 14.8 for the code which is defined for an end-point of a given line segment. The code uses the the minx (left), maxx (right), maxy (top) and miny (bottom) of the given bounding-box to compute a four-bit code. If the x-coordinate of the end-point is less than minx, then the code bit for left is set to one. Similarly for the other sides of the bounding box. The center region as shown in Figure 14.8 has code of 0b0000. This code gives a simple check for line clipping, if the two end-points of the line segment have a bit set in the same position, then the line segment lies completely outside the box and nothing needs to be done, else the line segment has to be clipped.

The code uses SIMD effectively and is shown in Listing 14.8.

```
///////////////////////////////////////////
// Polygon-Line Clipping
///////////////////////////////////////////
vec_uint4 OnOppSide(vec_float4 X1,
                    vec_float4 X2,
                    vec_float4 line) {
  // answer the Q is X1 X2 on same side of line
  vec_uint4 X1_line = spu_cmpgt(X1,line);
  vec_uint4 X2_line = spu_cmpgt(X2,line);
  return spu_xor(X1_line,X2_line);
}
vec_float4 YAtX(vec_float4 X1, vec_float4 Y1,
                vec_float4 X2, vec_float4 Y2,
```

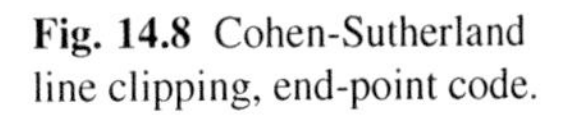

Fig. 14.8 Cohen-Sutherland line clipping, end-point code.

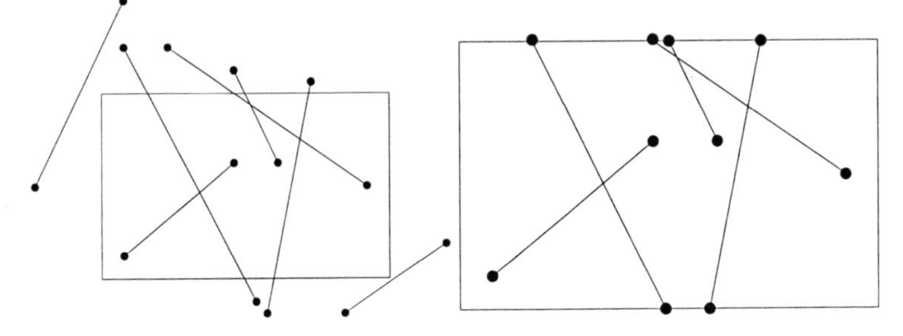

Fig. 14.9 Implementing Cohen-Sutherland line-clipper using SIMD on SPU.

```
                    vec_float4 x, vec_uint4 opp) {
15    // calculate y at x for these 4 lines
      vec_float4 dy = spu_sub(Y2,Y1);
      vec_float4 dx = spu_sub(X2,X1);
      vec_float4 m  = spu_mul(dy,spu_re(dx));
      vec_float4 ad = spu_sub(x,X1);
20    vec_float4 ay = spu_mul(ad,m);
      ay = spu_add(ay,Y1);
      return ay;
    }
    void PolygonLineClip(vec_float4* edges,
25                       int vn,
                         vec_float4 line_x,
                         vec_uint4* places,
                         vec_float4* values) {
      // clip polygon w.r.t vertical line_x
30    int i;
      int n = (((vn/4)+1)*2);
      for(i=0;i<=n;i+=2) {
        vec_float4 X1 = edges[i], X2 = edges[i+2];
        vec_uint4 opp = OnOppSide(X1,X2,line_x);
35      places[i] = opp; print_vector("opp\n",&opp);
        values[i] = YAtX(X1,edges[i+1],X2,edges[i+3],line_x,opp);
        fprint_vector(&values[i]);
      }
    }
```

Listing 14.8 Implementing Cohen-Sutherland line clipping using SIMD on SPU.

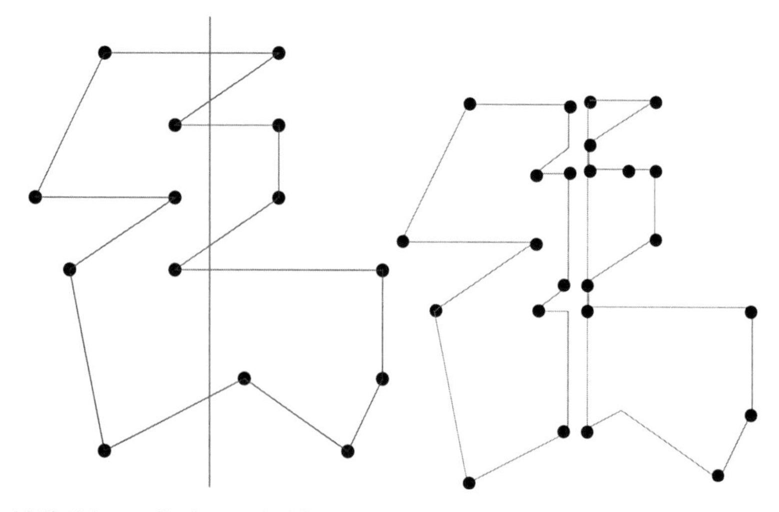

Fig. 14.10 Polygon clipping, vertical line clips polygon.

14.2.8 Implementation of parallel polygon clipper

Consider a simple polygon shown in Figure 14.10 which is intersected by a vertical line $x = \alpha$. The goal of the computation is to compute the separation of the polygon into two parts defined by the line clipping. A simple SIMD based line clipper for simple polygons is implemented in Listing 14.9.

```
////////////////////////////////////////////
// Polygon-Line Clipping
////////////////////////////////////////////
vec_uint4 OnOppSide(vec_float4 X1,
                    vec_float4 X2,
                    vec_float4 line) {
  // answer the Q is X1 X2 on same side of line
  vec_uint4 X1_line = spu_cmpgt(X1,line);
  vec_uint4 X2_line = spu_cmpgt(X2,line);
  return spu_xor(X1_line,X2_line);
}
vec_float4 YAtX(vec_float4 X1, vec_float4 Y1,
                vec_float4 X2, vec_float4 Y2,
                vec_float4 x, vec_uint4 opp) {
  // calculate y at x for these 4 lines
  vec_float4 dy = spu_sub(Y2,Y1);
  vec_float4 dx = spu_sub(X2,X1);
  vec_float4 m  = spu_mul(dy,spu_re(dx));
  vec_float4 ad = spu_sub(x,X1);
  vec_float4 ay = spu_mul(ad,m);
  ay = spu_add(ay,Y1);
  return ay;
}
void PolygonLineClip(vec_float4* edges,
                     int vn,
                     vec_float4 line_x,
                     vec_uint4* places,
                     vec_float4* values) {
  // clip polygon w.r.t vertical line_x
```

```
30    int i;
      int n = (((vn/4)+1)*2);
      for(i=0;i<=n;i+=2) {
        vec_float4 X1 = edges[i], X2 = edges[i+2];
        vec_uint4 opp = OnOppSide(X1,X2,line_x);
35      places[i] = opp; print_vector("opp\n",&opp);
        values[i] = YAtX(X1,edges[i+1],X2,edges[i+3],line_x,opp);
        fprint_vector(&values[i]);
      }
    }
```

Listing 14.9 Implementing polygon line-clipping using SIMD.

The main SPU function for unit testing of the computational geometry framework is shown in Listing 14.10.

```
    int main( unsigned long long spuid,
              unsigned long long argp ) {
      float perim,area;
      PrepData( 20 );
5     printf(" Hello from %x\n", spuid );
      perim = ComputePerimeter( EDGES, 20 );
      printf("\n Perimeter = %f\n", perim);
      area = ComputeArea( EDGES, 20 );
      printf("\n Area = %f\n", area);
10    TestDeterminant(EDGES);
      printf("\n");
      vec_float4 bbox = ComputeBoundingBox(EDGES,20);
      printf("\nBoundingBox = ");
      fprint_vector(&bbox);
15    printf("\n");
      TestClipCode();
      printf("\n");
      vec_float4 line_x = (vec_float4){1300,1300,1300,1300};
      PolygonLineClip(EDGES,20,line_x,S2,SCRATCH);
20    return (0);
    }
```

Listing 14.10 Geometry framework on SPU.

14.3 Conclusion

We have described the design and implementation of some high-performance floating point SIMD based computational geometry operations on the SPU. Additional algorithms which can be implemented on the SPU come from the domain of pixel-based image processing and include filters, convolutions and shape based functions.

Chapter 15
Optimizing SPU Programs

Abstract In this chapter we discuss some of the automated tools we can use to optimize code for the Cell. We begin with the venerable `gprof`, and GCC's branch probability framework. These include the command-line switches -fprobability-generate and the -fprobability-use in recent versions of GCC. These tools are applicable not only to the Cell, but any modern processor with deep pipelines. We extend the assembly language inspection to include cycle counting methods; manually as well as using the `spu-timing` tool. The Assembly visualizer tool from IBM is also presented and we explain its usage. We conclude by presenting `FDPRPRO`, the automated post-link optimization tool for POWER from IBM.

15.1 Introduction to Program Optimization for Cell

In this chapter we will investigate tools and techniques to measure, and then improve the performance of compute-intensive code for the Cell architecture. The first method we present is the venerable `gprof` GNU tool for *profiling*.

15.2 GNU `gprof`

Compile the program using `gcc -pg -g -O2`, where $-pg$ activates profiling, and the $-g$ flag preserves debugging information, especially symbol names in the binary. To get a correct profile I recommend compiling with optimizations, unless the objective is to generate static function call graphs for analysis. This compilation produces a binary which when run produces a file called `gmon.out`. The actual run of the program maybe slower by a factor of 2x. The file produced is called the *profile-file*. Next we use the `gprof` tool as `gprof -b <binary> gmon.out` to produce an execution profile. An example execution profile from our code on Sequence-Pairs is shown below:

S. Koranne, *Practical Computing on the Cell Broadband Engine*,
DOI: 10.1007/978-1-4419-0308-2_15, © Springer Science + Business Media, LLC 2009

```
Flat profile:

Each sample counts as 0.01 seconds.
  %   cumulative   self              self     total
 time   seconds   seconds    calls  us/call  us/call  name
 81.37   1018.50  1018.50  2000000   509.25   509.25  PermuteSequence
 10.05   1144.30   125.80  2000000    62.90    62.90  BF
  5.73   1216.01    71.71                             main
  3.10   1254.78    38.77  2000000    19.39    19.39  LongestPath
  0.00   1254.80     0.02   833576     0.02     0.02  UpdateTemperature
  0.00   1254.80     0.00        4     0.00     0.00  PrintSequence
     Call graph
granularity: each sample hit covers 2 byte(s) for 0.00% of 1254.80 seconds

index % time    self  children    called     name
                                                 <spontaneous>
[1]    100.0   71.71 1183.09                 main [1]
             1018.50    0.00 2000000/2000000     PermuteSequence [2]
              125.80    0.00 2000000/2000000     BF [3]
               38.77    0.00 2000000/2000000     LongestPath [4]
                0.02    0.00  833576/833576      UpdateTemperature [5]
                0.00    0.00       4/4           PrintSequence [6]
-----------------------------------------------
             1018.50    0.00 2000000/2000000     main [1]
[2]     81.2 1018.50    0.00 2000000         PermuteSequence [2]
-----------------------------------------------
              125.80    0.00 2000000/2000000     main [1]
[3]     10.0  125.80    0.00 2000000         BF [3]
-----------------------------------------------
               38.77    0.00 2000000/2000000     main [1]
[4]      3.1   38.77    0.00 2000000         LongestPath [4]
-----------------------------------------------
                0.02    0.00  833576/833576      main [1]
[5]      0.0    0.02    0.00  833576         UpdateTemperature [5]
-----------------------------------------------
                0.00    0.00       4/4           main [1]
[6]      0.0    0.00    0.00       4         PrintSequence [6]
-----------------------------------------------
Index by function name

[3] BF                     [2] PermuteSequence (sqp.c) [5] UpdateTemperature
[4] LongestPath (sqp.c) [6] PrintSequence (sqp.c)    [1] main
```

15.3 Branch probability extraction

The SPEs are heavily pipelined, making the penalty for incorrect branch prediction high, namely 18 cycles. In addition, the hardware's branch prediction policy is simply to assume that all branches (including unconditional branches) are not taken. Even on the PPE mis-predicted branches can be expensive. Consider an example run of the sequence-pair computation below; without any optimizations we get:

```
 0 1 7 4 2 9 5 10 6 8 3 11
 0 5 4 6 10 1 2 7 3 8 9 11
Best cost = 207

real    3m12.281s
user    3m12.148s
sys     0m0.008s
```

15.3.1 Using -fprofile-generate

We compile the program again, but this time we add the `-fprofile-generate` command-line to GCC. This instruments the code to collect branch probability statistics. These statistics are written to a file (with suffix .gcda) for every compilation unit of the original program. The program must terminate with `_exit` for this file to be correctly created. The gcda file is a binary file which contains information that gcc can use when invoked with the `-fprofile-use` switch. The gcda file is additive, hence, the instrumented binary can be run on many representative workloads and the branch probabilities will be accumulated.

15.3.2 Using -fprofile-use

Replacing `-fprofile-generate` with `-fprofile-use` compile the program again. This informs gcc that we have run the instrumented binary on some typical workloads and now the branch probabilities have been estimated in the gcda file (which should be located in the same place as the compilation unit). This compilation of gcc uses the information in the gcda file to produce a faster running program. We compiled our sequence-pair based floorplanner using this method and the runtime became:

```
 0 9 1 8 3 4 7 5 2 10 6 11
 0 2 6 4 1 10 8 7 3 5 9 11
Best cost = 221

real    1m52.079s
user    1m52.059s
sys     0m0.004s
```

We have manged to reduce the runtime of the program by 70 seconds. On larger examples this optimized binary gave even faster run-times.

The method we have presented above is broadly termed *profile guided optimization*, and for domains where the representative workload can be efficiently estimated (such as a floorplanner), this method gives spectacular improvements in run-time with only modest increase in build-times.

15.4 SPU Assembly Analysis

Consider a fragment of code (shown in Listing 15.1) taken from the line-of-sight example. We shall use this code to analyze the SPU pipeline performance using spu-timing and ASM visualization tools.

```
static void
SwapIfHSL(vec_uint4 hsl,
          vec_uint4 *a,
```

```
        vec_uint4 *b) {
5   vec_uint4 atemp = spu_sel( *b, *a, hsl );
    vec_uint4 btemp = spu_sel( *a, *b, hsl );
    *a = atemp;
    *b = btemp;
}
```

Listing 15.1 Example code for SPU pipeline analysis using spu-timing.

```
81:los_spe.c      **** static void
82:los_spe.c      **** SwapIfHSL(vec_uint4 hsl,
83:los_spe.c      ****           vec_uint4 *a,
84:los_spe.c      ****           vec_uint4 *b) {
75                     .LVL8:
76                     .LVL9:
77 0038 4020007F                nop     127
78 003c 35800009                hbr     .L11,$lr
79 0040 4020007F                nop     $127
85:los_spe.c **** vec_uint4 atemp = spu_sel(*b,*a,hsl);
81 0044 34000208                lqd     $8,0($4)
82 0048 4020007F                nop     $127
83 004c 34000286                lqd     $6,0($5)
84 0050 80E20303                selb    $7,$6,$8,$3
86:los_spe.c **** vec_uint4 btemp = spu_sel(*a,*b,hsl);
86 0054 80418403                selb    $2,$8,$6,$3
87:los_spe.c **** *a = atemp;
88 0058 24000207                stqd    $7,0($4)
88:los_spe.c **** *b = btemp;
90 005c 24000282                stqd    $2,0($5)
91                     .L11:
89:los_spe.c **** }
93 0060 35000000                bi      $lr
94                     .LFE17:
```

15.4.1 SPU-Timing tool

SPU-Timing is a static-timing analysis tool which produces annotated assembly listings with clock cycle, stalls, pipeline status and dual-issue information in its output. Consider the assembly language program shown above, when run through the spu-timing tool it produced the following annotation.

```
                          .align  3
                          .type   SwapIfHSL, @function
                          SwapIfHSL:
0d 0                          nop     127
1d ---345678                  hbr     .L11,$lr
0D     4                      nop     $127
1D     456789                 lqd     $8,0($4)
0D      5                     nop     $127
1D      567890                lqd     $6,0($5)
```

```
0        -----12              selb    $7,$6,$8,$3
0              23             selb    $2,$8,$6,$3
1               345678        stqd    $7,0($4)
1                456789       stqd    $2,0($5)
                       .L11:
1                 5678        bi      $lr
                       .size   SwapIfHSL, .-SwapIfHSL
```

The output of spu-timing can be explained as follows:
Column 0 contains a label of 0/1 to represent *even* or *odd* pipeline. This depicts the pipeline which will execute the instruction which *starts* at this address.

The next column informs us about the *dual-issue* status of the SPU; if the symbol *D* appears, then dual-issue instructions were issued, a lowercase *d* implies that dual-issue was *possible* but did not happen, either because of dependencies or other reasons. If there is no symbol, then dual-issue was not possible at all. The goal of SPU program optimization should be to maximize the *D* symbol.

The next 50 columns represent clock-cycles and they are printed as repeating digits from 0–9. Consider the `lqd` instruction which has a clock-cycle number 456789 before it. This implies that this 6-cycle instruction will issue at clock-cycle 4 and continue for the next 6 clock-cycles. A *n*-cycle instruction will therefore display *n*-digits, once each in the clock-cycle where it is executing. A – symbol in the clock-cycle tells of a *dependency stall*. The goal of the SPU-optimization should be reorder code, or rewrite it, to remove cycles lost due to stalls. Moreover, having long horizontal sections is also bad; thin vertical sections, preferably in both columns (which looks like a V shape) are good. We have run spu-timing on all of our time-critical SPU-kernels and have always found something to optimize. The remaining columns of the annotation are the actual assembly code itself.

15.4.2 IBM ASM Visualizer

IBM Assembly Visualizer works in a similar manner to spu-timing, but instead of producing an annotated listing, it shows the SPU pipeline information in a graphical manner. Clicking on any instruction slot shows the dependencies of that slot (both forward and backward). And the issuing instruction is highlighted to pinpoint the exact location where the instruction was issued. An example is shown in Figure 15.1 for the ASM visualization of our SPU code for line-of-sight. This example is the `SwapIfHsl` function discussed above.

Unfortunately, the clock cycle pane in the ASM VIS tool start off from 1, instead of 0, which is the number reported by `spu_timing`, hence, we have to remember to subtract 1 from the clock cycle when comparing the spu-timing output with ASM VIS tool. When you click on any instruction slot in the even or odd pipeline, the instruction is shown marked in black. The dependencies of the instruction are showed in red, and the

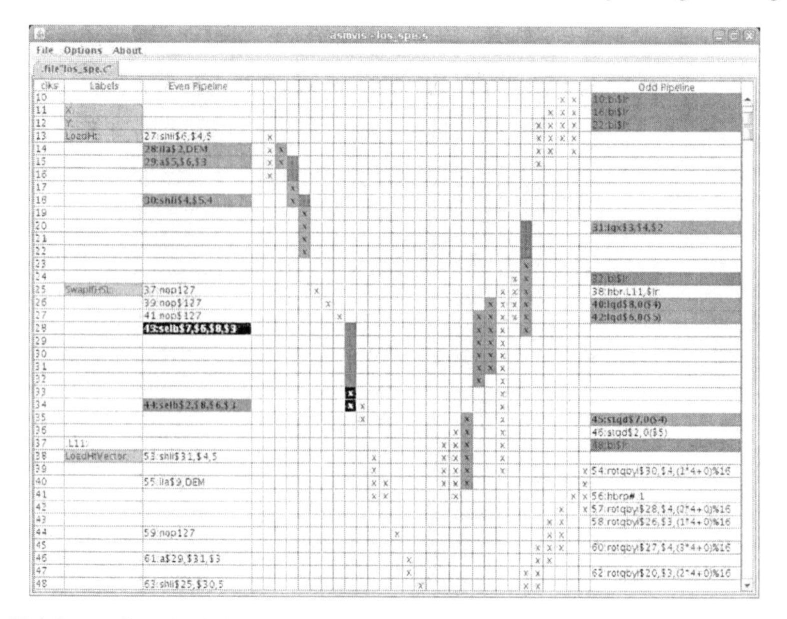

Fig. 15.1 Instruction scheduling in `SwapIfHsl`.

15.5 FDPRPRO: Optimization Tool

FDPRPRO (Postlink Optimization Tool for POWER) is a performance-tuning utility used to improve the execution time and the real memory utilization of user-level application programs. The tool optimizes the executable image of a program by collecting information on the behavior of the program while the program is used for some typical workload. It then re-analyzes the program (together with the collected profile), applies global optimizations (including program restructuring), and creates a new version of the program that is optimized for that type of workload. The new program generated by the optimizer typically runs faster than the original program. We have used FDPRPRO for SPU-kernel optimization, and have seen improvement upto 20% with automated program restructuring and optimization. Some common FDPRPRO usage options for its command line are given below, please see the man page for more detailed help. We describe the `instr` (instrument) and `opt` (optimize) actions only. To instrument the SPU code, run the following:

```
$fdprpro -a instr <spu-binary-file-name>
```

This step will produce an instrumented binary (with a .instr suffix) which can be run as usual on typical and representative workloads. When this binary it run it produces a .mprof file which needs to be brought back to the compilation directory, and then we can run fdprpro with its optimization action, as follows:

```
$fdprpro -a opt <spu-binary-file-name> -f <mprof-file>
```

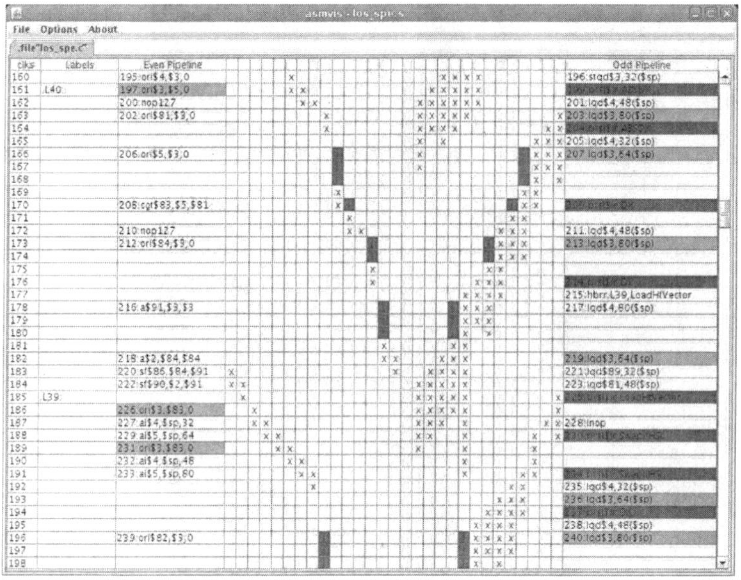

Fig. 15.2 Instruction scheduling in `CalculateLOS`.

Some of the other optimization options which can be specified are:

-A alignment –align-code alignment Align program so that hot code will be aligned on alignment-byte addresses,

-bf, –branch-folding Eliminate branch to branch instructions,

-bh factor, –branch-hint factor add branch hints to basic blocks that are hotter then the average by given (float) factor. This is a SPE specific optimization. Value of -1 (the default) disables this option,

-bp, –branch-prediction Set branch prediction bit for conditional branches according to collected profile,

-dce, –dead-code-elimination Eliminate instructions related to unused local variables within frequently executed functions. This is useful mainly after applying function inlining optimization,

-dp, –data-prefetch Insert data-cache prefetch instructions to improve data-cache performance,

-hr, –hco-reschedule Relocate instructions from frequently executed code to rarely executed code areas, when possible,

-i, –inline Same as –selective-inline with –inline-small-funcs ,

-lu aggressiveness_factor, –loop-unroll aggressiveness_factor Unroll short loops containing of one to several basic blocks according to an aggressiveness factor between (1,9), where 1 is the least aggressive unrolling option for very hot and short loops,

-nop, –nop-removal Remove NOP instructions from reordered code,

-O Switch on basic optimizations only. Same as -RC -nop -bp -bf,

-O2 Switch on less aggressive optimization flags. Same as -O -hr -pto -isf 8 -tlo -kr,

-O3 Switch on aggressive optimization flags. Same as -O2 -RD -isf 12 -si -dp -lro -las -vro -btcar -lu 9 -rt 0,

-O4 Switch on aggressive optimization flags together with aggressive function inlining. Same as -O3 -sidf 50 -ihf 20 -sdp 9 -shci 90 and -bldcg (for XCOFF files),

-rcaf aggressiveness_factor, –reorder-code-aggressivenes-factor Set the aggressiveness of code reordering optimization. Allowed values are [0 — 1 — 2], where 0 preserves original code order and 2 is the most aggressive. Default is set to 1. (applicable only with the -RC flag),

-RD, –reorder-data Perform static data reordering.

15.6 Conclusion

In this chapter we have presented tools for automatic measurement, analysis, and optimization of C/C++ programs running on the Cell Broadband Engine. Many of these tools are also applicable to other processors. The key idea to take-away from this chapter is that profile measurement should always be used to find out program hot-spots. Moreover, with the deep pipelining present in current processors including SPU, mispredicted branches can be very expensive. Using profile-guided optimization framework in GCC using -fprobability-generate and -fprobability-use can be very effective in reducing running time. For the Cell Broadband Engine, the spu-timing tool provides static analysis of the clock cycles. This can be used to see dual-pipeline issue efficiency of the written code. A visual tool, the IBM Assembly Visualizer presents the same information in a graphical manner. The FDPRPRO optimization tool was introduced which combines many of the ideas presented in this chapter in a single, easy to use tool. We have seen performance improvements ranging from 20 to 50% from using the methods presented in this chapter. However, some of the more aggressive optimizations should be regressed carefully, as both GCC and FDPRPRO have been known to produce incorrect code in certain corner cases.

Part III
Case Studies

In this part of the book we shall look at mapping various real life applications to the Cell Broadband Engine.

In Chapter 16 we present an implementation of an algorithm to perform line-of-sight estimation on 3d terrain. We use a novel under-sampled SIMD Bresenham's line-drawing algorithm with efficient SPU implementation to compute line-of-sight. We extend this method to perform watershed analysis for facility location (eg., cell-phone towers, wind-turbines).

The chapter (Chapter 17 on *ab-initio* methods for structure recovery using partial distance functions, is a compute intensive procedure to discover the structure of a 3d atomic or molecular structure given spectroscopic distance measurements. We have used statistical methods and have run this on the SPE with good performance.

We present a full polytope exploration system in Chapter 18. This implements a novel polytope enumeration and analysis algorithm which can generate and analyze catalogs of general n,d-polytopes.

Chapter 19 presents an implementation of a simplified (no motion correction, no registration correction) implementation of the core algorithms in Synthetic Aperture Radar. This technique of remote sensing has become vastly popular with increasing compute power, and the SIMD performance of SPEs on this task validates the argument of choosing Cell as the optimal platform for signal-processing applications.

The last couple of chapters are Cell implementations of classical VLSI CAD algorithms which are used in solving problems such as placement, channel-routing, FPGA switch-box routing, microword minimization, switching probability estimation, coupling length estimation, scheduling and BDD (binary decision diagram) analysis. We conclude in Chapter 24.

Chapter 16
Line-of-sight Computation

Abstract In this chapter we discuss the problem of calculating line-of-sight (LOS) between objects located at different points of a given digital-elevation-model. This problem finds practical application in many areas of civil engineering, military tactical command and control, cellular operators, wind-turbine positioning and even ray-tracing. We describe the problem in its general terms, and model the DEM using partitioned matrix layout. We present a novel under-sampled Bresenham's line-drawing algorithm implemented using 4-way SIMD on the SPU. The code has been optimized to support multi-observer, multi-target setup. Implementation issues on the SPU are discussed and example runs on simulated data is presented. We also present a novel concept of shielding which we have used to partition the data on the PPE, and execute parallel computations on the 6 SPEs. Correspondingly, we get upto 30x speed enhancements as compared to a multi-threaded execution on a conventional processor.

16.1 Introduction to the line-of-sight computation

This chapter discusses the problem of performing visibility analysis using parallel algorithms. A novel under-sampled Bresenham's line drawing algorithm is implemented using SIMD on the SPUs, to compute line-of-sight between four observers and four targets simultaneously. We combine the SIMD optimization with the six-way SPU parallelism present in the Cell to achieve near linear speedup of this problem. Visibility indices need to be computed for a variety of reasons, including, but not limited to, civil engineering projects, tactical military applications, cell-phone tower location and many more. The entities between which visibility is calculated are presented on a digital-elevation-model (DEM) or a digital-terrain-model (DTM).

With the advent of GPS and GIS computing, vast amounts of high-resolution spatial data is now available. This data can be used in a multitude of ways including site selection for housing projects. In this application, visibility indices of objects

S. Koranne, *Practical Computing on the Cell Broadband Engine*,
DOI: 10.1007/978-1-4419-0308-2_16, © Springer Science + Business Media, LLC 2009

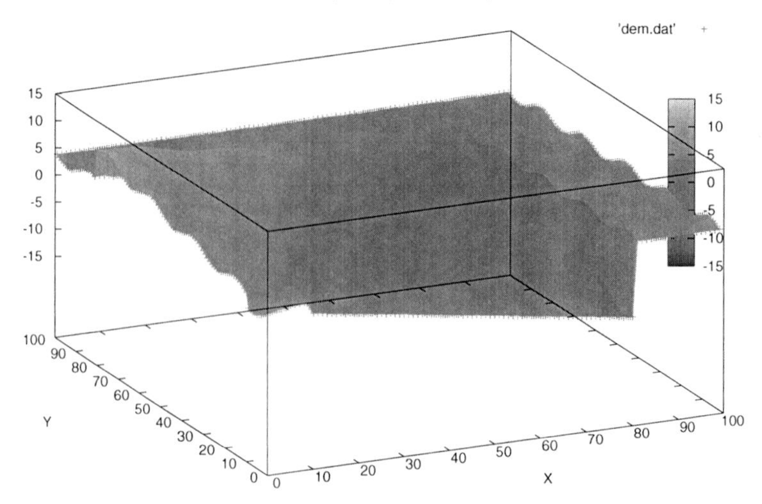

Fig. 16.1 Watershed analysis using SPU SIMD implementation of line-of-sight algorithm.

may need to be calculated to evaluate site candidates which offer less obtrusive views of offending objects in the distance. Conversely, visibility analysis also needs to be performed when site selection is being done to find areas to locate cell-phone towers which are visible from the maximum number of observer locations in an area. The area under consideration is called the *zone of visual influence* (ZVI), and can be thought of as a radius around observer or target.

The general visibility calculation problem is thus the problem of identifying location on the terrain which can be seen from an observer location. The terrain is most often represented as digital elevation model on a regular grid. A profile of the terrain from the target to the observer is constructed using a line-of-sight method. For every grid point on the line-of-sight, if the elevation is more than either the observer, or the target, then the target is not visible from the observer. This is shown in Figure 16.4. When we compute visibility index for all observer-target pairs, we in effect, calculate the watershed of the terrain. We have written the SPU implementation of our line-of-sight with SIMD and parallel processing, and in practice it can calculate a 2000x2000 grid under 2 minutes.

We have implemented a line-of-sight algorithm for determining the visibility indices of entities (or objects) such as elevation vertices, buildings, or road centrelines on a digital terrain model (DTM). This may be a requirement of site selection for a contentious development, especially if visibility, or more specifically, visual intrusion is likely to be a key factor in gaining planning approval. See Figure 16.2 for an example. In the figure, there are two computation scenarios, (i) the ideal location for the wind turbines was calculated using terrain data for ideal wind location, and/or (ii) the location of a housing development needs to be calculated and one of the

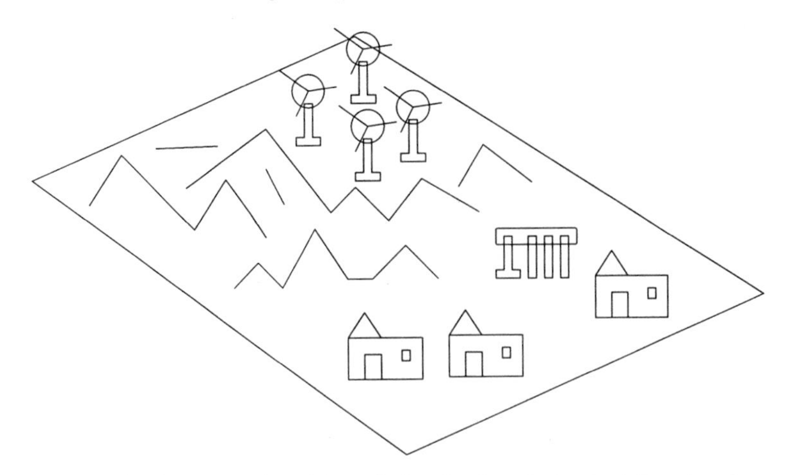

Fig. 16.2 Line-of-sight computation with terrain data for site-selection in civil engineering project. In this problem, the visibility index of every point in the development zone with-respect-to the far away, but visible wind-turbines is computed. Zones with low visibility indices will presumably fetch more money to the developer.

metrics being used in choosing the land is low visibility index (with-respect-to the wind-turbines). In both scenarios, the line-of-sight calculation with digital terrain model is indispensable.

As vast quantities of spatial data become available, particularly DTMs at larger scales and denser resolution, the demands for parallel processing will inevitably increase. Moreover, line-of-sight calculations are also an integral and often time-consuming part of tactical real-time target reconnaissance applications in defense. See Figure 16.3 for an illustration. The hilly terrain divides the battlefield, and line-of-sight, using optical, or microwave can be used to calculate target location and attack vectors. Some sensors can also see through foliage, and line-of-sight calculation with sensor uncertainty is an area of current research.

For general LOS (line-of-sight) computations, or visibility analysis, the problem is to identify the coordinates on a grid which can be seen from a given observer position. GIS data for terrain from satellite or hand held measurement is used to get the terrain data which is usually in the form of a regular grid digital elevation model (DEM) data. The run-time computation is to construct a profile of the terrain from the observer to each DEM vertex and to classify the location as observable or not-observable.

Consider the example shown in Figure 16.3. In this problem time-to-response is also equally important as final accuracy. When setting up defensive positions which can overlook as much terrain as possible, a fast turn-around time for the computation is required. When performing LOS on moving targets, obviously better than real-time response is desired, and the 6 SPUs are given the same problem but with different velocity vectors, and an average of the vector solution is returned.

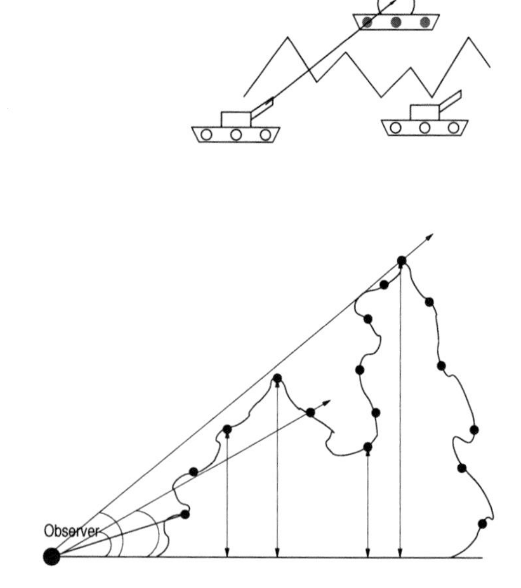

Fig. 16.3 Line-of-sight with tactical response time constraint.

Fig. 16.4 Observer position for line-of-sight with terrain height.

16.2 Basic equation for LOS calculation

A simple mathematical calculation can be used to find out which points in the terrain are visible from other points (see Figure 16.4). Typically, applications will define a *zone of visual influence* in which to calculate the visibility blocking. By calculating the elevation angles from the observer to the measured location, and recording the elevation angle of all (or some suitably sampled) intermediate points we can compute the visibility index of the location with-respect-to the observer. When multiple metrics are used a combined index (with-respect-to multiple observers) can be used (this is more time consuming as line-of-sight calculation needs to be independently processed for each observer).

In Figure 16.4, if $ht(x) > ht(o)$ or if $ht(x) > ht(t)$, where x is an interpolated point on the line-of-sight between the observer o and the target t, the t is deemed not visible from o. This equation forms the basis of LOS checks in our code.

The DEM can be modeled as a matrix as below:

$$M = \begin{bmatrix} 39 & 54 & 20 & 10 & 47 & 67 & 66 & 8 \\ 34 & 38 & 78 & 46 & 64 & 49 & 4 & 60 \\ 86 & 19 & 91 & 43 & 73 & 31 & 66 & 26 \\ 65 & 56 & 83 & 82 & 69 & 71 & 64 & 79 \\ 89 & 8 & 42 & 68 & 92 & 6 & 56 & 75 \\ 3 & 79 & 89 & 54 & 26 & 28 & 29 & 21 \\ 44 & 36 & 93 & 90 & 83 & 46 & 39 & 2 \\ 38 & 71 & 86 & 94 & 14 & 30 & 2 & 1 \end{bmatrix} \quad (16.1)$$

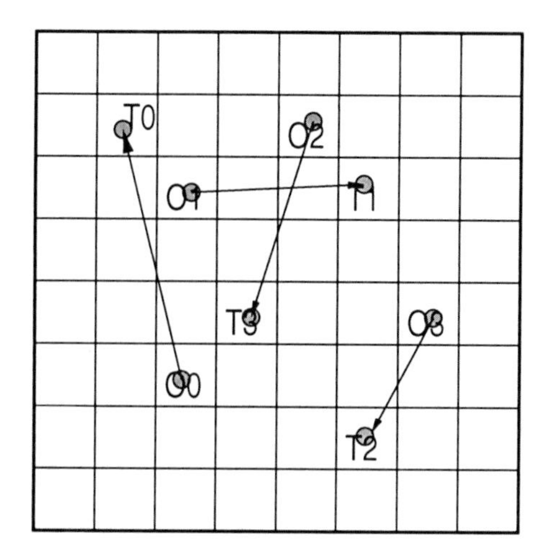

Fig. 16.5 Multi-observer, multi-target SIMD line-of-sight. Consider the 8×8 matrix shown. With observers located at (2,2),(2,5),(4,6) and (6,3), and targets located at (1,6),(5,5),(3,3) and (5,1). The line-drawing algorithm will interpolate each observer-target pair simultaneously in our SIMD implementation.

Consider the 4 Observers located at indices (2,2),(2,5),(4,6) and (6,3) as shown in Figure 16.5. The corresponding targets are located at (1,6),(5,5),(3,3) and (5,1). The LOS code will compare the DEM height at the following points calculated by the Bresenham's algorithm.

```
(2,2)->(2,2)->(1,6) (2,5)->(2,5)->(5,5) (4,6)->(4,6)->(3,3) (6,3)->(6,3)->(5,1)
(2,2)->(2,3)->(1,6) (2,5)->(3,5)->(5,5) (4,6)->(4,5)->(3,3) (6,3)->(6,2)->(5,1)
(2,2)->(2,4)->(1,6) (2,5)->(4,5)->(5,5) (4,6)->(3,4)->(3,3) (6,3)->(5,1)->(5,1)
(2,2)->(1,5)->(1,6) (2,5)->(5,5)->(5,5) (4,6)->(3,3)->(3,3) (6,3)->(5,1)->(5,1)
(2,2)->(1,6)->(1,6) (2,5)->(5,5)->(5,5) (4,6)->(3,3)->(3,3) (6,3)->(5,1)->(5,1)
```

The SIMD layout of the observer X,Y, target X,Y and the interpolated points calculated by the line-drawing algorithm are shown below.

$$Vis = \begin{bmatrix} X & [2,2,4,6] & [2,2,4,6] & [1,5,3,5] \\ Y & [2,5,6,3] & [2,5,6,3] & [6,5,3,1] \\ X & [2,2,4,6] & [2,3,4,6] & [1,5,3,5] \\ Y & [2,5,6,3] & [3,5,5,2] & [6,5,3,1] \\ X & [2,2,4,6] & [2,4,3,5] & [1,5,3,5] \\ Y & [2,5,6,3] & [4,5,4,1] & [6,5,3,1] \\ X & [2,2,4,6] & [1,5,3,5] & [1,5,3,5] \\ Y & [2,5,6,3] & [5,5,3,1] & [6,5,3,1] \\ X & [2,2,4,6] & [1,5,3,5] & [1,5,3,5] \\ Y & [2,5,6,3] & [6,5,3,1] & [6,5,3,1] \end{bmatrix} \tag{16.2}$$

16.3 Watershed analysis using LOS

Now consider a given observer located at (i,j) and a list of targets given by $(x_0,y_0),(x_1,y_1),\ldots,(x_n,y_n)$. The algorithm for watershed and line-of-sight for multiple targets can be given as:

```
   Data: Resolution and Coordinate reference
   Data: Number of rows m and columns n of DEM
   Data: DEM as a floating point matrix of m × n
   Data: List of observers, or full matrix
   Data: ZVI (zone-of-visual-influence) metric
   Result: Matrix of visibility (watershed)
1  Matrix WatershedAnalysis(m,n,zvi,DEM) begin
       Data: Matrix
2      Visibility;
3      for Each o ∈ Observer do
4          int count = 0;
5          for Each vertex v ∈ DEM ∈ ZVIofo do
6              count += LOS(o,v);
7          end
8          Matrix[o] = count;
9      end
10 end
```

Algorithm 23: Watershed or visibility index computation of a DEM.

16.4 LOS formulation using Bresenham's line-drawing for interpolation

As we have stated above, for multiple indices, the problem is inherently parallel, and even for a single observer there is a lot of computation which can be performed in parallel. The most basic formulation of this problem uses a line sampling method wherein linear or bilinear points on the straight-line joining the observer to the desired location are computed, and their terrain is looked up from the DTM. If any of these points has a higher elevation than either the observer or the measured point, then the line-of-sight does not exist and we can mark the location as not observable. If we assume that DTM is available at a suitable resolution and can fit into the local store memory (or can be accessed in a batch mode for certain coordinates) then the compute intensive part of the calculation is the calculation of which points lie on the straight-line and their linear interpolation. The first problem can be solved using Bresenham's line drawing algorithm which computes raster grid coordinates given $x1,y1$ and $x2,y2$ where $x1,y1$ and $x2,y2$ are known to lie on the grid and are the end-points of the line. Bresenham's algorithms is used as it does not involve division and most calculations in the algorithm can be performed using bit-wise arithmetic only. We would like to go a step further and implement the algorithm on SPEs so that we can perform 6-observer's LOS simultaneously. We would also, at the same time, like to use the 4-way SIMD capabilities of the SPEs.

Thus we modify the original Bresenham's algorithm to perform undersampling on the coordinate space (see Figure 16.7). We divide the available x resolution by 4 such that every coordinate can be assumed to lie within a quadword containing 4 single precision floats of terrain data. In the tactical mode, where we can sacrifice accuracy, the vector internal error is not corrected, and line-of-sight proceeds as if the vector data was constant average of all DTM at the 4 x locations subsumed in the vector.

When accuracy is needed, we perform shuffles in the vector to calculate the *exact* terrain height for the x location. We have also modified the algorithm so that repeated application of an extra step of the rasterization does not cause points to be generated which do not lie on the line. This is needed as when we use SIMD organization to calculate 4 LOS simultaneously, due to distance variations, some fields of the vector may already reach the end point, while other fields are still producing valid intermediate points. We *saturate* the calculation so that the end point of the line is repeatedly added to the list of output points till all LOS computations for that vector are complete. We check for terrain blocking during the loop and set selection bits when we have computed a non-observable coordinate, although this does not alter the program behavior in a large part, since all memory is already divided into 4 x locations, so even if we have identified a coordinate as non-observable, its DTM is part of an aligned memory fetch. We measured that it is faster to load aligned data than to write extra code which will disable the computation for a single vector element.

```
   Data: 4-Observer vector X,Y coordinates
   Data: 4-Target vector X,Y coordinates
   Data: DEM as a floating point matrix of m × n
   Result: Vector visibility count for each observer
 1 Vector SIMDBresenham(ox,oy,tx,ty,DEM) begin
 2     vector count;
 3     const vector MINUS1 = {-1,-1,-1,-1},ZERO = {0,0,0,0};
 4     const vector ONE = {1,1,1,1},TWO = {2,2,2,2};
 5     vector dx = tx − ox,dy = ty − oy;
 6     vector adx = |dx|,ady = |dy|;
 7     vector hsl = ady > adx,xstep = ONE, ystep = ONE;
 8     vector 2dy = dy + dy,2dx = dx + dx;
 9     vector 2dydx = 2dy − 2dx,err = 2dy − dx;
10     vector x = ox, y=oy,xp=ZERO, yp=ZERO;
11     vector progress = ONE,isvis = ONE;
12     vector oht = LoadHtVector(ox,oy);
13     vector tht = LoadHtVector(tx,ty);
14     SwapIfHSL(hsl,&ox,&tx);SwapIfHSL(hsl,&oy,&ty);
15     dx = tx − ox;dy = ty − oy;
16     vector is_dx_lt0 = dx < ZERO;
17     xstep = (is_dx_lt0) ? MINUS1 : xstep;
18     vector mdx = ZERO - dx;dx = (is_dx_lt0) ? mdx : dx;
19     vector is_dy_lt0 = dy < ZERO;
20     ystep = (is_dy_lt0) ? MINUS1 : ystep;
21     vector mdy = ZERO - dy;dy = (is_dy_lt0) ? mdy : dy;
22     for true do
23         xp = (hsl)? y : x, yp = (hsl)? x : y;
24         vector cht = LoadHtVector(xp,yp);
25         vector c_o = oht > cht,c_t = tht > cht;
26         isvis = isvis ∧ c_o ∧ c_t;
27         vector xdone = x == tx, ydone = y == ty;
28         vector notdone = NAND(xdone,ydone);
29         if AllBitsZero(notdone) then
30             break;
31         end
32         vector iserr = err > 0;
33         y = y + (iserr) ? ZERO : ystep;
34         err = err + (iserr) ? 2dy : 2dydx;
35         x = x + xstep;
36     end
37     count = (isvis) ? ONE : ZERO;
38     return(count);
39 end
```

Algorithm 24: SIMD presentation of Bresenham's line-drawing algorithm.

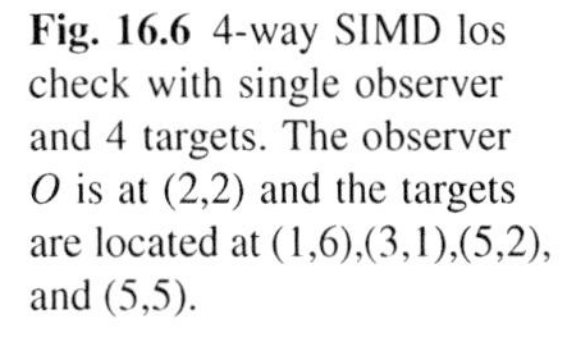
Fig. 16.6 4-way SIMD los check with single observer and 4 targets. The observer O is at (2,2) and the targets are located at (1,6),(3,1),(5,2), and (5,5).

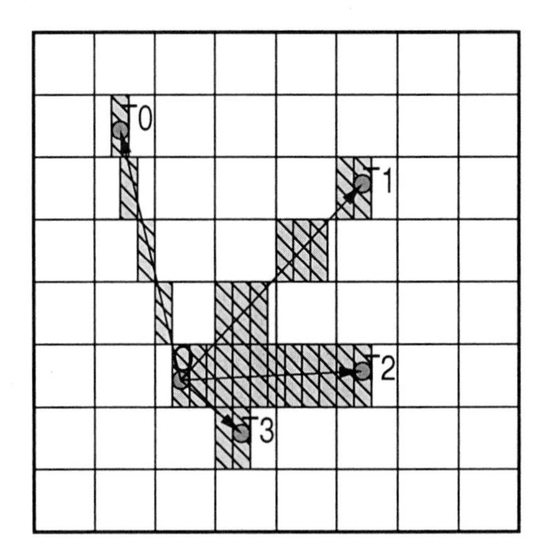

16.5 SIMD Implementation of line-of-sight algorithm

At first sight (pun intended), it looks (again), as if the computation is completely dependent on the position of both observer and target; thus if you are given a single observer and list of targets the best you can do is to do a pairwise los check between them. But as we show in this section we can do better, using SIMD we will perform 4-way los checks between a single observer and 4 targets. Consider the example shown in Figure 16.6

We setup vector registers containing the X and Y location of the observer and targets as follows. We create a single register which contains the X and Y location of the observer in all 4 slots using `spu_splats`, we create $N/4$ (where N is the number of total targets) X-vectors for the targets and specify their X location in each slot (respectively for Y as well). We now have:

```
vector float OX = spu_splats(2);
vector float OY = spu_splats(2);
vector float TX = {1,3,5,5};
vector float TY = {6,1,2,5};
```

We now perform an under-sampled Bresenham's line algorithm between O and T proviso, if any of the vector elements reach their end-points before the other, they saturate at that location. Thus the δx and δy computed by Bresenham's algorithm in this case for the pair $O \rightarrow T_0$ will be $(0,+1),(-1,+1),(0,+1),(0,+1)$, thus in 4 steps we have reached the target. But for target T_3, which is considerably closer to the observer we have the δx and δy as $(+1,-1)$. For the next 3 clock-cycles the delta values for T_3 will saturate at these values of $(+1,-1)$. For T_2, $(0,+1),(0,+1),(0,+1)$, and for T_1, $(+1,+1)^4$. We can see from this example, that the processing time was dominated by the target which is farthest away, and thus it makes sense to sort by max-manhattan distance the list of targets such that

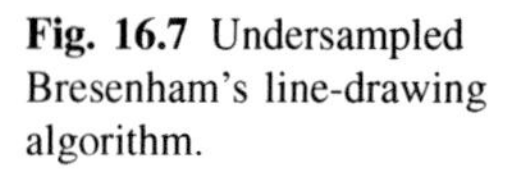

Fig. 16.7 Undersampled Bresenham's line-drawing algorithm.

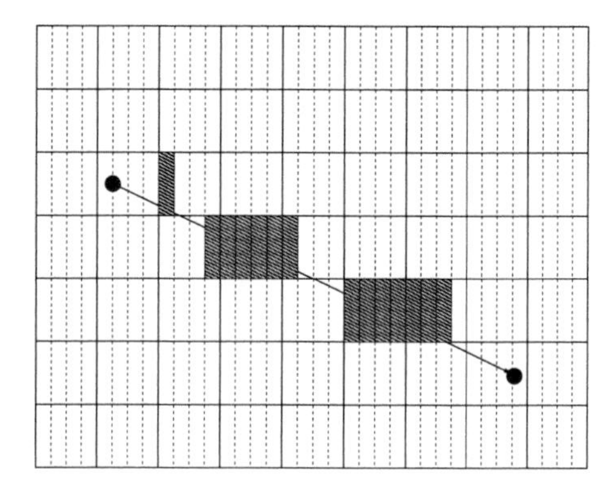

many targets are at the same distance so as to improve efficiency. A multi-observer, multi-target situation is shown in Figure 16.5. Our algorithm is able to compute line-of-sight (using SIMD Bresenham) for 4 observers, and 4 targets independently. For some calculations, the observer location can be identical in all 4 slots, but no special optimization is done for this case.

By using the saturation property of the line drawing algorithm we are able to process upto 4 targets simultaneously. In the line-of-sight computation, the the δx and δy values are used as table offsets in the digital-elevation model to compare elevation values of the $T_x + \delta x$, $T_y + \delta y$ against O_x, O_y and T_x, T_y. If the intermediate point at δx and δy has higher elevation than either observer or target, then that target is not visible from the observer.

16.6 Watershed Analysis

In the above problem the list of observers and targets was given as an input to the problem. A related problem is of *watershed analysis*, where given a DEM of size n^2, the problem is to compute the number of other points visible at every point in the matrix of n^2. In other words, we consider every point in the matrix an observer and calculate visibility to every other point. Even with trivial optimizations in place this is a compute intensive problem, but finds ready application in civil engineering, and tactical planning.

Consider yourself in the position of a battle-group commander who has just arrived in an enemy zone, and you are only given the digital-elevation-model of the area. You are tasked with planning the location of guard positions, and bunkers with which to *hold* the area. A near-real-time watershed analysis tool would surely come handy. With DEM models often approaching 10^3 grid size resolution we would like to use the SPU to calculate the watershed in parallel using SIMD. This is described below.

An example of a digital-elevation model is shown in Figure 16.1. The height data is generated by a program, and is stored as a n^2 matrix of floating point values. The SPU parallel watershed program reads this data in and computes the watershed for every point.

16.6.1 Watershed analysis on multiple SPUs

We partition the DEM data on multiple SPUs. We have done this to simplify the presentation, as external memory DEM computation can be performed by the following argument. Consider the situation of a far-away observer as shown in Figure 16.8. The square contains the pre-computed LOS data for a part of the DEM which resides in external memory (external with-respect-to the observer). Given the observer's location, the watershed analysis reports the visibility index of the marked sites on the periphery of the data as the visibility data for the observer. This is true in general, but when the observer is far away, sites on the periphery can be used as an approximation to the intersection of the ray from the observer to multiple targets in the data.

Thus in Figure 16.8, for every point in the square, a peripheral coordinate on the boundary of the square is computed; this depends on the points location and the observer's location. The observer is assigned the visibility index of the peripheral square (shown marked in the figure). This is a similar concept to *terminal propagation* which we used in graph partitioning (cf. Section 11.11.3).

16.7 SPU Implementation

The key component of our implementation is an SIMD line-of-sight calculator which calculates line-of-sight between 4 observers and 4 targets independently using an under-sampled Bresenham's algorithm. The major functions of our implementation are shown in Listing 16.1 and Listing 16.2.

```
   ////////////////////////////////////////
   // Program : LOS
   // Author  : Sandeep Koranne
   ////////////////////////////////////////
 5 #include <spu_mfcio.h>
   #include <stdio.h>
   #include <math.h>
   #include <spu_intrinsics.h>
   #include <simdmath.h>
10 #include <spu_timer.h>

   typedef struct _mbox_control_block {
     unsigned int N,M;
     unsigned long long addr;
15   unsigned int N2,M2;
     unsigned long long addr2;
   } LOSCB;
```

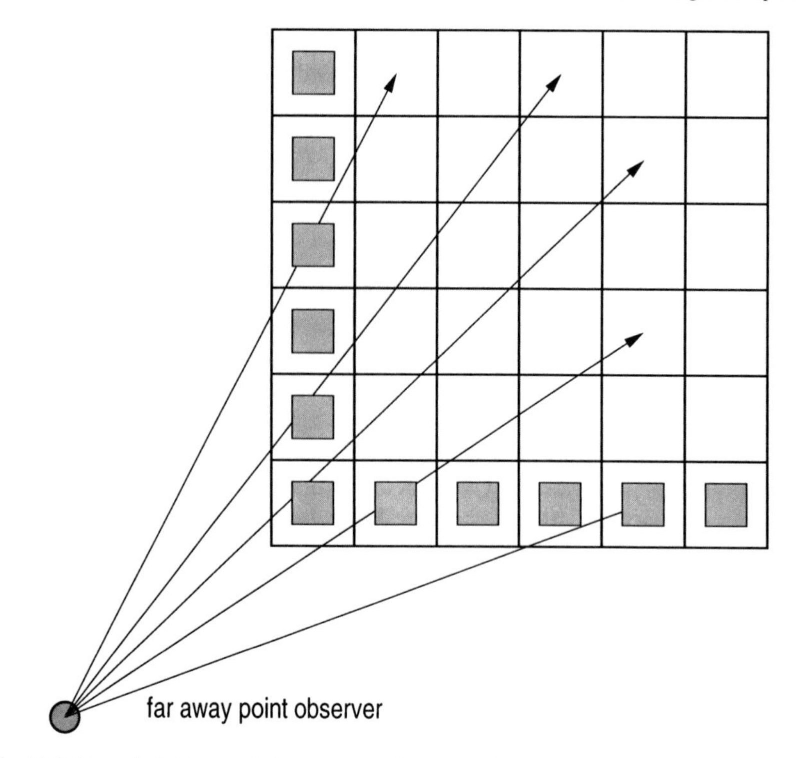

Fig. 16.8 Line-of-sight computation on external memory data.

```
   static LOSCB cb __attribute__ ((aligned(128)));
20 #define DIM_M 128
   #define DIM_N 128
   #define DIM_MN (DIM_M*DIM_N)
   #define NEL DIM_MN
   #define CHUNK 16384
25
   static vector float DEM[DIM_M][DIM_N/4] __attribute__ ((aligned(128)));
   static vector unsigned int WSHED[DIM_M][DIM_N/4]
                __attribute__ ((aligned(128)));
   unsigned long int total_tags = 0;
30 void ev_handler( int time_tag ) {
     total_tags++;
   }
   static void LoadDEM( LOSCB cb, int tag_id ) {
     int i;
35   for(i=0;i<4;++i) {
       mfc_get(DEM+(i*CHUNK), cb.addr+(i*CHUNK),
               sizeof(float)*NEL, tag_id, 0, 0);
       mfc_write_tag_mask(1<< tag_id);
     }
40   mfc_read_tag_status_all();
   }
   inline vec_uint4 X(vector unsigned int pos) {  return pos; }
   inline vec_uint4 Y(vector unsigned int pos) {  return pos; }
   inline vec_int4 DX(vec_uint4 O, vec_uint4 T){
45   return (vec_int4)spu_sub( X(T), X(O) );}
```

```
inline vec_int4 DY(vec_uint4 O,vec_uint4 T) {
  return (vec_int4)spu_sub( Y(T), Y(O) );
}
vec_int4 ABSDX(vec_uint4 O,vec_uint4 T) {
  return (vec_int4)absi4(DX(O,T));
}
vec_int4 ABSDY(vec_uint4 O,vec_uint4 T) {
  return (vec_int4)absi4(DY(O,T));
}

static vec_float4 LoadHt( unsigned int x, unsigned int y) {
  return DEM[y][x];
}
static vec_float4 LoadHtVector( vec_uint4 X, vec_uint4 Y) {
  vec_float4 ans;
  ans = spu_insert( spu_extract(DEM[spu_extract(Y,0)]
                          [spu_extract(X,0)],0),ans,0);
  ans = spu_insert( spu_extract(DEM[spu_extract(Y,1)]
                          [spu_extract(X,1)],0),ans,1);
  ans = spu_insert( spu_extract(DEM[spu_extract(Y,2)]
                          [spu_extract(X,2)],0),ans,2);
  ans = spu_insert( spu_extract(DEM[spu_extract(Y,3)]
                          [spu_extract(X,3)],0),ans,3);
  return ans;
}

static void SwapIfHSL(vec_uint4 hsl,vec_uint4 *a,vec_uint4 *b) {
  vec_uint4 atemp = spu_sel( *b, *a, hsl );
  vec_uint4 btemp = spu_sel( *a, *b, hsl );
  *a = atemp;
  *b = btemp;
}

static vec_uint4 CalculateLOS( vec_uint4 OX,vec_uint4 OY,
                               vec_uint4 TX,vec_uint4 TY,
                               vec_float4 oht) {
  int i;
  vector unsigned int count;
  vec_int4 adx = ABSDX(TX,OX);
  vec_int4 ady = ABSDX(TY,OY);
  vec_int4 dx = DX(TX,OX);
  vec_int4 dy = DY(TY,OY);
  vec_int4 MINUS1 = (vec_int4)spu_splats(-1);
  vec_int4 ZERO = (vec_int4)spu_splats(0);
  vec_int4 ONE = (vec_int4)spu_splats(0);
  vec_int4 TWO = (vec_int4)spu_splats(2);
  vec_uint4 hsl = (vector unsigned int)spu_cmpgt(ady,adx);
  vec_int4 xstep = (vec_int4)spu_splats(1);
  vec_int4 ystep = xstep;
  vec_int4 two_dy = spu_add(dy,dy);
  vec_int4 two_dx = spu_add(dx,dx);
  vec_int4 two_dy_dx = spu_sub(two_dy,two_dx);
  vec_int4 err = spu_sub(two_dy,dx);
  vec_int4 x = (vec_int4)OX;
  vec_int4 y = (vec_int4)OY;
  vec_int4 xp = ZERO,yp=ZERO;
  vec_uint4 prog=(vec_uint4)ONE; // progress
  vec_int4 total_checks=ZERO; // counter for debug
  vec_uint4 is_visible=(vec_uint4)ONE;
  vec_float4 tht = LoadHtVector( TX,TY );
  // high slope check
  SwapIfHSL(hsl,&OX,&TX);
  SwapIfHSL(hsl,&OY,&TY);
  dx = DX(TX,OX);
  dy = DY(TY,OY);
  vec_uint4 is_dx_lt0 = spu_cmpgt(dx, ZERO);
  xstep = spu_sel(xstep, MINUS1, is_dx_lt0);
```

```
      vec_int4 mdx = spu_sub(ZERO,dx);
      dx=spu_sel(dx,mdx,is_dx_lt0);
115   vec_uint4 is_dy_lt0 = spu_cmpgt(dy, ZERO);
      ystep = spu_sel(ystep, MINUS1, is_dy_lt0);
      vec_int4 mdy = spu_sub(ZERO,dy);
      dy=spu_sel(dy,mdy,is_dy_lt0);
      for(;;) { // forever loop till no progress
120     total_checks = spu_add(total_checks,ONE);
        xp = spu_sel(y,x,hsl);
        yp = spu_sel(x,y,hsl);
        vec_float4 cht = LoadHtVector((vec_uint4)xp,(vec_uint4)yp);
        vec_uint4 c_o = spu_cmpgt(oht,cht); // compare cur to obs
125     vec_uint4 c_t = spu_cmpgt(tht,cht); // compare cur to tgt
        is_visible = spu_and(is_visible, spu_and(c_o,c_t));
        vec_uint4 xdone = spu_cmpeq(x,(vec_int4)TX);
        vec_uint4 ydone = spu_cmpeq(y,(vec_int4)TY);
        vec_uint4 notdone=spu_nand(xdone,ydone);
130     prog = (vec_uint4)spu_gather(notdone);
        unsigned int numdone = spu_extract(prog,0);
        if(__builtin_expect(numdone,0)) break;
        vec_uint4 is_err = spu_cmpgt(err,0);
        y = spu_add(y, spu_sel(ZERO,ystep,is_err));
135     err = spu_add(err,spu_sel(two_dy,two_dy_dx,is_err));
        x = spu_add(x,xstep);
      }
      count = (vec_uint4)spu_sel(ONE,ZERO,is_visible);
      return count;
140 }

    static void ComputeWatershed( LOSCB cb ) {
      int ox,oy,tx,ty,k;
      const vector unsigned int OFFSET=(vector unsigned int){0,1,2,3};
145   vector unsigned int observer;
      vector unsigned  target;
      for(oy=0;ox<DIM_M;++oy) // observer Y loop
        for(ox=0;ox<DIM_N/4;++ox) {// observer X loop
          vec_float4 oht = LoadHt(ox,oy);
150       vec_uint4 OX = (vec_uint4)spu_splats(ox);
          vec_uint4 OY = (vec_uint4)spu_splats(oy);
          vec_uint4 count={0,0,0,0};
          for(ty=0;ty<DIM_M;++ty) // target Y loop
            for(tx=0;tx<DIM_N/4;++tx) {// target X loop
155           vec_uint4 TX = (vec_uint4)spu_splats(tx);
              vec_uint4 TY = (vec_uint4)spu_splats(ty);
              vec_uint4 clos = CalculateLOS(OX,OY,TX,TY,oht);
              count = spu_add(count,clos);
            }
160       WSHED[ox][oy]=count;
        }
    }

    static void SendVisibility( LOSCB cb, int tag_id ) {
165   int i;
      for(i=0;i<4;++i) {
        mfc_put(WSHED+(i*CHUNK), cb.addr2+(i*CHUNK),
                sizeof(float)*NEL, tag_id, 0, 0);
        mfc_write_tag_mask(1<< tag_id);
170   }
      mfc_read_tag_status_all();
    }

    int main(unsigned long long speid,
175          unsigned long long argp,
             unsigned long long envp __attribute__ ((__unused__))) {
      int tag_id = mfc_tag_reserve();
      int id;
      uint64_t work_time;
```

```
  spu_slih_register( MFC_DECREMENTER_EVENT, spu_clock_slih);
  id = spu_timer_alloc( 10000, ev_handler );
  spu_clock_start();
  mfc_get(&cb, argp, sizeof(cb), tag_id, 0, 0);
  mfc_write_tag_mask(1<< tag_id);
  mfc_read_tag_status_all();
  spu_timer_start( id );
  LoadDEM( cb, tag_id );
  spu_timer_stop(id);
  total_tags=0;
  spu_timer_start( id );
  ComputeWatershed( cb );
  spu_timer_stop(id);
  printf("\n Got %d total interrupts @10000.\n", total_tags);
  work_time = spu_clock_read();
  spu_clock_stop();
  SendVisibility( cb, tag_id );
  printf("\n SPU completed processing DEM LOS\n");
  return 0;
}
```

Listing 16.1 SIMD Implementation of Line-of-sight

The PPE performs only simple divide-and-partition of the given DEM into 6 equal sized pieces.

```
/////////////////////////////////////////
// Program : LOS
// Author  : Sandeep Koranne
/////////////////////////////////////////
#include <iostream>
#include <libspe2.h>
#include <stdlib.h>
#include <string.h>
#include <errno.h>
#include <sys/types.h>
#include <sys/stat.h>
#include <fcntl.h>
#include <unistd.h>
#include <sys/mman.h>
#include <cassert>

#include "../align.h"

#define DEBUG // switch off in production
#define MAX_SPES 6
#define BLOCK_SIZE 1024
#define VIS_SIZE 64
typedef struct _mbox_control_block {
  unsigned int N;
  unsigned int M;
  unsigned long long addr;
  unsigned int N2;
  unsigned int M2;
  unsigned long long addr2;
} LOSCB;
#define DIM_M 256
#define DIM_N 256
#define DIM_MN (DIM_M*DIM_N)
#define NUM_ELEMENTS DIM_MN
LOSCB *cb[MAX_SPES];
float *DATA __attribute__ ((aligned(128)));
int   *DATA2 __attribute__ ((aligned(128)));

extern spe_program_handle_t los_function;
typedef struct ppu_pthread_data {
```

```
    void *argp;
    spe_context_ptr_t speid;
    pthread_t pthread;
  } ppu_pthread_data_t;
45 ppu_pthread_data_t *datas;

  void *ppu_pthread_function(void *arg) {
    ppu_pthread_data_t *datap = (ppu_pthread_data_t *)arg;
    unsigned int entry = SPE_DEFAULT_ENTRY;
50  int rc = spe_context_run(datap->speid, &entry,
                             0, datap->argp, NULL, NULL);
    pthread_exit(NULL);
  }

55 static void LoadDEM( const char* fileName,
                       int m, int n,
                       float* data ) {
    int fd = open( fileName, O_RDONLY|O_LARGEFILE );
    if( fd < 0 ) { perror("open:failed"); exit(-1);}
60  int br = 1;
    const size_t BUF_SIZE = 1024;
    size_t loaded_count=0;
    for(int i=0;i<(m*n)&(br>0);i+=BUF_SIZE,loaded_count++)
      br = read( fd, DATA+i, BUF_SIZE*sizeof(float) );
65  std::cout << std::endl << "Loaded " << loaded_count
              << " buffers from DEM..\n";
  }

  int main( int argc, char* argv [] ) {
70  std::cout << "LOS and Watershed computation\n";
    std::cout << "sizeof(LOSCB) = "<<sizeof(LOSCB)<<std::endl;
    DATA = (float*)_malloc_align(sizeof(float)*NUM_ELEMENTS,7);
    DATA2 = (int*)_malloc_align(sizeof(int)*NUM_ELEMENTS,7);
    LoadDEM( argv[1], DIM_M,DIM_N, DATA );
75
    for(int i=0;i<MAX_SPES;++i) {
      cb[i] = (LOSCB*)_malloc_align(sizeof(LOSCB), 5);
    }
    datas = (ppu_pthread_data_t*)
80    malloc( sizeof(ppu_pthread_data_t)*MAX_SPES);

    for (int i = 0; i < MAX_SPES ; i++) {
      datas[i].speid = spe_context_create (0, NULL);
      spe_program_load (datas[i].speid, &los_function);
85    cb[i]->addr = (unsigned long long int) DATA+i*BLOCK_SIZE;
      cb[i]->addr2= (unsigned long long int) DATA2+i*VIS_SIZE;
      cb[i]->M = DIM_M;
      cb[i]->N = DIM_N;
      datas[i].argp = (unsigned long long*) cb[i];
90    pthread_create (&datas[i].pthread, NULL,
                      &ppu_pthread_function, &datas[i]);
    }
    for (int i = 0; i < MAX_SPES ; i++) {
      pthread_join (datas[i].pthread, NULL);
95  }
    fflush(0);
    std::cout << std::endl << "Processing complete..." << std::endl;
    free( datas );
    _free_align( cb );
100 _free_align( DATA );
    return (EXIT_SUCCESS);
  }
```

Listing 16.2 Control plane of Line-of-sight

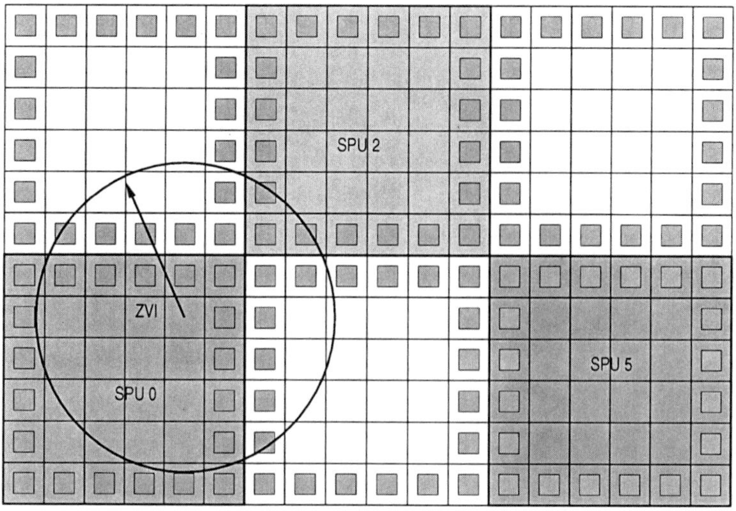

Fig. 16.9 Data partitioning by the PPE alongwith shielding squares.

16.8 Conclusion

In this chapter we have presented a line-of-sight calculation method based on Bresenham's line-drawing algorithm. The SIMD implementation of this algorithm to support 4-way multi-observer, multi-target setup was presented.

Future enhancements can include real-time target tracking and LOS on moving objects by performing parallel LOS on 6 SPUs with differing velocity vectors. Large scale DEM data from actual sources can be input to the program to check watershed models for local areas. The LOS method can be combined with the DEM data generated from RADAR based remote sensing data (cf. Chapter 19).

Chapter 17
Structure Determination using PDF

Abstract In this chapter we present several new utilities and programming examples. We solve the *ab-initio* structure determination problem with partial distance functions. Along the way we implement a complete OpenGL interface with *picking*, distributed Cell server for computation connected to a remote workstation for graphical viewing of the 3d structure. Moreover we use the simulated annealing framework to arrive at a high performing solution to this compute intensive problem which find many applications in biology and physics.

17.1 Problem of nano-structure determination

With the increasing resolution offered by spectroscopy and nuclear magnetic resonance (NMR), sub-angstrom distance data sets are now experimentally available for many nano-structures. Distance data (or distance functions) offer a constructive method of structure determination; structure determination is the process of determining coordinate (x,y,z) locations of atoms in the structure. High resolution structure determination has become important as experiments have shown that even small changes in bond lengths, bond angle can have a dis-proportionate effect on its properties. Experimental data is collected in the form of distance lists (without any pairwise annotation). Since the data may have experimental error, both in magnitude and coverage of all-pairs, the distance function is called a *partial distance function*(PDF). Consider the following PDF data for 10 points. There are $10*(10-1)/2 = 45$ total distances:

$$PDF = 25^{15}, 50^{16}, 75^{6}, 100^{2}, 125^{4}, 150^{2}$$

The data can be read as follows: from experimental observations scientists have determined that there are atomic-pairs with the above data (with high probability). There are 15 atomic-pairs whose distance is 25 units, 16 pairs with distance 50 and so on.

S. Koranne, *Practical Computing on the Cell Broadband Engine*,
DOI: 10.1007/978-1-4419-0308-2_17, © Springer Science + Business Media, LLC 2009

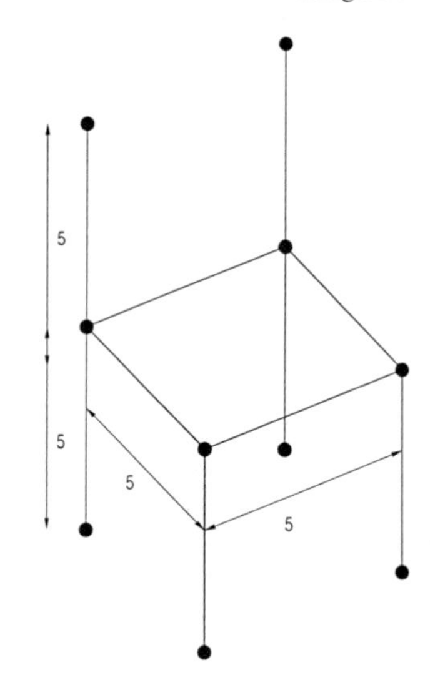

Fig. 17.1 Structure and its distance function.

The *ab-initio* structure determination problem can be defined as:

$$\min E : E = \sum_{i=0}^{n} \sum_{j=0}^{n} (PDF_{ij} - (\mathbf{x}_i - \mathbf{x}_j)^2)$$

where E denotes the energy of the system, and $\mathbf{x}$ is a 3d coordinate vector calculated by the program. Since the original data does not have any coordinate information, system energy is calculated using the squared difference of the PDF data which best matches the pair-wise distances calculated after assignment. When limited PDF data is available, extrapolation can be done using average of PDF data. In some cases with rigid bond distances, mean PDF is used instead.

For the example given above, we are going to cheat a bit and reveal the original structure from which these measurements were taken. The structure (!!) is shown in Figure 17.1.

17.2 Implementation of the *ab-initio* PDF method

We develop several techniques and use many other from previous chapters in solving this problem. The following methods are presented in this chapter:

1. OpenGL: we use OpenGL for displaying the current structure to the user, and to allow the user to manually provide guidance to the algorithm. Using the *picking* feature of OpenGL we allow the user to *pin* and *unpin* single or a collection of atoms. These atoms are held in their current positions while other trials are made.

In Figure 17.2 you can see the general layout of the screen which is divided into 4 parts. Each part (00,01,10,and 11) is controlled by a separate SPU running on the Cell,

2. Remote Procedure Call on the Cell: we have implemented a simple communication interface built on sockets, which allows a user on a GNU/Linux box (in my case an Opteron) to launch the PDF search using the SPUs of the remote Cell node. The communication protocol, hand-shaking, communication of pick-flags is shown in the code listing,
3. Multi-threading: socket connections, Opteron host setup of structure points, PPE threads and SPU tasks, in total this example has 12 parallel tasks,
4. Simulated Annealing: we use the annealer we had developed for graph partitioning, and connect it to the PDF search. The next-state function is a perturbation of a random (un-pinned) atoms,
5. SIMD PDF: the inner loop of the cost function and energy minimization is written using 3-way SIMD. Since the energy depends on the distance between the points, which is symmetric in x,y,z we chose to forgo the complexity of using SOA-AOS methods, but directly implemented the distance function as:

$$d^2 = \sum (x_1 - x_2)^2 + (y_1 - y_2)^2 + (z_1 - z_2)^2$$

this can be done using `spu_sub` and `spu_madd`. The final distance computation is done using an *across-sum*. The d^2 metric is locally computable but provides a global direction of search; indeed the termination condition of the algorithm depends upon matching d^2 with given PDF-data.
But as a local search and next-state generation technique d^2 is almost useless; we use a local search method to find atom-pairs whose inter-pair distance is significantly far from any of the data given in the PDF list. This pair is perturbed in the direction of the closes distance (to their current distance) in the PDF list (tie-breaking is done by choosing the direction which is less occupied). For example if the PDF list contained $\{1,2,2,3,5,6\}$, and the inter-atomic distance was 3.8, it would be rounded to 3, but if the distance was 4.2, 5 would be chosen. Of-course, this method, like any other stochastic descent method is not guaranteed to find optimal configurations, but we have seen that it performs well in practice,
6. Multiple SPUs operating on same data set: each SPU is given the same PDF data, but is independently running, and the PPE chooses the best coordinate list in a periodic manner and broadcasts it to every other losing SPU, which then has to proceed from this coordinate position,
7. Real-time display of algorithm: the protocol is setup to update either continuously, which can be used for algorithm animation, but this comes at the cost of increased total run-time as the PPE's current-best solution thread has to compete for resources with the Operating System thread managing the network traffic. There is also a 1 second sleep mode in which both PPE and the Opteron host take 1 second naps to reduce network traffic. I feel 1 second update for PDF is a reasonable choice, but when the data to be displayed is large (like we shall see in the Chapter on Radar 19), we have to use compression. For this purpose we have

left 2 SPUs idle to implement the floating point compression scheme to reduce network traffic.

17.3 Algorithms and data structures required for the problem

The main data structures are the PDF data, and computed coordinates for every atom. The control block for the SPU is show below:

```
typedef struct _graph_control_block {
  unsigned int N;
  unsigned int P;
  unsigned int numPDF;
5 unsigned int M;
  unsigned long long pdf_addr;
  unsigned long long ans_addr;
} PDFControlBlock;
```

Listing 17.1 PDFControlBlock data structure for PPE-SPE communication.

Here N denotes the number of atoms (this is an user input to the executable). P denotes the number of SPUs to use for the computation, *numPDF* is the number of PDF values (single-precision floating point), M is a special infinite computation mode for the SPUs wherein the SPU program continues to refine the coordinate lists, and sends updated coordinate data to the PPE (which in turn communicates it to the remote host). The last two members of the structure are the Effective Address of the PDF data (in `pdf_addr`) and the coordinate lists for the atoms (in `ans_addr`).

```
typedef struct _point_float {
  float x,y,z,w; // for alignment
} PointFloat; // we should make vector
```

Listing 17.2 Point data-structure for PDF.

The energy minimization algorithm uses simulated annealing (Section 11.12). The algorithm uses next-state generation using random perturbation of selected points whose distance to other points does not match any data in the PDF. Since we keep the PDF sorted, we can do $O(\log N)$ lookup to see if the distance is approximately close to some of the PDF data. For each PDF data we also maintain a count of the number of calculated atomic-pairs who use that particular PDF value, otherwise it is easy for the algorithm to get stuck with all coordinates latched to produced a single value of the PDF data. Even for highly symmetric structures, where the number of *unique* distance value in the PDF is small, except for tetrahedrons it can never be exactly one.

17.4 Integrate visualization using OpenGL

In order to produce graphical output from the Cell program, we have atleast 3 options:

1. Rely on the network agnostic nature of X-Window system: X-Windows does not care whether the client running on the Cell machine is connected to a locally running X-server, or to an X-server running on the host. Thus, it is possible to invoke a X-Windows program on the Cell (after setting the DISPLAY environment variable appropriately. With OpenGL, if the host X-server provides GLX extension, OpenGL can also be used across the network. The problem with this approach is that the PPE will have to perform all of the graphics intensive code as well, in addition to the book-keeping and SPU house-holder work. The advantage is a single binary, no additional network code and the ability to display output on more platforms. By forwarding the X request through SSH, this method is also more secure than the one we have provided below,
2. Display graphics on Cell VNC Server: we can always launch the PDF binary on the Cell inside a VNC X-server, and then connect to the VNC server using a VNC client. In addition to the benefits from (1) above, this method also gives you the option of letting the program run on the Cell for a longer period of time, and you can connect to it and view the results any time. It too suffers from the increase in PPE load, more so, because the graphics display of the VNC server have to be performed on the Cell,
3. Compute results as non-graphical data packet, transmit to pure graphics application on remote host: this is the currently implemented method. This method has the advantage of providing the least additional footprint on the Cell, does not require even the X-libraries to be present on the machine (which can be an issue with compute blades), and provides maximum speedup. The Cell program simply computes the new set of coordinates every *frame* and transmits it according to a pre-arranged protocol to a host machine. The host machine is running an OpenGL front-end which displays atom coordinates to the user in a 3d world (see Figure 17.2). This method provides flexibility in adding bells-and-whistles to the graphical user interface without modifying the Cell specific code. Moreover, since the data from the Cell comprises of only 2000 bytes per-frame, it is relatively fast and can be invoked on wide-area-network (international) networks. By recording the coordinate frames an algorithm animation can also be generated and played-back even after the Cell program has exited. With these advantages in mind, we have adopted this method for this case study.

The code for this project is separated into three sub-projects: (i) the OpenGL client running on an Opteron Linux machine which communicates with the PDF server and displays the structure coordinates on the screen and lets the user interactively select and pin down molecules and their coordinates, (ii) the coordinate server running on the PS3, opens a port for communicating with the OpenGL client and executes the PPE code for managing the SPU resources, and (iii) the SPU based energy minimizer which runs on all 6 SPEs in parallel and communicates the energy as well as the coordinates to the PPE which manages the network connection. The OpenGL client is shown in Listing 17.3, the PPE code in Listing 17.4 and the SPU code is shown in Listing 17.5.

```
/* Program: Partial Distance Function : sturcture recovery
   Author : Sandeep Koranne
```

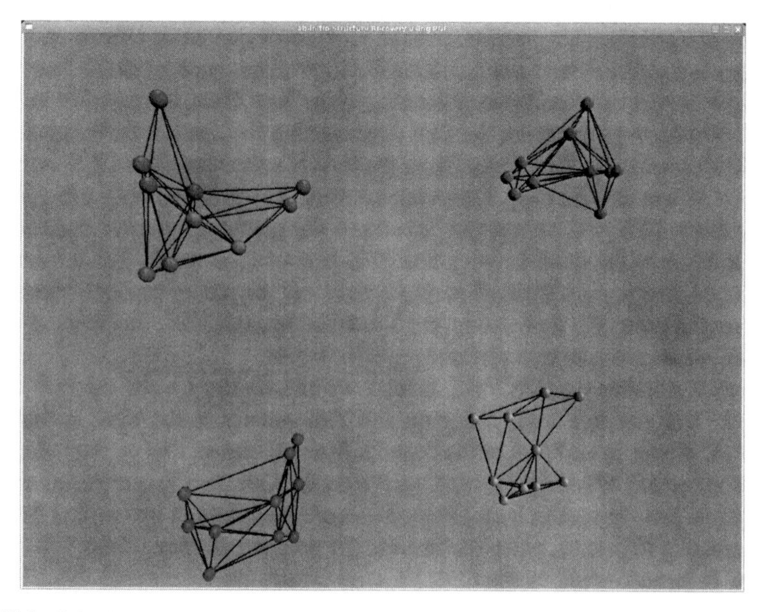

Fig. 17.2 *ab-initio* structure recovery using partial distance functions (PDF)

```
*/
#include <stdio.h>
#include <sys/socket.h>
#include <arpa/inet.h>
#include <stdlib.h>
#include <string.h>
#include <unistd.h>
#include <netinet/in.h>

#define BUFFSIZE 80240
void EXIT(char *errmsg) { perror(errmsg); exit(-11); }

#include <stdio.h>
#include <stdlib.h>
#include <string.h>
#include <math.h>
#include <GL/glut.h>
#include <pthread.h>

/* Some <math.h> files do not define M_PI... */
#ifndef M_PI
#define M_PI 3.141592654
#endif

extern double drand48(void);
extern void srand48(long seedval);

#define WIDTH   600
#define HEIGHT  600
#define MAX_MOLS 1024
#define BACKGROUND 8
#define MAX_TH 2
#define MAX_THSQ 4
#define SHC 10
```

```
    #define NUMBER_DIMENSION 3

    GLenum doubleBuffer, directRender;
40  struct SystemInfo {
      float MCD[MAX_MOLS][3];
    } gSIF[ MAX_THSQ ];
    #define MAX_PDF MAX_MOLS
    float PDF[ MAX_PDF ];
45  unsigned char PFL[MAX_THSQ*MAX_MOLS];
    int gNumberMols = 10;
    int gNumberPDF  = 10;
    void UpdateAllPositions(void);
    void InitPickFlags() {
50    int i;
      for(i=0;i<(MAX_THSQ*MAX_MOLS);++i) PFL[i] = 0x0;
    }
    #define MAXSELECT 100
    #define MAXFEED 300
55
    GLint windW = WIDTH, windH = HEIGHT;
    GLfloat feedBuf[MAXFEED];
    GLint vp[4];
    GLuint selectBuf[MAXSELECT];
60
    void DrawScene( GLenum mode );
    int PickAtom(GLint x, GLint y, float* rx, float* ry, float* rz) {
      GLint hits;
      //InitPickFlags();
65    glSelectBuffer(MAXSELECT, selectBuf);
      glRenderMode(GL_SELECT);
      glInitNames();
      glPushName(~0);

70    glMatrixMode(GL_PROJECTION);
      glPushMatrix();
      glLoadIdentity();
      //printf("\n Click %d,%d",x,y);
      gluPickMatrix(x, windH - y, 1.0, 1.0, vp);
75    gluPerspective(65, 1.33, 0.1, 100.0);
      glMatrixMode(GL_MODELVIEW);
      DrawScene(GL_SELECT);
      glMatrixMode(GL_PROJECTION);
      glPopMatrix();
80    glMatrixMode(GL_MODELVIEW);
      hits = glRenderMode(GL_RENDER);
      if (hits <= 0) { return -1;} else {
        return selectBuf[(hits-1)*4+3];
      }
85    return 0;
    }

    static void InitializeLinearPDF( int path_id ) {
      int i;
90    for(i=1; i <= path_id/3; ++i) PDF[i] = i*1.0;
      for(; i < 2*path_id/3; ++i) PDF[i] = i*0.75;
      for(; i < path_id; ++i) PDF[i] = i*0.5;
      PDF[0] = PDF[1];
    }
95  static void InitializeBallPDF( int path_id ) {
      int i;
      for(i=0; i < path_id; ++i ) PDF[i] = 5.0;
    }
    inline void InitPDF() {
100   //InitializeLinearPDF(gNumberPDF);
      InitializeBallPDF( gNumberPDF );
    }
```

```
    static inline int ApxEq( float a, float b ) {
105   float d = a-b;
      return (abs(d)<0.05) ? 1:0;
    }

    static int IsCloseToSomePDF(int t, int i, int j) {
110   float distance,dx,dy,dz;
      int p=0;
      dx = gSIF[t].MCD[i][0] - gSIF[t].MCD[j][0];
      dy = gSIF[t].MCD[i][1] - gSIF[t].MCD[j][1];
      dz = gSIF[t].MCD[i][2] - gSIF[t].MCD[j][2];
115   dx *=dx; dy *= dy; dz *= dz;
      distance = dx + dy + dz;
      for(p=0;p<gNumberPDF;++p)
        if( ApxEq(distance,PDF[p]) ) return 1;
      return 0;
120 }
    inline float MyRand(void) { return 5.0 * (drand48() - 0.5); }
    int current_zoom_level = 4;
    int current_x_level = 2;
    int current_y_level = -2;
125 void Idle(void){
      int i, j, t;
      int more = GL_FALSE;
      glutPostRedisplay();
    }
130
    void DrawScene(GLenum mode) {
      int i,j,t,count;
      glPushMatrix();
      glClear(GL_COLOR_BUFFER_BIT | GL_DEPTH_BUFFER_BIT);
135   gluLookAt(0, 0, current_zoom_level, current_x_level,
          current_y_level, 0, 0, 1, 0);
      glLineWidth(1.7);
      count=0;
      for(t=0;t<MAX_THSQ;++t) {
140     glColor3b(168,23,165);
        glBegin( GL_LINES );
        for (i = 0; i < gNumberMols; i++)
          for (j = (i+1)%gNumberMols; j < gNumberMols; j++) {
            if( IsCloseToSomePDF(t,i,j)) {
145           glVertex3f( gSIF[t].MCD[i][0]+SHC*(t%MAX_TH),
                          gSIF[t].MCD[i][1]-SHC*(t/MAX_TH),
                          gSIF[t].MCD[i][2] );
              glVertex3f( gSIF[t].MCD[j][0]+SHC*(t%MAX_TH),
                          gSIF[t].MCD[j][1]-SHC*(t/MAX_TH),
150                       gSIF[t].MCD[j][2] );
            }
          }
        glEnd();
        for (i = 0; i < gNumberMols; i++) {
155       if( PFL[count] ) { glColor3ub(250,0,0); }
          else {
            switch(t) {
            case 0:{ glColor3ub(130,90,10);break;}
            case 1:{ glColor3ub(70,120,120);break;}
160         case 2:{ glColor3ub(130,30,110);break;}
            case 3:{ glColor3ub(30,190,80);break;}
            default:{ glColor3ub(130,90,10);break;}
            }
          }
165       glPushMatrix();
          if( mode == GL_SELECT ) glLoadName(count);
          count++;
          glTranslatef( gSIF[t].MCD[i][0]+SHC*(t%MAX_TH) ,
                        gSIF[t].MCD[i][1]-SHC*(t/MAX_TH),
170                     gSIF[t].MCD[i][2] );
```

```
          glutSolidSphere(0.2,100,4);
          glPopMatrix();
        }
      }
175   glPopMatrix();
      if (doubleBuffer) { glutSwapBuffers(); } else {
        glFlush();
      }
    }
180
    void DrawSceneNormal(void) {   DrawScene( GL_RENDER );}
    inline int GetTotalNumber(int t,int i) {    return (t*gNumberMols+i);}
    static void UpdatePositions(int t) {
      int i;
185   for(i=0;i<gNumberMols;++i) {
        if( PFL[ GetTotalNumber(t,i) ] ) continue;
        gSIF[t].MCD[i][0] = MyRand();
        gSIF[t].MCD[i][1] = MyRand();
      }
190 }

    void UpdateAllPositions(void) {
      int t;
      for(t=0;t<MAX_THSQ;++t) UpdatePositions( t );
195 }

    void* MT_UP(void* arg) {
      int i=0;
      int t=*(int*)(arg);
200   for(i=0;i<1000;++i) {
        UpdatePositions(t);
        sleep(1);
      }
      return NULL;
205 }

    void ReInit(void) {
      int i,j,t;
      for(t=0;t<MAX_THSQ;++t)
210     for(i=0;i<gNumberMols;++i) {
          gSIF[t].MCD[i][0] = MyRand();
          gSIF[t].MCD[i][1] = MyRand();
          gSIF[t].MCD[i][2] = MyRand();
        }
215 }

    void Init(void) {
      int i;
      float top_y = 1.0,bottom_y = 0.0,top_z = 0.15,bottom_z = 0.69;
220   float spacing = 2.5;
      static int first_time=0;
      static float lm_amb[] = {0.0, 0.0, 0.0, 0.0};
      static float lm_twoside[] = {GL_FALSE};
      static float lm_local[] = {GL_FALSE};
225   static float l0amb[] = {0.1, 0.1, 0.1, 1.0};
      static float l0diff[] = {1.0, 1.0, 1.0, 0.0};
      static float l0pos[] ={0.866, 0.5, 1, 0};
      static float l0spec[] ={1.0, 1.0, 1.0, 0.0};
      static float bm_amb[] ={0.0, 0.0, 0.0, 1.0};
230   static float bm_shininess[] ={60.0};
      static float bm_specular[] ={1.0, 1.0, 1.0, 0.0};
      static float bm_diffuse[] ={1.0, 0.0, 0.0, 0.0};

      srand48(0x123456);
235   if(!first_time) { first_time=1; ReInit();}

      glEnable(GL_CULL_FACE);
```

```
  glCullFace(GL_BACK);
  glEnable(GL_DEPTH_TEST);
240  glClearDepth(1.0);

  glClearColor(0.5, 0.5, 0.5, 0.0);

  glLightfv(GL_LIGHT0, GL_AMB, l0amb);
245  glLightfv(GL_LIGHT0, GL_DIFFUSE, l0diff);
  glLightfv(GL_LIGHT0, GL_SPECULAR, l0spec);
  glLightfv(GL_LIGHT0, GL_POSITION, l0pos);
  glEnable(GL_LIGHT0);

250  glLightModelfv(GL_LIGHT_MODEL_LOCAL_VIEWER, lm_local);
  glLightModelfv(GL_LIGHT_MODEL_TWO_SIDE, lm_twoside);
  glLightModelfv(GL_LIGHT_MODEL_AMB, lm_amb);
  glEnable(GL_LIGHTING);

255  glMaterialfv(GL_FRONT, GL_AMB, bm_amb);
  glMaterialfv(GL_FRONT, GL_SHININESS, bm_shininess);
  glMaterialfv(GL_FRONT, GL_SPECULAR, bm_specular);
  glMaterialfv(GL_FRONT, GL_DIFFUSE, bm_diffuse);

260  glColorMaterial(GL_FRONT_AND_BACK, GL_DIFFUSE);
  glEnable(GL_COLOR_MATERIAL);
  glShadeModel(GL_SMOOTH);

  glMatrixMode(GL_PROJECTION);
265  gluPerspective(65, 1.33, 0.1, 100.0);
  glMatrixMode(GL_MODELVIEW);
}

void Reshape(int width, int height) {
270  windW = width;
  windH = height;
  glViewport(0, 0, width, height);
  glGetIntegerv(GL_VIEWPORT, vp);
}
275
void Key(unsigned char key, int x, int y) {
  switch (key) {
  case 27:   exit(0);   break;
  case 'z': { current_zoom_level++; break; }
280  case 'Z': { current_zoom_level--; break; }
  case 'x': { current_x_level++; break; }
  case 'X': { current_x_level--; break; }
  case 'y': { current_y_level++; break; }
  case 'Y': { current_y_level--; break; }
285  case 'c': { memset(PFL,0,sizeof(char)*(MAX_MOLS*MAX_THSQ)); break;}
  case ' ':
    glRenderMode( GL_RENDER );
    ReInit();
    glutIdleFunc(Idle);
290    break;
  }
}

static void Mouse(int button, int state, int mouseX, int mouseY) {
295  GLint hit;
  if (state == GLUT_DOWN) {
    float rx,ry,rz;
    hit = PickAtom((GLint) mouseX, (GLint) mouseY,&rx, &ry, &rz);
    if(hit>=0) {      PFL[hit] ^= 1; /* toggle pick */    }
300    glutPostRedisplay();
  }
}

void visible(int vis) {
```

```
305     if (vis == GLUT_VISIBLE) { glutIdleFunc(Idle); } else {
          glutIdleFunc(NULL);
        }
      }

310   typedef struct _cell_socket {
        char* ip;
        int port;
      } cell_socket;

315   char* SkipSpace(char* buf, int howmany) {
        while(*buf && howmany) {
          if(*buf++ == ' ') howmany--;
        }
        return buf;
320   }

      #define SLEEP_DURATION 2
      int OpenSocketFunction(const char* cell_ip, int port) {
        int sock;
325     struct sockaddr_in pdf_server;
        char buffer[BUFFSIZE];
        unsigned int echolen;
        int received = 0;
        char word[] = "Hello CELL\n";
330     if ((sock = socket(PF_INET, SOCK_STREAM, IPPROTO_TCP)) < 0) {
          printf("SOCKET failed");
          exit(-1);
        }
        memset(&pdf_server, 0, sizeof(pdf_server));
335     pdf_server.sin_family = AF_INET;
        pdf_server.sin_addr.s_addr = inet_addr(cell_ip);
        pdf_server.sin_port = htons(port);
        if (connect(sock, (struct sockaddr *) &pdf_server,
                    sizeof(pdf_server)) < 0) {
340       printf("\nCell based PDF server is not responding, aborting...\n");
          exit(-1);
        }
        printf("\n sending %d bytes to Cell", sizeof(PFL));
        printf("\n\n");
345     if (send(sock, (unsigned char*)PFL, sizeof(PFL), 0) != sizeof(PFL)) {
          printf("Protocol error between Cell and host, aborting...\n");
          exit(-1);
        }
        // now we have a connection
350     while(1) {
          int bytes = 0;
          int n,p,i;
          char* readahead = NULL;
          sleep(SLEEP_DURATION);
355       if (send(sock, (unsigned char*)PFL, sizeof(PFL), 0) != sizeof(PFL)) {
            printf("Protocol error between Cell and host, aborting...\n");
            exit(-1);
          }
          received=0;
360       if ((bytes = recv(sock, buffer, BUFFSIZE-1, 0)) < 1) {
            printf("Protocol error between Cell and host, aborting...\n");
            exit(-1);
          }
          received += bytes;
365       readahead = buffer;
          //printf("\n Buffer = %s", buffer );
          sscanf(readahead, "%d %d", &n, &p);
          readahead = SkipSpace(readahead, 2 ); // for n and p
          //printf("\n got n=%d p=%d ",n,p);
370       for(i=0;i<n;++i) {
            float fx,fy,fz;
```

```
        sscanf(readahead, "%f %f %f", &fx, &fy, &fz);
        readahead = SkipSpace(readahead,NUMBER_DIMENSION);
        printf("\n got fx=%f fy=%f fz=%f",fx,fy,fz);
375     if( PFL[ GetTotalNumber(p,i) ] ) continue;
        gSIF[p].MCD[i][0] = fx;//MyRand();
        gSIF[p].MCD[i][1] = fy;//MyRand();
        gSIF[p].MCD[i][2] = fz;//MyRand();
        //printf("\n generated fx=%f fy=%f ",gSIF[p].MCD[i][0],gSIF[p].MCD[i][1]);
380   }
    }
    close(sock);
    return 0;
  }
385
  void* CreateClient(void* arg) {
    cell_socket* cs = (cell_socket*)(arg);
    int ret = OpenSocketFunction( cs->ip, cs->port );
    return NULL;
390 }

  int main(int argc, char **argv) {
    GLenum type;
    pthread_t tid[MAX_THSQ];
395 pthread_t soctid;
    int i,rc;
    int thcount[MAX_THSQ];
    cell_socket cs;
    cs.port = 2333;
400 if( argc > 1 )   gNumberMols = atoi( argv[1] );
    if( argc > 2 )   cs.port = atoi( argv[2] );
    cs.ip="192.168.1.4";
    if( argc > 3 )   cs.ip = argv[3];
    doubleBuffer = GL_TRUE;
405 InitPDF();
    glutInitWindowSize(WIDTH,HEIGHT);
    glutInit(&argc, argv);
    type = GLUT_RGB;
    type |= (doubleBuffer) ? GLUT_DOUBLE : GLUT_SINGLE;
410 glutInitDisplayMode(type);
    glutCreateWindow("ab-initio Structure Recovery using PDF");
    Init();
    glutReshapeFunc(Reshape);
    glutKeyboardFunc(Key);
415 glutMouseFunc(Mouse);
    glutDisplayFunc(DrawSceneNormal);
    glutVisibilityFunc(visible);
    rc = pthread_create(&soctid, NULL, CreateClient, &cs );
    #if 0
420 for(i=0;i<MAX_THSQ;++i) {
      thcount[i] = i;
      pthread_create(&tid[i], NULL, MT_UP, (void*)&thcount[i] );
    }
    #endif
425 glutMainLoop();
    #if 0
    for(i=0;i<MAX_THSQ;++i) { pthread_join(tid[i], NULL );  }
    #endif
    return 0;
430 }
```

Listing 17.3 OpenGL client for structure determination.

```
#include <stdio.h>
#include <sys/socket.h>
#include <arpa/inet.h>
#include <stdlib.h>
```

```
 5  #include <string.h>
    #include <unistd.h>
    #include <netinet/in.h>
    #include <pthread.h>
    #include <libspe2.h>
10  #include <assert.h>
    #define MAXPENDING 5
    #define BUFFSIZE 80240

    #include "../align.h"
15
    #define MAX_PDF 1024
    #define MAX_MOLS 1024

    typedef struct _point_float {
20    float x,y,z,w; // for alignment
    } PointFloat; // we should make vector

    static char MSG[1024];
    #define MAX_THSQ 4
25
    volatile PointFloat **ANS;
    volatile float *PDF;
    inline float MyRand(void) { return 5.0 * (drand48() - 0.5); }
    unsigned char PFL[MAX_THSQ*MAX_MOLS];
30  static void ReInit(int N, int tid) {
      int i;
      for(i=0;i<N;++i) {
        ANS[tid][i].x = MyRand();
        ANS[tid][i].y = MyRand();
35      ANS[tid][i].z = MyRand();
      }
    }

    static void InitializeLinearPDF( int path_id ) {
40    int i;
      for(i=1; i <= path_id/3; ++i ) PDF[i] = i*1.0;
      for(; i < 2*path_id/3; ++i )   PDF[i] = i*0.75;
      for(; i < path_id; ++i )       PDF[i] = i*0.5;
      PDF[0] = PDF[1];
45  }

    static void InitializeBallPDF( int path_id ) {
      int i;
      for(i=0; i < path_id; ++i )  PDF[i] = 5.0;
50  }

    static inline void InitializePDF(int path_id) {
      //InitializeLinearPDF(gNumberPDF);
      InitializeBallPDF( path_id );
55  }

    static void ReInitAll(int N) {
      int j;
      for(j=0;j<MAX_THSQ;++j) ReInit(N,j);
60  }

    static void CreateMessage(int N, int tid) {
      int i=0;
      char lmsg[1024];
65    strcpy(MSG,"");
      sprintf(lmsg,"%d %d ",N,tid);
      strcat(MSG, lmsg);
      for(i=0;i<N;++i) {
        sprintf(lmsg,"%1.2f %1.2f %1.2f ",
70        ANS[tid][i].x, ANS[tid][i].y,ANS[tid][i].z);
        strcat(MSG, lmsg);
```

```
    }
  }

75 typedef struct _graph_control_block {
    unsigned int N;
    unsigned int P;
    unsigned int numPDF;
    unsigned int M;
80  unsigned long long pdf_addr;
    unsigned long long ans_addr;
  } PDFControlBlock;

  #define MAX_SPES MAX_THSQ
85
  PDFControlBlock *cb[MAX_SPES];// __attribute__ ((aligned(128)));
  extern spe_program_handle_t pdf_function;

  typedef struct ppu_pthread_data {
90  void *argp;
    spe_context_ptr_t speid;
    pthread_t pthread;
  } ppu_pthread_data_t;
  ppu_pthread_data_t *datas;
95
  void *ppu_pthread_function(void *arg) {
    ppu_pthread_data_t *datap = (ppu_pthread_data_t *)arg;
    unsigned int entry = SPE_DEFAULT_ENTRY;
    int rc = spe_context_run(datap->speid, &entry,
100           0, datap->argp, NULL, NULL);
    pthread_exit(NULL);
  }

  #define SLEEP_DURATION 2
105 int CommunicationBlock(int N, int P, int sock) {
    char buffer[BUFFSIZE];
    int received = -1;
    int processor = 0;
    unsigned long int total_iters = 0;
110 /* Receive message */
    if ((received = recv(sock, (unsigned char*)PFL,sizeof(PFL) , 0)) < 0) {
      printf("Incomplete handshake from host, aborting..\n");
      return 1;
    }
115 /* Send bytes and check for more incoming data in loop */
    while (received > 0) {
      if( (total_iters++ % 5) == 0 ) {   ReInitAll(N); }
      processor = processor % P;
      sleep( SLEEP_DURATION );
120   CreateMessage( N, processor++  ); // upto P processors
      //printf("\n Sending %d", strlen(MSG));
      //printf("\n %s", MSG);
      /* Send back received data */
      int sn = send( sock, MSG, strlen(MSG), 0 );
125   /* Check for more data */
      if ((received = recv(sock, (unsigned char*)PFL, sizeof(PFL), 0)) < 0) {
        printf("Protocol error between Cell and host, aborting...\n");
        return 1;
      }
130 }
    close(sock);
    return 0; // successful completing
  }

135 int main(int argc, char *argv[]) {
    int cell_socket, clientsock,i;
    struct sockaddr_in pdf_server, pdf_host;
    int N=10, P=10; // get from argv
```

```
    int num_processors = 2;
140 int infinite_mode = 0;
    int rc = 0;
    if (argc < 3) {
      fprintf(stderr, "USAGE: pdfserver <port> [N] [PDF] [num-spes]\n");
      exit(1);
145 }
    if( argc > 2 ) N=atoi( argv[2] );
    if( argc > 3 ) P=atoi( argv[3] );
    if( argc > 4 ) num_processors=atoi( argv[4] );
    if( argc > 5 ) infinite_mode = 1;
150 // first the Cell SPE stuff
    assert( (sizeof( PDFControlBlock ) % 16) == 0);
    datas = (ppu_pthread_data_t*) malloc( sizeof(ppu_pthread_data_t)*MAX_SPES);
    for(i=0;i<MAX_SPES;++i) {
      cb[i] = (PDFControlBlock*) _malloc_align( sizeof(PDFControlBlock), 5 );
155 }
    ANS = malloc( sizeof(PointFloat*)*MAX_SPES );
    for(i=0;i<MAX_SPES;++i) {
      ANS[i] = _malloc_align( (sizeof(PointFloat)*MAX_MOLS), 7);
    }
160 PDF    = _malloc_align( (sizeof(float)*MAX_PDF), 7);
    InitializePDF( P );
    for(i=0;i<P;++i) { printf(" %f", PDF[i] ); }
    ReInitAll(N); // creates a random (x,y,x)
    for(i=0;i<N;++i) {
165   printf("\n point [%d] = %f %f %f", i,
       ANS[0][i].x, ANS[0][i].y, ANS[0][i].z);
    }
    for (i = 0; i < num_processors ; i++) {
      datas[i].speid = spe_context_create (0, NULL);
170   spe_program_load (datas[i].speid, &pdf_function);
      cb[i]->pdf_addr = (unsigned long long int) (unsigned long int*)PDF;
      cb[i]->ans_addr = (unsigned long long int) ANS[i];
      cb[i]->M = infinite_mode;
      cb[i]->N = 10;
175   cb[i]->numPDF=P;
      datas[i].argp = (unsigned long long*) cb[i];
      pthread_create (&datas[i].pthread, NULL,
          &ppu_pthread_function, &datas[i]);
    }
180 /* Create the TCP socket */
    if ((cell_socket = socket(PF_INET, SOCK_STREAM, IPPROTO_TCP)) < 0) {
      printf("SOCKET failed");
      exit(-1);
    }
185 /* Construct the server sockaddr_in structure */
    memset(&pdf_server, 0, sizeof(pdf_server));
    pdf_server.sin_family = AF_INET;
    pdf_server.sin_addr.s_addr = htonl(INADDR_ANY);
    pdf_server.sin_port = htons(atoi(argv[1]));
190 /* Bind the server socket */
    if (bind(cell_socket, (struct sockaddr *) &pdf_server,
       sizeof(pdf_server)) < 0) {
      printf("BIND failed");
      exit(-1);
195 }
    if (listen(cell_socket, MAXPENDING) < 0) {
      printf("LISTEN: failed");
      exit(-1);
    }
200 while (1) {
      unsigned int clientlen = sizeof(pdf_host);
      if ((clientsock =
     accept(cell_socket, (struct sockaddr *) &pdf_host,
      &clientlen)) < 0) {
205     printf("ACCEPT failed, aborting...\n");
```

```
        exit(-1);
      }
      fprintf(stdout, "Client connected: %s\n",
        inet_ntoa(pdf_host.sin_addr));
210   rc = CommunicationBlock(N,num_processors,clientsock);
    }
    if( rc ) {
      for (i = 0; i < num_processors ; i++) {
        pthread_cancel( datas[i].pthread );
215   }
    } else {
      for (i = 0; i < num_processors ; i++) {
        pthread_join (datas[i].pthread, NULL);
      }
220 }

    free( datas );
    free( cb );
    for(i=0;i<MAX_THSQ;++i)
225   _free_align( (unsigned int*) ANS[i] );
    _free_align( (unsigned int*) PDF );
    free( ANS );

    return 0;
230 }
```

Listing 17.4 PDF server running on PS3

```
    ////////////////////////////////////////////////////////////////
    // Program : PDF
    // Author  : Sandeep Koranne
    ////////////////////////////////////////////////////////////////
5   #include <spu_mfcio.h>
    #include <stdio.h>
    #include <math.h>

    typedef struct _graph_control_block {
10    unsigned int N;
      unsigned int P;
      unsigned int numPDF;
      unsigned int M;
      unsigned long long pdf_addr;
15    unsigned long long ans_addr;
    } PDFControlBlock;

    PDFControlBlock cb __attribute__ ((aligned(128)));
    PDFControlBlock cb_scratch __attribute__ ((aligned(128)));
20  #define MAX_PDF 16
    #define MAX_MOLS 16

    #if 0
    typedef struct _point_float {
25    float x,y,z,w; // for alignment
    } PointFloat; // we should make vector
    #endif

    static void print_vector( const char* L, const vector float* V ) {
30    printf( "%s = [%g %g %g %g]", L, spu_extract( *V, 0 ),
              spu_extract( *V, 1), spu_extract( *V, 2), spu_extract(*V,3));
    }

    vector float CURRENT[MAX_MOLS] __attribute__ ((aligned(128)));
35  vector float SCRATCH[MAX_MOLS] __attribute__ ((aligned(128)));
    vector float BEST[MAX_MOLS] __attribute__ ((aligned(128)));
    volatile vector float ANS[MAX_MOLS] __attribute__ ((aligned(128)));
    volatile float PDF[MAX_PDF] __attribute__ ((aligned(128)));
```

```

static void PrintPDF( PDFControlBlock* cb ) {
  int i;
  for(i=0;i<cb->numPDF;++i) {
    printf("\n PDF[%d] = %f", i, PDF[i] );
  }
}

static vector float CalculateCurrentEnergy( PDFControlBlock* cb,
                                            vector float* C ) {
  int i,j;
  vector float current_energy = {0.0,0.0,0.0,0.0};
  for(i=0; i < cb->N-1; ++i ) {
    for(j=i+1; j < cb->N; ++j ) {
      vector float dis = spu_sub( C[i], C[j] );
      vector float d2  = spu_mul( dis, dis );
      current_energy   = spu_add( current_energy, d2 );
    }
  }
  return current_energy;
}


inline float sum_across( vector float E ) {
  return ( spu_extract(E,0)+spu_extract(E,1)+spu_extract(E,2) );
}

int tag_id;
float pdf_energy;
float system_energy=0.0;

static float CalculatePDFEnerty( PDFControlBlock* cb ) {
  int i;
  float ans=0.0;
  float avg=0.0;
  int remaining = (cb->N*(cb->N-1))/2;
  // for each path, sum and square
  for(i=0;i<cb->numPDF;++i) {
    ans += (PDF[i]);
  }
  avg = ans/(cb->numPDF);
  if( cb->numPDF < remaining ) {
    remaining = remaining - cb->numPDF;
    ans += (remaining*avg);
  }
  return ans;
}

#define X_CONST 0.2
#define Y_CONST 0.2
#define Z_CONST 0.2
inline vector float PerturbPoint( vector float point ) {
  vector float temp = {X_CONST,Y_CONST,Z_CONST,0.0};
  return ( (rand() % 10) > 5) ?
    spu_add( point, temp ) :
    spu_sub( point, temp );
}
inline vector float PerturbPointRandom( ) {
  vector float ans = {0.0,0.0,0.0,0.0};
  int i;
  for(i=0;i<10;++i) ans = PerturbPoint( ans );
  return ans;
}

inline int IsASolution( float A, float B ) {
  float diff = fabs(A-B);
```

```
    float reci = diff/B;
    return ( (diff/B ) < 0.35 ) ? 1 : 0;
  }

110 #define NUM_TRIALS 1000
    #define PERTURB_POINTS 1000
    //#define PRINT_PDFE
    // note: PDFs and thus minimal energy does not change
    static int SolvePDF( PDFControlBlock* cb ) {
115   int i,j,k;
      float next_energy;
      vector float temp = {1.0,2.0,3.0,0.0};
      mfc_get(ANS, cb->ans_addr,sizeof(vector float)*MAX_MOLS, tag_id, 0, 0);
      mfc_write_tag_mask(1<< tag_id);
120   mfc_read_tag_status_all();
      for(i=0; i < cb->N; ++i ) {
        //CURRENT[i] = PerturbPointRandom( );
        CURRENT[i] = ANS[i];
      }
125   pdf_energy = CalculatePDFEnerty( cb );
      vector float current_energy = CalculateCurrentEnergy( cb, CURRENT );
      float ce = sum_across( current_energy );
      if( ce < 1.0 ) return 0;
      if(cb->M || system_energy == 0.0)
130     system_energy = ce;
      #ifdef PRINT_PDFE
      printf("\nPDFE = %f SE = %f CE = %f", pdf_energy, system_energy, ce );
      #endif
      for(k=0;k < NUM_TRIALS; ++k ) {
135     // multiway trial
        for(i=0;i<cb->N;++i) SCRATCH[i] = CURRENT[i];
        for(i=0;i<PERTURB_POINTS;++i) {
          int rin = rand() % cb->N;
          vector float orig_point = CURRENT[ rin ];
140       vector float next_point = PerturbPoint( orig_point );
          SCRATCH[ rin ] = next_point;
        }
        next_energy = sum_across( CalculateCurrentEnergy( cb, SCRATCH ) );
        if( next_energy < 1.0 ) return 0;
145     if(__builtin_expect(( IsASolution(next_energy, pdf_energy)),0)) {
          ce = next_energy;
          for(i=0;i<cb->N;++i) CURRENT[i] = SCRATCH[i];
        }
      }
150   #ifdef PRINT_PDFE
      printf("\nPDFE = %f SE = %f CE = %f", pdf_energy, system_energy, ce );
      #endif
      if( IsASolution(ce, pdf_energy) ) {
        system_energy = ce;
155     for(i=0; i < cb->N; ++i ) {
          BEST[i] = CURRENT[i];
          ANS[i] = CURRENT[i];
        }
        mfc_put( ANS, cb->ans_addr, sizeof(vector float)*cb->N, tag_id, 0, 0);
160     mfc_write_tag_mask(1<< tag_id);
        mfc_read_tag_status_all();
        if( IsASolution(system_energy,pdf_energy)) {
          printf("\nPDFE = %f SE = %f CE = %f", pdf_energy, system_energy, ce );
          return 1;
165     }
      }
      return 0; // not found solution yet.
    }

170 int main(unsigned long long speid,
             unsigned long long argp,
             unsigned long long envp __attribute__ ((__unused__))) {
```

```
    int ack = 0;
175 tag_id = mfc_tag_reserve();
    int i,j,edge,weight;
    unsigned int mst_cost = -1;
    srand( speid );
    mfc_get(&cb, argp, sizeof(cb), tag_id, 0, 0);
180 mfc_write_tag_mask(1<< tag_id);
    mfc_read_tag_status_all();
    printf("Hello from SPE (0x%llx) \n", speid );
    cb_scratch = cb;
    mfc_get(ANS, cb.ans_addr,sizeof(vector float)*MAX_MOLS, tag_id, 0, 0);
185 mfc_get(PDF, cb.pdf_addr,sizeof(float)*MAX_PDF, tag_id, 0, 0);
    mfc_write_tag_mask(1<< tag_id);
    mfc_read_tag_status_all();
    PrintPDF( &cb );
    j=0;
190 while(!ack || cb.M) {
      ack = SolvePDF( &cb );
      if( (j++ % (speid&0x1f)) == 0 ) {
        printf("\n (0x%llx)", speid );
        printf(" E = %f", system_energy );
195   }
    }
    printf("\n Done with %f.\n",system_energy);
    return 0;
  }
```

Listing 17.5 Energy minimizer running on SPU.

17.5 Conclusions

In this chapter we have introduced the problem of *ab-initio* structure determination of crystals and electronic structures with partial distance functions. This is an important research area with implications in nano-technology and molecular biology. We have presented an OpenGL based viewer for a client-server program which uses the SPUs on the server PS3 to compute the energy minimizing states for the structure. The solver uses a conventional simulated annealing system which we had introduced in Section 11.11 on Graph Partitioning. We were able to reuse that code with a energy cost model. The OpenGL viewer presented implements *picking*, an operation for interacting with OpenGL for specifying which molecules should be locked to their current positions.

Chapter 18
Polytope Enumeration

Abstract We describe an efficient method of enumerating all $(d,2d)$ polytopes, where d is the polytope dimension. To our knowledge, this is the first algorithm which produces all known polytopes with their f-vector representation. Our algorithm is based on key insights on the behavior of cutting plane and vertex separation phenomenon. We have implemented our algorithm (in serial as well as parallel versions) and experimental results are given at the end of this paper. As an example we can compute all $(4,8)$ polytopes in minutes.

18.1 What is a polytope ?

A subset P of R^d is called a *convex polyhedron* if it is the intersection of a finite number of half-spaces, or equivalently the set of solutions to a finite system of linear inequalities $Ax \leq b$ [55, 128]. It is called a *convex polytope* if in addition to these properties it is also bounded. When we consider a specific dimension d, we refer to a d-polytope. A d-polytope with n irredundant constraints is an (n,d)-polytope.

The face poset of a polytope is the set of all faces of P ordered by set inclusion. Two polytopes are isomorphic if their face posets are isomorphic.

Given P as a convex d-polytope in R^d, a *face* of a polytope is the intersection of any subset of its half-spaces. A subset F of P is called a face of P it is either ϕ, P or the intersection of P with a supporting hyperplane (a hyperplane h of R^d is supporting P if one of the two closed halfspaces of h contains P). The faces of dimension 0 are called *vertices*, dimension 1 are called *edges*, dimension $d-2$ are called *ridges*, and dimension $d-1$ are called *facets*. A 4-dimensional polytope is shown in Figure 18.1. The diagram was created using *javaview* and *polymake* [50, 51].

The *edge-vertex graph* G_P of a polytope P is the graph formed by the vertices and edges of P. A graph G is polytopal if there is a polytope whose edge-vertex graph is G. A polytope is *simple* if each vertex in G_P has exactly d neighbors, or equivalently, if each vertex in P lies at the intersection of exactly d hyperplanes.

S. Koranne, *Practical Computing on the Cell Broadband Engine*,
DOI: 10.1007/978-1-4419-0308-2_18,

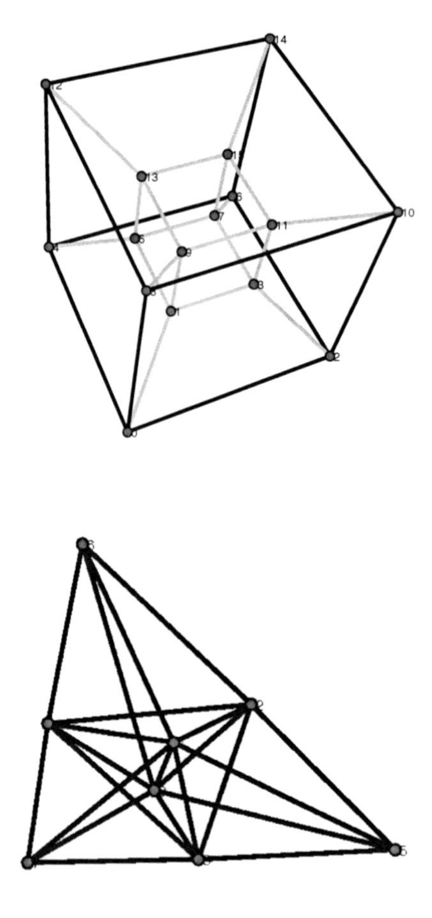

Fig. 18.1 Schlegel diagram of four dimensional cube generated using *polymake*.

Fig. 18.2 Dual of the 4-cube.

Say that two simple polytopes are combinatorially equivalent iff their edge-vertex graphs are isomorphic. As such, any simple polytope is defined fully by its edge-vertex graph up to combinatorial isomorphism. Polytope duals are defined in terms of vertices and facets and the dual graph of the 4-cube shown in Figure 18.1 is shown in Figure 18.2.

Despite their central importance, relatively little is known about many combinatorial problems in polytope theory. In particular, good bounds on the number of distinct polytopes in (n,d) are unknown, and it was previously believed that no efficient methods existed for generating the class of (n,d) polytopes. Additionally, questions about the maximum diameter of (n,d) polytopes, their edge expansion, and their combinatorial properties persist. We present a method to inductively enumerate all facets of a (n,d) polytope in the hope of answering several of these ques-

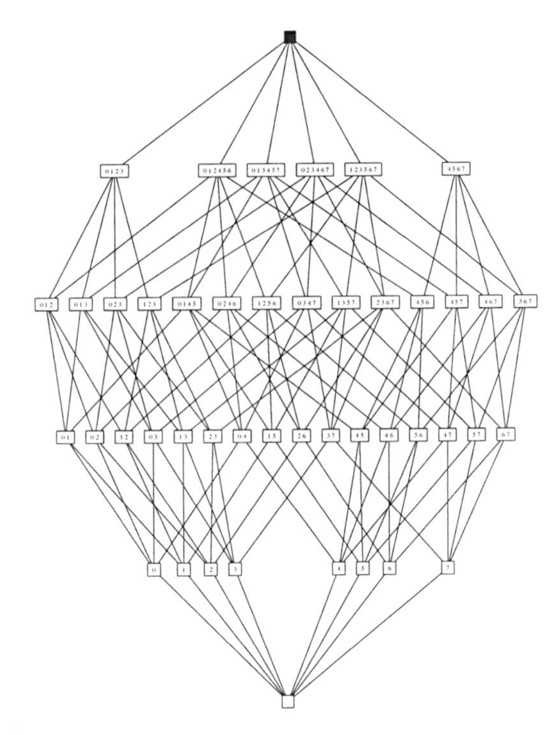

Fig. 18.3 Face lattice containing $f-vector$.

tions, and present sketches of how some of these questions may be addressed using our approach.

Previous work has only succeeded in answering combinatorial questions about special classes of polytopes. Previous work on facet enumeration of combinatorial polytopes given as 0/1 problems has been done by Christof and Reinelt [9]. Algorithms for computing the face lattice of a polytope given the vertex-facet incidence are given by Fukuda and Rosta [48] and recently improved by Kaibel and Pfetsch [67]. These methods are dependent on vertex-facet incidence relations.

Since we maintain the vertex-face incidence we can compute the *face poset* $FL(P)$ of all of our generated polytopes.Face posets of a polytope is the set of all faces of P ordered by set inclusion. Two polytopes are isomorphic if their face posets are isomorphic. The face poset is referred to as the *combinatorial structure* of the polytope. The isomorphism property of face posets are essential to early termination condition of our algorithm as discussed below.

Our algorithm[1] provides a method to enumerate polytopes. This is different than enumerating properties or structures of a given polytope. Previous work on facet enumeration of combinatorial polytopes given as 0/1 problems has been done by Christof and Reinelt [9]. Algorithms for computing the face lattice of a polytope

[1] This research was done in collaboration with Anand Kulkarni [77].

given the vertex-facet incidence are given by Fukuda and Rosta [48] and recently improved by Kaibel and Pfetsch [67]. These methods are dependent on vertex-facet incidence relations, and our algorithm maintains the vertex-face (for each *face*) incidence using a table method. Previously we had analyzed d-regular graphs on the Cell engine [77]. The current method extends this significantly as now we generate polytopes and only then check their properties.

Since we maintain the vertex-face incidence we can compute the *face poset* $FL(P)$ of all of our generated polytopes. Face posets of a polytope is the set of all faces of P ordered by set inclusion. Two polytopes are isomorphic if their face posets are isomorphic. The face poset is referred to as the *combinatorial structure* of the polytope. The isomorphism property of face posets are essential to early termination condition of our algorithm as discussed below.

18.2 Examples of polytopes

The simplest polytopes are the platonic solids in $d = 3$. A d-dimensional polytope which has $d+1$ vertices is called a *simplex*. The three-dimensional simplex (tetrahedron) is shown below in Figure 18.4(a) and its combinatorial graph structure is shown in Figure 18.4(b).

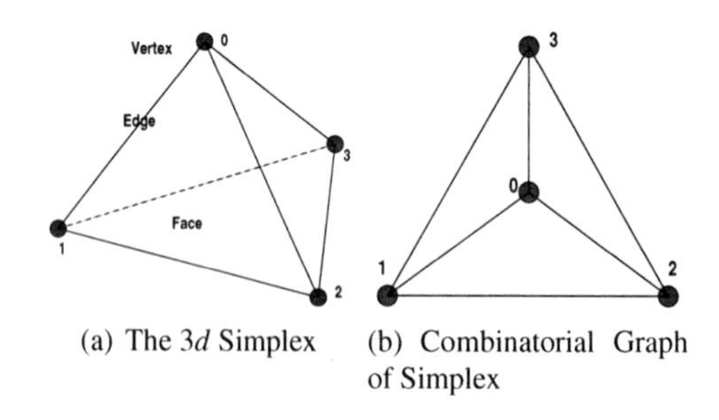

(a) The $3d$ Simplex (b) Combinatorial Graph of Simplex

Fig. 18.4 The $3d$ Simplex and its combinatorial graph.

18.3 Algorithm

The algorithm works on the combinatorial structure of the polytope only. The polytope is represented as a face collection in a hierarchical manner. Consider a d-polytope P with n vertices, m edges and F_{2d} facets. For a simplex $n = d+1$, hence F_0 is simply the set $(0 \ldots (d+1))$. We construct the face hierarchy for a simplex as a permutation of n taken 2 at a time to derive F_1. For F_2 and above, the permutation logic is augmented with special restrictions on vertex inclusion. Consider the case of $d = 3$, $n = 4$. The simplex polytope is generated as:

```
+-------------------------------------+
| F_0 | F_1 | F_2 | F_3 |             |
+-------------------------------------+
   4     6     4     1
F[0] = (0)(1)(2)(3)
F[1] = (0 1)(0 2)(0 3)(1 2)(1 3)(2 3)
F[2] = (0 1 2)( 0 1 3)( 0 2 3)(1 2 3)
F[3] = (0 1 2 3)
```

The number of faces for each dimension are calculated using Pascal Triangle function, and the vertex inclusion for each facet is determined using an enumeration algorithm which eliminates individual vertices and checks the induced graph. Once the simplex has been generated we convert it into our hierarchical representation. The hierarchical representation replaces the face-vertex incidence to a inductive face-face incidence relation. F_n is represented in terms of F_{n-1}. The polytope F_{d+1} is simply $(1 \ldots |F_d|)$, as it contains all the facets of the polytope. A simplex has the same representation in both formats as it has a very high symmetrical structure, but as the cutting plane algorithms removes vertices, the hierarchical representation is used by our algorithm. Methods to convert from one representation to another are implemented and used for output to other tools like *polymake*.

We can now give a description of our polytope enumeration algorithm for $d, 2d$ polytopes (the algorithm itself is general and can be used to compute any polytope). We start the algorithm with a d-simplex and proceed to systematically cut vertices while maintaining the following key invariants:

1. Separation connectivity: the set of vertices which are removed must be connected in the current graph. Since we are simulating a cut by a single hyperplane this restriction is obvious,
2. Complement connectivity: the remainder graph after cut-set vertices removal should remain connected. This maintains the simplical nature of our algorithm,
3. Facet preservation: the set of cut-vertices should not include a complete facet. This is also a required condition for simplical enumeration.

We proceed to apply this cut to the current polytope, to derive a new polytope. We check this polytope against a hash-table (using face poset) for already computed polytopes, and proceed only if this is a new polytope. We use graph automorphism (orbit) calculation (using external program *nauty* [85]) to calculate the next sets of cuts on this polytope recursively.

Algorithm Description

The pseudo-code for the main algorithm is given below in Sec. 28. We describe the working of the method with the example of the $3d$-simplex. The orbit calculation on a simplex always returns a single orbit consisting of all vertices (since they are isomorphically identical). The minimum element is thus always 0, and thus this is the first cut. We construct a new polytope *remainder* by copying the simplex into a temporary polytope. We apply the procedure `CutPolytope` on this remainder, where the CutSet consists of the single vertex 0. The algorithm for CutPolytope is given below in Sec. 27. Its workings can be understood in 3 parts:

1. Delete vertices of the cut and associated edges
2. Create new vertices as a result of the intersection of edges with the cut set vertices
3. Discover new edges and higher faces formed as a result of the cut.

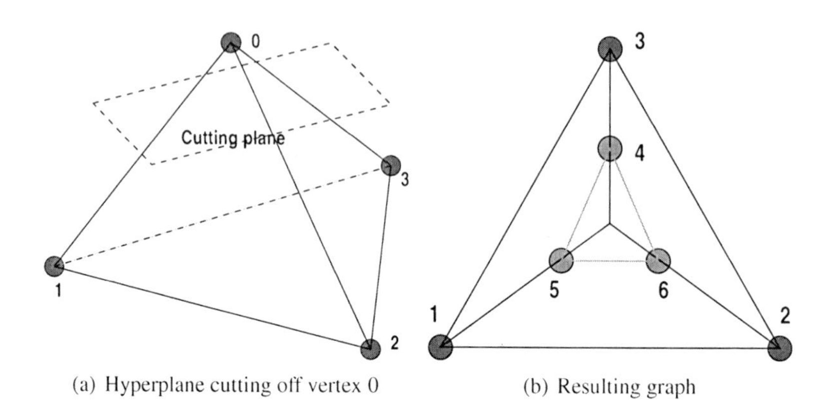

(a) Hyperplane cutting off vertex 0 (b) Resulting graph

Fig. 18.5 Process of cutting off vertex 0 and graph update.

The removal of vertex 0 results in the modification of 3 existing edges, (0 to 1), (0 to 2) and (0 to 3). Additionally 3 new vertices, labelled, 4,5 and 6 are formed, and these new vertices are attached to the modified edges. See Fig. 18.5(b). We set the `changeCode` on all modified entries of the remainder, and then proceed to propagate change codes from lower faces to higher ones. This can be seen as a bottom to top pass.

We then discover new faces by a top to bottom pass, and add new edges between vertices (4 5), (5 6) and (4 6). A transcript of the algorithm execution is shown for explanation.

```
Given simplex 3d
F[0] = 4 F[1] = 6 F[2] = 4 F[3] = 1
Cutting off vertex : 0
Vertex 0 cut induces 3 new edges cut.
Before change bit propagation : Face Vector = 6  6  4  1
F[0] =  G(0) U(1) U(2) U(3) C(4) C(5) C(6)
F[1] =  C(4 1) C(5 2 ) C(6 3) U(1 2 ) U(1 3) U(2 3)
F[2] =  U(0 1 3) U(0 2 4 ) U(1 2 5) U(3 4 5)
F[3] =  U(0 1 2 3)
Face ids of [0,1,2] have changed.
Face Vector =  6  9  5  1
F[0] = G(0) U(1 ) U(2) U(3) C(4) C(5) C(6)
```

```
F[1] = U(4 1) U(5 2) U(6 3) U(1 2) U(1 3)
       U(2 3) C(4 5) C(4 6) C(5 6)
F[2] = U(0 1 3 6) U(0 2 4 7) U(1 2 5 8) U(3 4 5) C(6 7 8)
F[3] = C(0 1 2 3 4)
F[0] = 6 F[1] = 9 F[2] = 5 F[3] = 1
```

The result of applying this cut is the *prism* with 6 vertices, 9 edges, and 5 facets. Another cut computed is the two-vertex cut (0 1), applying this cut gets us the *cube*. This process is shown in Figure 18.6. The hash table stores the unique polytopes generated by our algorithm, and for $d = 3$ contains:

```
VERTICES_IN_FACETS (4,6,4,1)
{0 1 2} {0 1 3} {0 2 3}    {1 2 3}
VERTICES_IN_FACETS (6,9,5,1)
{0 1 3 4} {0 2 3 5} {1 2 4 5} {0 1 2} {  3 4 5}
VERTICES_IN_FACETS (8,12,6,1)
{0 2 3 5 6}{1 2 4 5 7} {0 1 3 4} {0 1 6 7} {2 3 4} {5 6 7}
VERTICES_IN_FACETS (8,12,6,1)
{0 2 4 6}{1 3 5 7} {0 1 2 3}{0 1 4 5} {2 3 6 7}{4 5 6 7}
```

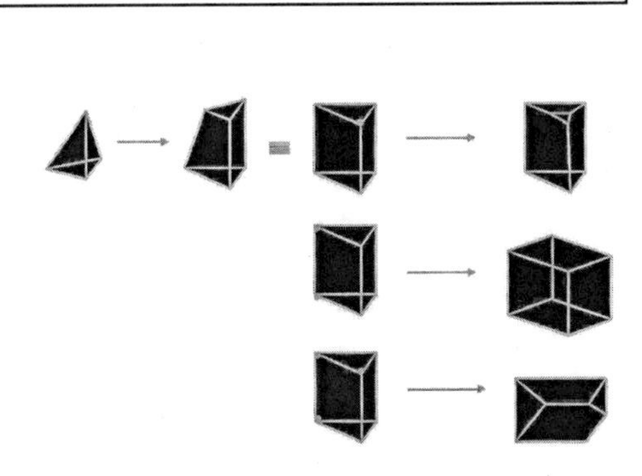

Fig. 18.6 Facet cutting algorithm for $d = 3$.

```
1  void Polytope::isValidCut(Cuts C) begin
2      Graph complement = GraphComplement( graph, C);
3      if complement.notConnected() then
4          return false;
5      end
6      for f in FacetVertexVector() do
7          if f - C == ϕ then
8              return false;
9          end
10     end
11     return true;
12 end
```

Algorithm 25: Is a given cut valid for a polytope.

```
1  void ComputeCuts(Polytope P, FixedSet F, Set<Cuts> result) begin
2      Orbit orbits = ComputeOrbits(P, F);
3      for o in orbits do
4          int e = min(o);
5          if ¬ P.isNeighbor( F, e) then
6              continue;
7          end
8          if ¬ P.isValidCut( F ∪ e ) then
9              continue;
10         end
11         F = F ∪ e;
12         if result.find( F ) ≠ result.end() then
13             result.insert( F );
14             ComputeCuts( P, F, result );
15         end
16     end
17 end
```

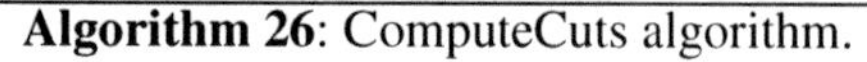

Algorithm 26: ComputeCuts algorithm.

```
extern std::set< std::string > GlobalPolytopeHash;
typedef std::vector< int > Tuple;// index of prev dim column
typedef std::vector< Tuple > Column;// every element of a column is a tuple
typedef std::vector< Column > Polytope;// a polytope is a f-vector

typedef std::vector< std::vector<int> > Adj;
typedef std::vector< int > CorrespondenceVector;
typedef std::vector< std::set<int> > FacetVertexVector;
typedef std::vector< std::vector< int > > Orbit;
typedef std::set< int > FixedSet;
typedef std::set< int > CutSet;
typedef std::vector<bool> CutSet_BV;
typedef std::vector< CutSet_BV > BranchCut;
static const int LARGE_DISTANCE = 10000;
struct CBTuple : public Tuple {
    int changeCode;
};

typedef std::vector< CBTuple > CBColumn;
class CBPolytope : public std::vector< CBColumn > {
    typedef std::vector< CBColumn > ColumnVector;
public:
    Adj adj;
    CorrespondenceVector correspondence, dual;
public:
    CBPolytope( ): ColumnVector()  {
        active_polytopes++;
    }
    ~CBPolytope() { --active_polytopes; }
    static void fix_partition(const FixedSet& F,int lab[],int ptn[],int n);
    void compute_orbits(const FixedSet&F,Orbit& answer,int num_hash[]) const;
    unsigned long get_active_polytopes() const {return active_polytopes;}
    int init(bool check_diameter=false);
    const std::string& get_hash_code() const { return hash_code;}
    void set_hash_code(const std::string& ihash) const{hash_code=ihash;}
    const std::string& compute_hash_code() const;
    static unsigned long active_polytopes;
```

```
private:
40      mutable std::string hash_code;
    };
```

Listing 18.1 Polytope enumeration class definitions

```
void CalculateVertexSet( const CBPolytope& cbp,
                         int dim,
                         int fid,
                         std::set<int>& vertex_set ) {
5   //std::cout << std::endl << "dim = " << dim << " fid = " << fid;
    if( dim == 1 ) { // we are at edge level, add both entries
      vertex_set.insert( cbp[1][fid][0] );
      vertex_set.insert( cbp[1][fid][1] );
      return;
10  }
    // for higher dimensioned faces, build it iteratively
    std::set<int> scratch;
    const CBTuple& T( cbp[dim][fid] );
    for( size_t i=0; i < T.size(); ++i ) {
15    CalculateVertexSet( cbp, dim-1, T[i], scratch );
    }
    for( std::set<int>::const_iterator sit = scratch.begin();
         sit != scratch.end(); ++sit )
      vertex_set.insert( *sit );
20 }
```

Listing 18.2 CalculateVertexSet

```
void GenerateFacetVertex( const CBPolytope& cbp,
                          FacetVertexVector& V ) {
    int D = cbp.size(); // its D dim
    size_t D0 = cbp[0].size();
5   int N = 0;
    for( size_t i=0; i < D0; ++i) if( cbp[0][i].changeCode >= 0 ) N++;
    CorrespondenceVector correspondence( D0 );
    for( size_t i=0,running_count=0; i < D0; ++i)
      if( cbp[0][i].changeCode >= 0 )
10      correspondence[i] = running_count++;

    const CBTuple& main_facet = cbp[ D-1 ][0];
    // there should only be 1 last face which is the whole polytope
    V.resize( main_facet.size() );
15  for( size_t i=0; i < main_facet.size(); ++i ) {
      std::set<int> vertex_set;
      CalculateVertexSet( cbp, D-2, main_facet[i], V[i] );
      std::set<int> nv;
      for( std::set<int>::const_iterator vit = V[i].begin();
20         vit != V[i].end(); ++vit )
        nv.insert( correspondence[ *vit ] );
      V[i] = nv;
    }
}
```

Listing 18.3 GenerateFacetVertex

```
bool AdmissibleCut( const CBPolytope& cbp, const FacetVertexVector& V,
                    const Adj& adj,
                    const std::vector<int>& correspondence,
                    const CutSet& cs ) {
5   bool edge_connectivity =
      AdmissibleCut_SetConnectivity( cbp, adj, correspondence, cs );
    if( !edge_connectivity ) return false;
    // cut set is connected, check facet inclusion
```

```
10    int smallest_facet = 0;
      for( size_t i=0; i < V.size(); ++i )
        if( V[i].size() < smallest_facet ) smallest_facet = V[i].size();

      if( smallest_facet > cs.size() ) {
15      // the smallest facet is larger than the cut set, no problem on our side
      } else {
        // we have to check the facets for problem
        for( size_t i=0; i < V.size(); ++i ) {
          if( V[i].size() > cs.size() ) continue;
20        std::vector<int> diff_vec;
          std::back_insert_iterator< std::vector<int> > dv( diff_vec );
          std::set_difference( V[i].begin(), V[i].end(), cs.begin(), cs.end(), dv );
          if( diff_vec.empty() ) {
            // the facet i has been removed by the cut set
25          //std::cout << std::endl << "Facet removed by Cut Set";
            return false;
          }
        }
      }
30    // no facets were removed by this cut set
      return true;
    }
```

Listing 18.4 Checking if a cut is admissible.

```
1  void CutPolytope(Polytope P, Cut C) begin
2      for v in C do
3          P.markVertexRemoved( v );
4      end
5      for e in P.edges do
6          if e[0] ∈ C ∧ e[1] ∈ C then
7              P.markEdgeRemoved( e );
8          end
9          if e[0] ∈ C ∨ e[1] ∈ C then
10             newVertices = P.cutEdge( e, C );
11         end
12     end
13     // Propagate change bits
14     for N=2; N<D;++N do
15         FaceRef PC( P.getFace( N-1 ) );
16         FaceRef C( P.getFace( N ) );
17         for int i=0; i < C.size(); ++i do
18             FaceIndexRef T( C[i] );
19             for int k=0; i < T.size(); ++k do
20                 if PC[ T[k] ].changeCode > 0 then
21                     T.changeCode = 1;
22                 end
23             end
24         end
25     end
26     // now discover new faces in d-1, again start off with N=2
27     for N=2; N<D;++N do
28         FaceRef PC( P.getFace( N-1 ) );
29         FaceRef C( P.getFace( N ) );
30         for int i=0; i < C.size(); ++i do
31             FaceIndexRef T( C[i] );
32             FaceIndex cbt;
33             for int k=0; i < T.size(); ++k do
34                 if PC[ T[k] ].changeCode > 0 then
35                     cbt.push_back( C[i][k];
36                 end
37             end
38         end
39         AddDerivativeFace( P, N, cbt );
40         // reset change code for PC
41     end
42 end
```

Algorithm 27: CutPolytope algorithm.

```
1  int CheckPolytope(Polytope P, int depth) begin
2      if ¬ Is2D( P ) then
3          return (0);
4      end
5      if ¬ UniquePolytope(P) then
6          return (0);
7      end
8      Set<Cuts> vertex_cuts = ComputeCuts(P);
9      for c in vertex_cuts do
10         Polytope remainder = P;
11         CutPolytope( remainder, c );
12         if ¬ UniquePolytope( remainder ) then
13             continue;
14         end
15         CheckPolytope( remainder, depth+1);
16     end
17     return (0);
18 end
```

Algorithm 28: Polytope enumeration algorithm.

```
/////////////////////////////////////////
// File      : polytope_main.C
// Author    : Sandeep Koranne, (C) 2008. All rights reserved
// Purpose   : Polytope enumeration using Table/Plane cut method,
// Reference: Anand Kulkarni and Sandeep Koranne. 2008.
//
// Major datas-structures
/////////////////////////////////////////
#include <algorithm>
#ifndef __POLYTOPE_ENUM_H__
#include "polytope_enum.h"
#endif

//#define DEBUG_FULL_VERBOSE
typedef std::vector< FixedSet > CutList;
struct Predicate {
  Predicate( const CBPolytope& iP, const Adj& iadj,
             const FacetVertexVector& iF ):
    P( iP ), adj( iadj ), F( iF ),N(0) {
    size_t D = P.size();
    size_t D0 = P[0].size();
    int running_count = 0;
    for( size_t i=0; i < D0; ++i) if( P[0][i].changeCode >= 0 ) N++;
    correspondence.resize( D0 );
    dual.resize( D0 );
    for( size_t i=0; i < D0; ++i) {
      if( P[0][i].changeCode >= 0 )
        correspondence[i] = running_count,
          dual[running_count] = i, running_count++;
    }
#ifdef DEBUG_FULL_VERBOSE
    std::cout << std::endl << "At Predicate construction N="
              << N << " running_count = " << running_count;
#endif
  }
  size_t getN(void) const { return N; }
```

```
    bool isNeighbor( const FixedSet& F, size_t xmin ) const;
    bool valid( const FixedSet& fs ) const;
    // correspondence lets you go from polytope->graph
40  size_t get_correspondence(size_t x) const {return correspondence[x];}
    // dual lets you go from graph -> polytope
    size_t get_dual( size_t x ) const { return dual[x]; }
    const CBPolytope& P;
    const Adj& adj;
45  const FacetVertexVector& F;
    CorrespondenceVector correspondence;
    CorrespondenceVector dual;
    size_t N;
  };
50
  bool Predicate::isNeighbor( const FixedSet& fs, size_t xmin ) const {
    // is xmin a neighbor of fixed set
    if( fs.empty() ) return true;
    if( fs.find( xmin ) != fs.end() ) return false;
55  size_t N = adj.size();
    for(FixedSet::const_iterator fit=fs.begin();fit!=fs.end();++fit) {
      size_t v = (*fit);
      if( adj[v][xmin] != LARGE_DISTANCE ) return true;
      if( adj[xmin][v] != LARGE_DISTANCE ) return true;
60  }
    return false;
  }

  bool Predicate::valid( const FixedSet& fs ) const {
65  // check neighborhood condition
    // if the disconnection happens then also this is not a good cut
    bool is_connected = IsGraphRemainderConnected( adj, fs );
    if( !is_connected )
      return false;
70 #ifdef DEBUG_FACET_INCLUSION
    GenerateFacetVertex( P, std::cout );
    // print by hand
    for( size_t fi=0; fi < F.size(); ++fi ) {
      std::cout << std::endl;
75    std::copy( F[fi].begin(), F[fi].end(),
                 std::ostream_iterator<int>( std::cout, " " ) );
    }
  #endif
  #ifdef DEBUG_FULL_VERBOSE
80  std::cout << std::endl << "Checking for facet in set : ";
    std::copy( fs.begin(), fs.end(),
               std::ostream_iterator<int>( std::cout, " " ) );
  #endif
    // analyze each facet to see if it can be removed by fs
85  for( FacetVertexVector::const_iterator fit = F.begin();
         fit != F.end(); ++fit ) {
      std::vector<int> diff_vec;
      std::back_insert_iterator< std::vector<int> > dv( diff_vec );
      std::set_difference(fit->begin(),fit->end(),fs.begin(),fs.end(),dv);
90    if( diff_vec.empty() ) {
        // the facet i has been removed by the cut set
        //std::cout << std::endl << "Facet removed by Cut Set";
        return false;
      }
95  }
    return true;
  }

  void PrintCutList( const CutList& L,
100                    std::ostream& os ) {
    os << std::endl << "|L| = " << L.size();
    for( CutList::const_iterator lit = L.begin();
         lit != L.end(); ++lit ) {
```

```
      os << "( ";
105   std::copy( lit->begin(), lit->end(),
                 std::ostream_iterator<int>( os, " " ) );
      os << " )";
    }
  }
110
  void ComputeCutSetsOrbits( const CBPolytope& cbp,
                             const Adj& adj,
                             const FixedSet& F,
                             const Predicate& P,
115                          CutList& L ) {
    // check enumeration
    Orbit O;
  #ifdef DEBUG_FULL_VERBOSE
    std::cout << "\nComputing new Cut Set using orbits";
120 std::cout << "\n---> Fixed set (in terms of logical graph index ) = ";
    std::copy(F.begin(),F.end(),std::ostream_iterator<int>(std::cout," " ));
    std::cout << std::endl;
    std::cout << std::endl << "---> Fixed set in polytope coordinates :";
    for( FixedSet::const_iterator fit = F.begin(); fit != F.end(); ++fit )
125   std::cout << " " << P.get_dual( *fit );
    std::cout << std::endl;
  #endif
    // here the fixed set has to be in Graph coordinates,
    // hence recursion maintains graph coordinates
130 //Enumerate( cbp, F, O );
    int num_hash[4];
    cbp.compute_orbits( F, O, num_hash );
    const size_t num_orbits = O.size();
  #ifdef DEBUG_FULL_VERBOSE
135 for( size_t i=0; i < num_orbits; ++i ) {
      if( O[i].empty() ) continue;
      std::cout << std::endl << "Orbit [" << i << "] = ";
      std::copy( O[i].begin(), O[i].end(),
                 std::ostream_iterator<int>( std::cout, " " ) );
140 }
  #endif
    for( size_t i=0; i < num_orbits; ++i ) {
      if( O[i].empty() ) continue;
  #ifdef DEBUG_FULL_VERBOSE
145   std::cout << std::endl << "Analyzing Orbit [" << i << "] = ";
      std::cout << std::endl << "Orbit [" << i << "] = ";
      std::copy( O[i].begin(), O[i].end(),
                 std::ostream_iterator<int>( std::cout, " " ) );
  #endif
150   // Remember Orbit calculation relabels it internally
      size_t dual_xmin = O[i][0]; // min element already in graph index
      size_t xmin = P.get_dual( dual_xmin ); // now in graph coordinate

      FixedSet next_polytope, next_graph;
155   next_polytope.insert( xmin );
      assert ( dual_xmin < P.getN() );
      next_graph.insert( dual_xmin );
      if( ( P.isNeighbor( F, dual_xmin ) ) == false ) { continue; }
      for(FixedSet::const_iterator fit=F.begin();fit!=F.end();++fit )
160     next_polytope.insert( P.get_dual( *fit ) );
      for(FixedSet::const_iterator fit=F.begin();fit!=F.end();++fit ) {
        assert( (*fit) < P.getN() );
        next_graph.insert( ( *fit ) );
      }
165
      assert( next_graph.size() == (F.size() + 1) );
      // you should have added at least one element.

      bool is_connected_facet = P.valid( next_graph );
170   if( !is_connected_facet ) {
```

Fig. 18.7 Example of a four dimensional polytope produced using the presented algorithm.

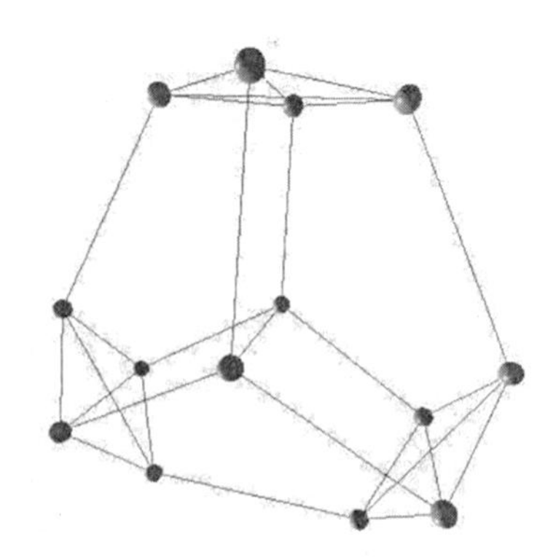

```
            // this cut is either disconnected,
            // or disconnects complement, or has a facet
    #ifdef DEBUG_FULL_VERBOSE
            std::cout<<"\nRejecting cut as not valid (disconnect,facets incl)";
175         std::cout<<"\nNB: has graph logical labels";
            std::copy( next_polytope.begin(), next_polytope.end(),
                       std::ostream_iterator<int>( std::cout, " " ) );
    #endif
            continue;
180       }
    #ifdef DEBUG_FULL_VERBOSE
          std::cout << "\nCandidate cut in polytope coordinates = : ";
          std::copy( next_polytope.begin(), next_polytope.end(),
                     std::ostream_iterator<int>( std::cout, " " ) );
185 #endif
          CutList::const_iterator already
            =std::find(L.begin(),L.end(),next_polytope);
          if( already != L.end() ) { continue;}
          L.push_back( next_polytope );
190       ComputeCutSetsOrbits( cbp, adj, next_graph, P, L );
        }
    }
```

Listing 18.5 Polytope enumeration driver program.

18.4 Experimental Results

Polytope catalogs for $d = 3,4,5,6$ are available from the authors. A sample four dimensional polytope is shown in Figure 18.7. External tools like *polymake* [50, 51], and mathematical analysis on the f-vectors have been used to revalidate the combinatorical structure of the polytopes. Visualization tools for various properties are also under development.

The code listing given in this chapter are generic C++ code which has been compiled and executed on Opteron as well as PPC Cell. The compute intensive sub-graph property checker is the main Cell based accelerator which computes the following:

1. Sub-graph diameter checking for Hirsch counter-example
2. Dehn-Sommerville f-vector inequalities for correctness
3. Euler-Poincare formula for correctness of polytopes
4. Graph connectivity for correctness
5. Integer relation search and inequality compaction for f-vectors

These functions when implemented on the main processor slow down the polytope enumeration code by a factor of 10 or more. Thus these have been implemented on the SPU in a round-robin manner. Every time a polytope has to be checked it is encoded as unsigned integer collection and the next free SPU is sent a message using mail box communication. The SPU then downloads the polytope from main memory using DMA and performs a number of compute intensive checks. If the polytope is correct (has no degeneracy or non-simple) then the SPU signals 0 to the PPE which continues processing the main enumeration code. If the SPU detects an error or an counter-example then it signals a 1 to PPE which stops the enumeration and prints the polytope to the log file.

```
VERTICES_IN_FACETS
{  0 1 2 3}
{  0 1 2 4}
{  0 1 3 4}
{  0 2 3 4}
{  1 2 3 4}

VERTICES_IN_FACETS
{  0 1 2 4 5 6}
{  0 1 3 4 5 7}
{  0 2 3 4 6 7}
{  1 2 3 5 6 7}
{  0 1 2 3}
{  4 5 6 7}

VERTICES_IN_FACETS
{  0 1 3 4 5 7 8 9}
{  0 2 3 4 6 7 8 10}
{  1 2 3 5 6 7 9 10}
{  0 1 2 4 5 6}
{  0 1 2 8 9 10}
{  3 4 5 6}
{  7 8 9 10}

VERTICES_IN_FACETS
{  0 2 3 4 6 7 8 10 11 12}
{  1 2 3 5 6 7 9 10 11 13}
{  0 1 2 4 5 6 8 9}
{  0 1 3 4 5 10 12 13}
{  0 1 7 8 9 11 12 13}
{  2 3 4 5}
{  6 7 8 9}
{  10 11 12 13}

VERTICES_IN_FACETS
{  1 2 3 5 6 7 9 10 11 13 14 15}
{  0 1 2 4 5 6 8 9 10 12}
```

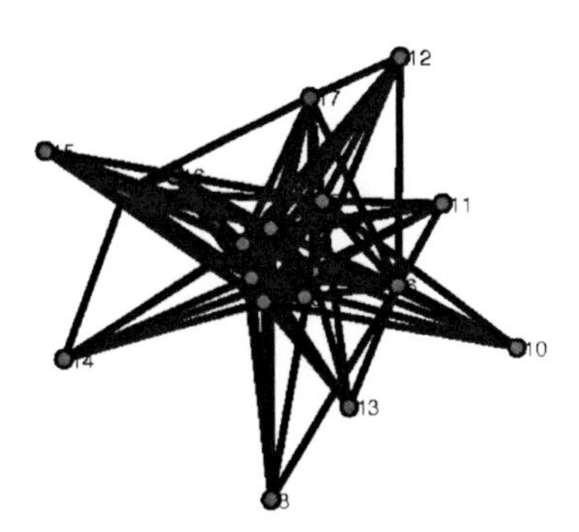

Fig. 18.8 Dual graph of a 6-dimensional polytope.

```
{  0 1 3 4 5 7 8 13 14 16}
{  0 2 3 4 9 11 12 13 15 16}
{  0 6 7 8 10 11 12 14 15 16}
{  1 2 3 4}
{  5 6 7 8}
{  9 10 11 12}
{  13 14 15 16}

VERTICES_IN_FACETS
{  0 1 2 4 5 6 8 9 10 12 13 14}
{  0 1 3 4 5 7 8 9 11 16 17 18}
{  0 2 3 4 6 7 12 13 15 16 17 19}
{  1 2 3 8 10 11 12 14 15 16 18 19}
{  5 6 7 9 10 11 13 14 15 17 18 19}
{  0 1 2 3}
{  4 5 6 7}
{  8 9 10 11}
{  12 13 14 15}
{  16 17 18 19}
```

Listing 18.6 Example of polytope catalog produced by algorithm.

18.5 Future Work

We plan to design an incremental update algorithm which can update a cut like (0 1) to (0 1 2) by perturbation of the (0 1) hyperplane. This should reduce the working set of the ComputeCut recursion, and may as well enable parallel computation of valid cuts given a fixed set. We are analyzing various properties of the generated polytopes, Hirsch conjecture among others.

18.6 Current Results

Consider the following 6-dimensional polytope produced by our algorithm.

The $f - vector$ for this polytope is:

$$f = 74, 222, 295, 220, 91, 18$$

The $h - vector$ can be constructed using Stanley's counting method

$$h = 1, 68, 92, 92, 92, 68, 1$$

and as can be seen it is unimodal.

The $f - vector$ polynomial

$$f := x^6 + 74x^5 + 222x^4 + 295x^3 + 220x^2 + 91x + 18$$

has complex roots. A collection of 6-dimensional face vectors has been generated and automated theorem proving methods are being used to generate new f-vector inequalities for $d = 6, 7, 8$ and higher.

18.7 Conclusion

In this chapter we have presented a combinatoric computing problem which maps well to the Cell architecture; we have used the presented algorithms to compute polytopes, analyze their properties and perform experiments on the $f - vector$ of hitherto unknown high-dimensional polytopes.

Chapter 19
Synthetic Aperture Radar

Abstract We describe the design and implementation of a novel SIMD data-parallel algorithm for real time SAR quick-look image generation using the Cell Broadband Engine as found in the consumer Playstation 3 gaming console. Our implementation of this chain on the CELL system processes 1 frame of image data, reading a 1024x3072 pixel of (simulated) RAW transmission data to produce 16-bit image data in 1.2 seconds. We present the following (a) describe fundamental principal behind Synthetic Aperture Radar, (b) signal processing functions required to solve image formation for SAR, (c) how to map these signal processing functions on the SPU, (d) data structure, communication and load balancing, (e) implementing image processing filters on SPU,

19.1 Introduction

In this chapter we develop a Synthetic Aperture Radar (SAR) quick-look generation system targeted for near-real time processing of Radar data collected by satellite. Microwave SAR has quickly become the method of choice for remote-sensing as it provides many advantages over conventional remote sensing techniques based on visual spectrum waves. Since typical satellite SAR systems are based on active microwaves, they generate their own illumination source and hence can operate in night as well as day. Moreover since microwave wavelength is not obstructed by clouds it can operate in all weather. This is very important for remote sensing in monsoon weather where cloud cover is to be expected. Additional references for SAR and other Radar systems can be found in [104, 116].

S. Koranne, *Practical Computing on the Cell Broadband Engine*,
DOI: 10.1007/978-1-4419-0308-2_19,

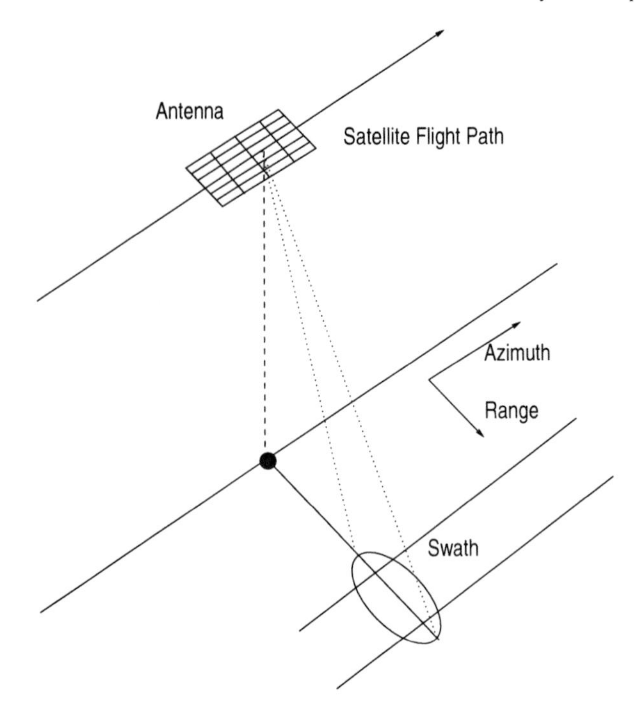

Fig. 19.1 SAR Schematic with azimuth and range directions shown.

19.2 Fundamental principal behind Synthetic Aperture Radar

In the case of space based SAR the radar is carried on a satellite moving at an uniform speed and altitude (which offers the advantage of not requiring motion compensation errors introduced when the radar is mounted on an airplane). The forward motion of the satellite in the direction of *azimuth* provides scanning in that direction while the radar looking angle is almost always orthogonal to the azimuth or track direction. This *sideway* looking radar angle is called the *squint* angle and the radar is also pointed down towards the surface of earth. See Figure 19.1 for details. The radar beam projects an ellipse shape on the surface of the earth.

The radar emits pulsating short duration signals with linear increasing in frequency called *chirps*. The echo of the radar pulse will be received at increasing range, and will overlap in time, but the shape of the radar return can be processed using matched filtering to resolve range with very high resolution.

Real aperture radar (with a physical antenna of limited size) achieves along track (azimuth) resolution by means of sharp beamwidth. To achieve high resolution the physical size of the antenna must be very large. Synthetic Aperture Radar does not have this limitation as the distance traveled by the antenna during the processing is used to create a *synthetic* antenna and image formation is done using coherent processing.

19.2.1 Operational Parameters

Table 19.1 Operational Parameters for SAR Image reconstruction.

Parameter	Value
Frequency	5.3 GHz (C-Band)
Antenna Type	Active Printed
Antenna Size	6m Azimuth × 2m Range
Side Lobe Level	-15 dB (Azimuth) and -18 dB (Elevation)
Pulse Width	20 μ sec
Chirp Bandwidth	18.75 MHz
Sampling Rate	20.83 MHz
Pulse Repetition Frequency	3000 Hz ± 200 Hz
Quantization	6 bit BAQ
MAX Data Rate	142 Mbits/sec

With the launch of Imaging Satellites space agencies have developed advanced capabilities for Disaster Management Support in estimating extent of damage over large areas in the event of natural disasters. Towards this end, generation of Synthetic Aperture Radar (SAR) images with rapid turn-around-time (TAT) is essential. In this paper we describe the design and implementation of an SIMD Data Parallel algorithm for real-time SAR image processing using the Cell Broadband Engine Processor (CBE). The system supports image generation using a modified Range-Doppler algorithm, and is currently processing data at 160 Giga FLOPS. The systems provides facilities for (i) OS file read and data-decryption, (ii) BAQ demodulation, (iii) AGC processing and caliberation, (iv) SLC complex data matrix formation, (v) Range walk followed by range processing, (vi) corner turn (implemented using self-tuned Eklundh method), (vii) Azimuth processing, (viii) Range Cell Migration Correction, (ix) Image formation, (x) filter processing, dynamic range compensation, (xi) watermarking and encryption, and (xii) OS file write. Our implementation of this chain on the CELL system processes 1 frame of image data, reading a 400 MB of (simulated) RAW transmission data to produce 8-bit image data in 1.2 seconds. The distribution of time for a single frame is shown in Figure 19.2.

19.3 Problem Statement for SAR Image Processing

A typical Satellite based SAR schematic is shown in Figure 19.1. The satellite flight path direction is referred to as *azimuth* and the direction of viewing is orthogonal (in a side looking configuration) and is called *range*. The point on the ground directly below the satellite is called *nadir*, and the angle of roll can be controlled to specify which *swath* should be looked at. The antenna is an active array microwave with system configuration as shown in Table 19.1. We first give operational parameters

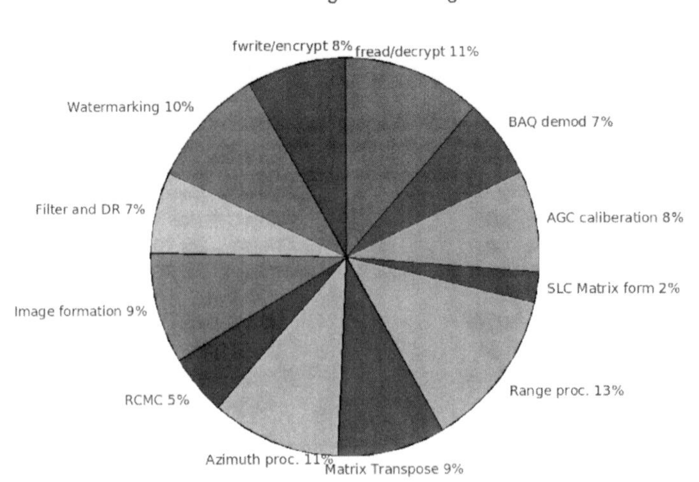

Fig. 19.2 Range Doppler Processing and Time-budget for real time processing.

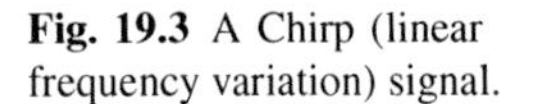

Fig. 19.3 A Chirp (linear frequency variation) signal.

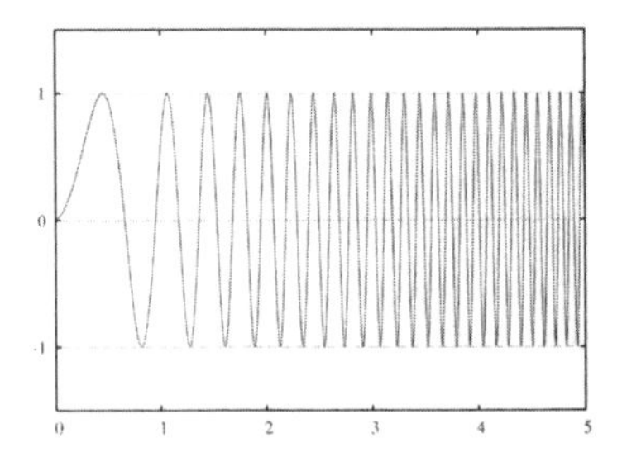

of the SAR image formation problem in Table 19.1. Consider a Chirp signal of the form shown in Figure 19.3.

Let the transmitted signal from satellite be of the form

$$f(t) = e^{i2\pi(f_0 t + 1/2at^2)}, 0 \leq t \leq T \tag{19.1}$$

where $T = 20\ \mu$ sec and the bandwidth a is 18.75.

19.4 Block Adaptive Quantization

The input data to the SAR processor is 6-bit complex I/Q raw signal data alongwith antenna characteristics. The data has to be converted to 16-bit complex numbers and corrected for antenna characteristics. We assume that the raw data packet is formatted to contain 64 I and 64 Q floating point data for every 128 signals. The antenna characteristics are appended as another 256 bits resulting in a total packet size of 128*32+128*6+256 = 5120 bits or 640 bytes. The PPE code reads in a file in chunks of 640 bytes and hands it off to another thread which builds up the BAQ tables, decodes the 6-bit signals into floating point I and Q signals and then hands this data (which is now 1024 bytes) alongwith the antenna characteristics to a free SPU for *antenna gain correction*(AGC). The result of AGC is passed on to the next step of range FFT. The code for BAQ processing on the SPE is shown in Listing 19.1.

```
static void CheckSatelliteData() {
  int i=0,j=0,ub=0;
  union ul {
    float f;
5   unsigned char cf[sizeof(float)];
  } U;
  for(i=0;i<64*4;i+=4) {
    U.cf[0]=SATDATA[i];U.cf[1]=SATDATA[i+1];
    U.cf[2]=SATDATA[i+2];U.cf[3]=SATDATA[i+3];
10  RE_TABLE0[i>>2] = U.f;
  }
  for(i=256;i<512;i+=4) {
    U.cf[0]=SATDATA[i];U.cf[1]=SATDATA[i+1];
    U.cf[2]=SATDATA[i+2];U.cf[3]=SATDATA[i+3];
15  IM_TABLE0[(i>>2) - 64] = U.f;
  }
  // the next data is 512 6-bit values in 384 bytes
  for(j=0;j<384;j+=3,ub+=4) { // process 3 bytes at a time
    unsigned char A = SATDATA[j+i],
20    B=SATDATA[j+i+1], C = SATDATA[j+i+2];
    unsigned char a,b,c,d;
    a = (A&0xFC) >> 2;b = ((A&3) << 4)|((B&0xF0)>>4);
    c = ((B&0x0F)<<2)|((C&0xC0)>>4);
    d = C&0x3F;
25  ACTUAL_RE_DATA0[ub+0] = RE_TABLE0[a];
    ACTUAL_RE_DATA0[ub+1] = RE_TABLE0[b];
    ACTUAL_RE_DATA0[ub+2] = RE_TABLE0[c];
    ACTUAL_RE_DATA0[ub+3] = RE_TABLE0[d];
  }
30 ub = 0;
  i += 384;
  for(j=0;j<384;j+=3,ub+=4) { // process 3 bytes at a time
    unsigned char A = SATDATA[j+i],
      B=SATDATA[j+i+1], C = SATDATA[j+i+2];
35  unsigned char a,b,c,d;
    a = (A&0xFC) >> 2;b = ((A&3) << 4)|((B&0xF0)>>4);
    c = ((B&0x0F)<<2)|((C&0xC0)>>4);
    d = C&0x3F;
    ACTUAL_IM_DATA0[ub+0] = IM_TABLE0[a];
40  ACTUAL_IM_DATA0[ub+1] = IM_TABLE0[b];
    ACTUAL_IM_DATA0[ub+2] = IM_TABLE0[c];
    ACTUAL_IM_DATA0[ub+3] = IM_TABLE0[d];
  }
  i += 384;
45 ub = i + 256;
  j = 0;
  for(;i<ub;i+=4) {
    U.cf[0]=SATDATA[i];U.cf[1]=SATDATA[i+1];
    U.cf[2]=SATDATA[i+2];U.cf[3]=SATDATA[i+3];
50  RE_TABLE1[j++] = U.f;
  }
  j=0;
```

```
    ub = i + 256;
    for(;i<ub;i+=4) {
      U.cf[0]=SATDATA[i];U.cf[1]=SATDATA[i+1];
      U.cf[2]=SATDATA[i+2];U.cf[3]=SATDATA[i+3];
      IM_TABLE1[j++] = U.f;
    }
    // the next data is 512 6-bit values in 384 bytes
    for(ub=0,j=0;j<384;j+=3,ub+=4) { // process 3 bytes at a time
      unsigned char A = SATDATA[j+i],
        B=SATDATA[j+i+1], C = SATDATA[j+i+2];
      unsigned char a,b,c,d;
      a = (A&0xFC) >> 2;b = ((A&3) << 4)|((B&0xF0)>>4);
      c = ((B&0x0F)<<2)|((C&0xC0)>>4);
      d = C&0x3F;
      ACTUAL_RE_DATA1[ub+0] = RE_TABLE1[a];
      ACTUAL_RE_DATA1[ub+1] = RE_TABLE1[b];
      ACTUAL_RE_DATA1[ub+2] = RE_TABLE1[c];
      ACTUAL_RE_DATA1[ub+3] = RE_TABLE1[d];
    }
    ub = 0;
    i += 384;
    for(j=0;j<384;j+=3,ub+=4) { // process 3 bytes at a time
      unsigned char A = SATDATA[j+i],
        B=SATDATA[j+i+1], C = SATDATA[j+i+2];
      unsigned char a,b,c,d;
      a = (A&0xFC) >> 2;b = ((A&3) << 4)|((B&0xF0)>>4);
      c = ((B&0x0F)<<2)|((C&0xC0)>>4);
      d = C&0x3F;
      ACTUAL_IM_DATA1[ub+0] = IM_TABLE1[a];
      ACTUAL_IM_DATA1[ub+1] = IM_TABLE1[b];
      ACTUAL_IM_DATA1[ub+2] = IM_TABLE1[c];
      ACTUAL_IM_DATA1[ub+3] = IM_TABLE1[d];
    }
    // here data is good to use
}
```

Listing 19.1 Block Adaptive Quantization unpacking.

Antenna gain correction (AGC) happens after BAQ dequantization and is performed based on actual caliberation data present in the satellite download header. Since we are simulating the satellite data we assume UNITY correction but nevertheless perform the multiplication as we want to measure the run-time impact of each and every step. The code for AGC is thus straightforward and is shown below in Listing 19.2.

```
static void PerformAGC() {
  // Perform Antenna Gain Correction based on SAT Data
  // this will use SIMD, so transfer the data to vector
  int i,j,k;
  memcpy(DATA, ACTUAL_RE_DATA0, 2048);
  memcpy(DATA+128,ACTUAL_IM_DATA0, 2048);
  memcpy(DATA+256,ACTUAL_RE_DATA1, 2048);
  memcpy(DATA+384,ACTUAL_IM_DATA1, 2048);
  // assume correction vector has been converted to SIMD
  for(i=0;i<128;i+=4) {
    DATA[i] = spu_mul( DATA[i], AGC_RE[0] );
    DATA[i+1]=spu_mul( DATA[i+1],AGC_RE[1]);
    DATA[i+2]=spu_mul( DATA[i+2],AGC_RE[2]);
    DATA[i+3]=spu_mul( DATA[i+3],AGC_RE[3]);
  }
  for(i=128;i<256;i+=4) {
    DATA[i] = spu_mul( DATA[i], AGC_IM[0] );
    DATA[i+1]=spu_mul( DATA[i+1],AGC_IM[1]);
    DATA[i+2]=spu_mul( DATA[i+2],AGC_IM[2]);
```

```
20        DATA[i+3]=spu_mul( DATA[i+3],AGC_IM[3]);
        }
      for(i=256;i<384;i+=4) {
        DATA[i] = spu_mul( DATA[i], AGC_RE[0] );
        DATA[i+1]=spu_mul( DATA[i+1],AGC_RE[1]);
25      DATA[i+2]=spu_mul( DATA[i+2],AGC_RE[2]);
        DATA[i+3]=spu_mul( DATA[i+3],AGC_RE[3]);
      }
      for(i=384;i<512;i+=4) {
        DATA[i] = spu_mul( DATA[i], AGC_IM[0] );
30      DATA[i+1]=spu_mul( DATA[i+1],AGC_IM[1]);
        DATA[i+2]=spu_mul( DATA[i+2],AGC_IM[2]);
        DATA[i+3]=spu_mul( DATA[i+3],AGC_IM[3]);
      }
    }
35
    static void CaliberateAGCTables() {
      int i;
      vec_float4 UNITY = {1.0,1.0,1.0,1.0};
      for(i=0;i<4;++i) {
40      AGC_RE[i] = UNITY;
        AGC_IM[i] = UNITY;
      }
    }
```

Listing 19.2 Antenna Gain Correction is a multiplication.

The main satellite processing function is responsible for getting data from PPE and unpacking and arranging it for efficient SIMD processing.

```
    #define USE_SYNCHRONIZATION
    int main(unsigned long long speid,
             addr64 argp,
             addr64 envp __attribute__ ((__unused__))) {
5     int tag_id = mfc_tag_reserve();
      mfc_get(&cb, argp.ui[1], sizeof(cb), tag_id, 0, 0);
      mfc_write_tag_mask(1<<tag_id);
      mfc_read_tag_status_all();
      printf("Hello from SPE (0x%llx)\n", speid);
10    mfc_get(&SATDATA,(unsigned int)cb.addr[1],4096,tag_id,0,0);
      mfc_write_tag_mask(1<<tag_id);
      mfc_read_tag_status_all();
      CheckSatelliteData();
      CaliberateAGCTables();
15    PerformAGC();
      TransmitData( cb,tag_id );
      #ifdef USE_SYNCHRONIZATION
      spu_write_out_mbox(1);
      #endif
20    return 0;
    }
```

Listing 19.3 Satellite unpacking function.

19.5 Signal processing functions for SAR

Range processing can be thought of as the convolution of range data with the chirp signal. The chirp signal is already processed using FFT and is available as an array of 1024 complex values. Using FFT of the I/Q data we convert the raw signals into complex numbers; this is a complex to complex FFT of 1024 number. Then we perform point-wise multiplication of complex number of the radar data and the chirp FFT. The result of this point-wise multiplication is sent to inverse FFT (which is again complex to complex on 1024 numbers) to get range processed data.

The range data is transposed (this is called *corner turn*) using SDK Matrix transpose function and sent to Azimuth processing as this data contains signal intensity but contains significant errors due to range cell migration (RCMC) which needs to be corrected before final image formation. The azimuth FFT is followed by RCMC interpolation and then by point-wise multiplication with range function. The output of this complex multiply is processed using inverse FFT to get the image intensity values.

The dynamic range of the image at this point is very large and thus dynamic range reduction has to be implemented alongwith some image smoothening filter to form the final image which is sent to the PPE to be written out to disk as 16-bit image.

The control logic of the signal processing is shown below in Listing 19.4.

```
int main(unsigned long long speid,
          addr64 argp,
          addr64 envp __attribute__ ((__unused__))) {
   int tag_id = mfc_tag_reserve();
 5 unsigned int ack=0;
   mfc_get(&cb, argp.ui[1], sizeof(cb), tag_id, 0, 0);
   mfc_write_tag_mask(1<<tag_id);
   mfc_read_tag_status_all();
   PrepData(cb);
10 printf("Hello from SPE (0x%llx)\n", speid);
   do {
     ack = spu_read_in_mbox();
   } while (ack != 1);
   printf("\nData has been computed and transferred to me.\n");
15 GatherData( cb,tag_id );
   printf("\nDone recieving.\n");
   ComputeTwiddleFactors(cb);
   PopulateChirp(cb);
   ComputeFFT(cb);
20 ComputeRangeProcessing(cb);
   ComputeInverseFFT(cb);
   DynamicRangeProcessing(cb);
   TransmitDataToPPE(cb,tag_id);
   return 0;
25 }
```

Listing 19.4 Signal Processing on Range.

19.6 Implementation Details

The main functions of the signal processor were shown above in Listing 19.4; in this section they are described in detail.

The *GatherData* implements receiver-initiated DMA as during experimentation this scheme was faster than sender initiated DMA, but the code is present using defines to try sender initiated DMA. The magic constant of 0x2e80 is actually the offset of the DATA variable in the sender SPE. This cannot be precomputed as the exact location depends not only on the number and size of *variables*, but also on the size of the code. The location of the variable in the other SPE (which is running a different SPE program), is needed to perform spe-to-spe DMA. The good thing about this arrangement is that the PPE does not need to participate. Moreover, this offset is not dependent on run-time data, as we are not allocating memory dynamically, thus we compute it at development time and hard-code it.

```
static void GatherData(ControlBlock cb, int tag) {
  int i;
  unsigned int ds = 0x2e80;
  unsigned int src_addr = (unsigned int)cb.addr[0];
  src_addr += ds;
  printf("\n Getting data from %x + %x", (unsigned int)cb.addr[0],ds);
  mfc_get(DATA,src_addr,N*sizeof(vec_float4),tag,0,0);
  mfc_write_tag_mask(1<<tag);
  mfc_read_tag_status_all();

  #if 0
  printf("\n Getting %d data", cb.row);
  for(i=0;i<cb.row/4;++i) {
    printf("\n %d] = ", i);
    fprint_vector( &DATA[i] );
  }
  #endif
}
```

Listing 19.5 Receiver initiated DMA for gathering satellite data.

Before we perform the Range processing we need to compute the Chirp waveform; infact as we know we are going to use the complex multiplication of FFTs for convolution it is better to compute the chirp data as the FFT of the original chirp waveform and then to use it directly for range processing. The chirp waveform is not dependent on positional data collected by the satellite and can be assumed to be a constant for the entire operation. This is shown in Listing 19.6.

```
int first_call = 0;
vec_float4 CHIRP[N];
static void PopulateChirp() {
  if(first_call) return;
  first_call=1;
  int i=0;
  for(i=0;i<N;i+=2) {
    float ai=i*M_PI*0.001, bi=ai+0.0005;
    vec_float4 A=(vec_float4)spu_splats(ai);
    vec_float4 B=(vec_float4)spu_splats(bi);
    CHIRP[i] = spu_mul(expf4(A),cosf4(B));
    CHIRP[i+1]=spu_mul(expf4(A),sinf4(B));
  }
}
```

Listing 19.6 Computing FFT of chirp signal analytically.

This chirp signal is the same signal that we used to convolve the original images with, as RADAR systems operate the same way, the actual waveform of the chirp is not important for our simulation as long as it is a pulsed waveform with linear increase in frequency.

Prior to computing the FFT we have to compute the *twiddle-factors* required by the FFT API. This is shown in Listing 19.7.

```
vec_float4 W[128];
static void ComputeTwiddleFactors(ControlBlock cb) {
  int i;
  for(i=0;i<256;i+=2) {
    vec_float4 sx,cx;
    float angle = i*2*M_PI/1024;
    sincosf4((vec_float4)spu_splats(angle),&sx,&cx);
    float c = spu_extract(cx,0), s=-spu_extract(sx,0);
    //printf("\n Angle = %f, cos = %f, sin = %f", angle,c,s);
    angle = (1+i)*2*M_PI/1024;
    sincosf4((vec_float4)spu_splats(angle),&sx,&cx);
    float d = spu_extract(cx,0), e=-spu_extract(sx,0);
    vec_float4 temp = spu_insert(c,temp,0);
    temp = spu_insert(s,temp,1);
    temp = spu_insert(d,temp,2);
    temp = spu_insert(e,temp,3);
    W[i/2] = temp;
  }
}
```

Listing 19.7 Computing twiddle factors for FFT on SPU.

Next the actual FFT API calls are presented.

```
static void ComputeFFT(ControlBlock cb) {
  int i;
  printf("\n Computing FFT on Signal Data....\n");
  fft_1d_r2(DATA,DATA,W,10);
}
static void ComputeInverseFFT(ControlBlock cb) {
  int i;
  printf("\n Computing Inverse FFT on Signal Data....\n");
  fft_1d_r2(DATA,DATA,W,10);
  vec_uint4 mask = (vec_uint4){-1,-1,0,0};
  vec_float4 scale = (vec_float4)spu_splats(1.0/1024);
  vec_float4 *start, *end, s0, s1, e0, e1;
  start = DATA, end = DATA+N;
  s0 = e1 = *start;
  for(i=0;i<1024/4;++i) {
    s1 = *(start+1);
    e0 = *(--end);
    *start++ = spu_mul(spu_sel(e0,e1,mask),scale);
    *end = spu_mul(spu_sel(s0,s1,mask),scale);
    s0 = s1, e1 = e0;
  }
}
```

Listing 19.8 Using the FFT API on SPE.

The wrapper functions for the complex multiplication to perform the convolution, inverse FFT and dynamic range reduction are shown below.

```
static void ComputeRangeProcessing(ControlBlock cb) {
  printf("\n Computing Range Processing....\n");
  int i;
  for(i=0;i<N;++i) {
    DATA[i] = ComplexMult( DATA[i], CHIRP[i] );
  }
}
static void TransmitDataToPPE(ControlBlock cb,int tag) {
  printf("\n Transmitting computed image to PPE...\n");
  mfc_put(DATA, (unsigned int)cb.addr[1],8192,tag,0,0);
  mfc_write_tag_mask(1<<tag);
  mfc_read_tag_status_all();
}
```

```
static void DynamicRangeProcessing(ControlBlock cb) {
  int i;
  vec_float4 DYN = (vec_float4)spu_splats(1e-14);
  for(i=0;i<N;++i) {
    vec_float4 d2 = spu_mul( DATA[i], DATA[i] );
    vec_float4 sh = spu_rlqwbyte( DATA[i], 4);
    vec_float4 arg= spu_add(d2,sh);
    arg = spu_mul(arg, DYN);
    DATA[i] = arg; // only 1st and 3rd value
  }
}
```

Listing 19.9 Convolution, inverse FFT and dynamic range reduction.

```

SAR Building Blocks on SPE

SPE-SPE Communication
 sizeof(ControlBlock) = 32
 (C) Satellite Information PacketVersion 0.01 Date April 06 2009.
 cb0=f7e73000 cb1=f7ed3000
 Hello from SPE (0x1002f128)
 Sending 0 @ 2e80 data to 3ff4800037688
 Size of data = 8192Hello from SPE (0x1002f308)
 Data has been computed and transferred to me.
 Getting data from f7ed3000 + 2e80
 Done recieving.
 Computing FFT on Signal Data....
 Computing Range Processing....
 Computing Inverse FFT on Signal Data....
 Transmitting computed image to PPE...

                 *  *                  **             *
         *  *                 **             *                   *     *
 *  *                 **             *                   *     *    ***
     ** **     **             *                  *     *    ***   *    *
  ***  * *      *     *    *             *     *    ***   *    *   ** *
*             *                  *     *                   ** *  *     **
      *   *      *        *     *    *    *    *   ** *  *     **    *
                  *     *    ***   *   *   **  *  *     **    *         *
          *     *    ***   *    *   * *  *     **    *         *  ** * **
  *     *          *    *   ** *  *    **    *         *  *  * ***     **
*    ***   *    *   ** *  *     **    *        *  ** * ***     ***     *
   *    *   ** *  *     **    *        *  ** *           *     *     **
*   ** *  *     **    *         *  ** * *** ***    ***     *     **    * **
  *     **    *         *  ** * ***     ***    *     **    * **
**    *         *  ** * ***     ***     *    **    * **            * **
        *  ** * ***     ***     *     **   * **            * **    *** *
PPE Done.
```

Listing 19.10 Range processing.

19.7 Experimental Results

The data in the following experiments was produced by simulating a radar scatter from 2d photographs using image intensity of 8 bits. The image intensity was converted to floating point radar back-scatter using a chirp function and FFT. This was subsequently sampled to 6-bit block-adaptive quantization as would happen on-board a satellite with broadband sampling. The data was formatted as a packet of 1024 samples in range and 2000 to 3000 samples in azimuth. Each packet comprised of 6-bit data with caliberation and other header sizes reserved.

This binary data is assumed to be streaming in and populating a directory on the PS3 where the application is running. The PPE reads the sequential file and holds it in a buffer *without transformation*. The Satellite decoding SPE is responsible for performing DMA to bring this formatted content into the SPE chain. The first SPE performs BAQ processing, AGC and sends the data to the second SPE in chain using SPE-to-SPE DMA.

```
int main(int argc, char* argv []) {
  std::cout << std::endl << "SAR Building Blocks on SPE\n";
  std::cout << std::endl << "SPE-SPE Communication\n";
  InitData();
5 cb[0].row = 1024;  cb[1].row = 1024;
  cb[0].cols = 1;  cb[0].cols = 1;
  datas[0].speid = spe_context_create(SPE_MAP_PS,NULL);
  datas[0].argp  = (unsigned long long*)&cb[0];
  datas[1].speid = spe_context_create(SPE_MAP_PS,NULL);
10 datas[1].argp  = (unsigned long long*)&cb[1];
  if( spe_program_load( datas[0].speid, &sender_function ) ) {
    perror( "Failed in spe_program_load");
    exit(1);  }
  if( spe_program_load( datas[1].speid, &recvr_function ) ) {
15  perror( "Failed in spe_program_load");
    exit(1);  }
  // now transfer the LS area
  cb[0].addr[0] = spe_ls_area_get( datas[1].speid );
  cb[1].addr[0] = spe_ls_area_get( datas[0].speid );
20 cb[0].addr[1] = SPU_MEM;
  cb[1].addr[1] = ComputedData;
  if( cb[0].addr[0] == NULL || cb[1].addr[0] == NULL ) assert(false);
  if(pthread_create(&datas[0].pthread,NULL,&ppu_thread_function,&datas[0])) {
    perror("pthread_create");
25  exit(1);  }
  unsigned int ack=0;
  do {
    spe_out_mbox_read(datas[0].speid,&ack,1);
  } while (ack <= 0 );
30 if(pthread_create(&datas[1].pthread,NULL,&ppu_thread_function,&datas[1])) {
    perror("pthread_create");
    exit(1);  }
  spe_in_mbox_write( datas[1].speid,&ack,1,SPE_MBOX_ANY_NONBLOCKING);
  pthread_join( datas[0].pthread, NULL );
35 pthread_join( datas[1].pthread, NULL );
  spe_context_destroy( datas[0].speid );
  spe_context_destroy( datas[1].speid );
  // print some sample image intensity
  for(int i=0;i<16;++i) {std::cout << std::endl;
40  for(int j=0;j<256;j+=2) {
      std::cout << (( ComputedData[i*16+j] > 2e4) ? "*" : " ");}}
  std::cout << std::endl << "PPE Done.\n";
  return (0);
}
```

Listing 19.11 Range processing on SPE.

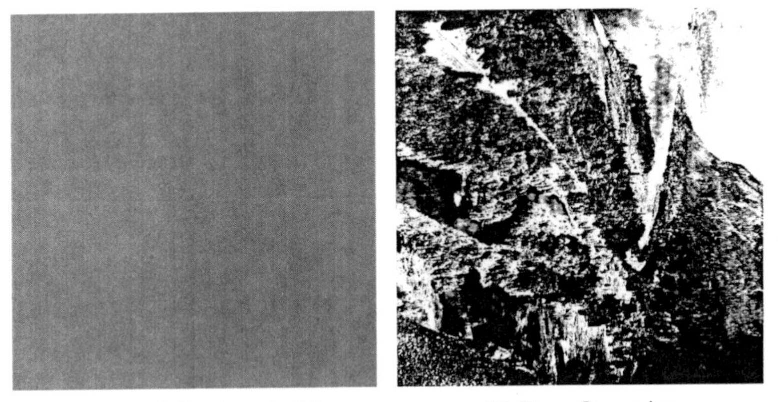

(a) Raw IQ data (magnitude) (b) Range Processing

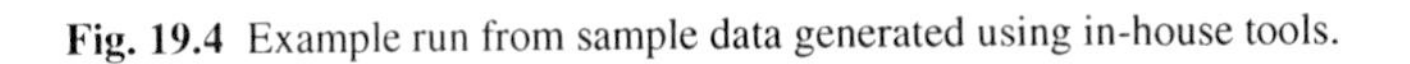

Fig. 19.4 Example run from sample data generated using in-house tools.

19.8 Conclusions

In this chapter we have presented a prototype SAR processing software written on the Cell Broadband Engine designed to use the computing power of the SPUs. We have used the FFT tools provided by the SDK and the matrix transpose for the corner-turn. Image processing and encryption was also used. We use the SPUs as a pipeline and have shown example of SPE-to-SPE DMA using LS address coherency.

Future work: We plan to enhance our system by validating it on real data (when publicly available). We plan to develop subsequent CELL algorithms with space-borne RADAR data applications including (i) rational polynomial coefficient data processing, (ii) data filtering, feature extraction and automatic classification, (iii) data compression of satellite data using arithmetic codes. We are also currently researching efficient SIMD implementations of phase unwrapping algorithms for interferometry applications.

Chapter 20
VLSI Design Automation

Abstract In this meta-chapter we introduce general VLSI terms and flow definitions which are used in the remainder of this book. VLSI design and its automation are now mature topics with hundreds of articles, books and dedicated journals to its credit. We only describe a small subset of the CAD universe in this chapter, mostly to introduce the subject to the non-specialist.

20.1 CAD of Very Large Scale Integrated Circuits

In this chapter we describe the various steps in the design flow of an integrated circuit, starting from the algorithmic description of the functionality, to its representation and model in a High Level *hardware description language*(HDL). Next we consider the important problems of scheduling, architecture optimization and synthesis. Weste and Eshraghian [125] is a good text book on VLSI design perspectives. For VLSI CAD the following references are useful [109, 115, 37, 38, 96]. We limit our attention to standard cell based design methodology, but we should point out that in addition to this style there are these other techniques which are also used:

1. Full-custom: the complete chip is directly drawn as a layout. This is useful for analog and radio-frequency (RF) block design. It is also used in custom bus design where the performance of the block is critical to the chip,
2. Data-path design: used in arithmetic blocks where similar operations happen on a wide-bus. In the Cell architecture, the SPU 32-bit single-precision floating-point unit is a data-path design, so is the vector processing unit in the PPE,
3. Memory array: memory arrays are often designed as a single *bit*-cell, which is then regularly placed by automated tools,
4. PLA design: control logic can be represented as *sum-of-products*(SOP), and PLA (programmable logic arrays) can be automatically drawn from SOP functions. The array is not run-time programmable, only that the choice of which products to sum is made by switches which are programmed when the PLA is drawn. The

S. Koranne, *Practical Computing on the Cell Broadband Engine*,
DOI: 10.1007/978-1-4419-0308-2_20,

SPU forwarding macro logic in the arithmetic unit and register file uses PLA design,
5. FPGA design:field-programmable gate arrays use arrays of logic elements (LEs) connected by a fabric of programmable routing. Some of the problems, such as floorplanning, and placement can be put in FPGA context as well.

VLSI design has been the study of automation and computer science, and many tools have been written to solve one or more problems associated with the VLSI design flow. Thus a number of standard (and not so standard) *file-formats* have emerged to communicate data from one step of the flow to another. A big hurdle to anyone writing CAD software is the parsing and generation of these myriad formats. In this chapter we present file formats as we use them, and also give the necessary software code, both on the PPE and SPE to parse, represent and generate the file formats.

We first describe the syntax of the Berkeley Logic Interchange Format, and present several tools to analyze BLIF data, as BLIF is a standardized logic interchange format for synthesis tools. Post synthesis the logic design has been converted into a *gate-level* netlist we discuss the problem of timing estimation, buffer-insertion and logic checking. This part of the flow is traditionally known as *front-end* of the VLSI flow, and the timing correct gate-level netlist was the accepted *handoff* to the physical design flow to follow. A pictorial depiction of the VLSI design flow is shown in Figure 20.1. The major design flow stages are (i) front-end design and HDL capture, (ii) synthesis to gate level netlist, (iii) floorplanning, (iv) placement, (v) global routing, (vi) detailed routing and (vii) mask level processing.

20.2 Algorithmic Design and HDL Capture

We state the title of this section as "algorithmic design", which refers to the task of solving a given problem by using an algorithm. Consider the problem of designing an elevator control module which services four elevators in a building with ten floors. The requirement analysis of this problem in detail may produces a list as follows:

1. Interact with sensors, lighting and control switches inside the elevator, in the floor and the main control room of the building:
2. Minimize the number of physical connections to the unit:
3. Minimize wait times:
4. Conserve power by maximizing idle item:
5. Guarantee maximum service time per request:
6. Offer emergency control to fire and safety personnel:
7. Integrate with building electronics and electrical facilities:
8. Comply with *STV 1.17.2 Elevator Control Standard*[1]:

[1] This is a made up standardization body, but most chip designs conform to one or more international standards.

A system architect will perform a Pareto analysis of various solutions to this problem. He may consider using an off-the-shelf micro-controller (e.g, 8051,or i486), program an FPGA, design a custom ASIC (chip). Depending on the number of units of this controller, its performance and power requirement, a specific design style will be chosen and implemented. If the design is based around an off-the-shelf component, then there is no VLSI CAD to be done (and you can close this chapter). But we shall assume that atleast for this time, the architect has chosen either FPGA (which also needs to perform HDL capture and synthesis), or an ASIC (which needs, in addition to HDL capture and synthesis, floorplanning, placement and routing).

The architect will then proceed to design an algorithm to solve the presented problem (within the constraints placed by the system requirements). This algorithm is then *captured* in an HDL for simulation and synthesis. Simulation is yet another area of VLSI CAD which is very important, but we have side-stepped, as writing a correct simulator for a modern HDL is non-trivial. By simulating the HDL of the elevator-controller module, its behavior under various conditions can be tested, and its properties measured. CAD tools to calculate expected power dissipation are also available (indeed, we shall present one such tool ourselves later in this chapter), and expected wait-times can be checked with random initial conditions.

The controller's response to emergency inputs under various states and its guarantee of a maximum service time per request can be checked formally using a technique known as *model checking*, which is derived from finite-state-machine state exploration. This is an area of continuing research, and again we have side-stepped this important component of the CAD flow, for brevity of exposition in the bigger context of presenting CAD tools on the Cell processor.

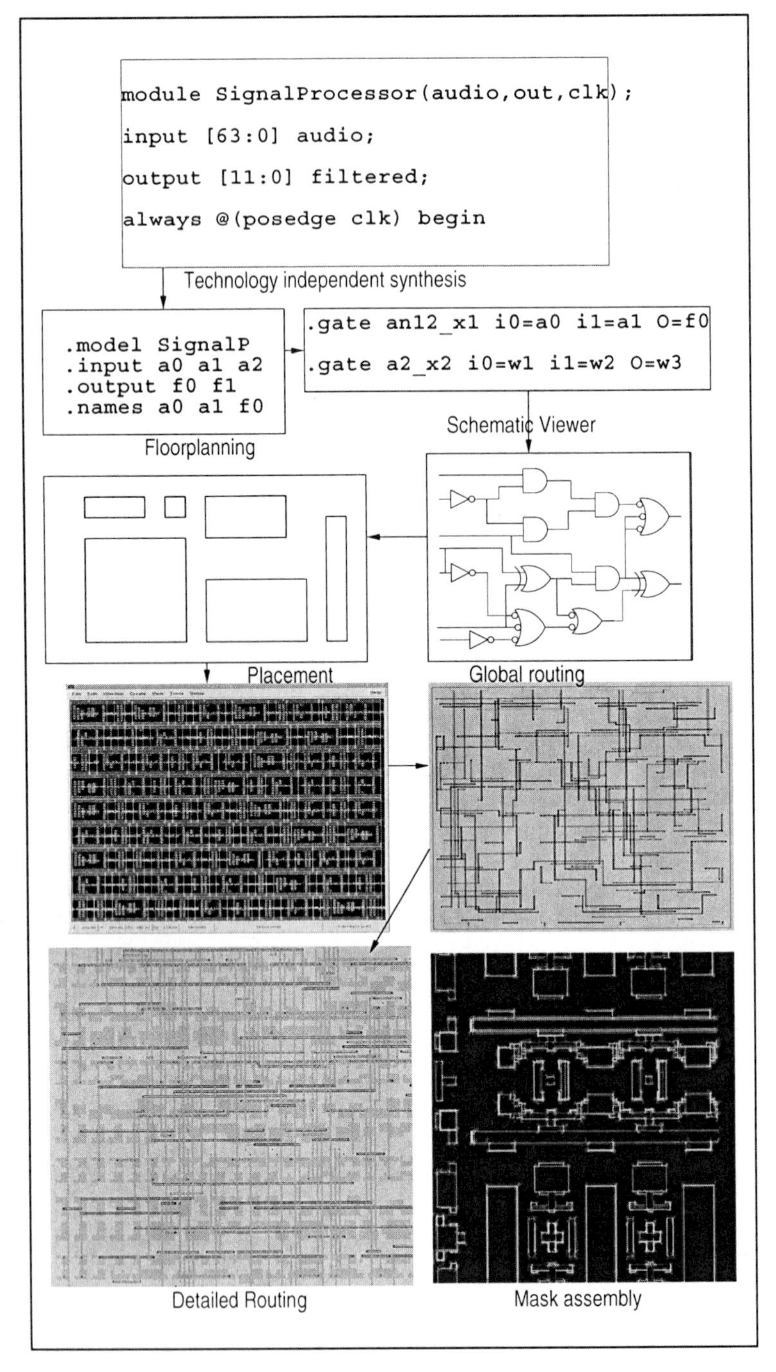

Fig. 20.1 A simplistic view of the VLSI CAD Flow.

20.2.1 HDL Capture

Common HDLs are Verilog and VHDL. Both offer similar features in terms of specifying the structure of the logic module, and its behavior using (i) combinational, and (ii) sequential processing elements. Combinational elements are logic circuits which have no state, and their outputs depend on their current inputs *only*. There may still be a logic-delay between the time an input changes, and the time this change is reflected in a change to the output. Combinational logic circuits are manipulated using Boolean algebra, and their properties have been extensively studied and understood. Sequential processing elements have *state*, and thus, their current output depends not only on their input, but also the current state of the element. This make sequential analysis more complex, but it also allows for a significant reduction in the size of the Boolean circuit needed to implement a given algorithm. For example, 16-bit multiplication would need thousands of logic elements to produce a valid output given only combinational element, but a sequential process combined with combinational logic can perform the same computation (albeit in multiple clock cycles) in only few hundred elements. Sequential elements are often modeled as *state-machines*, and are represented in Verilog/VHDL as encoded state-registers. Consider the following Verilog HDL fragment:

```
module ElevatorControl( clk, rst, swButton,
                        elvButton, emgButton,
                        sensor, opControl);
input    clk,rst;
input [31:0] swButton; //switches in floor, control room
input [63:0] elvButton;// switches inside  4 elevators
...
output [31:0] opControl;// motor control switch lights
endmodule;
```

Although automated tools to convert a given algorithmic description (which may have been done using another language) to HDLs exist, most of the time, the HDL is written by hand in conjunction with the system architect. Key system parameters, such as, interface registers, IO register locations, interface pin-diagrams, need to be created and frozen. Many of these physical parameters have no equivalence in the algorithmic description (what is the weight of an algorithm ?, the controller once manufactured will have weight and dimensions which need to be specified so that the rest of the system can be built around the controller).

As we mentioned earlier, HDL descriptions are simulated and analyzed for their properties. Once the designers are satisfied that the design meets the functional requirements, the design is *synthesized*. Synthesis is the process of taking HDL descriptions, converting them to Boolean logic expressions (using combinational and sequential elements), and finally writing out an equivalent Boolean description of the circuit with each logic elements *mapped* to a given primitive logic element. The set of primitives is also given to the synthesis tool, and this element set is called a *library*. Each library element has been designed to have a physical representation on the chip, and once the circuit is mapped onto the library elements, it can be manufactured purely as a connected graph of these library elements. Each library element provides the synthesis tool the *single* (for single-output standard cells) Boolean function it can implement. It is the task of the synthesis tool to *map* the given Boolean circuit using these primitives. Obviously, an universal primitive like two-input NAND can be used to map the complete combinational portion of the circuit, but it has been known that providing higher levels of Boolean primitives (for example, primitives which compute AND-OR-INV, or AOI, combinations of their input, not only reduce the number of total primitives used, but also reduce the maximum depth of the circuit). The maximum depth of the combinational circuit determines the clock cycle of the chip, as each clock cycle must leave enough time for the output of every combinational block in the chip to be computed.

Synthesis tools have a rich history, and UC Berkeley provided many of the fundamental results which are now part of every synthesis CAD tool. Thus it is only fitting that we present BLIF (Berkeley Logic Interchange Format) syntax before continuing on to the next steps of the VLSI flow.

20.2.2 BLIF Format in a nutshell

The BLIF format describes a circuit in terms of models. A model is an arbitrary combinational or sequential network of logic functions. A circuit is built up of modules connected together to form a connected, directed graph. Each cyclic edge in this network must go through a *latch* element. Each net has a single driver, which must be named without ambiguity. A model is declared in BLIF as

```
.model <name>
.inputs <input-list>
.outputs <output-list>
<command>
...
<command>
.end
```

The *command* is one of:

1. logic gate
2. generic latch
3. library-gate: library format is described below,
4. model reference:
5. subfile reference:
6. fsm-description:
7. clock constraint
8. delay constraint

A # begins a comment line which continues to the end of the line. A *logic gate* associates a logic function with a signal in the model, which can be used as an input to another logic function. A logic gate is declared as follows:

```
.names <in-1> <in-2> ... <in-n> <output>
<single-output-cover>
```

An example of a logic-gate is shown below:

```
.names x y z
1- 1
01 1
```

The single-output cover may include – to represent a don't care. Library gates are specified using *.gate* construct, where the syntax is as follows:

```
.gate <name> <formal-argument>
```

The argument list of the gate is looked up from the library file, and is matched by name.

Consider a very small model represented as a BLIF file:

```
.model small
.inputs ck  i0 i1 i2 i3
.outputs  o0 o1
.names  i0 i1 i2 i3 o0
0001 1
0000 1
0010 1
.names  i0 i1 i2 i3 o1
0001 1
0000 1
0010 1
.end
```

This model has 4 inputs (i0,i1,i2 and i3), and 2 outputs (o0 and o1). The truth table (or single-output cover) for o0 and o1 is given in the BLIF file.

Next we write a library description file containing the four gates of INV, BUF, 2-AND and 2-OR. This is shown below:

```
GATE inv_x1 1  q=!i;
     PIN    i  INV 1 999 1.00 0.00 1.00 0.00
GATE buf_x2 1  q=i;
     PIN    i  INV 1 999 1.00 0.00 1.00 0.00
GATE a2_x2  2  q=(i0*i1);
     PIN   i0  INV 1 999 1.00 0.00 1.00 0.00
     PIN   i1  INV 1 999 1.00 0.00 1.00 0.00
GATE o2_x2  2  q=(i0+i1);
     PIN   i0  INV 1 999 1.00 0.00 1.00 0.00
     PIN   i1  INV 1 999 1.00 0.00 1.00 0.00
```

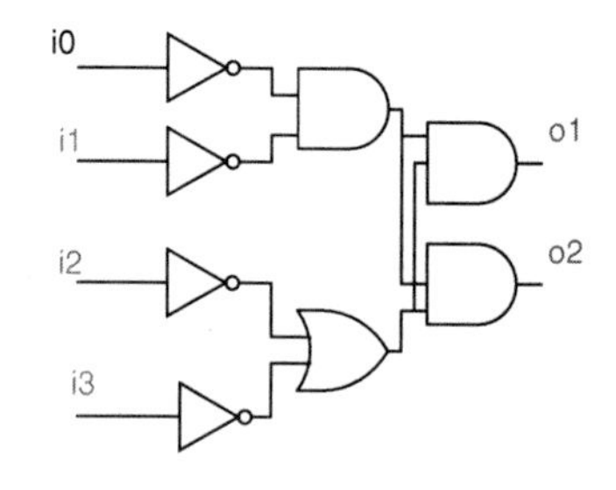

Fig. 20.2 Gate level netlist, schematic is shown after mapping onto library.

Using ABC Synthesis tool [121] we map the BLIF file using the library to produce a gate-level mapped netlist as shown below. The ABC commands necessary to do the mapping were: (i) read_blif (ii) read_library, (iii) map, and (iv) write_blif. We get the following BLIF file.

```
.model small
.inputs ck i0 i1 i2 i3
.outputs o0 o1
.gate inv_x1 i=i2 q=n7
.gate inv_x1 i=i3 q=n8
.gate o2_x2  i0=n8 i1=n7 q=n9
.gate inv_x1 i=i0 q=n10
.gate inv_x1 i=i1 q=n11
.gate a2_x2  i0=n11 i1=n10 q=n12
.gate a2_x2  i0=n12 i1=n9 q=o0
.gate a2_x2  i0=n12 i1=n9 q=o1
.end
```

Once the following gate is introduced in the library, the number of elements and the critical path get reduced.

```
GATE  noa22_x1   3    nq=((!i0+!i1)*!i2);
      PIN      i0 INV 1 999 1.00 0.00 1.00 0.00
      PIN      i1 INV 1 999 1.00 0.00 1.00 0.00
      PIN      i2 INV 1 999 1.00 0.00 1.00 0.00
```

We again map the original input BLIF using the augmented library to get the following BLIF file.

```
.model small
.inputs ck i0 i1 i2 i3
.outputs o0 o1
.gate o2_x2    i0=i1 i1=i0 q=n7
.gate noa22_x1 i0=i3 i1=i2 i2=n7 nq=o0
.gate noa22_x1 i0=i3 i1=i2 i2=n7 nq=o1
.end
```

Thus, it is obvious, how large an impact, choosing a good set of primitive elements in the library can have on the size of the chip and its critical path.

This gate level netlist is the traditional *hand-off* from front-end design tools to the *back-end* processing, also known as the *physical design flow*. In the physical design flow, the netlist is first divided into block regions at the chip level using an automated tool called the *floorplanner*. The input to this tool is the bounding box of each module of the netlist, and it calculates relative positions of the modules to minimize the overall size of this chip.

Once the floorplan is complete, we perform *cell-placement*. Standard cell placement has been an extensively researched topic, with recent publications (as late as Jan. 2009) reporting break-throughs in technology. We present simple, but effective and time-tested placement solutions based on simulated annealing, and spectral partitioning. Both methods are fully described earlier in the context of graph partitioning (Breuer had shown in the 1970s an interesting relationship between placement and partitioning). Parallel SPU optimized code for cell-placement based on these techniques is presented later in this chapter.

Placement is followed by *global routing*, a process, in which we assign nets to *channels*, which are defined as pathways on the grid of the chip with exclusive reservation of a limited number of nets (to avoid congestion when we perform final detailed routing). An obvious optimization criterion is the minimization of wirelength used to connect any net, but secondary objectives, such as minimizing via-counts (a *via* occurs whenever a wire changes direction on a net), minimizing coupling-length between long and *crosstalk sensitive* wires, and minimizing skew. In our formulation of global routing we have used wirelength as the primary objective, with via-minimization and coupling analysis and minimization as the secondary objectives. Our formulation of global routing is based on *flows* in the grid-graph.

Global routing is followed by detailed routing. Detailed routing has developed into its own industry as it needs to cater to foundry specific design rules, timing calculations, and yield analysis. Thus, we shall not present any detailed router in this text (it is fit to be the topic of a book by itself).

A completely routed design database is processed at the *mask shop* where image processing of the data is done to verify printability on the wafer. This is done using a combination of image processing tasks such as optical proximity correction, resolution enhancement and sub-resolution assist features. Like detailed routing, mask optimization is also an industry within its own right, and we defer on that topic as well. These flows are shown in Figure 20.1. In the coming chapters we shall discuss the efficient implementation of many of these steps on the Cell architecture.

20.3 Equation Processing

A simple equation parser is presented in Listing 20.1.

```
////////////////////////////////////////
// Simple .eqn file parser
// Sandeep Koranne
////////////////////////////////////////
#include <boost/spirit/core.hpp>
```

```
#include <iostream>
#include <string>
#include <stack>
#include <cassert>
#include <vector>
#include <set>
using namespace std;
using namespace boost::spirit;
typedef std::set<std::string> StringSet;
StringSet inSet, outSet, tempSet;
typedef std::stack<std::string> ExpressionStack;
ExpressionStack gStack;
struct Tree;
typedef enum { NONE, AND, OR, XOR, NEG,
               ASSIGN, LITERAL } Operation;
const char* opName[] = { "NONE", "AND", "OR", "XOR",
                         "NEG", "ASSIGN", "LIT" };
class Tree {
public:
  Operation op;
  Tree *left;
  Tree *right;
  std::string literal;
  Tree(Operation iop=NONE,const std::string& ilit=""):
    op(iop),left(NULL),right(NULL),literal(ilit) {}
private:
  Tree() {}
  Tree(const Tree&) {}
  Tree& operator=(const Tree&) { return *this;}
};

std::ostream& operator<<(std::ostream& os, const Tree& t) {
  switch(t.op) {
  case ASSIGN: {
    if(t.right)
      os << t.literal << " = " << t.right->literal << ";\n"; break; }
  case AND: {
    os << t.literal << " = " << t.left->literal
       << " * " << t.right->literal << ";\n"; break; }
  case OR: {
    os << t.literal << " = " << t.left->literal
       << " + " << t.right->literal << ";\n";  break; }
  case XOR: {
    os << t.literal << " = " << t.left->literal
       << " * !" << t.right->literal << " + "
       << "! " << t.left->literal << " * " << t.right->literal << " ;\n";
    break; }
  case NEG: {
    os << t.literal << " = !" << t.left->literal << ";\n";  break; }
  default:   break;
  }
  assert(t.left != &t); assert(t.right != &t);
  if(t.left && t.left != &t)   os << *t.left;
  if(t.right && t.right != &t) os << *t.right;
  return os;
}
struct Expression {
  std::string lhs;
  Tree* pTree;
};
typedef std::vector<Expression> ExpVector;

static void PrintExpressions(const ExpVector& e);
ExpVector gExpVector;
Tree* gCT = NULL;

typedef std::stack<Tree*> TreeStack;
```

```
    TreeStack treeStack;

75  static void ClearParserState(void) {
      while(!treeStack.empty()) treeStack.pop();
      while(!gStack.empty()) gStack.pop();
    }
    void do_int(char const* str, char const* end) {assert(false);}
80
    void do_literal(char const* str, char const* end) {
      string  s(str, end);
      gStack.push(s);
      Tree *pT = new Tree(LITERAL, s);
85    treeStack.push(pT);
    #ifdef DEBUG_FULL
      cerr << "PUSH LIT(" << s << ')' << endl;
    #endif
      inSet.insert(s);
90  }

    std::string GetName(void) {
      static int count = 0;
      char name[100];
95    sprintf(name,"T%d",count);
      count++;
      return name;
    }

100 void do_lhs(char const* str, char const* end) {
      string  s(str, end);
      Expression e;
      e.lhs = s;
    #ifdef DEBUG_FULL
105   cerr << "LHS = (" << s << ')' << endl;
    #endif
      gCT = new Tree(ASSIGN, s);
      e.pTree = gCT;
      assert(treeStack.empty());
110   gExpVector.push_back(e);
      outSet.insert(s);
    }

    void do_expr(char const* str, char const* end) {
115 #ifdef DEBUG_FULL
      string  s(str, end);
      cerr << "PUSH EXPR(" << s << ')' << endl;
    #endif
    }
120
    void do_add(char const* l, char const* r) {
    #ifdef DEBUG_FULL
      cerr << "ADD\n";
    #endif
125   // take the 2 top elements from the stack
      // and make their tree and add to sub tree
      Tree* pT = new Tree(OR,GetName());
      assert(!treeStack.empty());
      pT->left = treeStack.top();
130   treeStack.pop();
      pT->right = treeStack.top();
      treeStack.pop();
      assert(pT->left && pT->right);
      treeStack.push(pT);
135 }

    void do_xor(char const* l, char const* r) {
    #ifdef DEBUG_FULL
      cerr << "XOR\n";
```

```
140 #endif
      // take the 2 top elements from the stack
      // and make their tree and add to sub tree
      Tree* pT = new Tree(XOR,GetName());
      assert(!treeStack.empty());
145   pT->left = treeStack.top();
      treeStack.pop();
      pT->right = treeStack.top();
      treeStack.pop();
      assert(pT->left && pT->right);
150   treeStack.push(pT);
    }

    void do_mult(char const* l, char const* r) {
    #ifdef DEBUG_FULL
155   cerr << "AND\n ";
    #endif
      // take the 2 top elements from the stack
      // and make their tree and add to sub tree
      Tree* pT = new Tree(AND,GetName());
160   assert(!treeStack.empty());
      pT->left = treeStack.top();
      treeStack.pop();
      pT->right = treeStack.top();
      treeStack.pop();
165   assert(pT->left && pT->right);
      treeStack.push(pT);
    }

    void do_neg(char const* l, char const* r) {
170 #ifdef DEBUG_FULL
      cerr << "NEG\n ";
    #endif
      // take the 2 top elements from the stack
      // and make their tree and add to sub tree
175   Tree* pT = new Tree(NEG,GetName());
      assert(!treeStack.empty());
      pT->left = treeStack.top();
      pT->right = NULL;
      treeStack.pop();
180   assert(pT->left);
      treeStack.push(pT);
    }

    static bool isInput = false;
185 static bool isOutput = false;

    void do_input(char const* l, char const* r) { isInput = true; }
    void do_output(char const* l, char const* r){ isOutput = true;}

190 void do_final(char const* l, char const* r) {
      if(isInput || isOutput) {
        isInput = false;
        isOutput = false;
        return;
195   }
      assert(!treeStack.empty());
      gCT->right = treeStack.top(); treeStack.pop();
      assert(treeStack.empty());
    }
200 void do_subt(char const*, char const*)    { cerr << "SUBTRACT\n"; }
    void do_div(char const*, char const*)     { cerr << "DIVIDE\n"; }

    static void PrintExpressions(const Expression& e) {
      std::cout << std::endl << (*e.pTree);
205 }
    static void PrintExpressions(const ExpVector& e) {
```

```
  for(ExpVector::const_iterator eit = e.begin(); eit != e.end(); ++eit) {
    PrintExpressions(*eit);
  }
  std::cout << std::endl;
}

struct calculator : public grammar<calculator> {
  template <typename ScannerT>
  struct definition {
    definition(calculator const& /*self*/) {
      expression
        = str_p("INPUT()")[&do_input]
        | str_p("OUTPUT()")[&do_output]
        | term
        >> *(    ('#' >> term)[&do_add]
                |    (DOLLAR >> term)[&do_xor]
                |    ('-' >> term)[&do_subt]
                );
      term
        =    factor
        >> *(    ('&' >> factor)[&do_mult]
                |    ('/' >> factor)[&do_div]
                );
      factor
        =    lexeme_d[(+digit_p)[&do_int]]
        |    lexeme_d[(+(alnum_p|'_'|'['|']'))[&do_literal]]
        |    '(' >> expression[&do_expr] >> ')'
        |    ('!' >> factor)[&do_neg]
        |    ('#' >> factor);
      assign
        =    lexeme_d[(+(alnum_p|'_'|'['|']'))
                      [&do_lhs]] >> '=' >>
        expression[&do_final] >> ';'
        ;
    }
    rule<ScannerT> expression, term, factor, assign;
    rule<ScannerT> const&
    start() const { return assign; }
  };
};

int main(int argc, char* argv[]) {
  cerr << "///////////////////////////\n\n";
  cerr << "\t\tExpression parser...\n\n";
  cerr << "///////////////////////////\n\n";
  cerr << "Type an expression...or [q or Q] to quit\n\n";
  calculator calc;    //  Our parser
  string str;
  while (getline(cin, str)) {
    ClearParserState();
    if(str.empty()) continue;
    if(strstr(str.c_str(),"INPUT()")) continue;
    if(strstr(str.c_str(),"OUTPUT")) continue;
    if(strstr(str.c_str(),"TRI")) continue;
    if(strstr(str.c_str(),"DFFEAS"))continue;
    if (str[0] == '-' && str[1] == '-')continue;
    parse_info<> info = parse(str.c_str(), calc, space_p);
    if (info.full) {
      cerr << "Parsing succeeded\n";
    }else {
      cerr << "Parsing failed\n";
      cerr << "stopped at: \": " << info.stop << "\"\n";
    }
  }
  std::cerr << std::endl << "InSet  has " << inSet.size() << " inputs "
            << std::endl << "OutSet has " << outSet.size() << " outputs\n";
  size_t lm=0;
```

```
    const size_t LM_MAX = 50000;
275 std::cout << "INORDER = ";
    std::set_difference(inSet.begin(),inSet.end(),
                        outSet.begin(),outSet.end(),
                        std::ostream_iterator<std::string>
                        (cout," "));
280 #if 0
    for(StringSet::const_iterator sit = inSet.begin();
        sit != inSet.end(); ++sit) {
      if(lm++ > LM_MAX) break;
      std::cout << " " << *sit;
285 }
  #endif
    lm=0;
    std::cout << " ;\n";
    std::cout << "OUTORDER = ";
290 std::set_difference(outSet.begin(),outSet.end(),
                        inSet.begin(),inSet.end(),
                        std::ostream_iterator<std::string>(cout," "));
  #if 0
    for(StringSet::const_iterator sit = outSet.begin();
295     sit != outSet.end(); ++sit) {
      if(lm++ > LM_MAX) break;
      std::cout << " " << *sit;
    }
  #endif
300 std::cout << " ;\n";
    PrintExpressions(gExpVector);
    gExpVector.clear(); // mem leak if tree allocated
    cerr << "Bye... :-) \n\n";
    return 0;
305 }
```

Listing 20.1 Equation processing in VLSI CAD.

A sample session of the equation processor is shown below. The # symbol denotes the OR operation, the & symbol the AND, $ denotes XOR and ! denotes negation. Complicated expressions can be built from these primitives using parenthesis and internal variables.

```
Type an expression...or [q or Q] to quit
out0=t1#t2;
out1=t3&t4;
t1=(a&b)#(c&!d);
t2=(!a#b)&!(!c#d);
t3=a#b#c;
t4=!a#!b&!c;
```

The output that the program produces is given next. The output when evaluated in reverse order only requires a single evaluation per line.

```
INORDER = a b c d  ;
OUTORDER = out0 out1  ;
out0 = T0;T0 = t2 + t1;
out1 = T1;T1 = t4 * t3;

t1 = T5;T5 = T4 + T2;
T4 = T3 * c;T3 = !d;
```

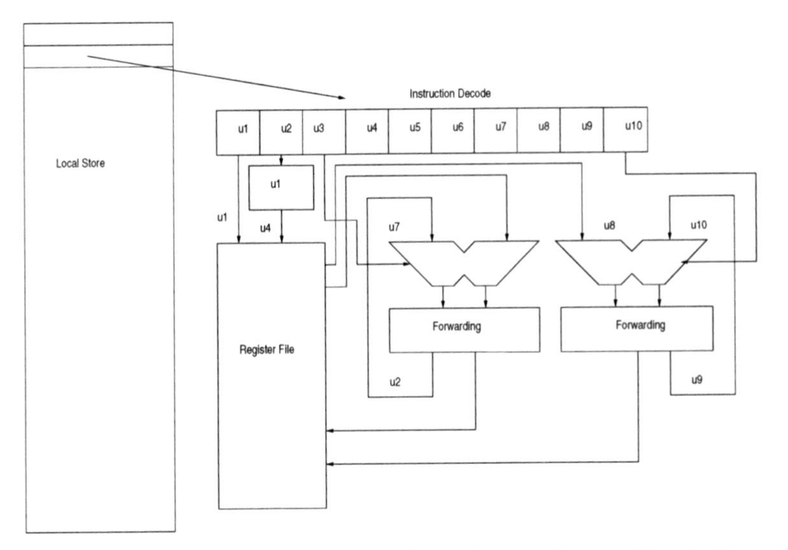

Fig. 20.3 Schematic of logic processor with micro-instructions coming out of decoded instruction.

```
T2 = b * a;

t2 = T11;T11 = T10 * T7;
T10 = !T9;T9 = d + T8;
T8 = !c;T7 = b + T6;
T6 = !a;

t3 = T13;T13 = c + T12;
T12 = b + a;t4 = T18;
T18 = T17 + T14;T17 = T16 * T15;
T16 = !c;
T15 = !b;
T14 = !a;
Bye... :-)
```

20.4 Microword Optimization using Local Search on SPU

The optimization of microword length is vital to microprogrammed controller design in a digital system. A microprogram can be stored in control memory by assigning the microoperations to word bits. The size of the control memory is determined by the size of the microprogram it stores. Any reduction in the microword length would reduce the control memory's silicon area and increase the system's real-time performance. Consider the schematic of a logic processor as shown in Figure 20.3.

Thus, microword length reduction techniques have been widely used in most existing automated synthesis systems, e.g., CMU-DA, MIMOLA and IDEAS. Some complex instruction set computers (CISC) have long microword lengths. The increasing number of microoperations per instruction stretches the control memory's size to its limits. In certain advanced computer architecture designs, such as a very long instruction word (VLIW) architecture, designers are now packing many microoperations into a single instruction.

There are basically two methods to encode microoperations into a control memory. One method, known as *direct control*, assigns one bit of memory word to each microoperation. This scheme provides maximum flexibility and concurrency, but leads to prohibitively large microword length for any reasonably sized microcode. Another method, known as *minimal encoding* , employs $\lceil \log_2 n \rceil$ bits to encode n microoperations.This scheme minimizes wordlength at the expense of speed and additional decoding logic. It provides no flexibility to accommodate microcode changes.

The above two methods are two extremes of a wide spectrum of trade-offs between the microcode length and the performance. An intermediate encoding scheme can be used to minimize the microword length without affecting the maximal concurrency. Such a scheme partitions the the microoperations into groups such that no more than one microoperation in each group is active at any given time. Then the microoperations in a single group can be encoded together. Such a group of microoperations defines a field. A separate decoder is required for each field. This encoding scheme provides a compromise between flexibility and the minimal wordlength.

We use the SPUs of the Cell Broadband Engine to implement bit level manipulation and computation to reduce the time complexity of the local search substantially. Our algorithm's time and space complexity is an improvement on the previous best known result.

20.4.1 Problem formulation

The microcode of a microprogrammed controller consists of a sequence of microinstructions $\mu_1, \mu_2, \cdots, \mu_M$, where each microinstruction is a set of microoperations $O_1, O_2, \cdots, O_n$. The set of microoperations in each microinstruction is conflict free in resource accessing. Thus, two microoperations O_i and O_j in a single instruction are executed concurrently and should not be encoded together. These microoperations are pairwise *incompatible*. Two microoperations are grouped together if they do not appear together in any microinstruction. Such a pair of microoperations is *compatible*. A *compatibility class* is a set of microoperations that are pairwise compatible and can be encoded together.

Let $\{C_1, C_2, \cdots, C_N\}$ be a set of compatibility classes for a given set of microinstructions such that each microoperation appears in only one of the compatibility classes. Thus, these compatibility classes are mutually exclusive and collectively exhaustive. Let $|C_i|$ be the number of microoperations in compatibility class C_i. We can encode the compatibility class C_i into a single field with $\lceil \log_2(|C_i| + 1) \rceil$ bits, where one bit is added to account for the NO-OP condition (i.e., none of the micro-operations in C_i are to be executed). The microword length L equals the total number of bits required to encode the compatibility classes $\{C_1, C_2, \cdots, C_N\}$:

$$L = \sum_{i=1}^{N} \lceil log_2(|C_i| + 1) \rceil$$

Definition 1: Given a set of microinstructions, the microword length minimization problem is to partition the set of microoperations into a collection of compatibility classes $\{C_1, C_2, \cdots, C_N\}$, such that the microword length L is minimum.

If we use a *compatibility graph* to represent the microoperations and the compatibility relationships, then we can formulate the above optimization problem in terms of graph theoretic descriptions. A microoperation O_i corresponds to node i in the graph. Two nodes i and j are connected by an undirected arc if the corresponding microoperations O_i and O_j are compatible. A compatibility class is a complete subgraph such that all its nodes are connected to each other, i.e., they are pairwise compatible. Table 20.1 gives a simple microcode example. The microoperations $\{O_0, O_1, \cdots, O_9\}$ are mapped to nodes $\{0, 1, \cdots, 9\}$ in the compatibility graph. Microoperations O_0, O_4, O_7 and O_9 do not appear together in the microinstructions $\mu_0, \mu_1, \cdots, \mu_3$, so they are pairwise compatible. Hence, $\{O_0, O_4, O_7, O_9\}$ forms a compatibility class. Mapping these nodes and compatibility relations onto the compatibility graph, nodes {0, 4, 7, 9 } are connected with each other and thus form a compatibility subgraph (shown in highlighted lines in Fig. 1). Similarly other compatibility classes can be derived from the graph.

Table 20.1 Microword Example

Micro Inst.	Micro Operations
μ_0	$O_0 O_1 O_5$
μ_1	$O_1 O_2 O_9$
μ_2	$O_3 O_4 O_8$
μ_3	$O_7 O_2 O_6$

20.4.2 Heuristic techniques & local search

Most techniques proposed for solving the microword length minimization problem build a search space by deriving all the maximal complete subgraphs; i.e., maximal compatibles, in the compatibility graph. The problem of extracting a maximal complete subgraph from an arbitrary undirected graph is known as the *maximal clique problem*. This problem was one of the first problems to be proved NP-complete [49].

A number of heuristics have been developed to solve the problem. A very common approach is to use *local search.*

Local Search

Local search was one of the early techniques proposed during the mid-sixties to cope with the overwhelming computational intractability of NP-hard combinatorial optimization problems. Given a minimization (maximization) problem with objective function f and feasible region R, a typical local search algorithm requires that, with each solution point $x_i \in R$, there is a predefined *neighborhood* $N(x_i) \subset R$. Given a current solution point $x_i \in R$, the set $N(x_i)$ is searched for a point x_{i+1} with $f(x_{i+1}) < f(x_i)$ $(f(x_{i+1}) > f(x_i))$. If such a point exists, it becomes the new current solution point, and the *local optimum* with respect to $N(x_i)$. Then, a set of feasible solution points is generated,and each of them is "locally" improved within its neighborhood. Our goal is to minimize the cost function L. It is costly to generate a maximal clique. The absolute maximality of the clique size derived from the compatibility graph does not necessarily contribute to a good practical solution. We show that minimizing L might require rearrangement of some microoperations (swap operation).

Minimizing L over maximal clique

We know that L is defined as the cost of the partition of the compatibility classes. Due to the form of L it is quite possible that the maximal clique might not contribute to the best practical solution. As an example consider the case when $|C_0|$ is 16, and $|C_1|$ is 5. Then the cost of the partition as calculated is 8 (i.e., L=8 , $\lceil \log_2(16+1) \rceil + \lceil \log_2(5+1) \rceil$). If we could identify $O_i \in C_0$, such that O_i is pairwise compatible with *all* microoperations in C_1, then we could swap O_i from C_0 to C_1 to get a reduction of 1 in L (i.e., L=7, $\lceil \log_2(15+1) \rceil + \lceil \log_2(6+1) \rceil$). This follows from the definition of L, which is a step function. In order to determine whether a microoperation O_i is pairwise compatible with a compatibility class C_j [2] we can use a "bit-bucket" scheme. This is explained further in the next section.

A "bit-bucket" data structure

Our goal is to devise a data structure so that the check whether O_i is pairwise compatible with C_j can be done efficiently (actually, by our scheme it can be done in

[2] By pairwise compatibility of O_i with compatibility class C_j we mean the O_i must be compatible with *all* microoperations in C_j

$O(1)$ time). For this purpose we encode the microoperations as a M bit word, with LSB at μ_0 and MSB at μ_M. Now, the check whether O_i is pairwise compatible with O_j is trivial and can be accomplished by checking the value of the logical AND of the two M bit words corresponding to O_i and O_j.We say

$$R \quad \leftarrow \quad W_i \quad \mathbf{AND} \quad W_j.$$

If O_i and O_j are active in the same microinstruction, say μ_p then both will have a 1 set in the pth bit-position. The logical AND of the two M bit words will give a non-zero result. Hence, if R is non-zero, we declare that O_i and O_j are *incompatible*, else they are compatible.

To determine whether O_i is pairwise compatible with a compatibility class C_j, we use a "bit-bucket." Corresponding to each compatibility class we maintain a number (max value is 2^M), which we call the *ORed Product* of the class. The ORed Product value is the value of the "bit-bucket" corresponding to a compatibility class. We shall prove that if O_i AND ORed Product (C_j) is zero, then the microoperation O_i is pairwise compatible with *all* microoperations in compatibility class C_j, or O_i is compatible with C_j, else it is incompatible. A combination of simulated annealing and bit-bucket local search has been implemented on the SPU.

20.5 Conclusion

In this chapter we have introduced the VLSI design flow. Obviously, VLSI is now a complex and very large subject hence in the limited space of few pages we can only discuss the key points of logic synthesis and placement which are related to the material we will be presenting in the next chapters. However we did present an Equation parser which was used to generate simple test cases for the placer and the microword optimizer. The interested reader is advised to read the references listed above for more details. The BLIF processing tool, parser and equation processors are used to produce, analyze and manipulate logic designs to create realistic test cases for the placement, and probabilistic activity estimator tools we develop in the next three chapters.

Chapter 21
Floorplanning: VLSI and other Applications

21.1 What is a VLSI floorplan

Once the circuit is available in the form of gate-level netlist, or even before at RTL if the area occupied by each block can be estimated or measured with some certainty alongwith the number of terminals required by each block, then the next step in completing the layout is to assign a specific shape to each block and arrange the blocks minimizing the total area of the chip. This problem of block level arrangement to minimize total chip area under the constraints of chip aspect ratio and block areas is called the VLSI Floorplanning problem. For further details please refer to [109, 115].

The input to the floorplanner is a set of blocks, the area and aspect ratio of each block, possible shapes of each block, and the number of terminals for each block. In this chapter we develop a floorplanner based on the *sequence-pair* method [91].

21.2 The sequence-pair data-structure

A sequence-pair is a pair of sequences of n elements representing a list of n blocks. Given a block placement (like the one given in Figure 21.1(a)) Murata *et al.* [91] describe a procedure for creating a linear order amongst the blocks using the concept of *positive step-line* of a module, and the *negative step-line* of a module. For more details on sequence-pair notation and its use in 2-d packing, please refer to [91, 122, 126, 127]. Using these a unique linear order on the blocks can be obtained. The module name sequence obtained by this method is denoted Γ_+ (Γ_-) for positive (negative) step-line order, resp. For Figure 21.1(a) we have $\Gamma_+ = [1,2,8,9,3,6,7,5]$, and $\Gamma_- = [3,8,2,9,6,1,5,7]$.

The sequence-pair of the block placement is precisely the pair (Γ_+,Γ_-). Given a sequence-pair (Γ_+,Γ_-), one can construct a 45 degree oblique grid as shown in Figure 21.1 (b). For every block, the plane is divided by the two crossing slope

S. Koranne, *Practical Computing on the Cell Broadband Engine*,
DOI: 10.1007/978-1-4419-0308-2_21,

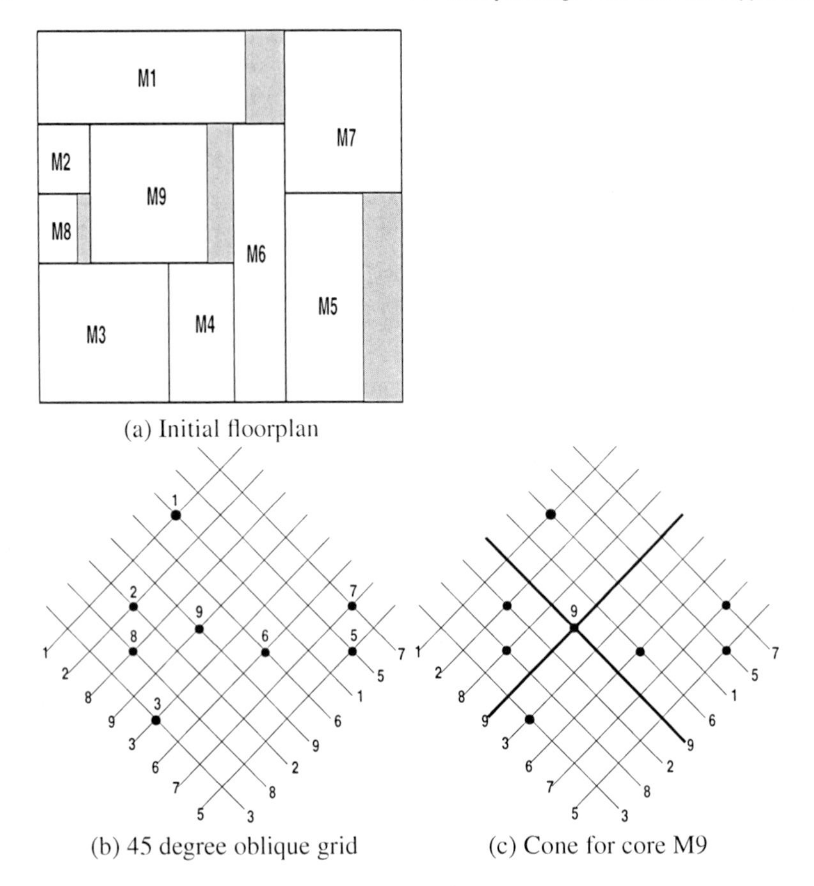

Fig. 21.1 Floorplan and its sequence-pair representation.

lines into four cones as shown in Figure 21.1 (c). In general, the sequence-pair imposes the following relationship between each pair of blocks:

$$(\langle \cdots b_i \cdots b_j \rangle, \langle \cdots b_i \cdots b_j \rangle) => b_i \textit{ is to the left of } b_j$$
$$(\langle \cdots b_j \cdots b_i \rangle, \langle \cdots b_i \cdots b_j \rangle) => b_i \textit{ is below } b_j$$

Given a sequence-pair representation for a block placement (test schedule, equiv.), the horizontal relationship amongst the cores follows a horizontal constraint graph $G_h(V,E)$, which can be constructed as follows:

1. $V = \{s_h\} \cup \{t_h\} \cup \{v_i | i = 1, \ldots, n\}$, where v_i corresponds to a core, s_h is the source node representing the left boundary and t_h is the sink node, representing the right boundary;
2. $E = \{(s_h, v_i) | i = 1, \ldots, n\} \cup \{(v_i, t_h) | i = 1, \ldots, n\} \cup \{(v_i, v_j) |$ core v_i is to the left of core v_j $\}$.

The vertical constraint graph $G_v(V,E)$ can be similarly constructed. The corresponding constraint graphs $G_h(V,E)$ and $G_v(V,E)$ for the example schedule shown in Figure 21.1(a) are shown in Figure 21.2(a) and (b), resp.

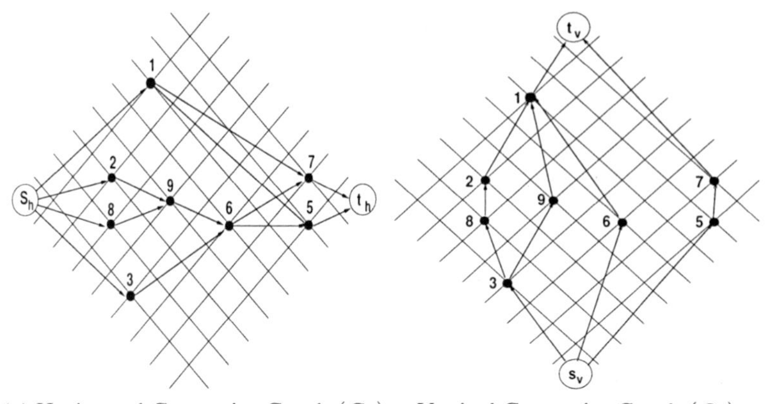

(a) Horizontal Constraint Graph (G_h) Vertical Constraint Graph (G_v)

Fig. 21.2 Graph representation of a floorplan (transitive edges have been omitted.)

In order to evaluate a block placement for the purpose of floorplanning, constraint graphs are vertex-weighted, according to the height (for $G_h(V,E)$) and width (for $G_v(V,E)$) of the block. Then a longest path algorithm is applied to find the x and y coordinates of each block. The coordinates of a block are the lower left corner of the block. The construction of constraint graphs $G_h(V,E)$ and $G_v(V,E)$ takes $\Theta(n^2)$ time. The longest path computation can be done in $O(n+m)$, where $n = |V|$ and $m = |E|$.

21.3 Use of a floorplan in non-VLSI applications

The VLSI floorplanning problem also finds application in the following domains:

1. Cloth & leather cutting: to minimize waste these industries would like to compact the desired shapes into smallest rectangles (usually with one dimension fixed). This problem is interesting in its own right as additional constraints of cloth-texture, cloth-print, leather-grain, have to be honored. A sequence-pair model of the simple (un-constrained) garment cutting problem follows immediately from the above discussion on VLSI floorplanning,
2. Newspaper layout: especially when laying out advertisement flyers where each advertisement block has adjacency relationships to many other blocks (think big-screen TV ad-space placed next to a digital video camera), VLSI floorplanning model can be used,

3. Online ad-placement on web-pages: with complex formatting being used in web-sites, dynamic advertisement placement in near-real-time is a challenge and the data-structures and algorithms we have described above can be used to solve them.

21.4 Implementation of a Sequence-pair based Floorplanner

Refer to the code shown in Listing 21.1.

```
/////////////////////////////////////////
// Program: Sequence Pair
// Author : Sandeep Koranne
/////////////////////////////////////////
#include <limits.h>
#include <stdio.h>
#include <stdlib.h>
#include <unistd.h>
#include <math.h>

#define INFTY 1000000

typedef struct {
  int source;
  int dest;
  int weight;
  int mark;
} Edge;

#define MAXNODES 1024

volatile int SP1_LOOKUP[ MAXNODES ] __attribute__ ((aligned(128)));
volatile int SP2_LOOKUP[ MAXNODES ] __attribute__ ((aligned(128)));
volatile int SP1[ MAXNODES ] __attribute__ ((aligned(128)));
volatile int SP2[ MAXNODES ] __attribute__ ((aligned(128)));
volatile int SP1C[ MAXNODES ] __attribute__ ((aligned(128)));
volatile int SP2C[ MAXNODES ] __attribute__ ((aligned(128)));
volatile int W[ MAXNODES ] __attribute__ ((aligned(128)));
volatile int H[ MAXNODES ] __attribute__ ((aligned(128)));
volatile int N;
volatile int GH[MAXNODES][MAXNODES] __attribute__ ((aligned(128)));
volatile int GV[MAXNODES][MAXNODES] __attribute__ ((aligned(128)));
#define MAX_EDGES (MAXNODES<<2)
volatile Edge EDGE_MATRIX[MAX_EDGES] __attribute__ ((aligned(128)));
volatile int DISTANCE[MAXNODES] __attribute__ ((aligned(128)));

int BF(volatile Edge edges[], int nE, int n, int source)
{
  int i, j;
  for (i=0; i < n; ++i) {DISTANCE[i] = INFTY;}
  DISTANCE[source] = 0;
  for (i=0; i < n; ++i) {
    int progress = 0;
    for (j=0; j < nE; ++j) {
      if (DISTANCE[edges[j].source] != INFTY) {
        int nd = DISTANCE[edges[j].source] + edges[j].weight;
        if (nd < DISTANCE[edges[j].dest]) {
          DISTANCE[edges[j].dest] = nd;
          progress=1; }}}
    if (!progress) break; }
  return -( DISTANCE[n-1] );
}
```

```
static int LongestPath( volatile int G[][MAXNODES], int n ) {
  int edge_count=0;
  int i,j;
  for(i=0;i<(n-1);++i)
    for(j=0;j<(n);++j)
      if( G[i][j] ) {
        EDGE_MATRIX[edge_count].source = i;
        EDGE_MATRIX[edge_count].dest = j;
        EDGE_MATRIX[edge_count].weight = -G[i][j];
        edge_count++; }
  if( edge_count >= MAX_EDGES ) {
    printf("\n Increase EDGE_MATRIX table size to %d", edge_count );
    exit(-1); }
  return BF( EDGE_MATRIX, edge_count, n, 0 );
}
static void ReadNodeInformation( const char* fileName ) {
  FILE* fp = fopen( fileName, "rt" );
  int i=0;  N = 0;
  fscanf( fp, "%d", &N );
  for(i=0; i < N; ++i) {
    int w,h;fscanf(fp, "%d %d",&w,&h);W[i]=w, H[i]=h;}
  fclose( fp );
}
static void PrintNodeInformation( ) {
  int i;
  printf("\n Total %d nodes.\n",N);
  for(i=0;i<N;++i)printf("\n Node %d : %d %d",i,W[i],H[i]);
  printf("\n");
}
static inline void InitializeSequence(volatile int *A,int n) {
  int i;for(i=0;i<n;++i) A[i]=i;
}
static void PermuteSequence(volatile int *A,int n, int trials) {
  int i=0; for(i=0; i<trials; ++i) {
    int p = rand() % n;
    int q = rand() % n;
    if( (p==0) || (q==0) || (p==(n-1)) || (q==(n-1))) continue;
    int temp_val = A[ p ];
    A[p] = A[q]; A[q] = temp_val;  }
}
static void PrintSequence( volatile int *A, int n ) {
  int i; printf("\n");for(i=0;i<n;++i) printf(" %d", A[i]);
}
static inline void CalculateRank( volatile int* S, volatile int* R, int n ) {
  int i=0;for(i=0;i<n;++i) R[ S[i] ] = i;
}
static void PrintGraph( volatile int G[][MAXNODES], int n ) {
  int i,j;
  printf("\n Graph:\n");
  for(i=0;i<n;++i) {
    for(j=0;j<n;++j) printf(" %d", G[i][j] );
    printf("\n"); }
}
static int AnalyzeGH( volatile int* SP1, volatile int* SP2, int n ) {
  // note that xi is left of xj when (xi,xj) in SP1 and (xi,xj) in SP2
  int i,j;
  for(i=0;i<n;++i)
    for(j=0;j<n;++j) GH[i][j]=0;
  for(i=1;i<n-1;++i) GH[0][i] = 1, GH[i][n-1] = W[i];
  for(i=1;i<n-2;++i) {
    for(j=i+1;j<n;++j) {
      int rcmp_sp1 = ( SP1_LOOKUP[i] < SP1_LOOKUP[j] );
      int rcmp_sp2 = ( SP2_LOOKUP[i] < SP2_LOOKUP[j] );
      if( rcmp_sp1 && rcmp_sp2 )GH[i][j] = W[i];
      rcmp_sp1 = ( SP1_LOOKUP[j] < SP1_LOOKUP[i] );
      rcmp_sp2 = ( SP2_LOOKUP[j] < SP2_LOOKUP[i] );
      if( rcmp_sp1 && rcmp_sp2 ) GH[j][i] = W[j];}
```

```
120     }
      return LongestPath( GH, n );
    }

    static int AnalyzeGV( volatile int* SP1, volatile int* SP2, int n ) {
125   // note that xi is left of xj when (xi,xj) in SP1 and (xi,xj) in SP2
      int i,j;
      for(i=0;i<n;++i)
        for(j=0;j<n;++j) GV[i][j]=0;
      for(i=1;i<n-1;++i) GV[0][i] = 1,GV[i][n-1] = H[i];
130   for(i=1;i<n-2;++i) {
        for(j=i+1;j<n;++j) {
          int rcmp_sp1 = ( SP1_LOOKUP[i] > SP1_LOOKUP[j] );
          int rcmp_sp2 = ( SP2_LOOKUP[i] < SP2_LOOKUP[j] );
          if( rcmp_sp1 && rcmp_sp2 ) GV[i][j] = H[i];
135       rcmp_sp1 = ( SP1_LOOKUP[j] > SP1_LOOKUP[i] );
          rcmp_sp2 = ( SP2_LOOKUP[j] < SP2_LOOKUP[i] );
          if( rcmp_sp1 && rcmp_sp2 )GV[j][i] = H[j]; }
      }
      return LongestPath( GV, n );
140 }

    static int PerformTrial( int TRIALS ) {
      int lph,lpv,cost;
      PermuteSequence( SP1, N, TRIALS );
145   PermuteSequence( SP2, N, TRIALS);
      CalculateRank( SP1, SP1_LOOKUP, N );
      CalculateRank( SP2, SP2_LOOKUP, N );
      lph = AnalyzeGH( SP1, SP2, N );
      lpv = AnalyzeGV( SP1, SP2, N );
150   cost = (lph*lpv);
      return cost;
    }
    static inline void CommitSequence( ) {
      int i;
155   for(i=0;i<N;++i)SP1C[i] = SP1[i],SP2C[i] = SP2[i];
    }
    static inline void UncommitSequence( ) {
      int i;
      for(i=0;i<N;++i) SP1[i] = SP1C[i],SP2[i] = SP2C[i];
160 }
    static int ConditionalAccept(int cost,int new_cost,float temperature) {
      float random_variable = (random()%1000)/(1000.0);
      if( temperature < 0.00001 ) return 0;
      return (temperature > random_variable);
165 }
    float UpdateTemperature( float temp ) {
      temp = temp*exp(-0.00005); return temp;
    }
    static void Usage( ) {
170   printf("\n sqp <dat-file> <loops> <trials>\n");
    }
    #define LOOP_COUNT 1000000
    int main(int argc, char* argv [] )
    {
175   int loops=0;
      int cost=0;
      int best_cost = INFTY;
      int trials=0;
      int new_cost = INFTY;
180   float temp=1000.0;
      if( argc > 1 ) {
        ReadNodeInformation( argv[1] );
        PrintNodeInformation();
      } else {
185     Usage();
        return 1;
```

```
        }
        trials = 2;
        loops = LOOP_COUNT;
190     if( argc > 2 ) loops = atoi( argv[2] );
        if( argc > 3 ) trials= atoi( argv[3] );
        InitializeSequence( SP1, N );
        InitializeSequence( SP2, N );
        PrintSequence( SP1, N );
195     PrintSequence( SP2, N );
        cost = INFTY;

        CommitSequence();
        srand( getpid() );
200
        for(loops=0; loops < LOOP_COUNT;++loops ) {
          new_cost = PerformTrial( trials );
          best_cost = (new_cost < best_cost) ? new_cost : best_cost;
          if( !(loops%10000) ) {
205         printf("\n Loop %d, Temp = %f, Cost = %d", loops, temp, cost);
          }
          if( new_cost < cost ) {
            cost = new_cost;
            CommitSequence();
210         continue;
          }
          temp=UpdateTemperature( temp );
          if( ConditionalAccept( cost, new_cost, temp ) ) {
            cost = new_cost;
215         CommitSequence();
            continue;
          }
          UncommitSequence();
        }
220     PrintSequence( SP1, N );
        PrintSequence( SP2, N );
        printf("\nBest cost = %d\n",best_cost);
        return 0;
      }
```

Listing 21.1 Sequence-pair based floorplanner

21.5 Conclusions

In this chapter we have presented the VLSI floorplanning problem. We introduced the sequence-pair notation and data-structure which is useful not only for floorplanning, but also the dual problem of scheduling and rectangle packing. The underlying solver for sequence-pair we have used is the general-purpose simulated annealer we had developed in Chapter 11 on graph-algorithms. The longest-path computation is performed using an implementation of Bellman-Ford [11]. Optimization of the Floorplanner code is performed using automated tools and branch elimination. This is discussed in detail in Chapter 15, pp. 301.

Chapter 22
VLSI Placement

Abstract In this chapter we introduce the VLSI standard-cell placement problem. We discuss how this problem can be modeled as a graph problem, and present our implementation to read gate-level BLIF netlists on the PPE. A simple method of generating BLIF designs with random output cover is also shown, so that the reader can generate example designs. These BLIF designs are mapped using ABC [121] producing gate-level netlists which are read by our PPE software, converted into a graph and then control is transferred to 6-SPUs which pull the graph from PPE using DMA and perform placement. The SPUs can model manhattan and euclidean metrics and implement simulated-annealing based placers with exponential cooling-schedule. The SPUs can run in a cooperative manner or competitive manner. In cooperative mode the SPUs solve different BLIF models, but in competitive mode, multiple SPUs solve the same problem, and the best solution is chosen by the PPE.

22.1 Introduction to VLSI Standard Cell Placement

In this chapter we introduce the VLSI Standard Cell Placement problem. In a previous chapter we had seen the VLSI physical design flow which includes floorplanning, placement and routing. In this chapter we focus on developing SPU based standard-cell placers. We use the simulated annealing framework we have developed (cf. Section 11.11). Our presentation of the cell placement problem is deliberately simple, but nevertheless captures the important notions of *node-placement* and *cost-functions*. For more details about cell-placement refer to the original survey by Shahookar and Mazumder in [114], and the book on VLSI Physical Design by Sherwani [115]. Another good introduction to Physical Design Automation is by Sarrafzadeh and Wong [109]. More recent information about VLSI placement can be obtained from any recent proceedings of the ISPD (International Symposium of Physical Design), and DAC (Design Automation Conference). A complete suite of free CAD tools for VLSI design from VHDL to layout is given by Alliance

S. Koranne, *Practical Computing on the Cell Broadband Engine*,
DOI: 10.1007/978-1-4419-0308-2_22, © Springer Science + Business Media, LLC 2009

CAD [123], and we have shown their layout viewer displaying standard-cell placement in Figure 22.1.

22.1.1 Input to the problem

The input to the cell-placement problem is a *netlist*, and a standard cell library. We describe a method to quickly generate realistic netlists with a given number of inputs, outputs and nodes. Consider the following Lisp program as shown in Listing 22.1. Using this program we can generate complex VLSI circuits based on covers of random Boolean functions.

```
;; Program: gen-blif, Author: Sandeep Koranne
(defun make-format-string (n) (format nil "~~~D,'0B 1~~%" n))
(defun print-random-function (fs n ni)
  (format fs (make-format-string ni) (random n)))
;; in addition to the inputs we can also use intermediate variables
;; the higher the op-count, the lesser and lesser (on avg) PI it
;; should get
(defconstant +TRUNC-FACTOR+ 2)
(defconstant +LAT-RAT+ 4)
(defun print-one-output (fs op-name n op-count)
  (let ((clip-n (floor (/ (- n (/ op-count +TRUNC-FACTOR+)) 1 ))))
    (format fs ".names ")
    (dotimes (i clip-n) (format fs " i~D" i))
    (dotimes (i (- n clip-n))
      (if (= 0 (mod i +LAT-RAT+))
          (format fs " lat_t_~D" (/ i +LAT-RAT+))
        (format fs " t_~D" i)))
    (format fs " ~A~%" op-name)))
(defun print-temp-one-output (fs op-name n)
  (format fs ".names ")
  (dotimes (i n)(format fs " t_~D" i))
  (format fs " ~A~%" op-name))
(defun print-header (fs model-name num-inputs num-outputs)
  (format fs ".model ~A" model-name)
  (format fs "~%.inputs ck ")
  (dotimes (i num-inputs)(format fs " i~D" i))
  (format fs "~%.outputs ")
  (dotimes (i num-outputs)(format fs " o~D" i))
  (format fs "~%"))
(defconstant +INTER-NODE+ 2)
(defun main (model-name num-input num-op sop)
  "Genrate a random combinational function"
  (let ((ks (expt 2 num-input))
        (op-done num-op)(temp-done 0)
        (fs (open (format nil "~A.blif" model-name)
                  :direction :output :if-exists :supersede)))
    (print-header fs model-name num-input num-op)
    (dotimes (f (* num-op +INTER-NODE+))
      (if (> (random 10) 3)
          (progn (print-one-output fs (format nil "t_~D" temp-done)
           num-input temp-done)
            (incf temp-done))
        (progn (when (> op-done 0)
   (print-one-output fs (format nil "o~D" (1- op-done))
           num-input (1- op-done))
   (decf op-done))))
     (dotimes (i sop) (print-random-function fs ks num-input)))
    ;; now if some primary-outputs are not covered
    ;; then generate for these ONLY
```

```
50      (format t "~%Covered ~D outputs using ~D temp variables.~%"
          op-done temp-done)
        (dotimes (i temp-done)
          (format fs ".latch t_~D lat_t_~D re ck 0~%" i i))
        (when (> op-done 0)
55        (let ((temp-ks (expt 2 temp-done)))
            (dotimes (f (+ 0 op-done))
              (print-temp-one-output fs (format nil "o~D" f) temp-done)
              (dotimes (i sop)
                (print-random-function fs temp-ks temp-done)))))
60      (format fs ".end")
        (close fs)))
    ;(main "b" 5 5 5)
```

Listing 22.1 Program to generate BLIF designs.

The output of this program is a BLIF netlist with a random output cover for every output as shown below:

```
    .model b
    .inputs ck  i0 i1 i2 i3 i4
    .outputs  o0 o1 o2 o3 o4
    .names  i0 i1 i2 lat_t_0 t_1 o4
5   11001 1
    10111 1
    00011 1
    11011 1
    00100 1
10  .names  i0 i1 i2 i3 i4 t_0
    11110 1
    .latch t_0 lat_t_0 re ck 0
    ..... file truncated to save space
    .end
```

Listing 22.2 Example BLIF design generated from Listing 22.1

We then map this file to a standard cell library containing primitive gates using ABC [121]. This results in a gate-level netlist, as shown below:

```
    # Benchmark "b" written by ABC on Wed Feb 25 02:31:59 2009
    .model b
    .inputs ck i0 i1 i2 i3 i4
    .outputs o0 o1 o2 o3 o4
5   .default_input_arrival 0 0
    .latch        n28    lat_t_1  0
    .latch        n33    lat_t_2  0
    .gate inv_x1         i=lat_t_0 nq=n35
    .gate inv_x1         i=i2 nq=n36
10  .gate a2_x2          i0=i3 i1=i2 q=n37
    .gate mx2_x2         cmd=i1 i0=n37 i1=n36 q=n38_1
    .gate a2_x2          i0=lat_t_0 i1=i3 q=n48_1
    .gate no2_x1         i0=lat_t_0 i1=i3 nq=n49
    .gate noa2a22_x1     i0=n49 i1=n46 i2=n48_1 i3=n47 nq=n50
15  .gate o2_x2          i0=n50 i1=i0 q=n51
    .... file truncated to save space
    .gate oa2ao222_x2    i0=n69 i1=n28 i2=n137 i3=n136 i4=n56 q=o4
    .end
```

Listing 22.3 Mapped gate-level BLIF netlist.

This gate-level netlist and the corresponding cell-library are the input to the placer which we develop in the next section.

22.2 Reading gate-level BLIF files and making a graph

Before we can use our graph based simulated annealing framework we need to convert this gate-level netlist into a graph. For every node we create a vertex, and for every connection in the netlist we add an edge into the graph. Since there maybe a net which contains more than 2 nodes in in, this creates a *hypergraph*. There are standard models for converting a hypergraph into a suitable graph for placement and we too have used a *clique* model. In this model a pairwise edge is added between every nodes present on an hyperedge. For the example considered above this results in a graph with $n = 116, m = 931$. The program to read in the gate-level BLIF and construct the graph is running on the PPE and some relevant part of that code is shown below in Listing 22.6. Once the graph has been constructed we launch the *placement* kernel on each of the 6-SPUs. The SPU based cell-placer is discussed in the next section.

22.3 SPU basic cell placer, using Simulated Annealing

If we refer back to our discussion on simulated-annealing, we see that we need atleast a *cost-function*, and a *cooling-schedule*. We have continued to use the exponential cooling schedule we developed for partitioning, but the cost function has to be developed with placement in mind. Consider an example VLSI placement as shown in Figure 22.1. The standard cell arrangement of the chip is organized in rows, with each row having a number of cells which abut to form contiguous metal lines for power supply on the top and bottom of each cell. Thus, each cell is exactly of the same height, but its width depends on the function of the cell. For simplicity in the annealer we start off with assuming that every cell has equal width as well. This yields a *checkerboard* placement problem where we are given an $W \times H$ checkerboard, and n cells to place on it, such that no two cells occupy the same square on the board. The cost of any given configuration is defined as follows:

$$C = \sum_{\forall e \in E} (|\Delta x| + |\Delta y|)$$

where Δx is defined as the *manhattan* distance between $n_1(x)$ and $n_2(x)$ with n_1 and n_2 being the endpoints of the edge. The nodes represent cells, and the edge represents a connection between them. The goal of the placer is to assign $n_i(x), n_i(y)$ for all $i \in n$, such that C is minimized. This is an NP-complete problem [49], thus simulated-annealing is one of the stochastic methods which has become popular in providing some solution.

Another cost function that can be considered is the *euclidean-metric* defined as:

$$C_{euc} = \sum_{\forall e \in E} ((\Delta x)^2 + (\Delta y)^2)$$

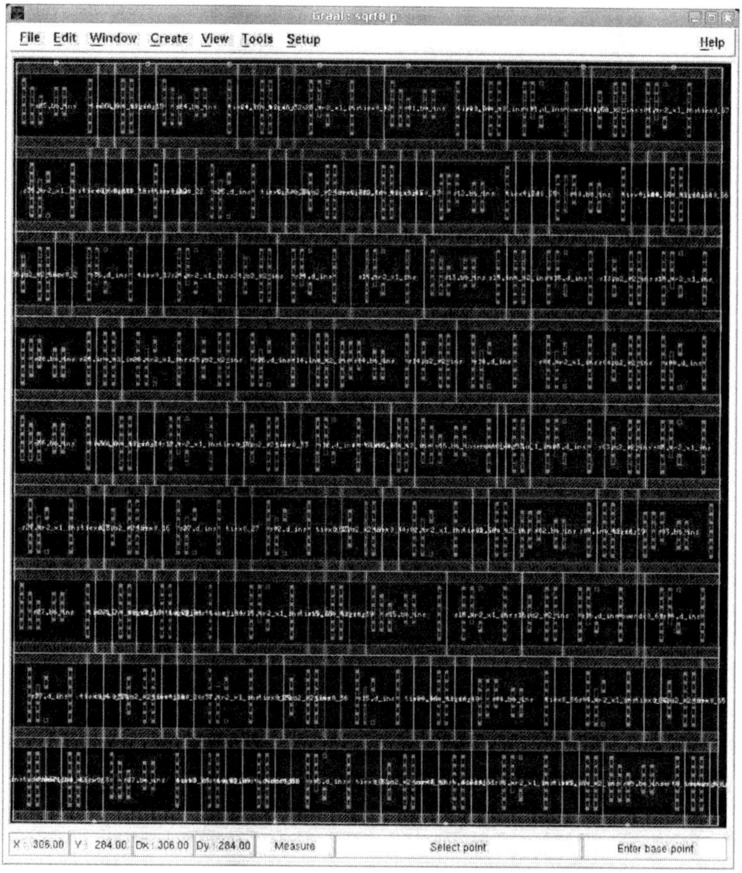

Fig. 22.1 Example of VLSI full chip placement.

We have implemented both in our SPU code as shown in Listing:22.8. Moreover, since we have 2 bits for our edge weight, we can add extra weight to timing-critical signals by selecting a small portion (less than 1%) of the nets to be timing critical. This information is recorded in the edge-weight bits of the edge.

```
////////////////////////////////////////
// Program: VLSI Placement
// Author : Sandeep Koranne
////////////////////////////////////////
#ifndef __NL_FWD_H__
#define __NL_FWD_H__

#include <string>
#include <vector>
#include <map>
#include <iostream>
#include "../align.h"

namespace Blif {
```

```
15     class BlifModel;
   }
   namespace Netlist  {
     class Blif::BlifModel;
     class Model;
20   typedef Model* ModelP;
     typedef std::vector<ModelP> ModelPVector;
     class Instance;
     class Net;
     class Port;
25   class Pin;
     class Cell;
     class GraphOperator;
     typedef std::map<std::string, Cell*> CellHash;
     class Library  {
30   public:
       Library();
       void setName(const std::string&);
       void addModel(const Blif::BlifModel&);
       Cell* addCell(const std::string&);
35     Cell* findCell(const std::string&) const;
       void ReadCellInformation(const std::string&);
       void PrintCellCatalog(std::ostream&) const;
       void WriteVstFormat(void) const;
       void CreateGraph(GraphOperator&) const;
40   private:
       std::string d_name;
       ModelPVector d_modelPVector;
       CellHash d_cellHash;
       Cell* d_modelCell;
45   };
     struct GraphOperator   {
      GraphOperator(size_t GSIZE) {
        m_graph = (unsigned int*)_malloc_align((GSIZE)*sizeof(int),7);
        std::cout << std::endl << "Mg = " << (unsigned long int)m_graph;
50      edge_count=0;n_count=0;
      }
      unsigned int getN(void) const { return n_count;}
      unsigned int getM(void) const { return edge_count; }
      unsigned int* get_m_graph(void) { return m_graph;  }
55    void ProcessConnection(int i, int j) const {
        n_count = (i>n_count) ? i : n_count;
        n_count = (j>n_count) ? j : n_count;
        unsigned int edge = (i<<20) | (j<<8);
        m_graph[edge_count++] = edge;
60    }
      unsigned int* m_graph;
      mutable unsigned int edge_count;
      mutable unsigned int n_count;
     };
65 } // end of Netlist namespace
   #endif
```

Listing 22.4 Netlist data structures, forwarding only.

```
   /////////////////////////////////////////
   // Program: VLSI Placement
   // Author : Sandeep Koranne
   /////////////////////////////////////////
 5 #ifndef __BLIF_H__
   #define __BLIF_H__

   #include <iostream>
   #include <fstream>
10 #include <string>
   #include <map>
```

```
#include <set>
#include <vector>
#include <cassert>
#include <algorithm>

#include "nlfwd.h"
namespace Blif  {
  struct Port  {
    std::string d_name;
    int d_id;
  };
  typedef std::vector<Port> PortVector;
  typedef std::pair<std::string, std::string> Argument;
  typedef std::vector<Argument> ArgumentVector;
  struct BlifInstance   {
    std::string d_cellName;
    int d_id;
    ArgumentVector d_argVector;
  };
  typedef std::vector<BlifInstance> InstanceVector;
  class BlifModel {
  public:
  BlifModel():numInstances(0){}
    std::string d_fileName;
    PortVector d_inputs;
    PortVector d_outputs;
    InstanceVector d_instances;
    std::string d_modelName;
    void createTerminalInstances(void);
    bool ShowInformation(void) const;
    size_t numInstances;
  };
  typedef std::vector<BlifModel> BlifModelVector;

  class BlifParser {
  public:
  BlifParser(const std::string& fileName)
    :d_fileName(fileName),d_library(NULL){}
    bool Read(void);
    void Init(void);
    bool ReadModel(std::ifstream&, const std::string&);
    bool ShowInformation(void) const;
    bool BuildNetlist(void);
    const Netlist::Library* getLibrary(void) const { return d_library; }
    Netlist::Library* getLibrary(void) { return d_library; }
  private:
    std::string d_fileName;
    BlifModelVector d_modelVector;
    Netlist::Library* d_library;
  };
} // end of Blif namespace
#endif
```

Listing 22.5 BLIF data structures, forwarding only.

```
// File : blif.cpp
// Author : Sandeep Koranne
// Purpose : A simple and fast BLIF reader

#include "blif.h"
#include <iostream>
#include <fstream>
#include <libspe2.h>
#include <stdlib.h>
#include <cassert>
#include <pthread.h>
```

```
#include <string.h>
#include <string>
#include <sstream>
15 #include <cstring>
#include "../SA/graph.h"

#define MAX_SPES 6
static bool min_cut_mode=true;
20 GraphControlBlock *cb[MAX_SPES];
extern spe_program_handle_t graph_function;

typedef struct ppu_pthread_data {
  void *argp;
25   spe_context_ptr_t speid;
  pthread_t pthread;
} ppu_pthread_data_t;
ppu_pthread_data_t *datas;

30 void *ppu_pthread_function(void *arg) {
  ppu_pthread_data_t *datap = (ppu_pthread_data_t *)arg;
  unsigned int entry = SPE_DEFAULT_ENTRY;
  int rc = spe_context_run(datap->speid, &entry,
                           0, datap->argp, NULL, NULL);
35   pthread_exit(NULL);
}

static void SetupSPES(void) {
  std::cout << "sizeof(CB) = "
40             << sizeof( GraphControlBlock) << std::endl;
  assert( (sizeof( GraphControlBlock ) % 16) == 0);
  for(int i=0;i<MAX_SPES;++i)
    cb[i] = (GraphControlBlock*)
      _malloc_align( sizeof(GraphControlBlock), 5 );
45   datas = (ppu_pthread_data_t*)
    malloc( sizeof(ppu_pthread_data_t)*MAX_SPES);
}
#define DP(token)
namespace Blif {
50   using namespace Netlist;
  bool BlifModel::ShowInformation(void) const {
    std::cout<<std::endl<<"Showing information for model "<<d_modelName
             <<" from reading file " <<d_fileName;
    std::cout<<std::endl<<"Model has "<<d_inputs.size()<<" inputs"
55             <<std::endl<<"Model has "<<d_outputs.size()<<" outputs"
             <<std::endl<<"Model has "<<d_instances.size()<<" instances.\n";
    for(InstanceVector::const_iterator vit=d_instances.begin();
        vit != d_instances.end();++vit) {
      std::cout<< std::endl << "Instance num " << vit->d_id
60               << " of cell " << vit->d_cellName
               << " has " << vit->d_argVector.size() << " ports.";
      for(ArgumentVector::const_iterator ait = vit->d_argVector.begin();
          ait != vit->d_argVector.end();++ait)
        std::cout << std::endl << "Port (local) " << ait->second
65                  << " -> (formal) " << ait->first; }
    return true;
  }
  void BlifModel::createTerminalInstances(void) {
    for(PortVector::const_iterator pit = d_inputs.begin();
70         pit != d_inputs.end(); ++pit) {
      BlifInstance binst;
      binst.d_id = numInstances++;
      binst.d_cellName = "ITERMINAL";
      binst.d_argVector.push_back(Argument("TEI",pit->d_name));
75       d_instances.push_back(binst);
    }
    for(PortVector::const_iterator pit = d_outputs.begin();
        pit != d_outputs.end(); ++pit) {
```

```
    BlifInstance binst;
    binst.d_id = numInstances++;
    binst.d_cellName = "OTERMINAL";
    binst.d_argVector.push_back(Argument("TRO",pit->d_name));
    d_instances.push_back(binst);
  }
}

bool BlifParser::ReadModel(std::ifstream&ifs,const std::string& fileName){
  BlifModel model;
  std::string token, fullLine;
  int numInputs=0, numOutputs=0;
  ifs >> model.d_modelName;
  model.d_fileName = fileName;
  getline(ifs, fullLine);
  if(ifs.fail()) return true;

  while(ifs) {
    ifs >> token;
    DP(token);
    if(token == "#") {
      getline(ifs, fullLine);
      continue;
    } else if(token == ".model") {
      std::cerr << std::endl << "Nested model definition found in file.\n";
      assert(false);
    } else if(token == ".inputs") {
    InputsAgain:
      std::string inputName;
      getline(ifs,fullLine);
      std::istringstream isfs(fullLine);
      isfs >> inputName;
      while(isfs) {
        Port p;p.d_name = inputName;
        p.d_id = numInputs++;
        model.d_inputs.push_back(p);
        isfs >> inputName;
        if(inputName == "\\") {
          goto InputsAgain;
        }
      }
    } else if(token == ".outputs") {
    OutputsAgain:
      std::string outputName;
      getline(ifs,fullLine);
      std::istringstream isfs(fullLine);
      isfs >> outputName;
      while(isfs) {
        DP(outputName);
        Port p;p.d_name = outputName;
        p.d_id = numOutputs++;
        model.d_outputs.push_back(p);
        isfs >> outputName;
        if(outputName == "\\") {
          goto OutputsAgain;
        }
      }
    } else if(token == ".latch") {
      BlifInstance inst;
      if(model.d_instances.empty())
        model.createTerminalInstances();
      inst.d_cellName = "sff1_x4";
      inst.d_id = model.numInstances++;
      Argument localArg;
      localArg.first = "ck";localArg.second = "ck";
      inst.d_argVector.push_back(localArg);
      localArg.first = "i";
```

```
        ifs >> localArg.second;inst.d_argVector.push_back(localArg);
        localArg.first = "q";
        ifs >> localArg.second; inst.d_argVector.push_back(localArg);
        model.d_instances.push_back(inst);
150   } else if(token == ".gate") {
        BlifInstance inst;
        if(model.d_instances.empty())
          model.createTerminalInstances();
        ifs >> inst.d_cellName; // an2v
155     inst.d_id = model.numInstances++;
      MoreArgs:
        getline(ifs,fullLine);
        std::istringstream isfs(fullLine);
        std::string arg;
160     isfs >> arg;
        while(isfs) {
          char formal[512];
          char local[512];
          //std::cout << std::endl << "Arg = " << arg;
165       // the arguments should not have space around the '=' operator
          // as we cannot parse it in that situation.
          const char* const eqPosition = std::strstr(arg.c_str(),"=");
          size_t numJump = eqPosition-arg.c_str();
          const char* const mainStr = arg.c_str();
170       if(eqPosition) {
            strncpy(formal,arg.c_str(), numJump); formal[numJump]='\0';
            strcpy(local,mainStr+numJump+1);
          }
          Argument localArg;
175       localArg.first = formal;
          localArg.second = local;
          inst.d_argVector.push_back(localArg);
          isfs >> arg;
          if(arg == "\\") {
180         goto MoreArgs;
          }
        }
        model.d_instances.push_back(inst);
      } else if(token == ".default_input_arrival") {
185     getline(ifs,fullLine);
        continue;
      } else if(token == ".end") {
        std::cout << std::endl << "Read model " << model.d_modelName;
        d_modelVector.push_back(model);
190     return true;
      }
    }
    assert(false);
    return false;
195 }

  bool BlifParser::Read(void) {
    std::ifstream ifs;
    std::string fullLine;
200 ifs.open(d_fileName.c_str(),std::ios::in);
    if(!ifs) {
      std::cerr << std::endl << "Cannot open file: " << d_fileName
                << " for reading.\n";
      exit(-1);
205 }
    std::string token;
    while(ifs) {
      ifs >> token;
      DP(token);
210   if(token == "#") { getline(ifs, fullLine); continue; }
      else if(token == ".model") {
        bool correctRead = ReadModel(ifs,d_fileName);
```

```
        if(!correctRead) {
          std::cerr << std::endl << "Unable to parse file.\n";
215       return false;}
      } else {
        std::cerr << std::endl << "Unknown token: " << token;
        assert(false);
      }
220   }
    return true;
  }

  bool BlifParser::ShowInformation(void) const {
225   for(BlifModelVector::const_iterator bmit = d_modelVector.begin();
          bmit != d_modelVector.end(); ++bmit) {
      const BlifModel& model = *bmit;
      model.ShowInformation();
    }
230   return true;
  }

  void BlifParser::Init(void) {
    if(!d_library) d_library = new Library();
235   assert(d_library);
    d_library->setName(d_fileName);
  }

  bool BlifParser::BuildNetlist(void){
240   assert(d_library);
    for(BlifModelVector::const_iterator bmit = d_modelVector.begin();
          bmit != d_modelVector.end();
          ++bmit) {
      const BlifModel& model = *bmit;
245     d_library->addModel(model);
    }
    return true;
  }
}
250
static size_t GetMemorySize(void) {
  std::ifstream ifs;
  ifs.open("/proc/self/statm",std::ios::in);
  size_t memoryUsed = 0;
255 ifs >> memoryUsed;
  return memoryUsed;
}

static void DisplayMemoryUsed(void) {
260 size_t memUsed = GetMemorySize();
  std::cout << std::endl << "Program consumed "<< memUsed << " kb.\n";
}

int main(int argc, char* argv[]) {
265 if(argc < 2) {
    std::cerr << std::endl << "Expecting blif file name\n";
    return -1;
  }
  std::string fileName(argv[1]);
270 using namespace Blif;
  BlifParser myParser(fileName);
  myParser.Read();
  std::cout << std::endl << "File parsed correctly.\n";
  //myParser.ShowInformation();
275 myParser.Init();
  myParser.getLibrary()->ReadCellInformation(argv[2]);
  myParser.BuildNetlist();
  myParser.getLibrary()->PrintCellCatalog(std::cout);
  myParser.getLibrary()->WriteVstFormat();
```

```
280   GraphOperator gobj(GRAPH_SIZE);
      myParser.getLibrary()->CreateGraph(gobj);
      std::cout<<std::endl<<"N="<<gobj.getN()<<" M="<<gobj.getM()<<std::endl;
      unsigned int TRIALS, TOPRUN, EXACT, SAVE;
      TRIALS=10000; TOPRUN=1000; EXACT=10; SAVE=0; // default,
285
      SetupSPES();
      for (int i = 0; i < MAX_SPES ; i++) {
        datas[i].speid = spe_context_create (0, NULL);
        spe_program_load (datas[i].speid, &graph_function);
290     cb[i]->addr = (long long unsigned int)gobj.get_m_graph();
        cb[i]->M = gobj.getM();
        cb[i]->N = gobj.getN();
        cb[i]->TRIALS = TRIALS;
        cb[i]->TOPRUN = TOPRUN;
295     cb[i]->EXACT  = EXACT;
        cb[i]->SAVE = SAVE;
        datas[i].argp = (unsigned long long*) cb[i];
        pthread_create (&datas[i].pthread, NULL,
                        &ppu_pthread_function, &datas[i]);
300   }
      for (int i = 0; i < MAX_SPES ; i++) {
        pthread_join (datas[i].pthread, NULL);
      }
      // now we can read the results
305   for (int i = 0; i < MAX_SPES ; i++) {
        unsigned int mbret;
        while( spe_out_mbox_status( datas[i].speid ) <= 0 ) ;
        spe_out_mbox_read( datas[i].speid, &mbret, 1 );
        printf("\n SAnnealing on (%x) done,cost = %d\n", datas[i].speid, mbret );
310     spe_context_destroy (datas[i].speid);
      }
      DisplayMemoryUsed();
      return 0;
    }
```

Listing 22.6 BLIF and PPE data structures.

```
    ////////////////////////////////////////
    // Program: VLSI Placement
    // Author : Sandeep Koranne
    ////////////////////////////////////////
5   #include <string>
    #include <cmath>

    #include "nlfwd.h"
    #include "blif.h"
10
    namespace Netlist {
      class Port {
      public:
        typedef enum { INPUT, OUTPUT, INPUT_OUTPUT } Direction;
15      Port(){}
        void Display(void) const;
        Direction d_dir;
        std::string d_name;
      };
20    typedef Port* PortP;
      typedef std::vector<PortP> PortPVector;
      typedef Pin* PinP;
      typedef std::vector<PinP> PinVector;
      typedef std::vector<const Pin*> ConstPinVector;
25    class Instance  {
      public:
        Instance(const Blif::BlifInstance& bid, const Cell* pCell)
          :d_id(bid.d_id),d_cell(pCell) {}
```

```
    void Display(void) const;
    float getWidth(void) const;
    float getHeight(void) const;
    void addPin(const Pin* p) const { d_pinVector.push_back(p);}
    int d_id;
    const Cell* d_cell;
    mutable ConstPinVector d_pinVector;
  };
  typedef Instance* InstanceP;
  typedef std::vector<InstanceP> InstanceVector;

  class Pin  {
  public:
    Pin(){}
    void Display(void) const;
    const Instance* d_pInstance;
    std::string d_portName;
    const Net* d_pNet;
    char getDirectionCode(void) const { return 'I';}
  };

  static std::string getVHDLName(const std::string& name) {
    return name;
    size_t len = name.size();
    const char* const mainString = name.c_str();
    char* vhname = new char[len+1];
    size_t count=0;
    for(size_t i=0; i < len; ++i) {
      if((! std::isalpha(*(mainString+i))) &&
         (! std::isdigit(*(mainString+i)))) {
        // dont add it
      }
      else {vhname[count++] = *(mainString+i);}
    }
    vhname[count] = '\0';
    std::string retStr = "sig" + std::string(vhname);
    delete [] vhname;
    return retStr;
  }

  class Net {
  public:
    typedef enum { TXT, VERILOG,VHDL } FormatType;
    Net(): is_primary( false ){}
    Pin *addPin(const Instance*, const std::string&);
    const std::string getName(FormatType t=TXT) const {
      switch(t) {
      case TXT:
      case VERILOG:
      default:
        return d_netName;
      case VHDL:{ return getVHDLName(d_netName);}
      }
    }
    void setName(const std::string& name)  { d_netName = name;}
    void Display(void) const;
    void WriteBookshelf(std::ostream&) const;
    const PinVector& getPinVector(void) const { return d_pinVector;}
    void SetPrimary( ) const { is_primary = true; }
    bool IsPrimary ( ) const { return is_primary; }
    template <typename T> void Operate(T& obj) const;
  private:
    PinVector d_pinVector;
    std::string d_netName;
    mutable bool is_primary;
  };
```

```
    typedef Net* NetP;
    typedef std::map<std::string, NetP> NetCollection;

    class Cell {
100 public:
      Cell():d_rootCell(false){}
      void createArgumentList(const Blif::ArgumentVector&);
      void createArgumentList(const Blif::BlifModel&);
      void createInterfaceNets(void);
105   Net* getNetHandle(const std::string&);
      void createModel(const Blif::BlifModel&, Library&);
      void Display(void) const;
      void WriteBookshelf(void) const;
      float getWidth(void) const;
110   float getHeight(void) const;
      size_t getNumberPins(void) const;
      void WriteEntity(std::ostream&) const;
      void WriteComponentDescription(std::ostream&) const;
      void WriteSignals(std::ostream&) const;
115   void WritePortMap(std::ostream&) const;
      void SetRootCell(bool i=true) { d_rootCell = i;}
      bool isRoot(void) const { return d_rootCell;}
      const NetCollection& getNetCollection() const { return d_netCollection;}
      const InstanceVector& getInstanceVector() const{ return d_instances;}
120   std::string d_name;
      int d_width,d_height;
      PortPVector d_portPVector;
    private:
      NetCollection d_netCollection;
125   InstanceVector d_instances;
      bool d_rootCell;
    };
  }

130 float Netlist::Instance::getWidth(void) const {
    return d_cell->getWidth();
  }
  float Netlist::Instance::getHeight(void) const {
    return d_cell->getHeight();
135 }
  Netlist::Library::Library()
    :d_modelCell(NULL) {}

  void Netlist::Library::ReadCellInformation(const std::string& fileName) {
140   std::ifstream ifs(fileName.c_str());
    assert(ifs);
    while(ifs) {
      std::string cellName;
      ifs >> cellName;
145   if(cellName == "") continue;
      Netlist::Cell* pCell = findCell(cellName);
      if(!pCell) pCell = addCell(cellName);
      int w,h;
      ifs >> w >> h;
150   pCell->d_width = w;
      pCell->d_height = h;
    }
  }
  void Netlist::Library::setName(const std::string& libName) {
155   d_name = libName;
  }
  void Netlist::Port::Display(void) const {
    std::cout << "(T) {"<<d_name<<","<<d_dir<<"} ";
  }
160 void Netlist::Instance::Display(void) const {
    std::cout << "(I) {"<<d_id<<"} ";
  }
```

```
void Netlist::Pin::Display(void) const {
  std::cout << "(P) {"<<d_pInstance->d_id << "," << d_portName
165            << "," << d_pNet->getName() << "} ";
}
void Netlist::Net::Display(void) const {
  std::cout << "(N) { " << d_netName
            << " [" << d_pinVector.size() << "]"
170            << "\t<";
  for(Netlist::PinVector::const_iterator pit = d_pinVector.begin();
      pit != d_pinVector.end();
      ++pit) {
    const Pin* pPin = *pit;
175    assert(pPin);
    pPin->Display();
  }
  std::cout << " > }\n";
}
180
template <typename T>
void Netlist::Net::Operate(T& obj) const {
  for(Netlist::PinVector::const_iterator pit = d_pinVector.begin();
      pit != d_pinVector.end();
185      ++pit) {
    for(Netlist::PinVector::const_iterator pjt = pit;
        pjt != d_pinVector.end();
        ++pjt) {
      if( pit == pjt ) continue;
190
      const Pin* iPin = *pit;
      const Pin* jPin = *pjt;
      int iid = iPin->d_pInstance->d_id;
      int jid = jPin->d_pInstance->d_id;
195      obj.ProcessConnection(iid,jid);
    }
  }
}

200 void Netlist::Cell::Display(void) const {
  std::cout << std::endl << "(C) {\n";
  for(Netlist::NetCollection::const_iterator nit = d_netCollection.begin();
      nit != d_netCollection.end();
      ++nit) {
205    const Net* pNet = nit->second;
    assert(pNet);
    pNet->Display();
  }
  std::cout << std::endl << "(TW)=" << d_instances.size();
210  std::cout << "}\n";
}

size_t Netlist::Cell::getNumberPins(void) const {
  size_t retval = 0;
215  for(Netlist::NetCollection::const_iterator nit = d_netCollection.begin();
      nit != d_netCollection.end();
      ++nit) {
    const Net* pNet = nit->second;
    assert(pNet);
220    retval += pNet->getPinVector().size();
  }
  return retval;
}

225 Netlist::Cell* Netlist::Library::findCell(const std::string& name) const {
  Netlist::CellHash::const_iterator findIt = d_cellHash.find(name);
  return (findIt == d_cellHash.end()) ? NULL : findIt->second;
}
```

```
230 Netlist::Cell* Netlist::Library::addCell(const std::string& name) {
      Netlist::Cell* pCell = new Netlist::Cell();
      pCell->d_name = name;
      d_cellHash[name] = pCell;
      return pCell;
235 }

    Netlist::Pin* Netlist::Net::addPin(const Netlist::Instance* pInst,
                                       const std::string& portName) {
      Netlist::Pin* pPin = new Netlist::Pin();
240   pPin->d_pInstance = pInst;
      pPin->d_portName = portName;
      pPin->d_pNet = this;
      d_pinVector.push_back(pPin);
      pInst->addPin(pPin);
245   return pPin;
    }

    void Netlist::Net::WriteBookshelf(std::ostream& os) const {
      os << "NetDegree :\t " << d_pinVector.size() << std::endl;
250   for(Netlist::PinVector::const_iterator pit = d_pinVector.begin();
          pit != d_pinVector.end();
          ++pit) {
        const Pin* pPin = *pit;
        assert(pPin);
255     os << "\t\t" << "U" << pPin->d_pInstance->d_id << "\t"
          << pPin->getDirectionCode() << " : " << " 0.5 0.5 " << std::endl;
      }
    }

260 void Netlist::Cell::WriteSignals(std::ostream& os) const {
      os << std::endl << "-- Starting to write signals \n";
      for(Netlist::NetCollection::const_iterator nit = d_netCollection.begin();
          nit != d_netCollection.end();
          ++nit) {
265     const Netlist::Net* pNet = nit->second;
        if( pNet->IsPrimary() ) continue;
        os << "signal " << pNet->getName(Netlist::Net::VHDL)
          << " : bit;" << std::endl;
      }
270   os << std::endl << "-- Completed write signals \n";
    }


    void Netlist::Cell::WriteComponentDescription(std::ostream& os) const {
275   // write this cell as
      /*
       * Component mx3_x2
       port (
       cmd0 : in      bit;
280    cmd1 : in      bit;
       i0   : in      bit;
       i1   : in      bit;
       i2   : in      bit;
       q    : out     bit;
285    vdd  : in      bit;
       vss  : in      bit
       );
       end component;
      */
290   //std::cout << std::endl << "Cell " << d_name
      //          << " has " << d_portPVector.size() << " ports.\n";
      if(d_portPVector.empty()) return;
      if(isRoot()) return;
      os << "\t component " << d_name << std::endl
295     << "\t\t port ( " << std::endl;
      size_t i=0;
```

```
  for(i=0; i < d_portPVector.size();++i) {
    const Netlist::Port* pPort = d_portPVector[i];
    assert(pPort);
300 //std::string represented_name( getVHDLName( pPort->d_name ) );
    std::string represented_name( pPort->d_name  );
    if(pPort->d_dir == Netlist::Port::INPUT)
      os << "\t\t\t" << represented_name << " : in\tbit;" << std::endl;
    else
305   os << "\t\t\t" << represented_name << " : out\tbit;" << std::endl;
  }
  os << "\t\t\tvdd : in\tbit;" << std::endl
     << "\t\t\tvss : in\tbit" << std::endl
     << "\t\t);" << std::endl
310  << "\t end component;" << std::endl;
}

void Netlist::Cell::WriteEntity(std::ostream& os) const {
  assert(isRoot());
315 os << "entity " << d_name << " is " << std::endl
     << "\t\t port ( " << std::endl;
  size_t i=0;
  for(i=0; i < d_portPVector.size();++i) {
    const Netlist::Port* pPort = d_portPVector[i];
320 assert(pPort);
    std::string represented_name( getVHDLName( pPort->d_name ) );
    if(pPort->d_dir == Netlist::Port::INPUT)
      os << "\t\t\t" << represented_name << " : in\tbit;" << std::endl;
    else
325   os << "\t\t\t" << represented_name << " : out\tbit;" << std::endl;
    NetCollection::const_iterator nit = d_netCollection.find( pPort->d_name );
    assert( nit != d_netCollection.end() );
    Net* const np = nit->second;
    np->SetPrimary();
330 }
  os << "\t\t\tvdd : in\tbit;" << std::endl
     << "\t\t\tvss : in\tbit" << std::endl
     << "\t\t);" << std::endl
     << "\t end " << d_name << ";" << std::endl;
335 }

void Netlist::Cell::WritePortMap(std::ostream& os) const {
  Netlist::InstanceVector::const_iterator iit = d_instances.begin();
  iit += d_portPVector.size();
340 for(;iit != d_instances.end();
      ++iit) {
    const Netlist::Instance* pInst = *iit;
    assert(pInst);
    os << "U" << pInst->d_id << " : "
345    << pInst->d_cell->d_name << std::endl
       << "\tport map (" << std::endl;
    Netlist::ConstPinVector& cpvec = pInst->d_pinVector;
    for(Netlist::ConstPinVector::const_iterator pit = cpvec.begin();
        pit != cpvec.end();
350     ++pit) {
      const Netlist::Pin* const pin = *pit;
      assert(pin);
      os << "\t\t" << pin->d_portName << " => "
         << pin->d_pNet->getName(Netlist::Net::VHDL) << "," << std::endl;
355 }
    os << "\t\tvdd => vdd, " << std::endl
       << "\t\tvss => vss " << std::endl
       << ");" << std::endl;
  }
360 }

void Netlist::Cell::createArgumentList(const Blif::ArgumentVector& arg) {
```

```
    // remember the second is the formal list which
365 // we want to add, direction is infered
    size_t num = arg.size();
    d_portPVector.reserve(num);
    Netlist::Port *pPort = NULL;
    for(size_t i=0; i < num-1; ++i) {
370   pPort = new Port();
      pPort->d_dir = Netlist::Port::INPUT;
      pPort->d_name = arg[i].first; // use actual local net name
      d_portPVector.push_back(pPort);
    }
375 pPort = new Port();
    pPort->d_dir = Netlist::Port::OUTPUT;
    pPort->d_name = arg[num-1].first;
    d_portPVector.push_back(pPort);
  }
380
  void Netlist::Cell::createArgumentList(const Blif::BlifModel& model) {
    d_portPVector.reserve(model.d_inputs.size() + model.d_outputs.size());
    for(Blif::PortVector::const_iterator pit = model.d_inputs.begin();
        pit != model.d_inputs.end(); ++pit) {
385   Port* pPort = new Port();
      pPort->d_dir = Netlist::Port::INPUT;
      pPort->d_name = pit->d_name;
      d_portPVector.push_back(pPort);
    }
390 for(Blif::PortVector::const_iterator pit = model.d_outputs.begin();
        pit != model.d_outputs.end(); ++pit) {
      Port* pPort = new Port();
      pPort->d_dir = Netlist::Port::OUTPUT;
      pPort->d_name = pit->d_name;
395   d_portPVector.push_back(pPort);
    }
  }

  float Netlist::Cell::getWidth(void) const {
400 return static_cast<float>(d_width);
  }
  float Netlist::Cell::getHeight(void) const {
    return static_cast<float>(d_height);
  }
405
  void Netlist::Cell::createInterfaceNets(void) {}

  void Netlist::Cell::WriteBookshelf(void) const {
    // write the <c>.nodes, <c>.nets files
410 const float PORT_WIDTH=10.0, PORT_HEIGHT=10.0;
    const std::string PORT_TERMINAL = "terminal";
    std::ofstream ofs;
    std::string fileName(d_name);
    fileName+=".nodes";
415 ofs.open(fileName.c_str(),std::ios::out);
    assert(ofs);
    ofs << "UCLA\tnodes\t1.0" << std::endl
        << "NumNodes : " << d_instances.size() << std::endl
        << "NumTerminals : " << d_portPVector.size() << std::endl;
420 size_t nt=0;
    for(Netlist::PortPVector::const_iterator pit = d_portPVector.begin();
        pit != d_portPVector.end();
        ++pit,++nt) {
      const Netlist::Port* pPort = *pit;
425   assert(pPort);
      ofs << "U" << nt << "\t" << PORT_WIDTH
          << "\t" << PORT_HEIGHT << "\t" << PORT_TERMINAL << std::endl;
    }
    // now write the instances
430 Netlist::InstanceVector::const_iterator iit = d_instances.begin();
```

```
      iit += d_portPVector.size();
      for(;iit != d_instances.end();
          ++iit) {
        const Netlist::Instance* pInst = *iit;
435     assert(pInst);
        ofs << "U" << pInst->d_id << "\t" << pInst->getWidth()
            << "\t" << pInst->getHeight() << std::endl;
      }
      ofs.close();
440
      fileName = d_name;
      fileName += ".corr";
      ofs.clear();
      ofs.open(fileName.c_str(), std::ios::out);
445   for(iit = d_instances.begin(); iit != d_instances.end(); ++iit) {
        const Netlist::Instance* pInst = *iit;
        ofs << "U" << pInst->d_id << "\t"
            << pInst->d_cell->d_name << std::endl;
      }
450   ofs.close();

      fileName = d_name;
      fileName += ".nets";
      ofs.clear();
455   ofs.open(fileName.c_str(),std::ios::out);
      assert(ofs);

      ofs << "UCLA\tnets\t1.0" << std::endl
          << "NumNets :\t " << d_netCollection.size() << std::endl
460       << "NumPins :\t " << getNumberPins() << std::endl;
      for(Netlist::NetCollection::const_iterator nit = d_netCollection.begin();
          nit != d_netCollection.end();
          ++nit) {
        const Netlist::Net* pNet = nit->second;
465     assert(pNet);
        pNet->WriteBookshelf(ofs);
      }
      ofs.close();

470   fileName = d_name;
      fileName += ".pl";
      ofs.clear();
      ofs.open(fileName.c_str(),std::ios::out);
      assert(ofs);
475
      ofs << "UCLA\tpl\t1.0" << std::endl;
      for(nt=0;nt < d_instances.size();++nt) {
        ofs << "U" << nt << "\t0 0 : N" << std::endl;
      }
480   ofs.close();

      fileName = d_name;
      fileName += ".wts";
      ofs.clear();
485   ofs.open(fileName.c_str(),std::ios::out);
      assert(ofs);
      ofs << "UCLA\twts\t1.0" << std::endl;
      for(nt=d_portPVector.size();nt < d_instances.size();++nt) {
        ofs << "U" << nt << "\t1" << std::endl;
490   }
      ofs.close();

      // calculate parameters for SCL file
      float denom = 2.0;
495   float bloat = 1.5;
      size_t totalInstances = d_instances.size();
      float totalArea = (float)totalInstances * 50.0;
```

```
    size_t numRows = 1 + static_cast<size_t>
      ((totalArea / ((1.0*50.0)*std::sqrt((float)totalInstances)))/denom);
500 std::cout << std::endl << "Total instances = "
              << totalInstances << " per row = "
              << static_cast<int>(denom*sqrt(totalInstances)) << std::endl;
    size_t verticalBump = 50;//static_cast<size_t>(75);
    fileName = d_name;
505 fileName += ".scl";
    ofs.clear();
    ofs.open(fileName.c_str(),std::ios::out);
    assert(ofs);
    ofs << "UCLA\tscl\t1.0" << std::endl
510     << "NumRows : " << numRows << std::endl;
    for(nt=0;nt<numRows;++nt) {
      ofs << "CoreRow Horizontal" << std::endl
          << "Coordinate : " << (50+nt*verticalBump) << std::endl
          << "Height : 50" << std::endl
515       << "SiteWidth : 1" << std::endl
          << "Sitespacing : 1" << std::endl
          << "Siteorient : 1" << std::endl
          << "Sitesymmetry : 1" << std::endl
          << "SubrowOrigin : 0 NumSites : "
520       << static_cast<int>(bloat * denom*sqrt(totalInstances)) << std::endl
          << "End" << std::endl;
    }
    ofs.close();

525 fileName = d_name;
    fileName += ".aux";
    ofs.clear();
    ofs.open(fileName.c_str(),std::ios::out);
    assert(ofs);
530 ofs << "RowBasedPlacement : " << d_name << ".nodes "
        << d_name << ".nets "
        << d_name << ".wts "
        << d_name << ".pl "
        << d_name << ".scl " << std::endl;
535 ofs.close();

  }

  void Netlist::Cell::createModel(const Blif::BlifModel& model, Library& library)
540 {
    createArgumentList(model);
    createInterfaceNets();
    d_instances.resize(model.d_instances.size());
    for(Blif::InstanceVector::const_iterator vit = model.d_instances.begin();
545     vit != model.d_instances.end(); ++vit) {
      const Blif::BlifInstance& instance = *vit;
      const std::string& cellName = instance.d_cellName;

      Netlist::Cell* pCell = library.findCell(cellName);
550   if(!pCell) pCell = library.addCell(cellName);
      if(pCell->d_portPVector.empty())
        pCell->createArgumentList(instance.d_argVector);
      Netlist::Instance* pInstance = new Netlist::Instance(instance,pCell);
      d_instances[instance.d_id] = pInstance;
555   for(Blif::ArgumentVector::const_iterator ait = instance.d_argVector.begin();
          ait != instance.d_argVector.end();
          ++ait) {
        Netlist::Net *pNet = getNetHandle(ait->second);
        pNet->addPin(pInstance,ait->first); }
560 }
  }

  Netlist::Net* Netlist::Cell::getNetHandle(const std::string& netName) {
    Netlist::NetCollection::const_iterator findIt = d_netCollection.find(netName);
```

```
565   if(findIt != d_netCollection.end())
        return findIt->second;
      Netlist::Net* pNet = new Netlist::Net();
      d_netCollection[netName] = pNet;
      pNet->setName(netName);
570   return pNet;
    }

    void Netlist::Library::addModel(const Blif::BlifModel& model) {
      std::cout << std::endl << "////////// Adding model "
575             << model.d_modelName << " to library.\n";
      d_modelCell = findCell(model.d_modelName);
      assert(d_modelCell == NULL);
      d_modelCell = addCell(model.d_modelName);
      d_modelCell->SetRootCell();
580   d_modelCell->createModel(model,*this);
      //d_modelCell->Display();
      d_modelCell->WriteBookshelf();
    }

585 void Netlist::Library::PrintCellCatalog(std::ostream& os) const {
      for(Netlist::CellHash::const_iterator cit=d_cellHash.begin();
          cit != d_cellHash.end(); ++cit) {
        //os << std::endl << cit->first;
        //const Netlist::Cell* pCell = cit->second;
590     //pCell->WriteComponentDescription(os);
      }
      os << std::endl;
      os << "Total cells in library = " << d_cellHash.size() << std::endl;
    }
595
    void Netlist::Library::WriteVstFormat(void) const {
      assert(d_modelCell);
      std::string vstName = d_modelCell->d_name + ".vst";
      std::ofstream ofs(vstName.c_str());
600   assert(ofs);
      d_modelCell->WriteEntity(ofs);
      ofs << std::endl
          << "architecture structural of " << d_modelCell->d_name
          << " is" << std::endl;
605   for(Netlist::CellHash::const_iterator cit=d_cellHash.begin();
          cit != d_cellHash.end(); ++cit) {
        const Netlist::Cell* pCell = cit->second;
        pCell->WriteComponentDescription(ofs);
      }
610   ofs << std::endl;
      d_modelCell->WriteSignals(ofs);
      ofs << "begin " << std::endl << std::endl;
      d_modelCell->WritePortMap(ofs);
      ofs << "end structural;" << std::endl;
615   ofs.close();
    }

    void Netlist::Library::CreateGraph(GraphOperator& gobj) const {
      assert(d_modelCell);
620   std::cout << std::endl << "Creating graph of model : "
                << d_modelCell->d_name << std::endl;
      const NetCollection& nc = d_modelCell->getNetCollection();
      const InstanceVector& iv = d_modelCell->getInstanceVector();
      std::cout << std::endl << " It has " << iv.size() << " instances.";
625   for(Netlist::NetCollection::const_iterator nit=nc.begin();
          nit!=nc.end(); ++nit) {
        nit->second->Operate(gobj);
      }
      std::cout << std::endl;
630 }
```

Listing 22.7 Netlist data structures, implementation.

```
/////////////////////////////////////////////////////////////////
// Program   : Placement
// Author    : Sandeep Koranne
// Purpose   : Implementation of Graph Algorithms on Cell Broadband

#include <spu_mfcio.h>
#include <stdio.h>
#include <math.h>
#include <simdmath.h>
#include "../SA/graph.h"
//#define COST_DEBUG
#define XY_DEBUG
//#define DEBUG_FULL
#ifdef DEBUG_FULL
  #define DEBUG
//#define DEBUG_SORT
  #define DEBUG_MST
#endif

static void print_vector( const vector float* V ) {
  printf( "VF = [%g %g %g %g]", spu_extract( *V, 0 ),
          spu_extract( *V, 1), spu_extract( *V, 2), spu_extract(*V,3));
}
static void print_vector_int(const char* s1, const vector unsigned int* V ) {
  printf( "%s = [%d %d %d %d]", s1, spu_extract( *V, 0 ),
          spu_extract( *V, 1), spu_extract( *V, 2), spu_extract(*V,3));
}
GraphControlBlock cb __attribute__ ((aligned (128)));
volatile int UNIVERSE[ GRAPH_SIZE ]
         __attribute__ ((aligned (128)));

#define NUM_EDGES 1024
#define NUM_NODES 512

volatile vector unsigned int EDGE_TABLE[ NUM_EDGES ];
volatile unsigned int NODE_TABLE[ NUM_NODES ];
const vector unsigned int N1M = {0xFE00,0xFE00,0xFE00,0xFE00};
const vector unsigned int N2M = {0xFE,0xFE,0xFE,0xFE};

static void InitializeUniverse( GraphControlBlock cb ) {
  int i; for(i=0; i < cb.N; ++i ) UNIVERSE[ i ] = -1;
}
static void ClearUniverse( GraphControlBlock cb ) {
  int i;  for(i=0; i < cb.N; ++i ) UNIVERSE[ i ] = 0;
}
static void PrintEdges( GraphControlBlock cb ) {
#ifdef DEBUG_FULL
  int i; for(i=0;i<cb.M/4;++i) {
    vector unsigned int edge = EDGE_TABLE[i];
    printf("%d ",i);print_vector_int("e", &edge );
  }
#else
  cb = cb;
#endif
}

static void RandomNodeAssignment( GraphControlBlock cb ) {
  unsigned int num_cols = cb.N/cb.EXACT;
  int i=0,j=0,ncount=0;
  for(i=0;i<num_cols;++i) {
    for(j=0;j<cb.EXACT;++j)
      NODE_TABLE[ncount++] = (i<<8) | j;
  }
}

```

```
static inline vector unsigned int LoadNodeDataN1( vector unsigned int E ) {
  vector unsigned int R = {0,0,0,0};
  unsigned int n0 = NODE_TABLE[ N1E( spu_extract( E, 0 ) ) ];
  unsigned int n1 = NODE_TABLE[ N1E( spu_extract( E, 1 ) ) ];
  unsigned int n2 = NODE_TABLE[ N1E( spu_extract( E, 2 ) ) ];
  unsigned int n3 = NODE_TABLE[ N1E( spu_extract( E, 3 ) ) ];
  R=spu_insert( n0, R, 0 );
  R=spu_insert( n1, R, 1 );
  R=spu_insert( n2, R, 2 );
  R=spu_insert( n3, R, 3 );
  return R;
}
static inline vector unsigned int LoadNodeDataN2( vector unsigned int E ) {
  vector unsigned int R = {0,0,0,0};
  unsigned int n0 = NODE_TABLE[ N2E( spu_extract( E, 0 ) ) ];
  unsigned int n1 = NODE_TABLE[ N2E( spu_extract( E, 1 ) ) ];
  unsigned int n2 = NODE_TABLE[ N2E( spu_extract( E, 2 ) ) ];
  unsigned int n3 = NODE_TABLE[ N2E( spu_extract( E, 3 ) ) ];
  R=spu_insert( n0, R, 0 );
  R=spu_insert( n1, R, 1 );
  R=spu_insert( n2, R, 2 );
  R=spu_insert( n3, R, 3 );
  return R;
}
//efine DEBUG_SWAP_NODES
static int SwapNodes( unsigned int n1,
          unsigned int n2 ) {
  unsigned int t1 = NODE_TABLE[n1];
  #ifdef DEBUG_SWAP_NODES
  printf("\n Swapping nodes %d and %d",n1,n2);
  #endif
  NODE_TABLE[n1] = NODE_TABLE[n2];
  NODE_TABLE[n2] = t1;
}
#define MANHATTAN_METRIC
vector unsigned int ALL_ONE={1,1,1,1};
vec_int4 L8 = (vec_int4){0xFF,0xFF,0xFF,0xFF};
vec_int4 U8 = (vec_int4){0xFF00,0xFF00,0xFF00,0xFF00};
#ifdef EUCLIDEAN_METRIC
static vec_float4 ComputeNodeCost( vec_uint4 n1, vec_uint4 n2 ) {
  vec_float4 ans=(vec_float4){0,0,0,0};
  vec_int4 t1 = spu_and((vec_int4)n1,L8);
  vec_int4 t2 = spu_and((vec_int4)n2,L8);
  t1 = spu_sub(t1,t2);
  vec_float4 dy = spu_convtf(t1,0);
  dy = spu_mul(dy,dy);
  t1 = spu_and((vec_int4)n1,U8); t1 = spu_rlmask( t1, -8);
  t2 = spu_and((vec_int4)n2,U8); t2 = spu_rlmask( t2, -8);
  t1=spu_sub(t1,t2);
  vec_float4 dx = spu_convtf(t1,0);
  dx = spu_mul(dx,dx);
  ans = spu_add(dx,dy);
  return ans;
}
#endif

inline vec_int4 GetAbs( vec_int4 A ) {
  const vec_int4 ZERO=(vec_int4){0,0,0,0};
  const vec_int4 ONE=(vec_int4){1,1,1,1};
  const vec_int4 M1=(vec_int4){-1,-1,-1,-1};
  vec_uint4 ispos = spu_cmpgt(A,ZERO);
  vec_int4 flipA = spu_sub(ZERO,A);
  A = spu_sel(flipA,A,ispos);
  return A;
}
```

```
    #ifdef MANHATTAN_METRIC
135 static vec_float4 ComputeNodeCost( vec_uint4 n1, vec_uint4 n2 ) {
      vec_float4 ans=(vec_float4){0,0,0,0};
      vec_int4 t1 = spu_and((vec_int4)n1,L8);
      vec_int4 t2 = spu_and((vec_int4)n2,L8);
      t1 = spu_sub(t1,t2);
140   t1 = GetAbs(t1);
      vec_float4 dy = spu_convtf(t1,0);

      t1 = spu_and((vec_int4)n1,U8); t1 = spu_rlmask( t1, -8);
      t2 = spu_and((vec_int4)n2,U8); t2 = spu_rlmask( t2, -8);
145   t1=spu_sub(t1,t2);
      t1=GetAbs(t1);
      vec_float4 dx = spu_convtf(t1,0);
      ans = spu_add(dx,dy); // |x|+|y|
      return ans;
150 }
    #endif

    static float ComputeCost(GraphControlBlock cb) {
      int cost=0;
155   int i=0;
      vec_float4 running_cost = (vec_float4){0,0,0,0};
      for(i=0;i<=cb.M/4;++i) {
        vector unsigned int e_1= EDGE_TABLE[i];
        vector unsigned int n1_data = LoadNodeDataN1( e_1 );
160     vector unsigned int n2_data = LoadNodeDataN2( e_1 );
        #ifdef COST_DEBUG
        print_vector_int("e1",&e_1); printf("\n");
        print_vector_int("n1",&n1_data); printf("\n");
        print_vector_int("n2",&n2_data); printf("\n");
165     #endif
        vec_float4 part_cost = ComputeNodeCost(n1_data,n2_data);
        running_cost = spu_add( part_cost, running_cost );
      }
      cost += ( spu_extract(running_cost,0)+spu_extract(running_cost,1)
170       +spu_extract(running_cost,2)+spu_extract(running_cost,3));
      return cost;
    }
    //vector float COOLING_SCHEDULE = (vector float){0.89,0.89,0.89,0.89};
    vector float COOLING_SCHEDULE = (vector float){0.99,0.99,0.99,0.99};
175 static inline int CoinToss( vector float temp ) {
      // temp = e^(-0.1*temp)
      unsigned int coint = (rand()%100);
      float fcoint = coint*0.01;
      //temp = spu_mul(temp, COOLING_SCHEDULE );
180   return ( fcoint < spu_extract(temp,0) );
    }
    int BEST_NODE_TABLE[ NUM_NODES ];
    float best_cost=(float)INT_MAX;

185 static void RunTrials( GraphControlBlock cb,int n,
              vector float temp ) {
      int i,j;
      unsigned int cost = INT_MAX,ncost;
      for(i=0;i<n;++i) {
190     unsigned int n1,n2;
        n1 = rand() % cb.N;
        n2 = rand() % cb.N;
        SwapNodes( n1,n2 );
        ncost = ComputeCost(cb);
195     if(__builtin_expect((ncost>cost),1)) {
          //printf("\n %d %d %d",i,n1,n2);
          SwapNodes( n1,n2 );
          continue; }
        cost=ncost;
```

```
200      if (__builtin_expect((ncost<best_cost),0)) {
           best_cost=ncost;
           if(cb.SAVE)
    for(j=0;j<cb.N;++j) { BEST_NODE_TABLE[j]=NODE_TABLE[j]; }
         }
205    }
     }

     static void PrintNodePartition( GraphControlBlock cb ) {
       int i;
210    int p1=0,p2=0;
       for(i=0;i<cb.N;++i) {
         if( NODE_TABLE[i] == 0 )
           p1++;
         else
215        p2++;
       }
       if( (p1+p2) != cb.N ) { printf("\n ERR"); }
       printf("\n |P1| = %d |P2| = %d\n", p1, p2);
     }
220
     #define COLUMN 30
     static void PrintNodeTable( GraphControlBlock cb ) {
       int i,j;
       int p1=0,p2=0;
225    for(i=0;i<cb.N;++i) {
         printf("\n [%4d] ", i);
         for(j=0;j<COLUMN;++j,++i) {
           printf(" %d", NODE_TABLE[i] );
           if(i>=cb.N) return;
230      }
       }
     }

     #define STTEMP 1e6
235  int main(unsigned long long speid,
              unsigned long long argp,
              unsigned long long envp __attribute__ ((__unused__))) {

       int ack = 0;
240    int tag_id = mfc_tag_reserve();
       int i,j,edge,weight;
       unsigned int mst_cost = -1;
       vector float temp = (vector float){STTEMP,STTEMP,STTEMP,STTEMP};
       srand( speid );
245    mfc_get(&cb, argp, sizeof(cb), tag_id, 0, 0);
       mfc_write_tag_mask(1<< tag_id);
       mfc_read_tag_status_all();
       printf("Hello from SPE (0x%llx) (0x%llx) \n", speid,cb );
       // now read the data as well but get it in 4096 * 4 bytes
250  #define CHUNK_OFFSET ((sizeof(unsigned int))*4096)
       for(i=0;i<GRAPH_SIZE/4096;++i) {
         mfc_get(EDGE_TABLE+(i*4096), cb.addr+(i*CHUNK_OFFSET),
           sizeof(int)*4096, tag_id, 0, 0);
         mfc_write_tag_mask(1<< tag_id);
255    }
       mfc_read_tag_status_all();
       ElementaryGraphProperties( cb, &i, &j );
       printf("\n N=%d M=%d MaxDeg=%d MinDeg=%d",cb.N,cb.M,i,j);
       RandomNodeAssignment( cb );
260    //RandomEdgeAssignment( cb );
       InitializeUniverse( cb );
       PrintUniverse( cb );
       PrintEdges( cb );
       PrintUniverse( cb );
265    i=ComputeCost(cb);
       printf("\n Cost = %d", i );
```

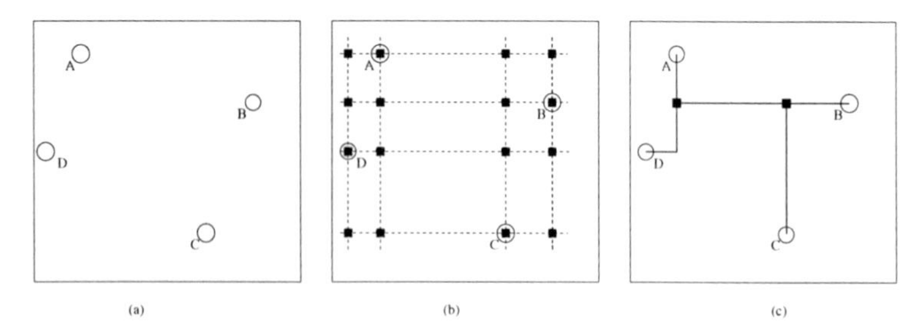

Fig. 22.2 Rectilinear Stiener tree using Hanan grid.

```
    for(i=1;i<cb.TOPRUN;++i) {
      RunTrials(cb,cb.TRIALS,temp);
      temp = spu_mul( temp, COOLING_SCHEDULE );
270   if(!(i%1000)) printf("\n %f %f", temp, best_cost);
    }
    mst_cost = best_cost;
    PrintNodePartition(cb);
    if(cb.SAVE) {
275   PrintNodeTable(cb);
    }
    i=ComputeCost(cb);
    printf("\n Final Cost = %d", i );

280 spu_write_out_mbox( mst_cost ); // put mst_cost
    return 0;
  }
```

Listing 22.8 SPU Placement Kernel.

22.4 What is the VLSI global routing problem

Once the placement of cells has been computed the electrical connections between pins has to be completed. This is done using metal wiring known as *routing* which uses multiple levels of metal routing to accomplish the connection. VLSI routing is an extremely complex problem and in this section we introduce the basic concepts of the problem. For further details we refer the reader to the book by Sapatnekar et al. [108], and the book by Sherwani [115]. Consider the terminals of a net located at points A,B,C and D as shown in Figure 22.2(a). Since VLSI routing is manhattan (or Orthogonal only), the Rectilinear Stiener Tree (RST) is used as the routing topology constructor for VLSI routing. RST is known to be NP-complete, and Hanan showed that all Stiener points must lie on a so-called *Hanan Grid* as shown in Figure 22.2(b) which is the intersection of horizontal and vertical projections of the terminals. The RST constructed by admitting these points is shown in Figure 22.2(c). Given n terminals the SPU can be used to find the Hanan-Grid and then to compute the RST.

The Hanan grid construction comes about in the problem of solving an instance of the VLSI full chip routing problem as shown in Figure 22.3. A cross-over is accomplished by switching the metal layer to form a *via*-connection. A related problem is of constructing routing which minimizes the number of vias for timing critical nets.

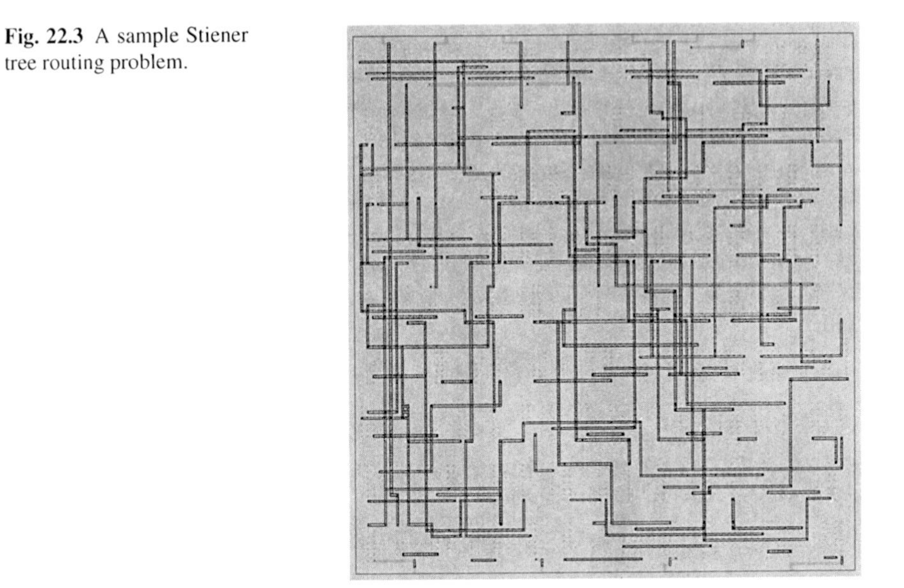

Fig. 22.3 A sample Stiener tree routing problem.

A completed routing solution computed by the Nero router of the Alliance CAD [123] framework is shown in Figure 22.4.

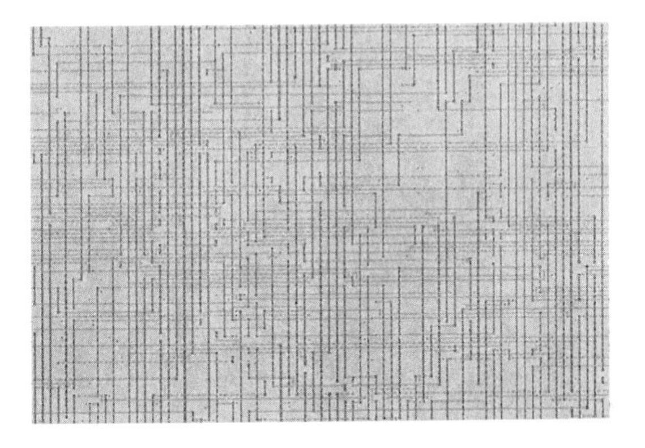

Fig. 22.4 A full chip routing example.

22.5 Conclusion

In this chapter we introduced the VLSI standard-cell placement problem. We discussed how this problem can be modeled as a graph problem, and presented our implementation to read gate-level BLIF netlists on the PPE. A Common Lisp program to generate BLIF designs with random output cover was also shown. The PPE software, reads in gate-level netlists in BLIF format, converts them into a graph and transfers control to SPUs using multi-threading. The 6-SPUs pull the graph from PPE using DMA and perform placement. The SPUs can model manhattan and euclidean metrics and implement simulated-annealing based placers with exponential cooling-schedule. The SPUs can run in a cooperative manner or competitive manner. In cooperative mode the SPUs solve different BLIF models, but in competitive mode, multiple SPUs solve the same problem, and the best solution is chosen by the PPE.

Chapter 23
Power Estimation for VLSI

Abstract We address the problem of estimating the average switching activity in digital CMOS circuits under random input sequences. We have used the a probabilistic approach and have given an efficient implementation for the estimation of average switching activity. We make use of the Parker-McCluskey heuristic and use bit-vector representation of Signal Probability Expressions to arrive at an efficient implementation. Experimental results on ISCAS '89 circuits show that reliable results can be obtained in considerable less time than required by exhaustive simulation.

23.1 Introduction

Estimation of switching activity has generated a lot of interest not only in the context of power analysis, but more recently, due to the emergence of a similar problem, that of *crosstalk analysis*. On the lines of power estimation techniques, the methods to calculate the switching activity can also be divided into two broad classes *dynamic* and *static* [84, 92]. Dynamic techniques explicitly simulate the circuit under a "typical" input stream. Since their results depend on the simulated sequence, the required number of simulated vectors is usually high. These techniques can provide a high level of accuracy, but the run time is very high. A few yeas ago, the static techniques came into the picture and demonstrated their usefulness by providing sufficient accuracy with low computational overhead. These techniques rely on probabilistic information about the input stream (e.g., switching activity on the input signals, temporal correlations, etc.) to estimate the internal switching activity of the target circuit. The main idea used in our approach is as follows. Based on some realistic assumptions about the transistor-level behavior, the problem of estimating switching activity in combinational circuits can be reduced to one of computing signal probabilities.

Common digital circuits exhibit many dependencies; the most known one is the dependency due to reconvergent fanout among different signal lines, but even struc-

S. Koranne, *Practical Computing on the Cell Broadband Engine*,

DOI: 10.1007/978-1-4419-0308-2_23,

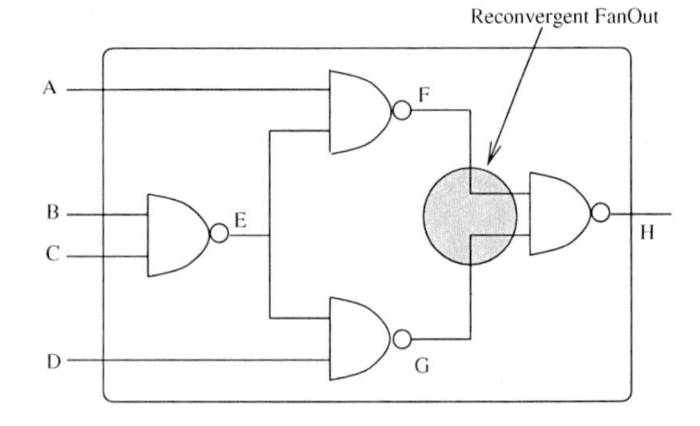

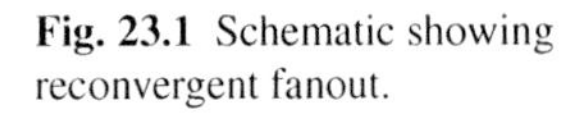
Fig. 23.1 Schematic showing reconvergent fanout.

turally independent lines may have dependencies (induced by the sequence of inputs applied to the circuit) which cannot be neglected. To date, only some dependencies have been considered and even then, only heuristics have been proposed. This is a consequence of the difficulty in managing complex data dependencies described above (also called *spatial correlations*, another type of correlation, namely *temporal correlation*, may also appear in digital circuits.

23.2 Static Probabilities

Consider the case of dynamic CMOS logic (e.g., Domino). At the beginning of each clock cycle, all the gates are precharged, and the gates make transitions only if their associated Boolean functions are satisfied. See for example Fig. 23.1. The three input AND-OR gate's Boolean function might be

$$z = (x \wedge c) \vee (c \wedge y)$$

Where, x,y and c are the primary inputs. In this case, the expected value of the number of transitions at this gate's output is

$$z(\text{transitions}) = 2 \times P(((x \wedge c) \vee (c \wedge y)) = 1)$$

where $P(x)$ is defined as the probability that x is true. The factor of two comes on account of the reset transition during precharge. To evaluate this expression it is necessary to determine the primary input probabilities. We assume that *primary inputs are uncorrelated* and that each is a waveform in time whose value is either zero or one, changing instantaneously at global clock edges. The probability of a particular input i_j being one at a given point in time, denoted by p_j^{one}, is given by

$$p_j^{one} = \lim_{N \to \infty} \frac{\sum_{k=1}^{N} i_j(k)}{N} \quad (23.1)$$

where N is the total number of global clock cycles, and $i_j(k)$ is the value of input i_j during clock cycle k. Clearly, the probability that i_j is zero at a given point, denoted p_j^{zero} is

$$p_j^{zero} = 1 - p_j^{one}$$

We refer to p_j^{one} and p_j^{zero} as *static probabilities*. Note that

$$P(((x \wedge c) \vee (c \wedge y)) = 1) \neq p_x^{one} p_c^{one} + p_c^{one} p_y^{one}$$

because the first and the second product terms are not independent. Rather

$$\begin{aligned} P(((x \wedge c) \vee (c \wedge y)) = 1) &= P(((x \wedge c) \vee (\neg x \wedge c \wedge y)) = 1) \\ &= p_x^{one} p_c^{one} + p_x^{zero} p_c^{one} p_y^{one} \end{aligned} \tag{23.2}$$

where the second equality holds because $x \wedge c$ is disjoint from $\neg x \wedge c \wedge y$. In this example, $(x \wedge c) \vee (\neg x \wedge c \wedge y)$ represents a disjoint cover for the logic function, and the terms $(x \wedge c)$ and $(\neg x \wedge c \wedge y)$ are referred to as *cubes* in the cover [37]. The equivalent logical expression $(x \wedge c) \vee (c \wedge y)$, does not represent a disjoint cover because $(x \wedge c \wedge y)$ is contained in both cubes $(x \wedge c)$ and $(c \wedge y)$.

In general, given a disjoint cover for a Boolean function of uncorrelated inputs described by static probabilities, it is easy to determine the probability of the function evaluating to a one.

23.2.1 Transition Probabilities

For static CMOS combinational logic, a gate output can only change when its inputs change, and then only if the Boolean function describing the gate evaluates differently. For example, a two-input AND gate's output will change between clock cycle t and $t+1$ if

$$(i_1(t) \cdot i_2(t)) \oplus (i_1(t+1) \cdot i_2(t+1)) \tag{23.3}$$

evaluates to one, where $i_1(t)$, $i_2(t)$ and $i_1(t+1)$, $i_2(t+1)$ are the inputs to the gate at clock cycle t and $t+1$ respectively. The disjoint cover for (23.3) is

$$\begin{aligned} & (i_1(t) \cdot i_2(t)) \cdot \neg(i_1(t+1)) \\ \vee \quad & (i_1(t) \cdot i_2(t)) \cdot (i_1(t+1) \cdot \neg i_2(t+1)) \\ \vee \quad & \neg i_1(t) \cdot (i_1(t+1) \cdot i_2(t+1)) \\ \vee \quad & (i_1(t) \cdot \neg i_2(t)) \cdot (i_1(t+1) \cdot i_2(t+1)) \end{aligned} \tag{23.4}$$

It is not possible to use the above method to evaluate the probability of (23.4), because an input $t+1$ is correlated to its behavior at time t (as in a sequential circuit). Instead, *transition probabilities* for the transition $0 \rightarrow 0$, $0 \rightarrow 1$, $1 \rightarrow 0$, and $1 \rightarrow 1$ must be used. We denote these transition probabilities by $p_j^{00}, p_j^{01}, p_j^{10}$, and p_j^{11} respectively, where for example p_j^{10} is defined by

$$p_j^{10} \quad = \quad \lim_{N \to \infty} \frac{\sum_{k=1}^{N} i_j(k) \neg i_j(k+1)}{N} \tag{23.5}$$

The other transition probabilities follow similarly.

Static probabilities can be computed from signal probabilities but *not* vice-versa, because of correlation between time frame and the next. So in general it is necessary to specify transition probabilities. The relations between static probabilities and transition probabilities follow directly from the definitions in Eq. 23.1 and Eq. 23.5, specifically

$$p_j^{zero} = p_j^{00} + p_j^{01} \tag{23.6}$$

$$p_j^{one} = p_j^{11} + p_j^{10} \tag{23.7}$$

Both static and transition probabilities are used to compute z(transitions) for static logic circuits, as can be seen from the expression for the probability that (23.4) evaluates to one

$$p_1^{10} \cdot p_2^{one} + p_1^{11} \cdot p_2^{10} + p_1^{01} \cdot p_2^{one} + p_1^{11} \cdot p_2^{01}. \tag{23.8}$$

For all primary inputs *we assume that successive input vectors are uncorrelated.* A one or a zero is equally likely, in which case all transition probabilities may be assumed to be 0.25, and all static probabilities to be 0.5. It is also possible that a one or a zero at a particular primary input are not equally likely[1], we assume that the user will provide these probabilities.

We generalize this notion in the next section.

23.3 Spatial Correlations

These correlations have two important sources :

- *structural dependencies* due to reconvergent fanout in the circuit;
- *input dependencies* i.e., spatial and/or temporal correlations among the input signals which result from the actual input sequence applied to the target circuit.

To take into account the exact correlation is practically impossible even for small circuits. To make the problem tractable, we allow only *pairwise correlated signals*, which is undoubtedly an approximation, but provides good results in practice.

Pairwise Conditional Probabilities : We define the conditional pairwise probabilities of a pair of signals (x,y) as :

$$p_{a,b}^{x,y} \quad = \quad p[\frac{(x_n \ = k \cap y_n \ = l)}{(x_{n-1} \ = i \cap y_{n-1} \ = j)}]$$

[1] In the case of power signals, for example.

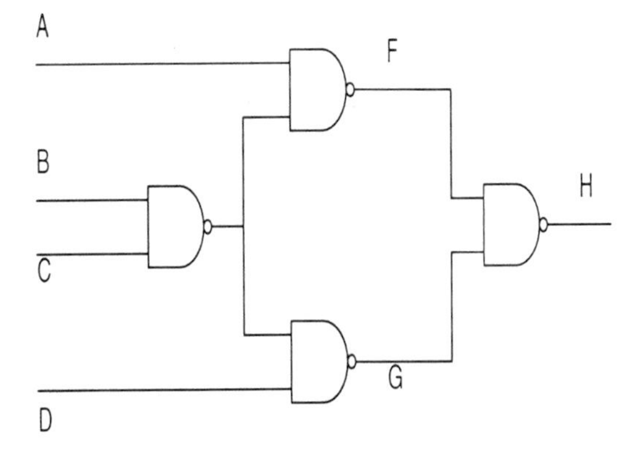

Fig. 23.2 McCluskey's Approximation

For the purpose of this study we have used the following approximation: we assume the *switching activity* of a net to be defined as :

$$sw(x) = 2\,p\,(1-p)$$

Where p is the *static signal probability* of net x.

In the next section we describe an efficient method to estimate the static signal probabilities of combinational circuits, keeping in mind the structural correlations arising due to reconvergent fanout.

23.4 A distributed algorithm for signal probability estimation for combinational circuits

In this section we describe an efficient method to estimate the signal probability for each net in the combinational circuit, given the static signal probability of primary inputs. The idea is to propagate the probabilities of the primary inputs in a breadth first fashion taking into account the error due to structural correlations due to reconvergent fanout.

23.4.1 The Parker-McCluskey's Approximation

Refer to Fig. 23.2 which shows a combinational circuit with four primary inputs A, B, C, D. Let p_x be the static signal probability of line x. For primary inputs the signal probability information is usually available from a knowledge of the kind of input vectors that are "usually" applied to the circuit. In the absence of any information, we shall assume that $p_x = 0.5$.

23.4.1.1 Computing static signal probabilities for internal lines

Given the signal probabilities of primary inputs, computing the probabilities of the internal lines for a *combinational circuit* is known to be NP-complete. The problem is rendered difficult by the presence of structural correlations in the form of reconvergent fanout. A simple approximation due to Parker-McCluskey can be applied to compute p_x. Given a two input gate g with inputs x and y, output z, we can relate $p_x = f(p_x, p_y)$ as shown in Table 23.1. We want to propagate the probabilities of the primary inputs in a breadth first fashion taking into account the error due to structural correlations due to reconvergent fanout using the Parker-McCluskey (PM) heuristic [98]. Using the PM approximation, we replace any literal of the form p_x^k by p_x in the *signal probability expression*. We have written a C++ program that computes the bit-pattern for all nets. The bit-patterns are stored as `0110+0011`, for the expression $ab + bc$. The MSB refers to the sign of the term. Note that the creation of the bit-pattern is a one time process, as it is based on the circuit alone. This paper describes a nested data-parallel method to evaluate such a bit-pattern, referred to as a *sequence*. The method described above has been implemented in Common Lisp in the tool called PRBC (for Polynomial Representation of Boolean Circuits). The tool first reads in a Lisp file containing a definition of primitive functions for each standard cell in the library of the design. The appropriate function is called on the input of each gate. As an example consider the definition of the NAND gate below.

```
;;; Definition of the gate-nand function
;;; Input = list of input polynomial
;;; Output= Supexe'd Output polynomial
(defun gate-nand (plist)
  (let ((ret-list (car plist)))
    (dolist (val (cdr plist))
      (setq ret-list (poly-and ret-list val)))
      (poly-not ret-list)))
```

Listing 23.1 Definition of gate-nand function.

The input to the program is a list containing the signal probability of the primary inputs and the circuit itself. An example is shown below from the MCNC ISCAS '89 benchmark circuit C17 :

```
((1.0 0.32 0.56 0.65 0.234 0.12       )
 ( gate-nand ( 1gat  3gat  ) 10gat  )
 ( gate-nand ( 3gat  6gat  ) 11gat  )
 ( gate-nand ( 2gat  11gat ) 16gat  )
 ( gate-nand ( 11gat 7gat  ) 19gat  )
 ( gate-nand ( 10gat 16gat ) 22gat  )
 ( gate-nand ( 16gat 19gat ) 23gat  ))
```

Listing 23.2 MCNC ISCAS'89 C17 Circuit.

The plots for these expressions are given below in Fig. 23.3.

Using the above table we can derive the following *probability polynomials* for the circuit shown in Fig 4.3, for the internal lines E, F and G.

$$p_E = (1 - p_B \cdot p_C) \tag{23.9}$$

$$p_F = (1 - p_A \cdot (1 - p_B \cdot p_C)) \tag{23.10}$$

Table 23.1 Signal probabilities for gate outputs.

Gate Type	Function	p_z
AND	$z = x \cdot y$	$p_x \cdot p_y$
OR	$z = x + y$	$p_x + p_y - p_x \cdot p_y$
XOR	$z = x \oplus y$	$p_x + p_y - 2 \cdot p_x \cdot p_y$
NOT	$z = \bar{x}$	1 - p_x

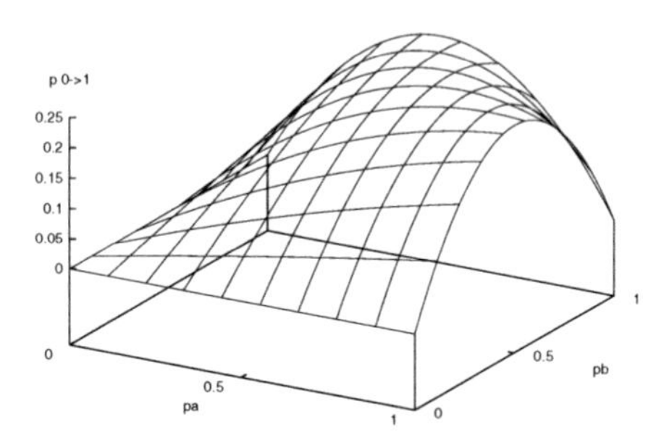

Fig. 23.3 Probability of Output Transition for 2-input AND gate.

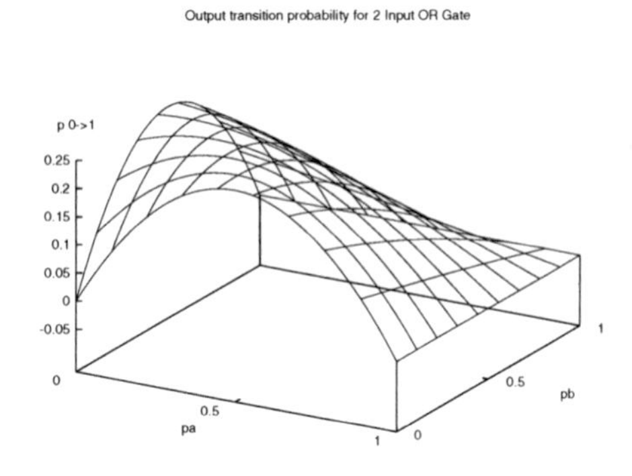

Fig. 23.4 Probability of Output Transition for 2-input OR gate.

$$p_G = (1 - p_D \cdot (1 - p_B \cdot p_C)) \tag{23.11}$$

We now have a reconvergent fanout situation at gate H. We have the following probability expression at H.

$$p_H = (1 - p_F \cdot p_G) \tag{23.12}$$

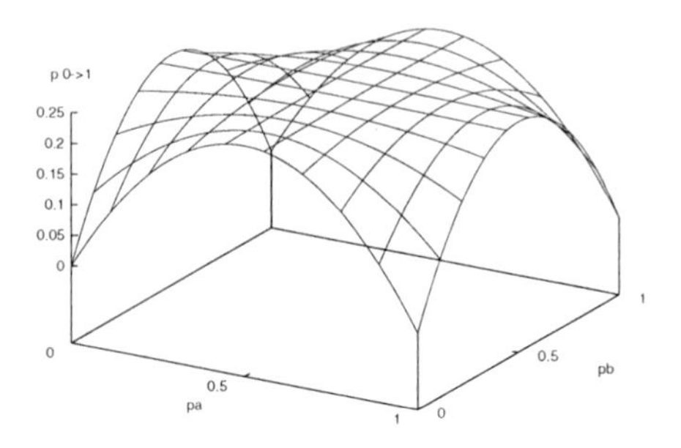

Fig. 23.5 Probability of Output Transition for 2-input XOR gate.

$$
\begin{aligned}
p_H = {} & 1 - p_A - p_D + p_A \cdot p_D + p_A \cdot p_B \cdot p_C \\
& + p_B \cdot p_C \cdot p_D - 2 \cdot p_A \cdot p_B \cdot p_C \cdot p_D \\
& + p_A \cdot p_B^2 \cdot p_C^2 \cdot p_D
\end{aligned}
\tag{23.13}
$$

When we substitute for p_F and p_G we must use the McCluskey approximation that replaces any literal of the form p_x^k by p_x. The Parker-McCluskey algorithm can be generalised to work with transition probabilities. Each input x_i has four probability variables associated with it, corresponding to the input staying low, making a rising transition, making a falling transition, and staying high. These are $x_i^{00}, x_i^{01}, x_i^{10}, x_i^{11}$, respectively. For each gate g, we now have four polynomials $P_g^{00}, P_g^{01}, P_g^{10}, P_g^{11}$ corresponding to the gate staying low, making a rising transition, making a falling transition, and staying high, respectively. Simulation tables can be used to obtain the basic rules for computing these polynomials for each type of logic gate. The Parker-McCluskey method has also been extended to handle gate delays. In other words, we suppress the exponents that are larger than 1 and change them to 1 before simplification. Using his rule we get :

$$
p_H = p_A + p_D - p_A \cdot p_D - p_A \cdot p_B \cdot p_C - p_D \cdot p_B \cdot p_C + p_A \cdot p_B \cdot p_C \cdot p_D
$$

In the next section we describe our data structure to store the signal probability polynomials.

23.5 Bit-vector implementation of probability polynomials

Let the number of primary inputs to the circuit be N, then we can write a polynomial as a linked list of *terms*, where each term is composed of exactly $N+1$ bits. The position of a bit is the representation of the the primary input in that term. The value of the bit i.e., whether it is true or false indicates whether that input is a part of that term or not. Let us explain it with an example. Let the primary inputs be A, B, C, D, then N is 4, and we have :

$$A = 01000 \quad B = 00100 \quad C = 00010 \quad D = 00001 \tag{23.14}$$

We can represent any term of the polynomial by simple bit-vectors, eg., $A \cdot B$ can be written as "01100". The MSB of the bit-vector is kept for the sign of the term and the special bit-vector of $000\ldots N+1$ is reserved for UNITY. We must have the concept of NEGATIVE UNITY also hence the bit-vector $1000\ldots N+1$ is reserved for -1. Using this simple notation we can build arbitrarily complex polynomials. From the above example we would write p_H as :

$$p_H = 01000 + 00001 + 11001 + 10111 + 11110 + 01111$$

To represent the polynomial we use a linked list of terms.

The significant advantage from using a bit-vector representation of the probability polynomials is apparent when we consider the effect of applying McCluskey's approximation on them. The AND operation is defined as :

$$p_z = p_x \cdot p_y$$

Followed by an exponent reduction phase :

$$\forall\, l \in p_z : l^k = l$$

Where l is a literal in the probability polynomial of z.

This operation can be written simply as :

$$p_z = p_x \vee p_y$$

For example, $p_x = p_A \cdot p_B \cdot p_C$ is represented as 01110 , and $p_y = p_B \cdot p_C \cdot p_D$ is represented as 00111. By the McCluskey approximation p_x AND p_y is $p_A \cdot p_B \cdot p_C \cdot p_D$. We can easily see that the INCLUSIVE OR of the bit-vector of the two probability polynomials give the correct representation. The INCLUSIVE OR operation is done on the [1..N] bits, the zeroth bit is computed independently, by the XOR function.

$$(0)1110 \vee (0)0111 = (0 \oplus 0)1111$$

For the case of the NOT operator we have, $p_z = (1 - p_x)$. We do this operation in two steps. First we reverse the sign bit of the polynomial representing p_x and then we ADD UNITY to it. For e.g., let $x = p_A \cdot p_B$, then $\bar{x}$ is represented as 000 + 111.

We have written a function which does these operations. To do breadth first search in the circuit we also store the depth of the polynomial. There is a hashing mechanism which returns the probability polynomial of a net based on its name. We have used this to write a switching activity estimator which reads in BLIF representation of the circuit and estimates the signal probabilities for each internal net.

The output of the Common Lisp based PRBC is a bit-vector representation of each net. The evaluation of this bit-vector w.r.t to a specific input probability list can be done external to the Lisp program. This is useful when we want to evaluate the bit-vector sequence for a specific net only. It also facilitates the use of distributed computing to evaluate such a sequence. Consider the following example :
x_1 = 0001+0010+1011+1110
x_2 = 0000+1001+0001+0000+1111
for a three input case. The evaluation of $N = \{x_1, x_2\}$ w.r.t $I = [0.5, 0.3, 0.8]$ can be done either sequentially or in a distributed fashion. A distributed implementation with SPUs to evaluate the bit-vectors has been implemented. On multiprocessor machines or vector machines we can speed the computation further by using *nested parallelism*. We can write it as :

```
for all n in {N} in parallel do
begin
  for all t in {n} in parallel do
  begin
  term[t] = 1.0;
  for all v in {t} in parallel do
    begin
      if v == 1 then term[t]*=I[index(v)];
    end
  end
  n[prob]=sum(term);
end
```

This exhibits *nested parallelism* and can be implemented on a variety of parallel computing platforms. Initial experiments with NESL : A Nested Data Parallel Language [6] are encouraging.

23.5.1 Approximation Scheme

The Parker McCluskey algorithm cannot be used on large circuits, since it involves *collapsing* the circuit into two levels. We use the method suggested by Costa *et al* to use interpolation to obtain a simpler polynomial that is closest to the original polynomial for large fraction of the input space. We generate some interpolation values by evaluating the original polynomial at several points in the input space. Initially the input probability is set to 0.5. For each primary input i, we evaluate the polynomial at different combinations of the probability value of i and $i+1$ at 6 selected points. The approximation polynomial we use is :

$$P' = a_0 + \sum_{i=1}^{N} (a_i P_i) \tag{23.15}$$

We then use a look up table based method to generate a polynomial which is closest to the original polynomial. When the number of terms in a polynomial exceeds a user defined parameter then the polynomial is evaluated at that point and added to the list of primary inputs. Although this loses some information, it has been shown to be a good approximation as the contribution of nets which reconverge at large depth is very less to the signal probability.

23.5.2 Virtual inputs at scannable FFs

For large dense combinational blocks within the design, the probability polynomial may become very complex. In order to reduce the complexity we can think of the outputs of scannable flip-flops which are driving the inputs of such a block (we assume that with good DfT such FFs will be nearby) as *virtual primary inputs*. We then calculate the probability polynomials of the nets inside the blocks assuming the inputs of the block to be primary inputs. The input probability list of such inputs can be easily calculated from the scan chain description and the scan out pin waveform. For example consider the scan chain to be consisting of $\{ff_{i1}, ff_{12}, ff_{o1}, ff_{i3}, ff_{o2}\}$. If the waveform at scan out is $so = [0101000..]$ and we know the scan enable signal waveform $se = [011111111100..]$ then we can calculate the output of each flip flop by tracing the scan out backwards. We get $ff_{i1} = [00..]$, $ff_{i2} = [000..]$, $ff_{o1} = [1000..]$, $ff_{i3} = [010000..]$, and $ff_{o2} = [1010000..]$. This is explained in Fig. 23.6. In circuits that implement full scan we can use the ATPG generated test patterns which are defined for each scannable flip flop. Instead of treating the whole combinational circuit as one design we partition the circuit at scannable flip flop boundaries. Then we can calculate the static signal probability for each output of a scannable flip flop (in test mode with ATPG test vectors). We can also annotate the scannable flip flop with a command to dump the value to a signal probability file. Then this flip flop can be inserted in the design as a part of the normal flow for scan insertion. With this we can analyse the design in functional simulation also. This method offers the advantage that simulation speed is not affected very much as only few cells are annotated and which require value dumping.

23.6 Applications of PRBC

Besides the obvious application in calculating switching activity for power estimation, we can also use PRBC to give an indication of the *cross talk* susceptance of a critical net w.r.t to its aggressors. A net which has higher switching activity is more likely to be an *aggressor* and hence will contribute more to crosstalk noise than a net which has lower switching activity. The $CrossTalkIndex\ XP\ (i \rightarrow j)$ of a aggressor(i)-victim(j) pair can be modelled as

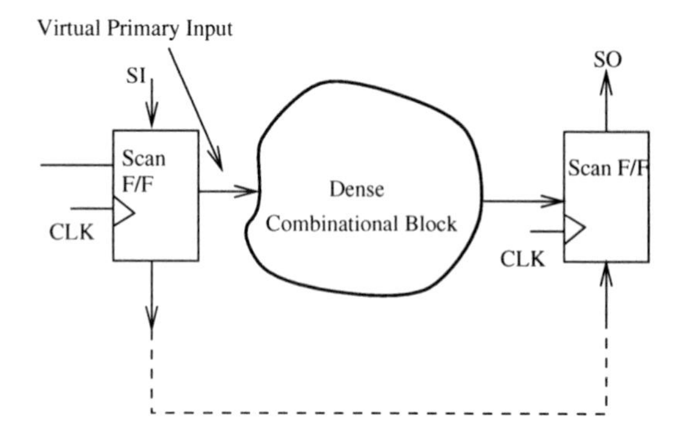

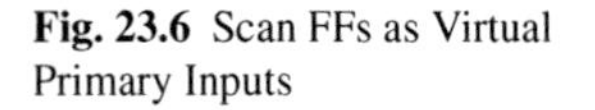
Fig. 23.6 Scan FFs as Virtual Primary Inputs

$$XP(i \rightarrow j) = w_1 \delta_i + w_2 \lambda(i, j) \tag{23.16}$$

where δ_i is the switching activity of the aggressor and $\lambda(i, j)$ is the mutual coupling length of nets i and j, w_1 and w_2 are experimentally calculated weights.

23.7 A Nested Data-Parallel Implementation of the Parker-McCluskey Method

We first describe the sequential implementation.

Sequential Implementation

Assume that the static signal probability for the primary inputs is given in `InVal`. The bit-patterns for the nets are stored in an array `SList`, which is an array of linked list of probability polynomials. As an example consider a 4 input circuit having 6 nets, then, $m = 4, n = 6$. A sample `SList` could be :

```
SList[0] = 01000
SList[1] = 00100
SList[2] = 00010
SList[3] = 00001
SList[4] = 00101 + 01010 + 10001
Slist[5] = 00110 + 01111
```

Assume `InVal` is : `InVal=[0.0,0.3,0.2,0.5,0.1];`
Using the above scheme we would get :
`OutVal=[0.3,0.2,0.5,0.1,0.07,0.103];`

23.7.1 The Parallel Implementation

We first prototyped the parallel algorithm using NESL which we had previously discussed in Chapter 12. The input to the NESL program is an integer sequence which contains the `SList`. NESL has powerful primitives that let us apply functions over each element of a sequence in parallel, and the parallelism can be nested. For example, consider the NESL code fragment below :

```
a = [1,-2,3,-4,5];
abs_a = {abs(i): i in a};
abs_a = [1,2,3,4,5];
```

This applies the `abs(x)`, function over each element of the sequence in parallel, and returns a sequence. This is written as `[int] → [int]`. The `'{' ...'}'` represents a *apply-to-each* construct, and it may be combined with conditions, and nested to an arbitrary level. Since NESL is a functional language, most of the implementation of our program is done by writing small functions.

The function to evaluate one term of the signal probability expression is written as :

```
function EvaluateTerm(a) =
{ InVal[i] : i in [1:#a] | a[i]==1 };
```

This function returns a sequence of floats in parallel when the condition written after the `'|'` evaluates to TRUE. The check for the sign is done in another function, which also calculates the value of this term.

```
function mult(a) =
if (#a==1) then a[0]
else let v={mult(x): x in bottop(a)};
     in v[0]*v[1];


function EvaluateProduct(a) =
let val = mult(EvaluateTerm(a)) in
   if(a[0]==1) then val = -val else val;
```

The `bottop` function splits the sequence into two parts, which are evaluated recursively. The signal probability for a net can be computed by applying this function over all its terms in parallel, and then summing the result.

```
function SignalProb(a) =
sum( { EvaluateProduct(x) : x in a });
```

The computation can be parallelized still further by doing the computation for *all nets* at the same time :

```
function SolveCircuit(a) =
{ SignalProb(x) : x in a };
```

Now, we have to read `SList` and make the call to `SolveCircuit(Slist);`. `SList` itself is a nested integer sequence :

Table 23.2 Work and Depth Complexities of the Functions

Function	Work	Depth
EvaluateTerm	$O(m)$	$O(1)$
mult	$O(m)$	$O(\lg m)$
EvaluateProduct	$O(m)$	$O(\lg m)$
SignalProb	$O(mL)$	$O(\lg m)$
SolveCircuit	$O(mnL)$	$O(\lg m)$
Total	$O(mnL)$	$O(\lg m)$

```
SList = [ [ [0,0,1,0,1],[0,1,0,1,0],[1,0,0,1,0] ],
        [ [0,0,1,1,0],[0,1,1,1,1] ] ];
InVal=[0.0,0.3,0.2,0.5,0.1];
OutVal = SolveCircuit(SList);
OutVal=[0.3,0.2,0.5,0.1,0.07,0.103];
```

23.8 Analysis of our method

The analysis of the sequential implementation is simple and yields an asymptotic time complexity of $O(nmL)$, where L is the average number of terms in a signal probability expression. L depends on the reconvergence and the depth of the circuit. The analysis of the parallel implementation requires the definition of the two complexities associated with all computations in NESL.

1. **Work complexity:** this represents the total work done by the computation, that is to say, the amount of time that the computation would take if executed on a serial random access machine.
2. **Depth complexity:** this represents the parallel depth of the computation, that is to say, the amount of time the computation would take on a machine with an unbounded number of processors.

The method for calculating the work, denoted $W(\texttt{exp(a)})$ and depth, $D(\texttt{exp(a)})$ for the apply-to-each construct can be written as :

$$W(\{\texttt{e1(a)} : \texttt{a in e2(b)}\}) = W(\texttt{e2(b)}) + \texttt{sum}(\{W(\texttt{e1(a)}) : \texttt{a in e2(b)}\}) \quad (23.17)$$

$$D(\{\texttt{e1(a)} : \texttt{a in e2(b)}\}) = D(\texttt{e2(b)}) + \texttt{max\_val}(\{D(\texttt{e1(a)}) : \texttt{a in e2(b)}\}) \quad (23.18)$$

Using Brent's rule [6] we get the time complexity for EREW PRAM model as :

$$T = O(W/P + D\lg P)$$

By substituting values from Table 23.2 we get :

$$T = O(mnL/P + \lg m \lg P)$$

23.9 PRBC Implementation

We first detail the flow which converts gate-level BLIF to signal probability polynomials. This has been written using Common Lisp, although an experimental C++ port has also been performed. The Lisp code is shown in Listing 23.3 and contains several functions which convert the gate-level netlist into polynomials with Parker-McCluskey heuristic applied for the exponent suppression. The result of this function is the representation of each net of the circuit as a vector of unsigned integer. In the example discussed below the number of primary inputs to the combinational logic block was 63, hence a 64-bit unsigned integer encodes the probability monomial. Refer to Section 13.4, pp. 251 for our discussion on monomials and polynomials. In this case the field of coefficients is binary, thus 0/1 is sufficient, and we have 63-variate monomials. The length of the polynomial is the number of monomials or equivalently (for this application), the number of unsigned integers. This is stored with the polynomial. A binary encoded file contains the number of primary inputs, their static signal probabilities. This is followed by the polynomial encoding for internal nets in topological order. These polynomials are to be analyzed and evaluated on the SPU as discussed in Section 23.10.

```
;; prbc.lisp
;; Polynomial Representation of Boolean Circuits
;; Sandeep Koranne
;;
(setf *gc-verbose* nil)                                          5
(proclaim '(speed 3 safety 0))
;; function read-circuit-file reads in a EDIF
;; like representation of the boolean
;; circuit. This must be prefixed by a list of the
;; Static Signal Probabilities of                                10
;; the primary inputs, as they occur in the source file.
;; eg (1.0 0.5 0.4 0.32 ....) the first value is not used.
;; This could be avoided by providing named list
;; like ((a 0.5) (b 0.4)) and so on
;; But we dont want that the user should be forced to find the PI.   15
;; First run should be used to fid the PI's.
;; Then the input-probability list should be given.
;; If input-list is nil : we can do a random run also.
;; (gate-nand ( a b ) c) == c <= gate-nand (a,b)
;; The circuit file must comprise of a net-list of nets,        20
;; where each (length nets) = 3
;;
(defun b2i (ibv)
  (let ((bl (length ibv))
        (bv (reverse ibv))                                      25
        (num 0))
                                        ;(format t "~%Len = ~D" bl)
    (dotimes (i bl)
      (when (= 1 (bit bv i))
        (setf num (logior num (ash 1 i)))))                     30
    num))

(defun num-pi (gate)
  (cond ((eql gate 'gate-nand) 2)
```

```
         ((eql gate 'gate-not) 1)
         (t 0)))

;; new format is ( gate-not ( n129013 n129014 ))
;;( gate-not ( n129016 n129017 )) ; input+ output

(defun convert-to-old-format (finput foutput)
  (let ((fsi (open finput :direction :input))
        (fso (open foutput :direction
         :output
         :if-exists :supersede))
        (circuit nil))
    (setf circuit (read fsi))
    (close fsi)
    (dolist (val circuit)
      (format fso "(~A (" (first val))
      (dotimes (i (num-pi (first val)))
        (format fso " ~A" (nth i (second val))))
      (format fso " ) ~A)~%" (nth (num-pi (first val)) (second val)))
      #+nil(format t "~%~A ~D" (first val) (num-pi (first val))))
    (close fso)))

(defun read-circuit-file (file-name)
  (let ((x nil) (in-stream (open file-name
         :direction :input)))
    (setq *input-prob* (read in-stream))
    (if (not *input-prob*)
        (format t  "~%Input list not given."))
    (setq x (read in-stream))
    (close in-stream) x))

;; Assert integrity of circuit file.
;; Side effect: Also set global variable *net-count* to be the number of nets
;; Integrity test is simply that all nets must have length 3
;; Later we shall add test for loops

(defvar *net-count* 0 "Number of nets in circuit")
(defvar *non-pi-list* nil "List of non-primary input nets")
(defvar *all-net-names-table* nil)
(defvar *all-net-names-list* nil "Contains the names of all nets")
(defvar *totalc-omputations* nil "Total number of non-pi nets")
(defvar *op-nodes* nil)
(defvar *non-pi-ht* nil)

(defun init-tables ()
  "Initialize all tables"
  (setf *op-nodes* (make-hash-table))
  (setf *non-pi-ht* (make-hash-table))
  (setf *all-net-names-table* (make-hash-table)))

(defun check-circuit-integrity (circuit-list)
  (dolist (x circuit-list)
    (if (/= 3 (length x))
        (error "Error : Circuit file not in correct format")
        (progn
          (push (nth 2 x) *non-pi-list*)
          (setf (gethash (nth 2 x) *non-pi-ht*) t); check if po exists
          (unless (nth-value 1 (gethash (nth 2 x) *op-nodes*))
            (setf (gethash (nth 2 x) *op-nodes*) t)) ; primary-op
          (dolist (val (second x))
            (setf (gethash val *op-nodes*) nil)
            (setf (gethash  val *all-net-names-table*) t))
          (setf (gethash (nth 2 x) *all-net-names-table*) t)
          (setq *net-count* (+ 1 *net-count*))))))

(defun populate-net-names ()
  (maphash #'(lambda (k v) (push k *all-net-names-list*))
```

```
            *all-net-names-table*))

    ;; Net Record data structure
105 ;; for each net-name we have to store the following information.
    ;; for each output node we store the computing function and the input list.
    ;; Also a LIST of Polynomials represnted as bit-vectors in "value"
    ;; and the final static signal probabiliy in "sp"
    ;;
110 (defstruct net-record child-list function-name value (sp 0))

    ;; Value is a List of Polynomials represented as a bit-vector.
    ;; The length of the bit-vector is the number of primary inputs+1
    ;; Make a global *net-name-hash* Hash Table.
115 ;; This will contain a net-record for each net-name
    (defvar *net-name-hash* (make-hash-table) "Net name hash table")

    (defun populate-net-hash-table (circuit-list)
      (let ((temp-struct nil))
120     (dolist (x circuit-list)
          (setq temp-struct (make-net-record :child-list (nth 1 x)
                :function-name (car x)))
          (setf (gethash (nth 2 x) *net-name-hash*) temp-struct))))

125 (defvar *primary-output-list* nil "List of primary outpputs")
    ;; Function calculate primary-output-nodes
    ;; This functio will calculate the primary-outputs
    ;; global variable *primary-output-list* will contain the list of PO nodes
    ;; A primary output will NOT be the member of child-list of any other node
130 ;; Side effect: changes the value of *primary-output-list*

    (defun calculate-primary-outputs (non-pi-list hash-table)
      (maphash
       #'(lambda (k v)
135        (when v (push k *primary-output-list*)))
       *op-nodes*))
    ;; Function to make a expression from hash-table.
    ;; given a net-name it will return a list containing
    ;; the function to evaluate it
140 ;; and the child list.
    (defun make-expression (net-name hash-table)
      (let ((temp-struct (gethash net-name hash-table)))
        (list (net-record-function-name temp-struct)
              (net-record-child-list temp-struct) net-name)))
145 (defvar *new-order-list* nil "New ordered circuit list")
    ;; Function reorder circuit.
    ;; Reorder means that we traverse the circuit netlist
    ;; such that all  nodes that are
    ;; evaluated before they occur in a child-list
150
    (defun circuit-reorder (cir po-list hash-table)
      (dolist (val cir)
        (dolist (net-name (second val))
          (when (gethash net-name *non-pi-ht*)
155         (let ((vexp (make-expression net-name hash-table)))
              (when vexp (push vexp *new-order-list*))))))
      (setf *new-order-list* (reverse *new-order-list*)))

    ;; List of primary inputs
160 ;;
    ;; Given *non-pi-list* and *all-net-names-list*
    ;; pi-list =  *all-net-names-list* - *non-pi-list*
    ;;
    (defvar *pi-list* nil "List of primary inputs")
165 (defvar *number-primary-inputs* 0 "Number of primary inputs")
    (defvar *number-primary-outputs* 0 "Number of primary outputs")
    (defvar *number-nets* 0 "Number of nets")
```

```
    (defun calculate-primary-inputs (non-pi-list all-names-list)
170   (dolist (val1 all-names-list)
        (unless (gethash val1 *non-pi-ht*) (push val1 *pi-list*))))

    (defun collect-statistics (file)
      (let ((cir (read-circuit-file file)))
175     (init-tables)
        (format t "~%Checking Circuit Integrity")
        (check-circuit-integrity cir)
        (format t "~%Completed checking Circuit Integrity")
        (terpri)
180     (populate-net-names)
        (populate-net-hash-table cir)
        (calculate-primary-outputs *non-pi-list* *net-name-hash*)
        (circuit-reorder cir *primary-output-list* *net-name-hash*)
        (calculate-primary-inputs *non-pi-list* *all-net-names-list*)
185     (setq *number-nets* (length *all-net-names-list*))
        (setq *number-primary-inputs* (length *pi-list*))
        (format t "~%~D primary-inputs" *number-primary-inputs*)
        (setq *total-computations* (length *non-pi-list*))
        (if (not *input-prob*)
190         (progn
              (format t "~%Entering random simulation mode~%")
              (dotimes (j (+ 1 *number-primary-inputs*))
                (push 0.50 *input-prob*))))
        (setq *number-primary-outputs* (length *primary-output-list*))))
195 ;; For each primary input create a Value bit-vector
    ;; corresponding to its bit-position
    ;; A=0100, B=0010 and C=0001
    ;; The MSB is the sign bit.

200 (defun populate-pi-net-hash-table (pi-list)
      (dolist (x pi-list)
        (setq temp-struct (make-net-record ))
        (setf (gethash x *net-name-hash*) temp-struct)))

205 (defun create-pi-values (pi-list hash-table)
      (let ((count 0))
        (dolist (val pi-list)
          (progn
            (setq count (+ 1 count))
210         (setq bit-vec (make-array (+ 1 *number-primary-inputs*)
              :element-type 'bit))
            (setf (aref bit-vec count) 1)
            (setq temp-struc (gethash val hash-table))
            (setf (net-record-value temp-struc) (list bit-vec))))))
215
    (defun print-values (in-list)
      (dolist (val in-list)
        (progn
          (setq temp-struc (gethash val *net-name-hash*))
220       (setq value-list (net-record-value temp-struc))
          (print value-list))))

    (defun primitive-compute-sp (input-prob in-vec)
      (let ((prod 1.0))
225     (dotimes (j (- ( length in-vec) 1))
          (if (> (aref in-vec (+ 1 j)) 0)
        (setq prod (* prod (nth  (+ 1 j) input-prob)))))
        (if (> (aref in-vec 0) 0) (- prod)
            prod)))
230
    ;; Given a list of bit-vectors
    ;; compute their signal probabilities using the function
    ;;   primitive-compute-sp
    (defun primitive-compute-sum (input-prob in-list)
235   (let ((sum 0))
```

```
      (dolist (val in-list)
        (setq sum (+ sum (primitive-compute-sp input-prob val)))) sum))

    (defun compute-sp (input-prob in-list hash-table)
240   (let ((sum 0))
        (dolist (val in-list)
          (progn
            (let* ((temp-struc (gethash val hash-table))
                   (temp-list (net-record-value temp-struc)))
245           (setq sum (primitive-compute-sum input-prob temp-list))
              (setf (net-record-sp temp-struc) sum))))))

    (defun print-sp (in-list)
      (dolist (val in-list)
250     (progn
          (setq temp-struc (gethash val *net-name-hash*))
          (setq temp-sp (net-record-sp temp-struc))
          (format t "~%Net ~S has SP ~S" val temp-sp))))
    ;; Functions for operating on polynomials.
255 ;; We require the following functions to operate on a list of polynomials
    ;; AND => (a+b) AND (b+c) = ab+ac+b+bc using the Parker-McCluskey Method
    ;; OR => (a.b) OR (b.c) = ab + bc simple list append operation
    ;; NOT => NOT a = 1 - a. Note: That UNITY is represented by 000000..
    ;; a string of all zeroes. So NOT (p) = Invert all sign bits and Push UNITY
260 ;; We shall first define low level functions for bit-wise operations.
    ;; We can safely assume that the size of all vectros
    ;; is (+ 1 *number-primary-inputs*)
    ;; For convinience we define this as *vectro-size*

265 (defun print-vec (in-vec)
      (dotimes (j *vector-size* )
        (print (aref in-vec j))))
    (defun print-vec-list (in-vec-list)
      (dolist (val in-vec-list)
270     (progn
          (format t "~%----------")
          (print-vec val))))
    ;; "and" and "or" work as planned for bit-vectors.
    (defun our-bit-not (x)
275   (if (= x 1) 0 1))
    (defun my-bit-xor (x y)
      (if (/= x y) 1
          0))
    (defun bit-my-ior (vec1 vec2)
280   (let ((ret-vec))
        (setq ret-vec (bit-ior vec1 vec2))
        (setf (aref ret-vec 0) (my-bit-xor (aref vec1 0) (aref vec2 0)))
        ret-vec))
    (defun poly-and (plist1 plist2)
285   (let ((ret-plist nil))
        (dolist (val1 plist1)
          (dolist (val2 plist2)
            (push (bit-my-ior val1 val2) ret-plist))) ret-plist))
    (defun poly-or (plist1 plist2)
290   (append plist1 plist2))

    (defun ret-rev-vec (vec)
      (let ((ret-vec (copy-seq vec)))
        (setf (aref ret-vec 0) (our-bit-not (aref ret-vec 0)))
295     ret-vec))

    (defun reverse-signs (plist)
      (let ((ret-list) (val-copy))
        (dolist (val plist)
300       (progn
            (setq val-copy (ret-rev-vec val))
            (push val-copy ret-list)))
```

```
        ret-list))

305  (defun poly-not (plist)
       (let ((ret-list nil))
         (setq ret-list (reverse-signs plist))
         (push (make-array *vector-size* :element-type 'bit) ret-list)))

310  ;; Now the gate-level implementation
     ;; This is called directly from the Hash-table symbol
     ;; Function gate-nand : takes a list of polynomials as input.
     ;; These symbols are guranteed to have a "value" as a list

315  (defun gate-nand (plist)
       (let ((ret-list (car plist)))
         (dolist (val (cdr plist))
           (setq ret-list (poly-and ret-list val)))
         (poly-not ret-list)))
320

     (defun gate-and (plist)
       (let ((ret-list (car plist)))
         (dolist (val (cdr plist))
325        (setq ret-list (poly-and ret-list val)))
         ret-list))

     (defun gate-or (plist)
       (let ((ret-list nil) (val-copy))
330      (dolist (val plist)
           (progn
             (let ((val-copy val))
               (push (poly-not val-copy) ret-list))))
         (gate-nand ret-list)))
335

     (defun gate-nor (plist) (poly-not (gate-or plist)))
     (defun gate-not (plist) (poly-not (car plist)))
     (defun gate-buff (plist) plist)
340
     (defun return-plist-from-net-name (net-name hash-table)
       (let* ((temp-struc (gethash net-name hash-table))
              (ret-list (net-record-value temp-struc)))
         (if (> (length ret-list) 10)
345          ;;(error "Too long plist here for Net: ~S" net-name)
             (progn
               (rplacd (nthcdr 9 ret-list) nil)
               (format t "~% Doing polynomial compaction..~S" (length ret-list))
               (setf (net-record-value temp-struc) ret-list))
350          ret-list)))

     ;; Main loop starts here
     ;; For each element in the *new-ordered-list* calculate the
     ;; We apply the function as a symbol.
355  ;; Function PRBC-comp takes as input the ordered netlist and
     ;; the hash-table and the input probability list.
     ;; It calculates the Static Signal Probability Polynomial
     ;; and stores it in the hash.
     ;; Side effect : modifies the "value" field
360
     (defvar *percent-count* 1)
     (defun calculate-percent (number)
       (round (* 100.0 (/ number *total-computations*))))

365  (defun PRBC-comp (in-list hash-table input-prob)
       (format t "~% === ~D ==== " (length in-list))
       (dolist (val in-list)
         (format t "~%Checking for ~A" val)
         (if (= (net-record-sp (gethash (nth 2 val) hash-table)) 0)
```

```
370        (let* ((temp-val nil)
                  (func-list nil))
             (dolist (temp-input (nth 1 val))
               (setq temp-val
           (return-plist-from-net-name temp-input hash-table))
375            (push temp-val func-list))
             (let ((will-compute (funcall (car val) func-list)))
               (format t "~%FFH ~A ~D [" (nth 2 val) (length will-compute))
               (dolist (wc will-compute)
                 (format t " ~D" (b2i wc)))
380            (format t " ]"))
             (setf (net-record-value (gethash (nth 2 val) hash-table))
                   (funcall (car val) func-list))
             (compute-sp *input-prob* (list (nth 2 val)) hash-table)
             (format t "~% Computing for : ~S with SP ~S : ~S\% completed "
385                  (nth 2 val)
           (net-record-sp (gethash (nth 2 val) hash-table))
                     (calculate-percent *percent-count*))
             (setq *percent-count* (+ 1 *percent-count*))))))

390
    ;; Set all global variables to nil to clear the states
    ;; in case we want to do multiple circuit analysis
    ;; function : PRBC-initialize : takes a dummy argument.
    ;; side effect : clear *net-name-hash*
395 ;;               clear *all-net-names-list*
    ;;               clear *pi-list*, *primary-output-list*
    ;;               clear *non-pi-list*, *input-prob*
    ;;               clear *new-order-list*

400 (defun PRBC-initialize (x)
      (setq *net-name-hash nil)
      (setq *all-net-names-list* nil)
      (setq *pi-list* nil)
      (setq *primary-output-list* nil)
405   (setq *non-pi-list* nil)
      (setq *input-prob nil)
      (setq *new-order-list* nil))

    ;; The MOST important function of all.
410 ;; The banner function.
    ;;

    (defun PRBC-banner (x)
      (format t "~% Polynomial Representation of Boolean Circuits ")
415   (format t "~% (C)       Sandeep Koranne"))

    (defvar *vector-size* 0)

    (defun PRBC-start (file-name )
420   (progn
        (collect-statistics file-name)
        (populate-pi-net-hash-table *pi-list*)
        (format t "~% Accumulated ~D net names" (length *all-net-names-list*))
        (create-pi-values *pi-list* *net-name-hash*)
425     (format t "There are ~D nodes in the circuit:~S"*number-nets* file-name)
        (format t "~%List of all nodes in circuit")
        (terpri)
        (format t "Primary inputs of circuits")
        (print *pi-list*)
430     (terpri)
        (format t "List of primary-outputs")
        (print *primary-output-list*)
        (terpri)
        (format t "Non-primary nodes")
435     ;;(print *non-pi-list*)
        (terpri)
```

```
      ;;(format t "Newly ordered circuit list")
      (setq *vector-size* (+ 1 *number-primary-inputs*))
      ;;(print *new-order-list*)
440   (create-pi-values *pi-list* *net-name-hash*)
      (PRBC-comp *new-order-list* *net-name-hash* *input-prob*)
      (compute-sp *input-prob* *pi-list* *net-name-hash*)
      #+nil(print-sp *all-net-names-list*)))

445 (defun PRBC-prompt (x)
      (PRBC-initialize 1)
      (PRBC-banner 1)
      (format t "~%~%~%PRBC> Enter circuit file name :")
      (setq file-name (string-downcase (read )))
450   (time (PRBC-start file-name))
      (format t "~%Bye.~%"))

    (PRBC-prompt 1)
    ;;;cat pmap.blif | awk '{printf "( " 2"("5 " " 8""11 "))\n"}' > t
455 (quit)
```

Listing 23.3 Probability Polynomial Representation

Our SPU based implementation of the probability polynomial evaluator is described below.

23.10 Probability Polynomial Evaluation on SPU

As discussed above the polynomials are vectors of unsigned integers (64-bit for this example). Using DMA these are transferred to SPUs which operate in parallel in analyzing multiple signals. The analysis may involve sensitivity analysis of a specific signal with-respect-to a primary input. The goal may be to identify *temporal correlation* between signals x and y for crosstalk analysis; this can be accomplished by sweeping primary input probabilities from 0 to 1 and measuring the probabilities for x and y. We had earlier discussed the NESL implementation and we use that as a prototype to write the SPU code. The SPU code for evaluation is shown in Listing 23.4.

```
   /////////////////////////////////////////
   // Program: Probability Polynomial
   // Author : Sandeep Koranne
   /////////////////////////////////////////
5  #include <stdio.h>
   #include <spu_intrinsics.h>
   #include "../util.h"
   #define MAX_PI 128
   #define N 1024
10 float PI_DATA[8][N];
   vec_uint4  PDATA[N];
   float TABLE[8][256];

   float CalculateProbability(int i, int j) {
15   float ret = 1.0;
     int k=0;
     for(k=0;k<8;++k) if(j & (1<<k)) ret *= PI_DATA[i][j];
     return ret;
   }
20 void ConstructProbabilityTables() {
```

```
    int i,j;
    for(i=0; i<8;++i)
      for(j=0;j<256;++j)
        TABLE[i][j] = CalculateProbability(i,j&0xFF);
25 }

  vec_float4 EvaluatePolynomial(vec_uint4 x) {
    int i;
    vec_uchar16 vp = (vec_uchar16)x;
30  float t[16];
    for(i=0;i<16;++i) t[i] = TABLE[0][ (spu_extract(vp,i))&0x7F ];
    vec_float4 *vt = (vec_float4*)t;
    // how fast can you multiply 16 floats as 4x4
    vec_float4 ans1 = spu_mul(*vt,*(vt+1));
35  vec_float4 ans2 = spu_mul(*(vt+2),*(vt+3));
    vec_float4 ans  = spu_mul(ans1,ans2);
    return ans;
  }

40 void PrepData() {
    int i,j;
    for(i=0;i<8;++i)
      for(i=0;j<N;++i)
        PI_DATA[i][N] = 0.5;
45 }
  int main( unsigned long long spuid,
            unsigned long long argp ) {
    PrepData();
    printf(" PRBC program running on  %x\n", spuid );
50  ConstructProbabilityTables();
    vec_uint4 data=(vec_uint4){0,102888,913838,8838811};
    EvaluatePolynomial(data);
    return (0);
  }
```

Listing 23.4 SPU code for evaluation of probability polynomial.

The PPE code is given in Listing 23.5.

```
// See SPU code.
```

Listing 23.5 PPE code for evaluation of probability polynomial.

23.11 Experiments

Consider the partial circuit given below:

```
  ()
  ((GATE-NOT ( I34 ) N160)
   (GATE-NOT ( I18 ) N161)
   (GATE-NOT ( I17 ) N162)
5  (GATE-NOT ( I30 ) N163)
   (GATE-NOT ( I20 ) N164)
   (GATE-NAND ( I19 I16 ) N165)
   (GATE-NOT ( N165 ) N166)
   (GATE-NAND ( N166 I12 ) N167)
10 (GATE-NOT ( N167 ) N168)
   (GATE-NOT ( I50 ) N169)
   (GATE-NOT ( I53 ) N170))
```

Listing 23.6 Circuit in CIR format

The output of this circuit by the Probability Polynomial representation is:

```
FFH N165 2 [ 0 1048 ]
FFH N166 3 [ 0 24 1024 ]
FFH N167 4 [ 0 1028 1052 4 ]
```

Listing 23.7 Encoded polynomials representing signal probability.

The detailed log is shown in Listing 23.8. As can be seen the probability polynomial is written using expanded bit-wise form, in Listing 23.7 this bitvector was compactly encoded as an unsigned integer. The unsigned integer form is used when evaluating this polynomial on the SPU as discussed below.

```
; Loading #P"/tmp/prbc.lsp".
Polynomial Representation of Boolean Circuits
(C)       Sandeep Koranne
PRBC> Enter circuit file name : n.cir
5  10 primary-inputs
Entering random simulation mode
There are 21 nodes in the circuit:  "n.cir"
Primary inputs of circuits
(I34 I18 I17 I30 I20 I19 I16 I12 I50 I53)
10 List of primary-outputs
(N170 N169 N168 N164 N163 N162 N161 N160)
Checking for (GATE-NAND (I19 I16) N165)
FFH N165 2 [ #*00000000000 #*10000011000 ]
Computing for : N165 with SP 0.75 : 9% completed
15 Checking for (GATE-NOT (N165) N166)
FFH N166 3 [ #*00000000000 #*00000011000 #*10000000000 ]
 Computing for : N166 with SP 0.25 : 18% completed
Checking for (GATE-NAND (N166 I12) N167)
FFH N167 4 [ #*00000000000 #*10000000100 #*10000011100 #*00000000100 ]
20  Computing for : N167 with SP 0.875 : 27% completed
```

Listing 23.8 Detailed log of running PRBC on example.

23.12 Conclusion

Parallel processing techniques can reduce the time taken for signal probability computation. Nested data-parallel techniques circumvent many of the problems of a simple parallel approach, like load distribution and irregular loops, to a great extent. With the advent of high-level language support in the form of NESL, vector processors can be used to exploit the natural nesting of parallel constructs in problems like logic simulation and fault simulation. This study shows the feasibility of such an approach and ease with which problems that have irregular loops can be modeled using NESL. We prototyped our system in NESL, and then coded it on the Cell SPU using SPU bit-operations and intrinsics. Using our parallel implementation we can also simulate the probability polynomial with real input data and intermediate sequential traces coming from flip-flops of the chip.

Chapter 24
Concluding Remarks

Abstract We present concluding remarks about the material presented in this book.

24.1 Progress in small steps

From our simple functions to calculate the Collatz count of an integer and calculating value of π, we have covered a lot of ground in understanding the basics of programming on the Cell Broadband Engine. The first part of this book concentrated on the implementation of Cell Broadband Engine itself as well as the PPE and SPE architecture and instruction set. Although much of the information presented in part one is also available in Journal papers and other archived papers, my intention was to explain not only how a certain aspect of the chip worked, but also why it was designed like that. It is my belief that only when the application writer merges the application requirements with the chip design goals, can the true potential of the application execution on the Cell Broadband Engine come forth.

24.2 Use of Common Lisp

By this time you must have seen the substantial number of examples which were coded up in Common Lisp prior to SPU execution. The goal of prototyping with Common Lisp was simple; to identify the algorithmic and implementation issues at the highest level possible without investing too much time into the implementation. A C or C++ implementation would have required the same detail of thought as the SPU implementation and would have been sufficiently close to the SPU so as to warrant the shipping of the first prototype itself.

Another advantage when writing with Common Lisp was the generation of test data which then could be directly sent to the SPU for comparison. Since Common Lisp is a high level language more time could be spent on choosing the right problem

S. Koranne, *Practical Computing on the Cell Broadband Engine*,
DOI: 10.1007/978-1-4419-0308-2_24,

space and simulation parameters than otherwise feasible. A direct compilation of Common Lisp to SPU instruction set is a goal that the author would like to work on after this book and that would alleviate the need to rewrite the complete code base over from Lisp to C/C++.

24.3 Future Work

I hope this book has been fun and instructive to read and that it gives you several ideas to improve upon or start afresh on the many problem areas we have discussed in this book. Particularly molecular structure determination, radar processing and VLSI CAD are areas of current research with significant potential for performance improvement.

Due to a lack of space and time there were areas which although I would have liked to cover, but was unable to. These include:

1. Digital Sound Processing: the SDK provides an example of normalizing WAV sound samples. With the increase of artificial intelligence in human-computer interaction digital sound processing to understand human speech as well as speech synthesis are becoming important problems. The SPUs with their SIMD instruction set are uniquely suited to contribute to this domain,
2. Digital Image Processing: the box-filter and dynamic range reduction filters we wrote for the Synthetic Aperture Radar are examples of the power of the SPU when computing on digital images. More advanced computations, e.g., removal of motion-blur, registration, geo-encoding, compression can also be implemented,
3. Data Compression: although originally I had planned on presenting material on Huffman codes, arithmetic coding and floating-point compression, for lack of space this material had to be omitted,
4. Data Encryption: for similar reasons material on data encryption using Blowfish had to be deferred,
5. Matrix Algebra: although we considered some Matrix applications namely matrix multiplication and matrix transpose using the SDK, still the field of matrix algebra is rich with examples of algorithms which can be elegantly mapped on the SPU. Two examples come to my mind:

    a. Rank of a Matrix: Mulmuley's Algorithm,
    b. Transitive closure computation on graphs using matrices.

6. Communication Framework: an analysis of MPI, PVM, CellSS, Cell Messaging Layer should be conducted with results for commonly used operations along with achieved bandwidth made available. We have preliminary data available on the book's website but this needs more experimentation with varied workloads,
7. SPU Overlays: since the code presented in this book was written to solve a specific problem at a time we did not run into the problem of limited SPU memory

space for instructions. SPU Overlaying is a method which can solve this problem by *swapping* out functions from SPU and replacing them with new functions within the same SPU context. When developing a full system with dynamic function call (as opposed to the presented scheme of static SPU function calls, where we knew exactly which program to load into the SPU ahead of time), SPU Overlays become essential,

8. SPU Isolation Mode: the SDK provides a mechanism wherein the SPU is *locked-up* in a sense; this is to ensure data and instruction integrity on the SPU. Since none of the applications we have written need this functionality we did not address it, but some applications notably encryption this ability may be necessary to provide additional security,
9. OpenGL Graphics: I had planned for a full chapter on OpenGL graphics interacting with programs running on the SPU, but now the example provided in Chapter 17 will have to suffice. Interesting applications can be developed with the SPU computing high-level display lists which are executed by the server connected to an OpenGL window,
10. Linear Programming: integration with *glpk*, as we have seen in Chapter 18, moving some of the discrete combinatoric part of the function evaluation on the SPU would accelerate the computation significantly,
11. Integration with GNU/Octave: since GNU/Octave already links to LAPACK, BLAS and other standardized linear algebra libraries, making a PS3 run as a server to communicate results of computation to an Octave process running on a workstation should be simple. The PS3 server could use Cell optimized versions of math kernel libraries (especially when dealing with single-precision floating point numbers) to enhance performance,
12. Binary Decision Diagrams (BDD): along with the chapter on polytopes I had planned for a chapter on BDDs to explain VLSI synthesis and its implementation on the SPU. But practically considering the size of the circuit being synthesized nowadays, the SPUs will be used in cases where design space exploration is required and not for full chip synthesis. Nevertheless an implementation of BDD on the SPU is planned,
13. Scheduling: in addition to partitioning, synthesis, floorplanning, and placement scheduling of tasks and operator mapping are important problems in VLSI design. Since the number of tasks (not the size of the operator, but the number of unique operators) can be small, this is an area where the SPUs could analyze multiple schedules in parallel and the best schedule under a variety of cost considerations could be adapted and used for synthesis.
14. Traditional high-performance computing tasks: since this area has been well researched I did not present any substantial data on problems such as:

    a. Solving large systems of linear equations
    b. Fast Fourier Transforms and its variants
    c. Hydrodynamical Schemes and Computational Fluid Dynamics
    d. Operations on Singular Matrices and Groups
    e. Semiconductor process simulation

24.4 Conclusions

It took longer than I had expected to write this book. What had started as a 3 month project is only now nearing completion (after almost 4 years), so I too am guilty of under-estimating the time and effort it takes to research and put down on paper what seems common knowledge at times. I hope this book proves useful to the reader who has stuck till this last page, pp. 464.

24.5 Website Information

The website `http://www.practical-cell-programming.info`, is a companion to this text. The site contains almost all the source code, and sample-data for the examples presented in this text. The Website will also be continuously updated with further examples, errata and information about latest development around the Cell Broadband Engine.

Appendix A
Introduction to Common Lisp

A.1 Introduction to Common Lisp

Common Lisp is a modern, object-oriented, multi-paradigm programming language which is a dialect of *Lisp*, the List Processing language designed by McCarthy in the 1950s for symbolic computation. Common Lisp is an ANSI standard and supports many data-types, functions, and macros. In this book we use only a fraction of its capabilities and this section describes the used subset for easy reference.

All data and programs in Common Lisp are *S-expressions*, or data which is parenthetically enclosed, for example:

```
(+ 5 3)
'(a e i o u)
(exp 1.01)
```

Common Lisp has a simple rule for evaluating S-exps, the first element of any S-exp is treated as a function call, and the remaining elements are arguments to that function. Once you get used to this notation you can appreciate the difference between typing:

```
(setf sum (+ 1 2 3 4 5)) ; vs
sum = 1 + 2 + 3 + 4 + 5;
```

Moreover, in Lisp, the function call syntax uniform for all types, builtin and otherwise. Whereas in C/C++, user defined functions (without operator overloading) are called using prefix notation:

```
sum = f(x) + g(x);
```

while operators such as '+', are called *infix*. In Common Lisp the function name (or symbol) always comes first. Literals (such as numbers, strings, and floating-point constants) are passed without evaluation, and they evaluate to themselves when used as a function argument. These are called *atoms*. Lists can be passed without evaluation using `quote`, which is mapped (using a read-table) to the ' symbol. Thus a list can be written as:

```
(setf len (length '(good this is never evaluated)))
```

In the above listing, the literal list is not evaluated (otherwise we would get an error as function `good` is not defined), but the **length** function returns the number of elements in the list, returning 5.

A.2 Functions in Common Lisp

We use the following functions often in this text:

defvar: declares a variable, used for global variables, e.g., it is traditional to put a "*" around global variable names in Common Lisp,

```
(defvar *matrix* nil "global matrix")
```

defconstant: declares a symbol with constant semantics, it is traditional to put a "+" around constants in Common Lisp. The string following the initialization value is called a *documentation string* and can be looked up using `(describe '*matrix*)`. If you forger the ' and do instead `(describe *matrix*)`, the **describe** function will get the evaluated `*matrix*` value, and not the name of the variable to describe.

```
(defconstant +NUM-POINTS+ 64 "number of sample points")
```

format: Common Lisp equivalent (and more) of C printf. The **format** function takes a destination stream as the first argument, a format control string as the second argument, and processes the remaining argument in the context of the control string. Some examples are given below:

```
(format t "Hello, World~%")
(format t "~D + ~D = ~D" 3 2 (+ 3 2))
(format t "(reverse of ~A) = ~A" '(1 2 3) (reverse '(1 2 3)))
```

defun: creates an user-defined function of the given name, and argument list. In the following example, a function called `add-two` is created, which returns the sum of its argument and the number 2,

```
(defun add-two (x) (+ x 2))
```

dotimes: loops using integers using variable given in the first argument. We use simple **dotimes** constructs in this book, although, the construct is very powerful.

```
(dotimes (i 10) (format t "~%i = ~D" i))
```

dolist: loops over a list using variable given in the first argument. We use simple **dolist** constructs in this book, although, the construct is very powerful.

```
(dolist (val '(alpha beta gamma)) (format t " in greek we say ~A" val))
```

`make-array`:

```
(setf *matrix* (make-array (list 3 3) :element-type
      'float :initial-element 0.0))
```

`aref`: references an element in an array; can be used as *l-value* with `setf`,

```
(dotimes (i n)
  (setf (aref *matrix i i) 1.0))
```

A.3 File I/O

Files in Common Lisp are represented as streams of data. In this book files are generated using Common Lisp test generators, sometimes with random data, sometimes programmatically generated sequence with some known or measured properties. The `open` and `close` functions take the file name and direction as parameters and return a valid *stream*. Consider the following program fragment:

```
(defun write-data (filename matrix)
 "Open file and write matrix data"
  (let (( fs (open filename :direction :output :if-exists :supersede)))
  ...))
```

A.4 Conclusion

This concludes our mini-tutorial of the Common Lisp language. We have only scratched the surface of this very powerful and complex programming language. Many of the programs written in this text were prototyped first with Common Lisp. Lisp was also used to generate test data for many of the programs and applications and was used to verify the correctness of the Cell implementation. For further readings on Common Lisp we recommend you to read Guy Steele's CLTL text [119].

Index

References

1. Thomas William Ainsworth and Timothy Mark Pinkston. Characterizing the cell eib on-chip network. *IEEE Micro*, 27(5):6–14, 2007.
2. Cilk Arts. Cilk. http://www.cilk.com.
3. David A. Bader, Virat Agarwal, Kamesh Madduri, and Seunghwa Kang. High performance combinatorial algorithm design on the cell broadband engine processor. *Parallel Comput.*, 33(10-11):720–740, 2007.
4. Kevin J. Barker, Kei Davis, Adolfy Hoisie, Darren J. Kerbyson, Mike Lang, Scott Pakin, and Jose C. Sancho. Entering the petaflop era: the architecture and performance of roadrunner. In *SC '08: Proceedings of the 2008 ACM/IEEE conference on Supercomputing*, pages 1–11, Piscataway, NJ, USA, 2008. IEEE Press.
5. Paul Biggar, Nicholas Nash, Kevin Williams, and David Gregg. An experimental study of sorting and branch prediction. *J. Exp. Algorithmics*, 12:1–39, 2008.
6. Guy E. Blelloch and Parallel Ram Model. Nesl: A nested data-parallel language. Technical report, Carnegie Mellon University, 1993.
7. J.-D. Boissonnat and M. Yvinec. *Algorithmic geometry.* Cambridge University Press, UK, 1998.
8. K. J. Bowers, B. J. Albright, B. Bergen, L. Yin, K. J. Barker, and D. J. Kerbyson. 0.374 pflop/s trillion-particle kinetic modeling of laser plasma interaction on roadrunner. In *SC '08: Proceedings of the 2008 ACM/IEEE conference on Supercomputing*, pages 1–11, Piscataway, NJ, USA, 2008. IEEE Press.
9. Thomas Christof and Gerhard Reinelt. Decomposition and parallelization techniques for enumerating the facets of combinatorial polytopes. *Int. J. Comput. Geom. Appl*, 11:2001, 1998.
10. J. B. Conway. *Functions of One Complex Variable-I.* Springer-Verlag, New York, 1978.
11. Thomas H. Cormen, Charles E. Leiserson, Ronald L. Rivest, and Clifford Stein. Introduction to algorithms, second edition, 2001.
12. International Business Machines Corporation. Gnu/linux standard base core specification for ppc 3.0. (http://www.linuxbase.org/spec).
13. International Business Machines Corporation. *PowerPC Compiler Writers Guide.* IBM Corp., Riverton, NJ, USA, 1996.
14. International Business Machines Corporation. *64-bit PowerPC ELF Application Binary Interface Supplement.* IBM Corp., Riverton, NJ, USA, 2004.
15. International Business Machines Corporation. *PowerPC User Instruction Set Architecture, Book I–III.* IBM Corp., Riverton, NJ, USA, 2005.
16. International Business Machines Corporation. *Basic Linear Algebra Subprograms (BLAS) Programmer's Guide and API Reference.* IBM Corp., Riverton, NJ, USA, 2007.
17. International Business Machines Corporation. *Cell Broadband Engine Architecture.* IBM Corp., Riverton, NJ, USA, 2007.
18. International Business Machines Corporation. *Cell Broadband Engine CMOS SOI 65 nm Hardware Initialization Guide.* IBM Corp., Riverton, NJ, USA, 2007.
19. International Business Machines Corporation. *Cell Broadband Engine Programming Handbook.* IBM Corp., Riverton, NJ, USA, 2007.
20. International Business Machines Corporation. *Cell Broadband Engine Registers.* IBM Corp., Riverton, NJ, USA, 2007.
21. International Business Machines Corporation. *Data Communication and Synchronization Library for Cell Broadband Engine Programmer's Guide and API Reference.* IBM Corp., Riverton, NJ, USA, 2007.
22. International Business Machines Corporation. *Example Library API Reference.* IBM Corp., Riverton, NJ, USA, 2007.
23. International Business Machines Corporation. *Linear Algebra Package Library (LAPACK) Programmer's Guide and API Reference.* IBM Corp., Riverton, NJ, USA, 2007.

24. International Business Machines Corporation. *Monte Carlo Library Programmer's Guide and API Reference*. IBM Corp., Riverton, NJ, USA, 2007.
25. International Business Machines Corporation. *SIMD Math Library Specification for Cell Broadband Engine Architecture*. IBM Corp., Riverton, NJ, USA, 2007.
26. International Business Machines Corporation. *Software Development Kit for Multicore Acceleration ver 3.0: Accelerated Library Framework for Cell Broadband Engine Programmer's Guide and API Reference*. IBM Corp., Riverton, NJ, USA, 2007.
27. International Business Machines Corporation. *Software Development Kit for Multicore Acceleration ver 3.0: Programmer's Guide*. IBM Corp., Riverton, NJ, USA, 2007.
28. International Business Machines Corporation. *SPE Runtime Management Library Version 2.2*. IBM Corp., Riverton, NJ, USA, 2007.
29. International Business Machines Corporation. *SPU Application Binary Interface Specification*. IBM Corp., Riverton, NJ, USA, 2007.
30. International Business Machines Corporation. *SPU Assembly Language Specification*. IBM Corp., Riverton, NJ, USA, 2007.
31. International Business Machines Corporation. *SPU Timer Library Programmer's Guide and API Reference*. IBM Corp., Riverton, NJ, USA, 2007.
32. International Business Machines Corporation. *Synergistic Processor Unit Instruction Set Architecture*. IBM Corp., Riverton, NJ, USA, 2007.
33. International Business Machines Corporation. *C/C++ Language Extensions for Cell Broadband Engine Architecture*. IBM Corp., Riverton, NJ, USA, 2008.
34. David A. Cox, John Little, and Donal O'Shea. *Ideals, Varieties, and Algorithms: An Introduction to Computational Algebraic Geometry and Commutative Algebra, 3/e (Undergraduate Texts in Mathematics)*. Springer-Verlag New York, Inc., Secaucus, NJ, USA, 2007.
35. Zbigniew J. Czech, George Havas, and Bohdan S. Majewski. Perfect hashing. *Theor. Comput. Sci.*, 182(1-2):1–143, 1997.
36. M. de Berg, M. van Kreveld, M. Overmars, and O. Schwarzkopf. *Computational Geometry: Algorithms and Applications*. Springer-Verlag, Berlin, 1997.
37. G. DeMichelli. *Synthesis and Optimization of Digital Circuits*. McGraw-Hill Science and Engineering, 1994.
38. S. Devadas, A. Ghosh, and K. Keutzer. *Logic Synthesis*. McGraw-Hill Professional, 1994.
39. W. E. Donath and A. J. Hoffman. Lower bounds for the partitioning of graphs. *IBM J. Res. Dev.*, 17:420–425, 1973.
40. A. E. Eichenberger, J. K. O'Brien, K. M. O'Brien, P. Wu, T. Chen, P. H. Oden, D. A. Prener, J. C. Shepherd, B. So, Z. Sura, A. Wang, T. Zhang, P. Zhao, M. K. Gschwind, R. Archambault, Y. Gao, and R. Koo. Using advanced compiler technology to exploit the performance of the cell broadband enginetm architecture. *IBM Syst. J.*, 45(1):59–84, 2006.
41. A. E. Eichenberger et al. Using advanced compiler technology to exploit the performance of the cell broadband engine architecture. *IBM Systems Journal*, 45(1):59–84, 2006.
42. B. Flachs et. al. Microarchitecture and implementation of the synergistic processor in 65-nm and 90-nm soi. *IBM J. Res. Dev.*, 51(5):529–544, 2007.
43. K. Asanovic et al. The landscape of parallel computing research: A view from berkeley. http://www.eecs.berkeley.edu/Pubs/TechRpts/2006/EECS-2006-183.html.
44. Richard Fateman. Comparing the speed of programs for sparse polynomial multiplication. *SIGSAM Bull.*, 37(1):4–15, 2003.
45. Roger Ferrer, Marc González, Federico Silla, Xavier Martorell, and Eduard Ayguadé. Evaluation of memory performance on the cell be with the sarc programming model. In *MEDEA '08: Proceedings of the 9th workshop on MEmory performance*, pages 77–84, New York, NY, USA, 2008. ACM.
46. Philippe Flajolet and G. Nigel Martin. Probabilistic counting algorithms for data base applications. *J. Comput. Syst. Sci.*, 31(2):182–209, 1985.
47. M. J. Flynn. *Computer Architecture, Pipelined and Parallel Processor Design*. Jones and Bartlett Publishers, 1995.
48. K. Fukuda and V. Rosta. Combinatorial face enumeration in convex polytopes. *Computational Geometry*, 4:191–198, 1994.

49. M. R. Garey and D. S. Johnson. *Computers and Intractability*. W. H. Freeman and Company, 1979.
50. Ewgenij Gawrilow and Michael Joswig. polymake: a framework for analyzing convex polytopes. In Gil Kalai and Günter M. Ziegler, editors, *Polytopes — Combinatorics and Computation*, pages 43–74. Birkhäuser, 2000.
51. Ewgenij Gawrilow and Michael Joswig. polymake: an approach to modular software design in computational geometry. In *Proceedings of the 17th Annual Symposium on Computational Geometry*, pages 222–231. ACM, 2001. June 3-5, 2001, Medford, MA.
52. Buğra Gedik, Rajesh R. Bordawekar, and Philip S. Yu. Cellsort: high performance sorting on the cell processor. In *VLDB '07: Proceedings of the 33rd international conference on Very large data bases*, pages 1286–1297. VLDB Endowment, 2007.
53. Jr. George L. Cain. A method for locating zeros of complex functions. *Commun. ACM*, 9(4):305–306, 1966.
54. Anwar Ghuloum, Eric Sprangle, Jesse Fang, Gansha Wu, and Xin Zhou. Ct:a flexible parallel programming model for tera-scale architectures. http://techresearch.intel.com/articles/Tera-Scale/1514.htm.
55. Branko Grünbaum. *Convex Polytopes*. Springer-Verlag, 2003.
56. M. Gschwind. Chip multiprocessing and the cell broadband engine. In *CF '06: Proceedings of the 3rd conference on Computing frontiers*, pages 1–8, New York, NY, USA, 2006. ACM.
57. Michael Gschwind, David Erb, Sid Manning, and Mark Nutter. An open source environment for cell broadband engine system software. *Computer*, 40(6):37–47, 2007.
58. Michael Gschwind, Fred Gustavson, and Jan F. Prins. High performance computing with the cell broadband engine. *Sci. Program.*, 17(1-2):1–2, 2009.
59. Richard W. Hamming. *Numerical Methods for Scientists and Engineers*. McGraw-Hill, Inc., New York, NY, USA, 1973.
60. Sung-Chul Han, Franz Franchetti, and Markus Püschel. Program generation for the all-pairs shortest path problem. In *Parallel Architectures and Compilation Techniques (PACT)*, pages 222–232, 2006.
61. K. Imamura. "A method for computing addition tables in $GF(p^m)$". *IEEE Trans. on Information Theory*, 26(3):367–369, May 1980.
62. Hiroshi Inoue, Takao Moriyama, Hideaki Komatsu, and Toshio Nakatani. Aa-sort: A new parallel sorting algorithm for multi-core simd processors. In *PACT '07: Proceedings of the 16th International Conference on Parallel Architecture and Compilation Techniques*, pages 189–198, Washington, DC, USA, 2007. IEEE Computer Society.
63. J. Siek et al. *Boost Graph Library*. 2008.
64. D. Jimenez-Gonzalez, X. Martorell, and A. Ramirez. Performance analysis of cell broadband engine for high memory bandwidth applications. *Performance Analysis of Systems and Software, IEEE International Symmposium on*, 0:210–219, 2007.
65. J. A. Kahle, M. N. Day, H. P. Hofstee, C. R. Johns, T. R. Maeurer, and D. Shippy. Introduction to the cell microprocessor. *IBM J. Res. Dev.*, 49(4/5):589–604, 2005.
66. Jim Kahle. Cell architecture: key physical design features and methodology. In *ISPD '07: Proceedings of the 2007 international symposium on Physical design*, pages 1–1, New York, NY, USA, 2007. ACM.
67. Volker Kaibel and M. E. Pfetsch. Computing the face lattice of a polytope from its vertex-facet incidence. *Preprint*, 2008.
68. David R. Karger and Clifford Stein. A new approach to the minimum cut problem. *Journal of the ACM*, 43:601–640, 1996.
69. S. Kirpatrick, C.D. Gelatt, Jr., and M.P. Vecchi. "Optimization by Simulated Annealing". *Science*, 220:671–680, May 1983.
70. Michael Kistler, Michael Perrone, and Fabrizio Petrini. Cell multiprocessor communication network: Built for speed. *IEEE Micro*, 26(3):10–23, 2006.
71. Donald E. Knuth. *Seminumerical Algorithms*. Addison-Wesley, 1981.
72. S. Koranne. A distributed algorithm for k-way graph partitioning. volume 2, pages 446–448 vol.2, 1999.

73. Sandeep Koranne. A high performance simd framework for design rule checking on sony's playstation 2 emotion engine platform. In *ISQED '04: Proceedings of the 5th International Symposium on Quality Electronic Design*, pages 371–376, Washington, DC, USA, 2004. IEEE Computer Society.
74. D. Kozen. *The Design and Analysis of Algorithms*. springer, 1991.
75. P. Kravanja, T. Sakurai, and M. van Barel. On locating clusters of zeros of analytic functions. *BIT Numerical Mathematics*, 39(4):646–684, ???? 1999.
76. P. Kravanja and M. Van Barel. A derivative-free algorithm for computing zeros of analytic functions. *Computing*, 63(1):69–91, 1999.
77. Anand Kulkarni and Sandeep Koranne. Large-scale polytope diameter experiments using the cbe processor: Towards a resolution to hirsch's conjecture. March 2007.
78. A. Kunimatsu, N. Ide, T. Sato, Y. Endo, H. Murakami, T. Kamei, M. Hirano, F. Ishihara, and H. Tago. "Vector Unit Architecture for Emotion Synthesis". *IEEE Micro*, 20(2):40–47, April 2000.
79. Jakub Kurzak, Wesley Alvaro, and Jack Dongarra. Optimizing matrix multiplication for a short-vector simd architecture - cell processor. *Parallel Comput.*, 35(3):138–150, 2009.
80. John Lakos. *Large-scale C++ software design*. Addison Wesley Longman Publishing Co., Inc., Redwood City, CA, USA, 1996.
81. Lewis. *Foundations of Parallel Programming: A Machine-Indepedent Approach*. IEEE Press, 1994.
82. Richard Linderman. Early experiences with algorithm optimizations on clusters of playstation 3's. In *HPCMP-UGC '08: Proceedings of the 2008 DoD HPCMP Users Group Conference*, pages 468–471, Washington, DC, USA, 2008. IEEE Computer Society.
83. Michael Luby. A simple parallel algorithm for the maximal independent set problem. *SIAM J. Comput.*, 15(4):1036–1055, 1986.
84. Diana Marculescu, Radu Marculescu, and Massoud Pedram. Theoretical bounds for switching activity analysis in finite-state machines. In *ISLPED '98: Proceedings of the 1998 international symposium on Low power electronics and design*, pages 36–41, New York, NY, USA, 1998. ACM.
85. Brendan McKay. nauty: program for computing automorphism groups of graphs and digraphs. http://cs.anu.edu.au/people/bdm/nauty/.
86. Michael H. Meylan and Lutz Gross. A parallel algorithm to find the zeros of a complex analytic function. *ANZIAM J.*, 44(E):E236–E254, February 2003.
87. Jayadev Misra. Powerlist: a structure for parallel recursion. *ACM Transactions on Programming Languages and Systems*, 16:1737–1767, 1994.
88. Alexander Morgan and Vadim Shapiro. Box-bisection for solving second-degree systems and the problem of clustering. *ACM Trans. Math. Softw.*, 13(2):152–167, 1987.
89. K. Mulmuley. *Computational Geometry: An Introduction Through Randomized Algorithms*. Prentice Hall, Englewood Cliffs, NJ, 1994.
90. Ketan Mulmuley, Umesh V. Vazirani, and Vijay V. Vazirani. Matching is as easy as matrix inversion. In *STOC '87: Proceedings of the nineteenth annual ACM symposium on Theory of computing*, pages 345–354, New York, NY, USA, 1987. ACM.
91. H. Murata, K. Fujiyoshi, S. Nakatake, and Y. Kajitani. "VLSI Module Placement Based on Rectangle-Packing by Sequence Pair". *IEEE Trans. on CAD*, 15(12):1518–1524, December 1996.
92. Farid N. Najm. Power estimation techniques for integrated circuits. In *ICCAD '95: Proceedings of the 1995 IEEE/ACM international conference on Computer-aided design*, pages 492–499, Washington, DC, USA, 1995. IEEE Computer Society.
93. W. M. Newman and R. F. Sproull. *Principles of Interactive Computer Graphics*. McGraw-Hill, New York, NY, 1979.
94. Bradford Nichols, Dick Buttlar, and Jacqueline Proulx Farrell. *Pthreads programming*. O'Reilly & Associates, Inc., Sebastopol, CA, USA, 1996.
95. M. Oka and M. Suzuoki. "Designing and Programming the Emotion Engine". *IEEE Micro*, 19(6):20–28, December 1999.

96. Samir Palnitkar. *Verilog®hdl: a guide to digital design and synthesis, second edition.* Prentice Hall Press, Upper Saddle River, NJ, USA, 2003.
97. Victor Y. Pan. Optimal (up to polylog factors) sequential and parallel algorithms for approximating complex polynomial zeros. In *STOC '95: Proceedings of the twenty-seventh annual ACM symposium on Theory of computing*, pages 741–750, New York, NY, USA, 1995. ACM.
98. K. P. Parker and E. J. McCluskey. Probabilistic treatment of general combinational networks. *IEEE Trans. Comput.*, 24(6):668–670, 1975.
99. Roman Pearce and Michael Monagan. A maple library for high performance sparse polynomial arithmetic. *ACM Commun. Comput. Algebra*, 41(3):110–111, 161.
100. Alex Pothen and Horst D. Simon. Spectral algorithms for ordering sparse matrices in parallel. In *Proceedings of the Fourth SIAM Conference on Parallel Processing for Scientific Computing*, pages 120–121, Philadelphia, PA, USA, 1990. Society for Industrial and Applied Mathematics.
101. Robert Preis. time 1/2-approximation algorithm for maximum weighted matching in general graphs. In *in General Graphs, Symposium on Theoretical Aspects of Computer Science, STACS 99*, pages 259–269. Springer, 1999.
102. F. P. Preparata and M. I. Shamos. *Computational Geometry: An Introduction.* Springer-Verlag, 3rd edition, October 1990.
103. RapidMind. Rapidmind. http://www.rapidmind.net/technology.php.
104. Mark Richards. *Fundamentals of Radar Signal Processing.* McGraw-Hill, Inc., New York, NY, USA, 2005.
105. M. W. Riley, J. D. Warnock, and D. F. Wendel. Cell broadband engine processor: design and implementation. *IBM J. Res. Dev.*, 51(5):545–557, 2007.
106. C. R.Johns and D. A. Brokenshire. Introduction to the cell broadband engine architecture. *IBM J. Res. Dev.*, 51(5):503–520, 2007.
107. P. Sanders and S. Winkel. Super scalar sample sort, 2004.
108. Sachin S. Sapatnekar, Prashant Saxena, and Rupesh S. Shelar. *Routing Congestion in VLSI Circuits: Estimation and Optimization (Series on Integrated Circuits and Systems).* Springer-Verlag New York, Inc., Secaucus, NJ, USA, 2007.
109. Majid Sarrafzadeh and C. K. Wong. *An Introduction to VLSI Physical Design.* McGraw-Hill Higher Education, 1996.
110. Tateaki Sasaki. A theorem for separating close roots of a polynomial and its derivatives. *SIGSAM Bull.*, 38(3):85–92, 2004.
111. M. J. Schaefer and T. Bubeck. A parallel complex zero finder. *Reliable Computing*, 1(3):317–324, 1995.
112. R. Seidel. "A simple and fast incremental randomized algorithm for computing trapezoidal decompositions and for triangulating polygons". *Comput. Geom. Theory Appl.*, 1(1):51–64, 1991.
113. Freescale Semiconductors. *Programming Environments Manual for 32-bit Implementations of the PowerPC Architecture.* Freescale Semiconductors, Arizona, 2005.
114. K. Shahookar and P. Mazumder. Vlsi cell placement techniques. *ACM Comput. Surv.*, 23(2):143–220, 1991.
115. N. Sherwani. *Algorithms for VLSI Physical Design Automation.* kluwer, 1995.
116. M. Skolnik. *Handbook of Radar Sysntems.* McGraw-Hill, Inc., New York, NY, USA, 1995.
117. David M. Smith. Algorithm 786: multiple-precision complex arithmetic and functions. *ACM Trans. Math. Softw.*, 24(4):359–367, 1998.
118. Daniel A. Spielman and Shang-Hua Teng. Spectral partitioning works: Planar graphs and finite element meshes. In *IEEE Symposium on Foundations of Computer Science*, pages 96–105, 1996.
119. G. Steele. *Common Lisp, the Language.* Digital Press, 1990.
120. Sriram Swaminarayan, Kai Kadau, Timothy C. Germann, and Gordon C. Fossum. 369 tflop/s molecular dynamics simulations on the roadrunner general-purpose heterogeneous supercomputer. In *SC '08: Proceedings of the 2008 ACM/IEEE conference on Supercomputing*, pages 1–10, Piscataway, NJ, USA, 2008. IEEE Press.

121. Berkeley Logic Synthesis and Verification Group. Abc: A system for sequential synthesis and verification. http://www.eecs.berkeley.edu/ alanmi/abc/.
122. X. Tang, R. Tian, and D. F. Wong. "Fast Evaluation of Sequence Pair in Block Placement by Longest Common Subsequence Computation". *IEEE Trans. on CAD*, 20(12):1406–1413, December 2001.
123. Alliance CAD Team. Alliance cad. http://www-asim.lip6.fr/recherche/alliance/.
124. M. Teillaud. "Union and Split Operations on Dynamic Trapezoidal Maps". *Comput. Geom. Theory Appl*, 17:153–163, 2000.
125. N. H. E. Weste and K. Eshraghian. *Principles of CMOS VLSI design*. Addison-Wesley, Reading, MA, 2nd edition, 1994.
126. J. Xu, P. Guo, and C. Cheng. "Sequence-Pair Approach for Rectilinear Module Placement". *IEEE Trans. on CAD*, 18(4):484–493, April 1999.
127. F. Y. Young, C. C. N. Chu, W. S. Luk, and Y. C. Wong. "Handling Soft Modules in General Non-slicing Floorplan using Lagrangian Relaxation". *IEEE Trans. on CAD*, 20(5):687–692, May 2001.
128. Gunter M. Ziegler. *Lectures on Polytopes*. Springer-Verlag, 1995.

Breinigsville, PA USA
11 November 2010
249128BV00010B/34/P